VOLUME ONE

ADVANCES IN FUEL CELLS

VOLUME ONE

ADVANCES IN FUEL CELLS

Edited by

T. S. ZHAO (Editor-in-chief)
The Hong Kong University of Science & Technology
Hong Kong, China

K.-D. KREUER
Max Planck Inst Festkorperforsch, Germany

TRUNG VAN NGUYEN
University of Kansas, USA

ELSEVIER

Amsterdam • Boston • Heidelberg • London • New York • Oxford
Paris • San Diego • San Francisco • Singapore • Sydney • Tokyo

ELSEVIER B.V.
Radarweg 29
P.O. Box 211,
1000 AE Amsterdam
The Netherlands

ELSEVIER Inc.
525 B Street, Suite 1900
San Diego,
CA 92101-4495
USA

ELSEVIER Ltd
The Boulevard, Langford Lane
Kidlington,
Oxford OX5 1GB
UK

ELSEVIER Ltd
84 Theobald's Road
London WC1X 8RR
UK

First edition 2007

ISBN-13: 978-0-08045-394-1
ISBN-10: 0-080-45394-5

ISSN: 1752-301X

Typeset by Charon Tec Ltd (A Macmillan Company), Chennai, India
www.charontec.com
Printed in Great Britain.

07 08 09 10 10 9 8 7 6 5 4 3 2 1

Contents

CONTRIBUTORS

Frank A. de Bruijn
Energy Research Centre of the Netherlands, P.O. Box 1, 1755 ZG Petten, The Netherlands.
University of Technology Eindhoven, P.O. Box 513, 5600 MB Eindhoven, The Netherlands.

Patric Jannasch
Department of Polymer Science & Engineering, Lund University, P.O. Box 124, SE-221 00 Lund, Sweden.

Gaby J. M. Janssen
Energy Research Centre of The Netherlands, P.O. Box 1, 1755 ZG Petten, The Netherlands.

Dong Hyun Kim
Department of Chemical Engineering, Kyungpook National University, 1370 Sankyukdong, Daegu 702-701, Republic of Korea.

Yu Seung Kim
MST-11, Electronic and Electro-chemical Devices, Los Alamos National Laboratory, Los Alamos, NM 87545.

A. A. Kulikovsky
Institute for Materials and Processes in Energy Systems (IWV–3), Research Center "Jülich", D–52425 Jülich, Germany.

Arunabha Kundu
Department of Chemical Engineering, Yonsei University, 134 Shinchon Dong, Sudeamun-Ku, Seoul 120-749, Republic of Korea.

Benoît Lafitte
Department of Polymer Science & Engineering, Lund University, P.O. Box 124, SE-221 00 Lund, Sweden.

Xianguo Li
Fuel Cells and Green Energy Research Group, Department of Mechanical Engineering, University of Waterloo, Waterloo, Ontario, Canada.

Robert C. Makkus
Energy Research Centre of the Netherlands, P.O. Box 1, 1755 ZG Petten, The Netherlands.

Ronald K. A. M. Mallant
Energy Research Centre of The Netherlands, P.O. Box 1, 1755 ZG Petten, The Netherlands.

John Newman
Lawrence Berkeley National Laboratory and Department of Chemical Engineering, University of California, Berkeley, CA 94720-1462.

Bryan S. Pivovar
MST-11, Electronic and Electro-chemical Devices, Los Alamos National Laboratory, Los Alamos, NM 87545.

Yong Gun Shul
Department of Chemical Engineering, Yonsei University, 134 Shinchon Dong, Sudeamun-Ku, Seoul 120-749, Republic of Korea.

Adam Z. Weber
Lawrence Berkeley National Laboratory and Department of Chemical Engineering, University of California, Berkeley, CA 94720-1462.

Preface

Fuel cells have been recognized to be destined to form the cornerstone of energy technologies in the twenty-first century. The rapid advances in fuel cell system development have left current information available only in scattered journals and Internet sites. To be truly useful to both present and future workers in the field, this prodigious outpouring of new information needs to be brought together, deposited in one central resource and conveyed from a more global perspective. The establishment of this new book series, *Advances in Fuel Cells*, will answer the need and provide the forum.

Advances in Fuel Cells is intended to fill the information gap between regularly scheduled journals and university level textbooks by providing in-depth coverage over a broader scope than in journals or texts. Contributions by leading experts working in both academic and industrial research and development, will serve as a central source of reference for the fundamentals and applications of fuel cells, establishing the state of the art, disseminating the latest research discoveries, and providing potential textbooks to senior undergraduate and graduate students.

The present volume provides informative chapters on thermodynamic performance of fuel cells, macroscopic modeling of polymer-electrolyte membranes, the prospects for phosphonated polymers as proton-exchange fuel cell membranes, polymer electrolyte membranes for direct methanol fuel cells, materials for state of the art proton-exchange membrane fuel cells, and their suitability for operation above 100°C, analytical modeling of direct methanol fuel cells, and methanol reforming processes.

The editorial board expresses their appreciation to the contributing authors of Volume 1, who have opened up the high standards associated *Advances in Fuel Cells*. Last, but not least, the editors acknowledge the efforts of the professional staff at Elsevier for providing invaluable editorial assistance.

T. S. Zhao
The Hong Kong University of Science & Technology, Hong Kong, China

K.-D. Kreuer
Max Planck Inst Festkorperforsch, Germany

Trung Van Nguyen
University of Kansas, USA

CHAPTER ONE

Thermodynamic Performance of Fuel Cells and Comparison with Heat Engines

Xianguo Li

Contents

Fuel Cells and Green Energy Research Group, Department of Mechanical Engineering, University of Waterloo, Waterloo, Ontario, Canada

Advances in Fuel Cells
0080453945

Abstract

Heat engines based on fossil fuel combustion produce harmful pollutants and greenhouse gas emissions. Environmental concerns and sustainable development call for new technology for energy conversion and power generation, which is more efficient, environmentally friendly and compatible with alternative fuels and renewable energy sources and carriers. Fuel cells meet all these requirements, and are being developed as one of the primary energy technologies of the future. In this chapter, the thermodynamic performance of fuel cells is analyzed, energy conversion efficiency of fuel cells and heat engines is studied and compared, and misconceptions about fuel cell efficiency clarified. It is shown that both fuel cells and heat engines have the same maximum theoretical efficiency, which is equivalent to the Carnot efficiency, when operating on the same fuel and oxidant. However, fuel cells are free from the high temperature limit imposed by materials on heat engines and less irreversibilities associated with heat rejection. As a result, fuel cells can have higher practical efficiencies.

1.1 INTRODUCTION

Progress in human society, and especially modern civilization, has been marked by ever-increasing energy consumption and power requirements. The majority of the energy needs have been provided by combustion of fossil fuels since the industrial revolution. Heat engines utilizing fossil fuel combustion have resulted in severe local air pollution, threatening the health of millions of people living in many of the world's urban areas. They continue to contribute significantly to the increase in the atmospheric carbon dioxide concentrations, thus intensify the prospect of global warming. In addition to the health and environmental concerns, a steady depletion of the world's limited fossil fuel reserves and the very survival of humankind call for new generation technology for energy conversion and power generation, which is more efficient than the conventional heat engines with minimal or no pollutant emissions, and also compatible with renewable energy sources and carriers for sustainable development and energy security. Fuel cell has been identified as the most promising and potential energy conversion technology, which meets all of the above requirements. In fact, fuel cell technology has been successfully used in many specific areas, notably in space explorations, where fuel cell operates on pure hydrogen and oxygen with over 70% thermal to electrical energy efficiency and the only byproduct water constitutes the sole source of drinking water for the crew of the spacecraft. There are now several hundred fuel cell units for terrestrial applications, from stationary cogeneration, mobile transportation to portable applications, operating in over a dozen of countries, impressive technical progress has been achieved, and is driving the development of competitively priced fuel cell-based power generation systems with advanced features.

Besides being efficient, clean and compatible with future energy sources and carriers, fuel cell also offers many additional advantages for both mobile and stationary applications. Fuel cell is an electrochemical device and has no moving components except for peripheral compressors and motors. As a result, its operation is very quiet, and virtually without vibration and noise, thus capable of being sited at the premises of the consumer to eliminate power transmission lines. Its inherent modularity allows for simple construction and operation with possible applications for dispersed, distributed and portable power generation, because it may be made in any size from a few watts to megawatt scale plant with equal efficiency. Its fast response to the changing load condition while maintaining high efficiency makes it ideally suited to load following applications. Its high efficiency represents less chemical, thermal and carbon dioxide emissions for the same amount of energy conversion and power generation.

At present, fuel cell is being used routinely in space applications, and has been under intensive development for terrestrial use, such as for utilities and zero emission vehicles. There exist a variety of fuel cells, and they can be classified based on their operating temperature such as low and high temperature fuel cells, the type of ion migrating through the electrolyte, etc. However, the choice of electrolyte defines the properties of a fuel cell. Hence, fuel cell is often named by the nature of the electrolyte used. There are presently six major fuel cell technologies at varying stages of development and commercialization. They are alkaline, phosphoric acid, polymer electrolyte membrane, molten carbonate, solid oxide and direct methanol fuel cells. Their electrochemical reactions, operation fundamentals, construction and design, application and state of the art technology can be found elsewhere [1–7].

Although numerous studies aiming at developing fuel cell technology as a practical source of power have been conducted, some confusion and misconception exists about the thermodynamic performance of fuel cells and its comparison with heat engines [2,8–13]. For example, it is often said that fuel cells are not limited by the Carnot efficiency, therefore they have, or can achieve, higher energy efficiency than the conventional heat engines, e.g., Ref. [9]. Some even go so far as to state that fuel cells are not subject to "thermodynamic limitations." It is also often the practice that in proving a fuel cell has better performance, a reversible fuel cell is compared with the irreversible heat engine [9,11,13] – certainly an unfair comparison since a reversible energy conversion system is assured to have better performance than the irreversible energy system based on the second law of thermodynamics. Further, many analyses used the specific fluids involved, e.g., Ref. [11], and assumed constant properties, e.g., Refs. [2,8,11], leaving to questions regarding the generality of the conclusions reached. Furthermore, some studies even show the possibility of fuel cells that could achieve over 100% efficiency – a result some even cite as a proof that fuel cell is more efficient than conventional heat engine since heat engine is not possible to achieve over 100% efficiency. On the other hand, Li [14,15] only uses the thermodynamic laws without consideration of specific fluids involved and their property variations, and demonstrate that both fuel cells and heat engines have the same maximum possible efficiency when operating on the same fuel and oxidant, and this maximum efficiency and the Carnot efficiency are equal. But practical limitations (irreversibilities) result in fuel cells having higher practical efficiency than heat engines. Unique fuel cell operation at much lower temperatures also yields less or no pollutant emissions, thus leading to environmentally clean energy conversion. Finally, the riddle about the possibility of over 100% energy conversion efficiency for some reversible fuel cells was clarified. Li's approach has also been used by others in similar studies, e.g. [12].

In this chapter, the general approach by Li [14,15] will be followed in the analysis of fuel cell performance. The method of analysis will be general, without considering specific working fluids involved or their specific variations of properties with temperature, pressure and concentration, so that the results obtained and conclusions reached will be generally applicable. Specifically, the reversible cell potential will be derived by using the first and second laws of thermodynamics, and its variations with the operating conditions such as temperature, pressure and reactant concentrations in the reactant streams will be obtained with general thermodynamic relations. The issue of energy conversion efficiency will be presented with the help of the first and second laws of thermodynamics. The maximum possible efficiency for fuel cells will be investigated; a comparison will be made with Carnot efficiency, which is the maximum possible efficiency for heat engines against which fuel cells are competing for commercial success. Then the possibility of over 100% efficiency for fuel cells is examined. The energy conversion efficiency for a fuel cell system comprising fuel cells and auxiliary equipments will be considered, and efficiency loss mechanism for operating fuel cells will be discussed. It is the intention of preparing this chapter that the thermodynamic performance of fuel cell and its comparison with its rival heat engine is properly dealt with so that any future misunderstandings and misconceptions can be avoided.

1.2 Reversible Cell Potential

In a fuel cell, the chemical energy of a fuel and an oxidant is converted directly into electrical energy, which is exhibited in terms of cell potential and electrical current output. The maximum possible electrical energy output and the corresponding electrical potential difference between the cathode and anode are achieved when the fuel cell is operated under the thermodynamically reversible condition. This maximum possible cell potential is called reversible cell potential, one of the significantly important parameters for

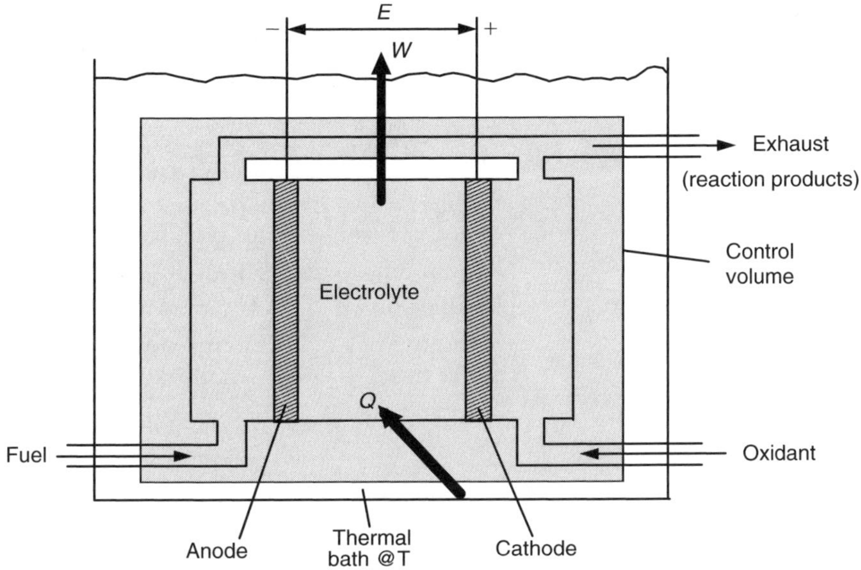

Figure 1.1 A thermodynamic model of fuel cell system.

fuel cells. We shall apply fundamental thermodynamic principles to derive the reversible cell potential in this section.

A thermodynamic system[1] model is shown in Figure 1.1 for the analysis of fuel cell performance. It is a control-volume system for the fuel cell to which fuel and oxidant streams enter and product or exhaust stream exits. The fuel cell is located inside a thermal bath in order to maintain the desired system temperature T. The reactant streams (fuel and oxidant) and the exhaust stream are considered to have the same temperature T and pressure P. It is assumed that the fuel and oxidant inflow and the exhaust outflow are steady; the kinetic and gravitational potential energy changes are negligible. Further, the overall electrochemical reactions occurring inside the fuel cell system boundary is described as follows:

$$\text{Fuel (e.g., } H_2) + \text{Oxidant (e.g., } O_2) \Rightarrow \dot{W} + \dot{Q} + \text{Product} \quad (1.1)$$

where $\dot{W}$ is the rate of work done by the system and $\dot{Q}$ the rate of heat transferred into the system from the surrounding constant temperature thermal bath, which may, or may not, be in thermal equilibrium with the fuel cell system at the temperature T and pressure P. For hydrogen/oxygen

[1] A thermodynamic system, or simply system, is in thermodynamics a collection of matter under study (or analysis); whereas the jargon "fuel cell system" in fuel cell literature usually denotes the fuel cell power plant that consists of fuel cell stack(s) and auxiliary equipment. In this chapter, a fuel cell system may imply both meanings. However, the context will tell which it is meant to be.

fuel cells, the reaction product is usually water. Then, the first and second laws of thermodynamics can be written, respectively, for the present fuel cell system, as

$$\frac{dE_{C.V.}}{dt} = \left[\left(\dot{N}h + KE + PE\right)_F + \left(\dot{N}h + KE + PE\right)_{Ox}\right]_{in} - \left[\left(\dot{N}h + KE + PE\right)_{Ex}\right]_{out} + \dot{Q} - \dot{W} \tag{1.2}$$

$$\begin{pmatrix}\text{Increase in}\\ \text{system energy}\end{pmatrix} = \begin{pmatrix}\text{Energy in}\\ \text{by mass flow}\end{pmatrix} - \begin{pmatrix}\text{Energy out}\\ \text{by mass flow}\end{pmatrix} + \begin{pmatrix}\text{Energy in}\\ \text{as heat}\end{pmatrix} - \begin{pmatrix}\text{Energy out}\\ \text{as work}\end{pmatrix}$$

and

$$\frac{dS_{C.V.}}{dt} = \left[\left(\dot{N}s\right)_F + \left(\dot{N}s\right)_{Ox}\right]_{in} - \left[\left(\dot{N}s\right)_{Ex}\right]_{out} + \frac{\dot{Q}}{T} + \dot{\wp}_s$$

$$\begin{pmatrix}\text{Increase in}\\ \text{system entropy}\end{pmatrix} = \begin{pmatrix}\text{Entropy in}\\ \text{by mass flow}\end{pmatrix} - \begin{pmatrix}\text{Entropy out}\\ \text{by mass flow}\end{pmatrix} + \begin{pmatrix}\text{Entropy in by}\\ \text{heat transfer}\end{pmatrix} + \begin{pmatrix}\text{Entropy}\\ \text{generated}\end{pmatrix} \tag{1.3}$$

where $\dot{N}$ is the molar flow rate, h (absolute) enthalpy per unit mole, s the specific entropy on a mole basis and $\dot{\wp}_s$ the rate of entropy generation due to irreversibilities. The subscript "$_F$", "$_{Ox}$" and "$_{Ex}$" stand for fuel, oxidant and exhaust stream, respectively. "KE" and "PE" denote kinetic and gravitational potential energy that are being carried in and out of the system by the mass flow.

For a steady process, there are no temporal changes in the amount of energy $E_{C.V.}$ and entropy $S_{C.V.}$ within the control-volume system, hence, $dE_{C.V.}/dt = 0$ and $dS_{C.V.}/dt = 0$. Further, the changes in the kinetic and gravitational potential energy are negligible for the present process. Therefore, Eqs. (1.2) and (1.3) can be simplified as follows:

$$\dot{N}_F\left(h_{in} - h_{out}\right) + \dot{Q} - \dot{W} = 0 \tag{1.4}$$

$$\dot{Q} = -T\dot{\wp}_s - \dot{N}_F T\left(s_{in} - s_{out}\right) \tag{1.5}$$

where

$$h_{in} = \left(h_F + \frac{\dot{N}_{Ox}}{\dot{N}_F}h_{Ox}\right)_{in} \quad \text{and} \quad h_{out} = \frac{\dot{N}_{Ex}}{\dot{N}_F}h_{Ex} \tag{1.6}$$

h_{in} is the amount of enthalpy per mole of fuel carried into the system by the reactant inflow and h_{out} is the amount of enthalpy per mole of fuel taken out of the system by the exhaust stream. Similarly,

$$s_{in} = \left(s_F + \frac{\dot{N}_{Ox}}{\dot{N}_F} s_{Ox} \right)_{in} \quad \text{and} \quad s_{out} = \frac{\dot{N}_{Ex}}{\dot{N}_F} s_{Ex} \tag{1.7}$$

are the amount of entropy per mole of fuel brought into the system by the reactant inflow, and the amount of entropy per mole of fuel carried out of the system by the outgoing exhaust stream containing the reaction products, respectively.

Substitution of Eq. (1.5) into (1.4) yields

$$\dot{W} = \dot{N}_F \left(h_{in} - h_{out} \right) - \dot{N}_F T \left(s_{in} - s_{out} \right) - T\dot{\wp}_s \tag{1.8}$$

Now let

$$w = \frac{\dot{W}}{\dot{N}_F}; \quad q = \frac{\dot{Q}}{\dot{N}_F}; \quad \text{and} \quad \wp_s = \frac{\dot{\wp}_s}{\dot{N}_F} \tag{1.9}$$

represent, respectively, the amount of work done, heat transferred and entropy generated per mole of fuel, Eqs. (1.5) and (1.8) then become

$$q = -T\wp_s - T\left(s_{in} - s_{out} \right) = T\Delta s - T\wp_s \tag{1.10}$$

$$w = \left(h_{in} - h_{out} \right) - T\left(s_{in} - s_{out} \right) - T\wp_s \tag{1.11}$$

Because the enthalpy and entropy change for the fuel cell reaction is defined as

$$\Delta h = h_{out} - h_{in} \quad \text{and} \quad \Delta s = s_{out} - s_{in} \tag{1.12}$$

Eq. (1.11) can also be expressed as

$$w = -\Delta h + T\Delta s - T\wp_s = -\left[\left(h - Ts \right)_{out} - \left(h - Ts \right)_{in} \right] - T\wp_s \tag{1.13}$$

From the definition of the Gibbs function (per mole of fuel) $g = h - Ts$, Eq. (1.11) or (1.13) can also be written as

$$w = -(g_{out} - g_{in}) - T\wp_s = -\Delta g - T\wp_s \tag{1.14}$$

Because by the second law of thermodynamics, entropy can be generated, but can never be destroyed, we know $\wp_s \geqslant 0$, and also the absolute temperature (in Kelvin scale) $T > 0$ by the third law of thermodynamics, the maximum possible work (i.e., useful energy) output from the present system occurs when $\wp_s = 0$, or under the thermodynamically reversible condition, since the change in the Gibbs function is usually negative for useful fuel cell reaction. Therefore, from Eq. (1.14) it is clear that the maximum possible work output from the present fuel cell system is equal to the decrease in Gibbs function, or

$$w_{max} = -\Delta g \tag{1.15}$$

for all reversible processes, regardless of the specific type of fuel cells involved. In fact, it might be pointed out that in the derivation of Eqs. (1.14) and (1.15), no specifics about the control-volume system have been stipulated; hence they are valid for any energy conversion systems.

For a fuel cell system, the electrical energy output is conventionally expressed in terms of the cell potential difference between the cathode and the anode. Since (electrical) potential is the (electrical) potential energy per unit (electrical) charge, its SI unit is J/Coulomb, which is more often called volt or simply V. Potential energy is defined as the work done when charge is moved from one location to another in the electrical field, normally refers to external circuits. For the internal circuit of fuel cells, such as the one shown in Figure 1.1, electromotive force is the terminology often used, which is also defined as the work done by transferring one Coulomb positive charge from a low to a high potential. Hence, electromotive force also has the SI unit of J/Coulomb, or volt. We shall adopt the terminology of cell potential, instead of electromotive force, from now on; and we shall use the notation E to represent the cell potential. Because normally electrons are the particles transferred that carry electrical charge, we may express the work done by a fuel cell as follows

$$w \text{ (J/mole fuel)} = E \times \text{(Coulomb of electron charge transferred/mole fuel)}$$

or

$$w = E \times (n\ N_0 e) = E \times (nF) \tag{1.16}$$

where n is the number of moles of electrons transferred per mole of fuel consumed, N_0 the Avogadro's number ($= 6.023 \times 10^{23}$ number of electrons/mole electron) and e the electric charge per electron ($= 1.6021 \times 10^{-19}$ Coulomb/electron). Since $N_0 e = 96{,}487$ Coulomb/mole electron $= F$ is often known as the Faraday constant, the cell potential becomes, from Eq. (1.14)

$$E = \frac{w}{nF} = \frac{-\Delta g - T\wp_s}{nF} \tag{1.17}$$

Hence, the maximum possible cell potential, or the reversible cell potential E_r, becomes

$$E_r = -\frac{\Delta g}{nF} \tag{1.18}$$

From the reversible cell potential given above, Eq. (1.17) can also be rewritten as

$$E = E_r - \frac{T\wp_s}{nF} = E_r - \eta \tag{1.19}$$

where

$$\eta = \frac{T\wp_s}{nF} \tag{1.20}$$

is the cell voltage loss due to irreversibilities (or entropy generation). Clearly, the actual cell potential can be calculated by subtracting the cell voltage loss from the reversible cell potential. Alternatively, the amount of entropy generation per mole fuel consumed can be determined as

$$\wp_s = \frac{nF\eta}{T} = \frac{nF\left(E_r - E\right)}{T} \tag{1.21}$$

Thus, the amount of entropy generation, representing the degree of irreversibilities (the degree of deviation from the idealized reversible condition), for the fuel cell reaction process can be measured once the cell potential E and the cell operating temperature T are known.

Note that the Gibbs function is a thermodynamic property, determined by state variables such as temperature and pressure. Hence, the change in the Gibbs function for the fuel cell reaction discussed here can be written as, considering the temperature throughout the fuel cell system shown in Figure 1.1 is uniform and constant

$$\Delta g = \Delta h - T\Delta s \tag{1.22}$$

and it is also a function of the system temperature T and pressure P, so is the reversible cell potential. The specific effect of the operating conditions, such as temperature, pressure and reactant concentrations, on the reversible cell potential will be presented in the next section. If the reaction occurs at the standard reference temperature and pressure (25°C and 1 atm), the resulting cell potential is usually called the standard reversible cell potential E_r°,[2] or

$$E^\circ_r(T_{ref}) = -\frac{\Delta g(T_{ref}, P_{ref})}{nF} \tag{1.23}$$

If pure hydrogen and oxygen are used as reactants to form product water, then E_r° (25°C) = 1.229 V for the product water in liquid form, and

[2] In literature, the superscript "°" sometimes denotes the value at the standard reference condition of 25°C and 1 atm; and sometimes it also refers to parameters evaluated at 1 atm. To avoid confusion, the latter meaning has been adopted in the present chapter.

Table 1.1 Standard enthalpy and Gibbs function of reaction for candidate fuels and oxidants, and corresponding standard reversible cell potential as well as other relevant parameters (at 25°C and 1 atm)[a]

Fuel	Reaction	N	$-\Delta h$ (J/mol)	$-\Delta g$ (J/mol)	E°_r(V)	η (%)[b]
Hydrogen	$H_2 + 0.5O_2 \rightarrow H_2O(\ell)$	2	286.0	237.3	1.229	82.97
	$H_2 + Cl_2 \rightarrow 2HCl(aq)$	2	335.5	262.5	1.359	78.33
	$H_2 + Br_2 \rightarrow 2HBr(aq)$	2	242.0	205.7	1.066	85.01
Methane	$CH_4 + 2O_2 \rightarrow CO_2 + 2H_2O(\ell)$	8	890.8	818.4	1.060	91.87
Propane	$C_3H_8 + 5O_2 \rightarrow 3CO_2 + 4H_2O(\ell)$	20	2221.1	2109.3	1.093	94.96
Decane	$C_{10}H_{22} + 15.5O_2 \rightarrow 10CO_2 + 11H_2O(\ell)$	66	6832.9	6590.5	1.102	96.45
Carbon monoxide	$CO + 0.5O_2 \rightarrow CO_2$	2	283.1	257.2	1.333	90.86
Carbon	$C(s) + 0.5O_2 \rightarrow CO$	2	110.6	137.3	0.712	124.18[c]
	$C(s) + O_2 \rightarrow CO_2$	4	393.7	394.6	1.020	100.22[c]
Methanol	$CH_3OH(\ell) + 1.5O_2 \rightarrow CO_2 + 2H_2O(\ell)$	6	726.6	702.5	1.214	96.68
Formaldehyde	$CH_2O(g) + O_2 \rightarrow CO_2 + H_2O(\ell)$	4	561.3	522.0	1.350	93.00
Formic acid	$HCOOH + 0.5O_2 \rightarrow CO_2 + H_2O(\ell)$	2	270.3	285.5	1.480	105.62[c]
Ammonia	$NH_3 + 0.75O_2 \rightarrow 1.5\ H_2O(\ell) + 0.5N_2$	3	382.8	338.2	1.170	88.36
Hydrazine	$N_2H_4 + O_2 \rightarrow 2H_2O(\ell) + N_2$	4	622.4	602.4	1.560	96.77
Zinc	$Zn + 0.5O_2 \rightarrow ZnO$	2	348.1	318.3	1.650	91.43
Sodium	$Na + 0.25H_2O + 0.25O_2 \rightarrow NaOH(aq)$	1	326.8	300.7	3.120	92.00

[a] Taken from A.J. Appleby, Chapter 5 in Fuel Cell Systems, ed. by L.J.M.J. Blomen and M.N. Mugerwa, Plenum Press, New York, 1993.
[b] Energy Conversion Efficiency.
[c] There is a conceptual problem with these efficiency data; as detailed later in efficiency section.

$E_r^\circ(25°C) = 1.185$ V if the product water is in vapor form. The difference in E_r° corresponds to the energy required for the vaporization of water. It might be pointed out that any fuel containing hydrogen (including hydrogen itself, hydrocarbons, alcohols and to a lesser extent coal) has two values for Δg and Δh, one higher (called higher heating value (HHV)) and one lower (called lower heating value (LHV)), depending on whether the product water is in the form of liquid or vapor. Hence, care should be taken when referring to reversible cell potential and energy efficiency to be discussed later.

From Eq. (1.23) the standard reversible cell potential, E_r°, can be determined for any electrochemical reactions. Some of the potential fuel cell reactions and the resulting E_r° obtained are shown in Table 1.1 along with other relevant parameters. From this table, it might be noted that E_r° should be approximately above 1 V in order for the reaction to be realistic for fuel cell application. This is because if E_r° is much less than 1 V, and considering the cell voltage loss that inevitably occur in practical fuel cells due to irreversibilities, the actual cell potential might become too small to be useful for practical applications. Therefore, the rule of thumb is, for any proposed fuel and oxidant, to calculate E_r° and to see if E_r° is on the order of 1 V or larger before proceeding to any further work on it.

1.2.1 A General Formulation

The calculation of reversible cell potential may be expressed in a more general form that would be convenient for computer programming. For a generalized electrochemical reaction:

$$\underbrace{\sum_{i=1}^{N} v_i' M_i}_{\text{Reactant}} \rightarrow \underbrace{\sum_{i=1}^{N} v_i'' M_i}_{\text{Product}} \tag{1.24}$$

where M_i is the chemical formula for species i; v_i' and v_i'' are the number of moles for species i in the reactant and product mixture, respectively; and N is the total number of species in the chemically reacting system. Take H_2 and O_2 reaction to form water e.g.:

$$H_2 + ½O_2 \rightarrow H_2O$$

Then, we have

$M_1 = H_2;$	$M_2 = O_2;$	$M_3 = H_2O;$
$v'_{H_2} = v'_F = 1;$	$v'_{O_2} = 0.5;$	$v'_{H_2O} = 0$
$v''_{H_2} = 0;$	$v''_{O_2} = 0;$	$v''_{H_2O} = 1$

The enthalpy and entropy at the inlet and outlet are equal to those of reactant and product, respectively

$$h_{in} = h_R = \frac{1}{\nu'_F}\sum_{i=1}^{N}\nu'_i h_{M_i}; \qquad h_{out} = h_P = \frac{1}{\nu'_F}\sum_{i=1}^{N}\nu''_i h_{M_i} \tag{1.25}$$

$$s_{in} = s_R = \frac{1}{\nu'_F}\sum_{i=1}^{N}\nu'_i s_{M_i}; \qquad s_{out} = s_P = \frac{1}{\nu'_F}\sum_{i=1}^{N}\nu''_i s_{M_i} \tag{1.26}$$

The enthalpy and entropy changes for the generalized reaction, Eq. (1.24), are

$$\Delta h = h_P - h_R = \frac{1}{\nu'_F}\sum_{i=1}^{N}(\nu''_i - \nu'_i)h_{M_i} \tag{1.27}$$

$$\Delta s = s_P - s_R = \frac{1}{\nu'_F}\sum_{i=1}^{N}(\nu''_i - \nu'_i)s_{M_i} \tag{1.28}$$

Then, the change in the Gibbs function can be determined as

$$\Delta g = \Delta h - T\Delta s \tag{1.29}$$

or alternatively

$$\Delta g = g_P - g_R = \frac{1}{\nu'_F}\sum_{i=1}^{N}(\nu''_i - \nu'_i)g_{M_i} \tag{1.30}$$

Finally, the reversible cell potential can be calculated according to Eq. (1.18).

1.3 The Effect of Operating Conditions on Reversible Cell Potential

The most important operating conditions that influence fuel cell performance are the operating temperature, pressure and reactant concentrations. Before analyzing these influences, we will first develop some thermodynamic relations that will be useful for our analysis. Recall that Gibbs function and enthalpy are defined as

$$g = h - Ts \tag{1.31}$$

$$h = u + Pv \tag{1.32}$$

where u is the internal energy and v is the specific volume. Combining the above equations together and taking the differential, we obtain

$$dg = du + Pdv + vdP - Tds - sdT \tag{1.33}$$

Another fundamental relation in thermodynamics is the Gibbs equation for a simple compressible substance, which is

$$Tds = du + Pdv = dh - vdP \tag{1.34}$$

Substituting Eq. (1.34) into (1.33) results in

$$dg = vdP - sdT \tag{1.35}$$

Therefore, we arrive at two important relations for fuel cell analysis

$$\left(\frac{\partial g}{\partial T}\right)_P = -s; \quad \left(\frac{\partial g}{\partial P}\right)_T = v \tag{1.36}$$

Applying the above two equations to Gibbs function change for a particular fuel cell reaction, such as the generalized one given in Eq. (1.24), we finally obtain

$$\left(\frac{\partial \Delta g}{\partial T}\right)_P = -\Delta s \tag{1.37}$$

$$\left(\frac{\partial \Delta g}{\partial P}\right)_T = \Delta v \tag{1.38}$$

where Δs and Δv are the respective changes in the entropy and specific volume between the products and reactants. It might be emphasized that the above two relations are obtained without making any specific assumptions such as ideal gas approximation; hence they are valid for any substances undergoing chemical reactions. They will be utilized for our following analysis of temperature and pressure effect on the reversible cell potential.

1.3.1 Effect of Temperature on Reversible Cell Potential E_r

The reversible cell potential, E_r, given in Eq. (1.18), is a function of temperature, because the change in Gibbs function depends on the fuel cell operating temperature and pressure. Hence,

$$E_r(T, P) = -\frac{\Delta g(T, P)}{nF} \tag{1.18}$$

Then the change of the reversible cell potential with temperature can be expressed as, incorporating Eq. (1.37)

$$\left(\frac{\partial E_r(T, P)}{\partial T}\right)_P = -\frac{1}{nF}\left(\frac{\partial \Delta g(T, P)}{\partial T}\right)_P = \frac{\Delta s(T, P)}{nF} \tag{1.39}$$

Clearly, the variation of E_r with temperature depends on the change in entropy for the particular fuel cell reaction, and three possible situations may arise:

1. If $\Delta s < 0$, like $H_2 + \frac{1}{2}O_2 \rightarrow H_2O(g)$, the reversible cell potential will decrease with cell operation temperature.
2. If $\Delta s > 0$, then the reversible cell potential will increase with temperature, e.g., for the reaction $C(s) + \frac{1}{2}O_2 \rightarrow CO$, the entropy change is about $+89\,J/K$.
3. If $\Delta s = 0$, then the reversible cell potential will be independent of temperature, like the reaction $CH_4 + 2O_2 \rightarrow CO_2 + 2H_2O(g)$.

For many useful electrochemical reactions, the entropy change is negative and is almost constant with the change of temperature to a good approximation, provided the temperature change $T - T_{ref}$ is not too large. Then Eq. (1.39) may be integrated from the standard reference temperature, $T_{ref} = 25°C$, to the arbitrary fuel cell operating temperature T, while keeping pressure P constant,

$$E_r(T, P) = E_r(T_{ref}, P) + \left(\frac{\Delta s(T_{ref}, P)}{nF} \right)(T - T_{ref}) \tag{1.40}$$

Alternatively, we can expand the reversible cell potential expression, Eq. (1.18), in Taylor series in terms of temperature, T, around the reference temperature, T_{ref}, keeping P = constant

$$E_r(T, P) = E_r(T_{ref}, P) + \left(\frac{\partial E_r(T_{ref}, P)}{\partial T} \right)_P (T - T_{ref})$$

Considering Eq. (1.39), the above equation becomes identical to Eq. (1.40).

It must be emphasized that the expression given in Eq. (1.40) is an approximation. Strictly speaking, the reversible cell potential at any temperature and pressure should be determined from Eq. (1.18) by calculating first the property changes for the particular fuel cell reaction involved. Such a procedure has been followed for the hydrogen and oxygen reaction to form gaseous water, and the results are presented in Figure 1.2. Clearly, the reversible cell potential indeed decreases almost linearly as temperature is increased over a large temperature range. However, it is noticed that the reversible cell potential is larger for product water as liquid at low temperatures, but it decreases much faster than the gaseous water as product when temperature is increased. So that at temperatures slightly above about 373 K, the reversible cell potential for liquid water product actually becomes smaller. This may seem curious, but it is because at such high temperatures pressurization is necessary to keep the product water in liquid

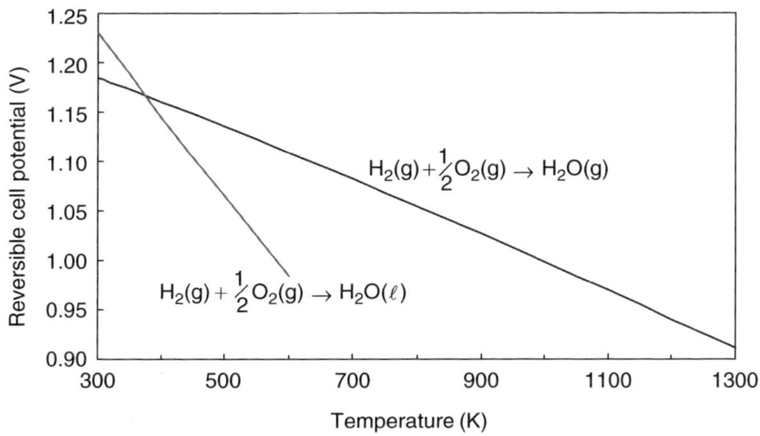

Figure 1.2 Effect of temperature on the reversible cell potential of a hydrogen–oxygen fuel cell for the reaction of $H_2 + ½O_2 \rightarrow H_2O$ at the pressure of 1 atm.

form as the reactants, hydrogen and oxygen, are fed at 1 atm. Also notice that the critical temperature for water is about 647 K, beyond which distinct liquid state does not exist for water, hence the shorter curves for the liquid water case shown in Figure 1.2.

As pointed out earlier, the entropy change for most of fuel cell reactions is negative; consequently the reversible cell potential decreases as temperature is increased as shown in Figure 2. However, for some few reactions such as

$$C(s) + ½O_2(g) \rightarrow CO(g)$$

the entropy change is positive, e.g., $\Delta s = 89\,J/(\text{mole fuel} \cdot K)$ at the standard reference temperature and pressure. As a result, the reversible cell potential for this type of reactions will increase with temperature.

Suppose N_P and N_R represent the number of moles of products and reactants, respectively, which are in gaseous state and on a per mole fuel basis, and $\Delta N = N_P - N_R$ represents the change, per mole fuel, in the number of moles of gas species during the reaction, then as a rough rule of thumb, it might be stated that:

1. $\Delta s > 0$ for $\Delta N > 0$ (due to increasing disorder because of more molecules in the product), and reversible cell potential increases with temperature.
2. $\Delta s < 0$ for $\Delta N < 0$ (due to decreasing disorder because of less molecules in the product), and reversible cell potential decreases with temperature.
3. $\Delta s \approx 0$ for $\Delta N = 0$, and reversible cell potential is almost independent of temperature.

Figure 1.3 shows the reversible cell potential as a function of temperature for a number of important fuel cell reactions. The aforementioned three trends of

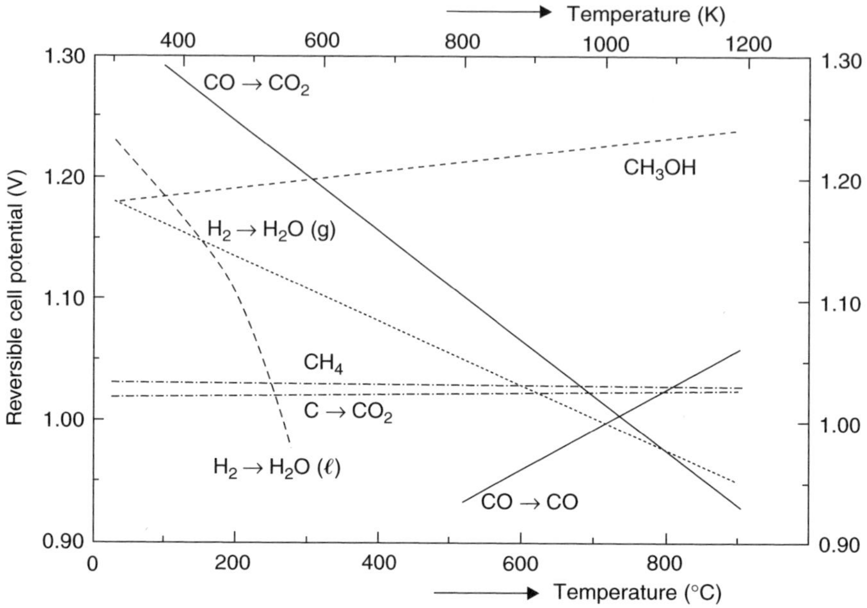

Figure 1.3 Standard reversible cell potential, E_r, as a function of temperature for the most important fuel cell reactions at the pressure of 1 atm [16].

variation for the reversible cell potential versus temperature can be clearly seen in the figure. For example, for methane reaction with oxygen,

$$CH_4(g) + 2O_2(g) \rightarrow CO_2(g) + 2H_2O(g)$$

and for solid carbon, C(s), reaction with oxygen,

$$C(s) + O_2(g) \rightarrow CO_2(g)$$

the change in the number of moles for the gaseous species is zero, and the reversible cell potential for these two reactions is almost a horizontal line, independent of temperature.

1.3.2 Effect of Pressure on Reversible Cell Potential E_r

Taking partial derivative of Eq. (1.18) with respect to pressure while keeping temperature constant, we obtain

$$\left(\frac{\partial E_r}{\partial P}\right)_T = -\frac{1}{nF}\left(\frac{\partial \Delta g}{\partial P}\right)_T = -\frac{\Delta v}{nF} \tag{1.41}$$

where

$$\Delta v = v_P - v_R \tag{1.42}$$

represents the volume change of all the gaseous species in the reaction, on a per mole fuel basis; the volume of the solids and liquids is much smaller than the volume of gas species, and can be neglected for the present purpose; v_P and v_R are the specific volume (per mole fuel) of gas products and reactants, respectively. Consider all the reactant and product gases can be treated as ideal gases, Eq. (1.42) can be expressed as

$$\Delta v = v_P - v_R = \frac{N_P \Re T}{P} - \frac{N_R \Re T}{P} = \frac{\Delta N \Re T}{P} \tag{1.43}$$

where $\Re$ is the universal constant. Combining Eqs. (1.41) and (1.43) yields

$$\left(\frac{\partial E_r}{\partial P} \right)_T = -\frac{\Delta N \Re T}{nF} \frac{1}{P} \tag{1.44}$$

Equation (1.44) indicates that

1. If $\Delta N > 0$, meaning product contains more moles of gas species than reactant, the reversible cell potential will decrease with pressure, such as for the reaction $C(s) + ½O_2(g) \rightarrow CO(g)$.
2. If $\Delta N < 0$, which is the case for most of useful fuel cell reactions, the reversible cell potential will increase with pressure.
3. If $\Delta N = 0$, the reversible cell potential will not change with pressure.

Note for (1) and (2) above, the amount of change for reversible cell potential decreases gradually when pressure is increased. Higher-pressure operation results in mechanical problems, such as mechanical strength of the cell components, cell sealing problem, corrosion, etc. This implies that the performance gain in E_r at high-pressure operation diminishes, and may become undesirable from system design point of view.

Now integrating Eq. (1.44) from the standard reference pressure, $P_{ref} = 1$ atm, to an arbitrary pressure, P, while keeping temperature fixed, results in

$$E_r(T, P) = E_r(T, P_{ref}) - \frac{\Delta N \Re T}{nF} \ln \left(\frac{P}{P_{ref}} \right) \tag{1.45}$$

This indicates that the pressure dependence of the reversible cell potential is a logarithmic function, and hence the dependence becomes weaker as pressure P is increased. Further, the pressure effect is small on the reversible cell potential at low temperatures, but this effect at high temperatures increases significantly because the pressure effect coefficient, $-\Delta N \Re T/(nF)$, is directly proportional to temperature. Figure 1.4 shows the pressure dependence of the reversible cell potential at two different temperatures for the reaction $H_2(g) + ½O_2(g) \rightarrow H_2O(g)$.

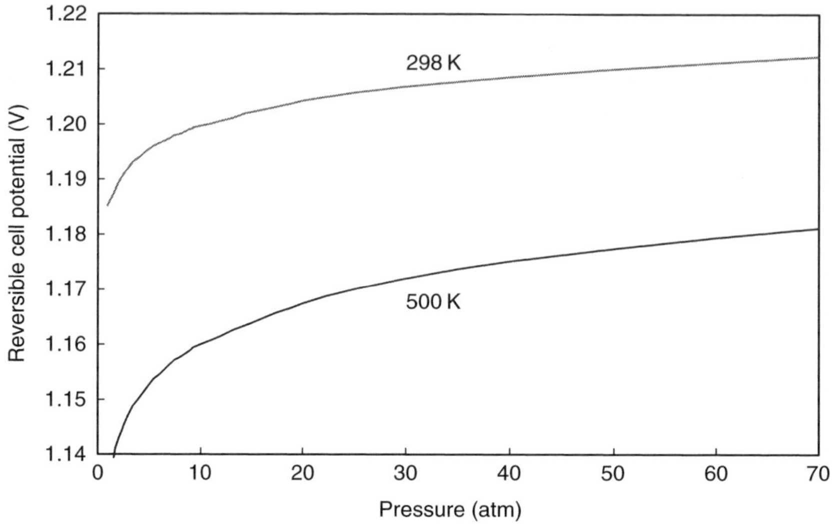

Figure 1.4 Standard reversible cell potential, E_r, as a function of pressure for the fuel cell reaction of $H_2(g) + ½O_2(g) \rightarrow H_2O(g)$ at the temperature of 298 K and 500 K, respectively.

It might be pointed out that for high temperature fuel cells the dependence of the actual cell potential E on the pressure follows closely the results given in Eq. (1.45), whereas a significant deviation occurs for the low temperature fuel cells. The difference arises from the fact that at high temperatures, the reaction kinetics is very fast and pressurization primarily increases the reactant concentration, hence better performance directly. At low temperatures, the reaction kinetics is slow and higher reactant concentration does not yield a proportional increase in the cell potential due to the significant cell potential loss associated with the slow kinetics.

1.3.3 Effect of Concentration on Reversible Cell Potential

All the results obtained so far for the reversible cell potential are, strictly speaking, valid for pure fuel and pure oxidant in the reactant streams for the fuel cell reaction; and pure reaction product in the exhaust stream. In reality, both fuel and oxidant streams are diluted by many other species for a variety of reasons. For example, H_2 as fuel is often humidified with water vapor before entering proton exchange membrane fuel cells for the hydration of the proton-conducting membrane; if H_2 is obtained from hydrocarbon fuel by reforming process (either steam reforming or partial oxidation reforming), then the fuel stream is a mixture of H_2 (about 50–70%), CO_2, $H_2O(g)$, CO, etc. On the other hand, oxygen in air is usually used as oxidant for most of commercial applications, and the oxidant stream is also required to contain sufficient amount of water vapor before entering the proton exchange membrane fuel cells, again for the hydration of the electrolyte membrane. Fuel cell performance, including reversible cell potential, is affected by the presence of the chemically inert diluents, which lowers the concentration of fuel and oxidant in the anode and cathode stream, and increases resistance to mass transport to the active sites. The effect of the chemically active species on the cell performance (such as CO poisoning of the polymer electrolyte membrane fuel cells) will not be considered for the analysis presented in this subsection and interested readers are referred to current literature, e.g., Refs [17,18].

Consider that the fuel, oxidant and exhaust streams, respectively, consist of a mixture of solids, liquids and ideal gases. Refer back to the schematic thermodynamic system for fuel cell shown in Figure 1.1, the first and second law analyses leading to Eqs. (1.14)–(1.23) are still valid because all inert diluents do not participate in the fuel cell reaction for energy conversion. However, the change for the Gibbs function, Δg, should be evaluated at the mixture temperature and the partial pressure of the fuel, oxidant and reaction product for the fuel, oxidant and exhaust stream, respectively.

Let P_i be the partial pressure of a component i in an ideal gas mixture of temperature T and total pressure P, then the Gibbs function of the component is $g_i(T, P_i)$ by definition, and also

$$g_i(T, P_i) = h_i(T) - Ts_i(T, P_i) \tag{1.46}$$

From Gibbs equation, Eq. (1.34), we have for component i

$$ds_i = \frac{dh_i}{T} - \frac{v}{T}dP = \frac{dh_i}{T} - \frac{\Re}{P}dP \tag{1.47}$$

where the equation of state for ideal gases, $pv = \Re T$, has been used to obtain the second equation in Eq. (1.47). Integration of the above equation from the partial pressure, P_i, to the mixture total pressure, P, while keeping the temperature fixed, yields

$$s_i(T, P) - s_i(T, P_i) = -\Re(\ln P - \ln P_i)$$

or

$$s_i(T, P_i) = s_i(T, P) - \Re \ln\left(\frac{P_i}{P}\right) \tag{1.48}$$

Substitution of Eq. (1.48) into (1.46) results in

$$g_i(T, P_i) = g_i(T, P) + \Re T \ln\left(\frac{P_i}{P}\right) \tag{1.49}$$

where

$$g_i(T, P) = h_i(T) - Ts_i(T, P) \tag{1.50}$$

is the Gibbs function for component i at the mixture temperature T and total pressure P.

Now consider the generalized electrochemical reaction given in Eq. (1.24)

$$\underbrace{\sum_{i=1}^{N} v_i' M_i}_{\text{Reactant}} \rightarrow \underbrace{\sum_{i=1}^{N} v_i'' M_i}_{\text{Product}} \tag{1.24a}$$

where N is the total number of species in the reacting system, including solids, liquids and gas species, and both electrochemically reacting and inert species as well. Since solids and liquids are treated as incompressible, the pressure has no effect on the value of the Gibbs function, or $g_i(T, P_i) = g_i(T, P)$ for incompressible substances.

Then the Gibbs function change for the above generalized reaction can be written, according to Eqs. (1.30) and (1.49), as

$$\Delta g(T, P_i) = \frac{1}{v_F'} \sum_{i=1}^{N} (v_i'' - v_i') g_i(T, P_i)$$
$$= \frac{1}{v_F'} \sum_{i=1}^{N} (v_i'' - v_i') g_i(T, P) + \frac{1}{v_F'} \sum_{i=1}^{N_g} (v_i'' - v_i') \Re T \ln\left(\frac{P_i}{P}\right)$$

or

$$\Delta g(T, P_i) = \Delta g(T, P) + \Re T \ln K \tag{1.51}$$

where $\Delta g(T, P)$ is the Gibbs function change for fuel, oxidant and exhaust streams at the system temperature T and pressure P; similarly, $\Delta g(T, P_i)$ is the Gibbs function change for fuel, oxidant and exhaust streams at the temperature T and respective partial pressure P_i; and

$$K = \prod_{i=1}^{N_g} \left(\frac{P_i}{P}\right)^{(v_i''-v_i')/v_F'} \tag{1.52}$$

is defined similarly to the equilibrium constant for partial pressure (but they are not the same); N_g is the total number of gas species in the reacting system, excluding the solid and liquid species; and for ideal gases the pressure ratio can be expressed in terms of the molar fraction X_i as follows:

$$X_i = \frac{P_i}{P} \tag{1.53}$$

Now substituting Eq. (1.51) into Eq. (1.18), we obtain the expression for reversible cell potential when inert diluents exist in the fuel, oxidant, and exhaust streams

$$E_r(T, P_i) = E_r(T, P) - \frac{\Re T}{nF} \ln K \tag{1.54}$$

This is the general form of the Nernst equation, representing the effect of the reactant and product concentrations on the reversible cell potential. When the reactant streams contain inert diluents for a given operating temperature and pressure, the diluents will cause a voltage loss for the reversible cell potential, and the amount of the loss is generally called the Nernst loss, and its magnitude is equal to the second term on the right-hand side of Eq. (1.54).

1.4 Energy Conversion Efficiency

There are several definitions for efficiency to describe the performance of a device, and usually two definitions for efficiency are used in thermodynamics; the first one is based on the First Law of thermodynamics, comparing the useful energy output with the total energy input. Using this as standard of reference, devices with different energy conversion processes can be compared against each other. The second definition is based on the Second Law of thermodynamics, which compares the actual performance of a device to the maximum possible work output. This indicates where there is room for improvement. By using the concept of exergy, the Second Law is a measure of efficiency relative to the maximum work potential of a system, hence the second definition of efficiency is also often called exergy efficiency. We shall show that both types of efficiency definition have been in use in fuel cells, as detailed below.

1.4.1 Definition of Energy Conversion Efficiency

The efficiency for any energy conversion process or system is often defined as

$$\eta = \frac{\text{Useful energy obtained}}{\text{Energy available for conversion that's an expense}}{}^{3} \tag{1.55}$$

Based on this definition, it is well known that 100% energy conversion efficiency is possible by the first law of thermodynamics, but is not possible by the second law of thermodynamics for many energy conversion systems that

[3] Note that in literature η is commonly used as efficiency in thermodynamics; whereas it is also conventionally used as overpotential, or voltage loss for fuel cell analysis, as in electrochemistry. In the present book, η is used for both in order to be consistent with the convention, and its meaning would become clear from the context.

produce power output by means of thermal energy, such as steam and gas turbines, internal combustion engines, which involve irreversible losses of energy. These thermal energy conversion systems are often referred to as heat engines.

On the other hand, there exist thermal energy conversion systems that have 100% or even higher efficiency. One of the examples is the refrigeration system such as heat pumps, air conditioners, etc. For these systems, often other performance measures are used, such as coefficient of performance (COP), instead of efficiency. Further, these systems are typically power consuming, instead of power generation.

Therefore, for the rest of the book, we will focus on the energy conversion efficiency of power generation systems, especially fuel cell; and how it compares with the efficiency of heat engine, a widely used existing and mature technology against which fuel cell is being compared and competing for the commercial application. Within this section, we will derive the reversible energy conversion efficiency, the best possible efficiency for fuel cells. This efficiency will be compared with the Carnot efficiency, which is the best possible efficiency for heat engines. Then we will demonstrate that these forms of efficiency are really identical, they are the different forms of the best possible efficiency under the same condition that is dictated by the second law of thermodynamics. The misconception that over 100% fuel cell efficiency is possible will be clarified, and other efficiencies associated with a fuel cell power plant will be introduced. Finally, additional energy loss mechanism in operating fuel cells will be described briefly. During this process, the amount of waste heat generation in the fuel cells will be determined, which is equal to the amount of cooling required for operating fuel cells.

1.4.2 Reversible Energy Conversion Efficiency for Fuel Cells

For the present fuel cell system described in Figure 1.1, the energy balance equation, Eq. (1.4), can be written as, on a per mole fuel basis

$$h_{in} - h_{out} + q - w = 0 \tag{1.56}$$

which indicates that the enthalpy change, $-\Delta h = h_{in} - h_{out}$, provides the energy available for conversion into the useful energy exhibited as work here, and it is the expense to be paid for the useful work output. At the same time, waste heat, q, is also generated, which would represent a degradation of energy. The amount of waste heat generated can be determined from the second law expression, Eq. (1.5) or (1.10), as

$$q = T\Delta s - T\wp_s \tag{1.57}$$

and the useful energy output as work is, from Eq. (1.14) or combining Eqs. (1.56) and (1.57)

$$w = -\Delta g - T\wp_s \tag{1.58}$$

Therefore, the energy conversion efficiency for the fuel cell system described in Figure 1.1 becomes, according to Eq. (1.55)

$$\eta = \frac{w}{-\Delta h} = \frac{\Delta g + T\wp_s}{\Delta h} \tag{1.59}$$

Note that both Δh and Δg are negative for power generation systems. By the second law, the entropy generation per unit mole of fuel is

$$\wp_s \geq 0 \tag{1.60}$$

and the equality holds for all reversible processes whereas entropy is always generated for irreversible processes. Therefore, the maximum possible efficiency allowed by the second law is, when the process is reversible (i.e., $\wp_s = 0$)

$$\eta_r = \frac{w_{max}}{-\Delta h} = \frac{\Delta g(T, P)}{\Delta h(T, P)} \tag{1.61}$$

Since both the enthalpy and Gibbs function change depend on the system temperature and pressure, so is the energy conversion efficiency. It should be pointed out that in the above derivation, no assumption specifically related to fuel cell has been made, and the only assumption made is that the energy conversion system for power production is reversible for all processes involved. Thus, Eq. (1.61) is valid for any power production system, be it electrochemical converter like fuel cells or conventional thermal energy converter like heat engines, as long as the process is reversible. Hence, it may be called the *reversible efficiency*, since it is the maximum possible efficiency that is allowed by the second law of thermodynamics. In what follows we will demonstrate that the maximum possible efficiency for conventional heat engines, the well-known Carnot efficiency, is really the reversible efficiency applied specifically to the conventional thermal power cycles, thus is equivalent to Eq. (1.61).

1.4.3 Reversible Energy Conversion Efficiency for Heat Engines

Consider a heat engine operating between two temperature thermal energy reservoirs (TER), one at a high temperature T_H and the other at a low temperature T_L, as shown in Figure 1.5. The heat engine obtains energy from the high temperature TER in the form of heat with the quantity q_H, a portion of this heat is converted to work output w and the remainder is rejected to the low temperature TER in the amount of q_L as waste heat. Applying the first and second laws to the 2T heat engine, we have

First law:

$$w = q_H - q_L \tag{1.62}$$

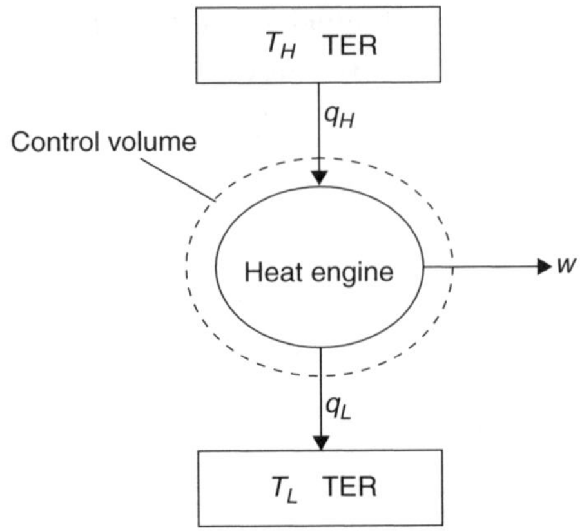

Figure 1.5 Thermodynamic system model of heat engines operating between two temperature TER.

Second law:
$$\wp_{s,\mathrm{HE}} = \frac{q_L}{T_L} - \frac{q_H}{T_H} \tag{1.63}$$

where $\wp_{s,\mathrm{HE}}$ represents the amount of entropy production during the energy conversion process by means of the heat engine. From Eq. (1.63), the amount of heat rejection can be determined as

$$q_L = \frac{T_L}{T_H} q_H + T_L \wp_{s,\mathrm{HE}} \tag{1.64}$$

The efficiency for the heat engine is, by the definition of Eq. (1.55)

$$\eta = \frac{w}{q_H} \tag{1.65}$$

Substituting Eqs. (1.62) and (1.64) into (1.65) yields

$$\eta = 1 - \frac{T_L}{T_H} - \frac{T_L}{q_H} \wp_{s,\mathrm{HE}} \tag{1.66}$$

As pointed out earlier, the second law of thermodynamics dictates that the entropy generation within the heat engine can never be negative, at most, it can vanish under the thermodynamically reversible condition. Therefore, the maximum possible efficiency for the heat engine is achieved if the process is reversible ($\wp_{s,\mathrm{HE}} = 0$):

$$\eta_{r,\mathrm{HE}} = 1 - \frac{T_L}{T_H} \tag{1.67}$$

This is the familiar Carnot efficiency, giving the upper bound for the efficiency of all 2T heat engines. Because $T_L < T_H$, the low temperature $T_L \neq 0$ by the third law of thermodynamics, and the high temperature T_H is finite, 100% efficiency is not possible by the second law for any energy conversion system that produces power output using heat engines, such as steam and gas turbines, internal combustion engines, etc. because of the second law requirement that the entropy generation term must never be negative. In contrast, 100% efficiency is always possible by the first law, which merely states the principle of energy conservation.

1.4.4 Equivalency of Carnot and Fuel Cell Efficiency

As shown above, both the Carnot efficiency and the reversible energy conversion efficiency for fuel cell, Eq. (1.61), are the maximum possible efficiency allowed by the second law, hence, they are called the irreversible efficiency; the former is applied specifically to heat engines, while the latter is derived for fuel cells. Therefore, they must be related somehow as they both are the maximum possible efficiency dictated by the second law. In this subsection we demonstrate that they are actually equivalent, just expressed in a different form, under a suitable condition for the comparison.

Suppose for the heat engine the high temperature TER is maintained at T_H by the combustion of a fuel with an oxidant, both reactants are originally at the temperature of T_L, as shown schematically in Figure 1.6. It is assumed that both the fuel and oxidant are the same as used in Figure 1.1 for the derivation of fuel cell performance, the combustion process is carried out at the same system pressure P in a controlled manner such that the combustion products leave the TER at the pressure P and temperature T_L. Neglecting the changes in the kinetic and gravitational potential energy, the first and second laws become for the high temperature TER

First law:
$$q_H = h_R - h_P = -\Delta h(T_L, P) \tag{1.68}$$

Second law:
$$\wp_{s,\mathrm{TER}} = (s_P - s_R) + \frac{q_H}{T_H} = \Delta s(T_L, P) + \frac{q_H}{T_H} \tag{1.69}$$

After rearranging, Eq. (1.69) gives the temperature T_H resulting from the combustion process

$$T_H = \frac{q_H}{\wp_{s,\mathrm{TER}} - \Delta s(T_L, P)} \tag{1.70}$$

Substitution of Eqs. (1.68) and (1.70) into (1.66) leads to

$$\eta = \frac{\Delta g(T_L, P)}{\Delta h(T_L, P)} + \frac{T_L}{\Delta h(T_L, P)}(\wp_{s,\mathrm{HE}} + \wp_{s,\mathrm{TER}}) \tag{1.71}$$

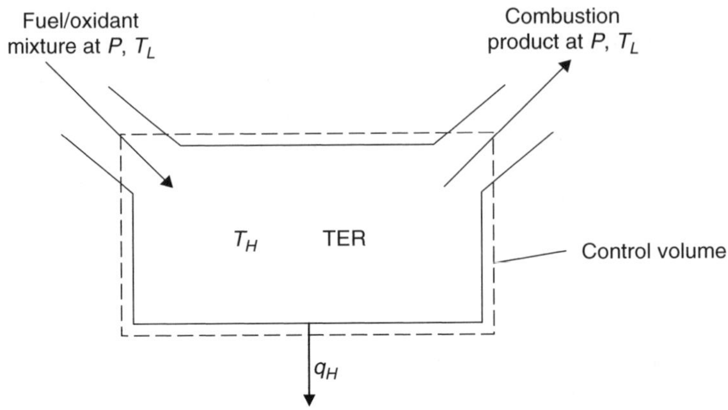

Figure 1.6 Thermodynamic system model of high temperature TER maintained by combustion process of a fuel/oxidant mixture.

where $\Delta g(T_L, P) = \Delta h(T_L, P) - T_L \Delta s(T_L, P)$ is the change in the Gibbs function between the reaction product and reactant. If all the processes within the heat engine and high temperature TER are reversible ($\wp_{s,\mathrm{HE}} = 0$ and $\wp_{s,\mathrm{TER}} = 0$), then Eq. (1.71) reduces to

$$\eta_r = \frac{\Delta g(T_L, P)}{\Delta h(T_L, P)} \tag{1.72}$$

which is exactly the same as the efficiency expression derived for fuel cells, Eq. (1.61). Note that in order for the combustion process to be reversible (i.e., $\wp_{s,\mathrm{HE}} = 0$) theoretically, there should be no product dissociations and no incomplete combustion products or byproducts (such as pollutants) formed, and the perfect combustion products should consist of stable chemical species only, as would be obtained from an ideal and complete stoichiometric reactions. Therefore, it may be stated that any reversible heat engine operating under the maximum temperature limit allowed by a "perfect" combustion of a fuel/oxidant mixture has the same efficiency as that of a reversible isothermal fuel cell using the same fuel and oxidant and operating at the same temperature as that of the low temperature TER. Or simply stated, the maximum possible efficiency is the same for both fuel cells and heat engines [14].

As a result, the often-heard statement that fuel cell is not subject to the Carnot efficiency and therefore it is more efficient [9,10] is partly "correct" and partly "incorrect," depending on the angle of viewpoint. It is "correct" because fuel cells do not require two temperature TERs to operate, and there are no T_H and T_L involved. Clearly the Carnot efficiency does not apply to fuel cell operations – this statement is preferable rather than the statement that a fuel cell is not subjected to the Carnot efficiency limitation. On the other hand, it is "incorrect" because the Carnot efficiency is in essence a specific form of the reversible efficiency which imposes upper limit on the performance of any energy conversion systems including both fuel cells and heat engines.

From the preceding analysis, it is clearly shown that both heat engines and fuel cells are subject to the same second law limitation. Then why do we often hear the statement that fuel cells have higher energy conversion efficiency than the corresponding heat engines? The reason may be several folds:

- The theoretical combustion temperature T_H is not achieved in practice: since flame temperature is usually above 3000 K for a perfect combustion, product dissociation occurs and incomplete combustion products are formed, such as CO, NO, C (particulates), OH, O, N, etc., leading to a lower temperature T'_H (usually about 2200 K for hydrocarbon fuels). In other words, combustion process is invariably irreversible, resulting in the degradation of useful energy. That is, a "perfect" combustion is unachievable in practice.
- Due to material (metallurgical) consideration, even T'_H has to be lowered in heat engines so that metal components can have sufficient mechanical strength. For example, the maximum allowable temperature for gas turbines is about 1000 K[4] (with good cooling of turbine blades).
- The heat rejected to the lower temperature TER is not achieved in practice at the temperature T_L, rather at a temperature T'_L which is higher than the ambient atmospheric temperature T_L, e.g., $T'_L \cong 550$ K[5] for gas turbines instead of about 300 K. Referring to Figure 1.6, it is not possible in reality for the hot combustion product to leave the high temperature TER at the low temperature of T_L, instead it is more likely at the high temperature T_H.

Therefore, the actual maximum efficiency

$$\eta'_r = 1 - T'_L/T'_H \cong 1 - 550/1000 = 45\%$$

which is only about half of the theoretical value $\eta_r = 1 - T_L/T_H \cong 1 - 300/3000 = 90\%$. Considering other related losses such as frictions, heat engines usually have lower than about 40% energy efficiency. For automobile engines, a rough estimate is that of chemical energy of fuels released as heat:

- 1/3 is lost to the cooling water (metallurgical requirement);
- 1/3 is lost to the environment by exhaust stream;
- 1/3 is converted to useful work output.

That is, the actual efficiency is only about 33%, much less than the Carnot efficiency – the maximum possible efficiency allowed by the second law.

[4] The maximum temperature for gas turbines has increased significantly in the last decade to about 1300–1400°C.

[5] The exhaust temperature for gas turbines is typically in the range of 300–600°C.

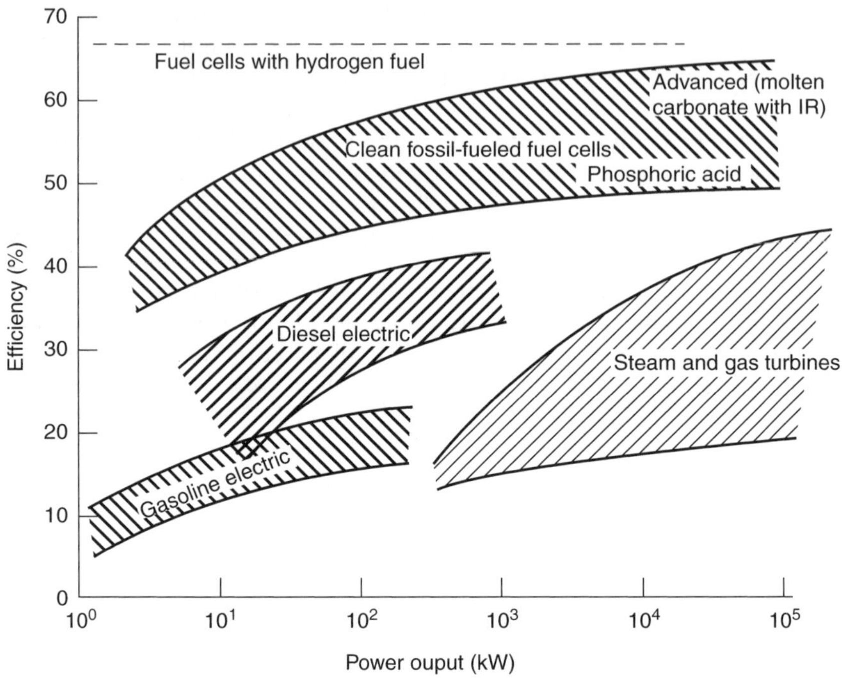

Figure 1.7 Comparison of practical energy conversion efficiency (based on LHV) of different technologies as a function of scale [5].

On the other hand, fuel cells can operate isothermally at a temperature sufficiently low so that no limitation imposed by materials and sufficiently close to the atmospheric temperature so that the degree of irreversibility arising from cooling requirement is much less than the corresponding heat engines. For example, proton exchange membrane fuel cells can be operated at $T \cong 80°C$, at which q_L is rejected to the ambient atmosphere at $T_L \cong 298\,K$ (or 25°C), compared to gas turbines of $T'_L \cong 550\,K$! Even though irreversibilities do occur for fuel cells, but at a lesser degree, so that fuel cells have a higher practical efficiency than the conventional heat engines. Figure 1.7 shows a comparison of the practical energy conversion efficiency for different power generation technologies as a function of scale [5], which clearly illustrates the superior performance of fuel cells over the conventional heat engines. Therefore, the important point about fuel cells is not that they are not Carnot efficiency limited, rather they are (i) free from incomplete reaction or product dissociation because of the much lower operating temperatures; (ii) free from the high temperature limit imposed by materials on any heat engines and (iii) less irreversibilities associated with heat rejection process.

On the other hand, it should be emphasized that just like any other energy conversion system, fuel cells can never achieve, although it is quite possible to achieve very closely, the maximum possible efficiency allowed by the second law. The various mechanisms of irreversible losses in fuel cells will be described later in this section.

Note the commercially available combined cycle (based on natural gas) has the chemical to electrical energy efficiency of around 58% or higher, and the total energy efficiency is much higher if cogeneration is considered [19].

1.4.5 The Possibility of Over 100% Fuel Cell Efficiency: Is It Real or Hype?

It is well known that no heat engine could have efficiency of 100% or more, including the ideal Carnot efficiency, as discussed earlier. However, it has been reported that the ideal fuel cell efficiency, η_r, according to Eq. (1.61), could be even over 100% in principle for some special fuel cell reactions, e.g., Appleby and Foulkes [2] and Appleby [8], even though it is unachievable in practice. This has also sometimes been used as evidence that fuel cells could have higher energy efficiency than the competing heat engines. Is this realistic even under the thermodynamically reversible condition? The answer is negative! With the following analysis, we can show that this is really due to a conceptual error in stretching the application of Eq. (1.61) beyond its range of validity.

Consider the thermodynamic model system used for fuel cell analysis shown in Figure 1.1. The amount of heat transfer from the surrounding thermal bath to the fuel cell system is given in Eq. (1.57) for practical fuel cells. Under the thermodynamically reversible condition, the amount of heat transfer becomes

$$q = T\Delta s = \Delta h - \Delta g \tag{1.73}$$

For most of fuel cell systems, Δs is negative (i.e., $\Delta s < 0$ just like Δh and Δg), indicating that heat is actually transferred from the fuel cell to the ambient environment, or heat is lost from the fuel cell system, rather than the other way around. Hence, the second law efficiency, according to Eq. (1.61)

$$\eta_r = \frac{\Delta g}{\Delta h} = \frac{\Delta h - T\Delta s}{\Delta h} = 1 - \frac{T\Delta s}{\Delta h} < 1 \tag{1.74}$$

is less than 100%, as it should be by the common perception of the parameter called efficiency.

However, for some special reactions, such as

$$C(s) + \tfrac{1}{2}O_2(g) \rightarrow CO(g) \tag{1.75}$$

the entropy change Δs is positive. Physically it indicates that the fuel cell absorbs heat from the ambiance and converts it completely into electrical energy along with the chemical energy of the reactants. This is equivalent to that the less useful form of energy – heat is converted completely into the more useful form of energy – electric energy without the generation of entropy (i.e., reversible condition) during the conversion process when Eq. (1.61) is used for the efficiency calculation – such a process is clearly a violation of the second law. Therefore, the second law efficiency for this particular fuel cell reaction becomes larger than 100%, i.e., an unphysical result, when Eq. (1.61) is utilized for the efficiency calculation for this type of fuel cell reactions. In fact, for the reaction shown in Eq. (1.75) and according to Eq. (1.61), the reversible fuel cell efficiency would be equal to $\eta_r = 124\%$ at the standard temperature and pressure, 163% at 500°C and 1 atm, and 197% at 1000°C and 1 atm!

The root of the problem from the straight forward application of Eq. (1.61) leading to the physically impossible result of over 100% energy efficiency is as follows. At atmospheric temperature for fuel cell operations, the energy from the thermal bath (or the atmosphere) as heat may be free. But at elevated temperatures, external means must be employed to keep the thermal bath at temperatures above the ambient atmospheric temperature, which constitutes an expense. Therefore, the heat from the thermal bath to the fuel cell system is no long a free energy input; rather it is part of the energy input that has to be paid for. By the definition, Eq. (1.55), the efficiency definition for fuel cells has to be modified accordingly, such that the ideal reversible efficiency will be no longer over 100% for fuel cells. Further, heat represents microscopically disorganized form of energy, while work or electrical energy is fully organized, by the second law of thermodynamics it is not possible to completely convert heat into work or electrical energy continuously in fuel cells. Thus, we conclude that the reversible fuel cell efficiency shown in Eq. (1.61) is only valid for fuel cell reactions where the entropy change between the product and reactant is negative (hence, heat is lost from the fuel cell), and it cannot be applied for reactions with positive entropy change, such as the one given in Eq. (1.75).

1.5 Practical Fuel Cell Efficiency and Energy Loss Mechanisms

From the preceding analysis, it is clear that energy loss in fuel cells occurs under both reversible and irreversible conditions. We will discuss each type of energy loss mechanisms and associated expression for energy conversion efficiency in fuel cells. Since this chapter is devoted to thermodynamic analysis, the reversible loss mechanism has been and will be presented in details, whereas the irreversible loss mechanisms will be described briefly only in this chapter.

1.5.1 Reversible Energy Loss and Reversible Energy Efficiency

The energy loss in fuel cells under reversible condition is equal to the heat transferred (or lost) to the environment, as given in Eq. (1.73), because of the negative entropy change for the fuel cell reaction. The associated energy conversion efficiency, which has been called the reversible efficiency, has been derived and given in Eq. (1.61) or (1.74). Combining Eq. (1.73) with (1.74) yields

$$\eta_r = \frac{\Delta g}{\Delta h} = \frac{\Delta g}{\Delta g + T\Delta s} \tag{1.76}$$

Dividing the numerator and the denominator by the factor (nF), and utilizing Eqs. (1.18) and (1.39), Eq. (1.76) becomes

$$\eta_r = \frac{E_r}{E_r - T\left(\frac{\partial E_r}{\partial T}\right)_P} \tag{1.77}$$

Therefore, when the entropy change is negative, as described earlier, the reversible efficiency, η_r, is less than 100% and the reversible cell potential decreases with temperature; and according to Eq. (1.77), the reversible efficiency, η_r, also decreases with temperature. For example, for H_2 and O_2 reaction forming gaseous water at 1 atm pressure

$$\left(\frac{\partial E_r}{\partial T}\right)_P = -0.2302 \times 10^{-3}\,\text{V/K}$$

at 25°C, and the reversible efficiency is about 95% at 25°C, and it becomes 88% at 600 K and 78% at 1000 K. Figure 1.8 illustrates the reversible efficiency as a function of temperature for the hydrogen and oxygen reaction with

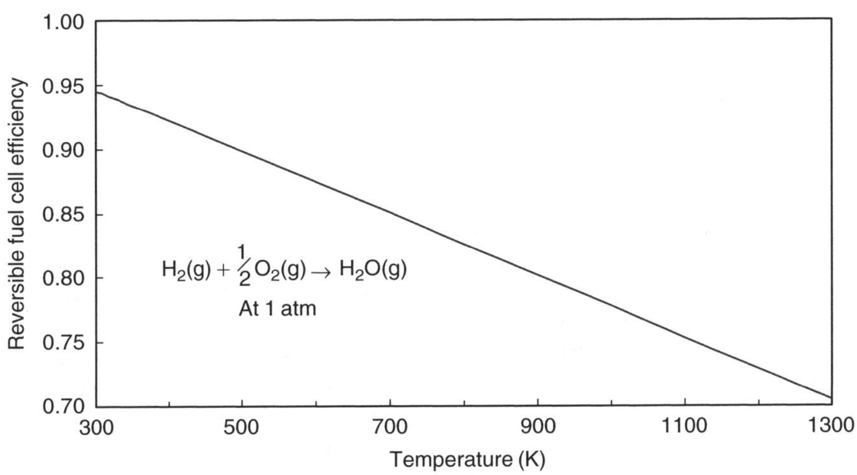

Figure 1.8 The reversible fuel cell efficiency (based on LHV) as a function of temperature for the reaction of $H_2 + ½O_2 \rightarrow H_2O(g)$ occurring at 1 atm pressure.

gaseous water as the reaction product. It is seen that the reversible efficiency decreases almost linearly. For most fuel cell reactions

$$\left(\frac{\partial E_r}{\partial T}\right)_P = -(0.1 \sim 1.0) \times 10^{-3}\,\mathrm{V/K}$$

at 25°C and 1 atm, hence the reversible efficiency is typically around 90%.

However, for the reaction of carbon and oxygen to form carbon monoxide, as shown in Eq. (1.75), the entropy change is positive, and the reversible cell potential increases with temperature, as presented previously; hence, the reversible efficiency will also increase with temperature, according to Eq. (1.77). But as discussed previously, the efficiency expression, Eq. (1.76) or (1.77), is not really valid for such reactions.

From the reversible energy efficiency, Eq. (1.61), and dividing the numerator and denominator by the factor (nF), we have, after utilizing the reversible cell potential, Eq. (1.18)

$$\eta_r = \frac{E_r}{(-\Delta h/nF)} = \frac{E_r}{E_{tn}} \tag{1.78}$$

where

$$E_{tn} = -\frac{\Delta h}{nF} \tag{1.79}$$

is called thermoneutral voltage (or potential), a voltage a fuel cell would have if all the chemical energy of the fuel and oxidant is converted to electric energy (i.e., 100% energy conversion into electricity). For example, for the reaction

$$H_2(g) + \tfrac{1}{2}O_2(g) \rightarrow H_2O(\ell)$$

$E_{tn} = 1.48\,\mathrm{V}$ and the corresponding reversible efficiency is $\eta_r = 83\%$ at 25°C and 1 atm. Whereas at the same temperature and pressure, for the reaction

$$H_2(g) + \tfrac{1}{2}O_2(g) \rightarrow H_2O(g)$$

$E_{tn} = 1.25\,\mathrm{V}$ and the corresponding reversible efficiency is $\eta_r = 95\%$.

From the above discussion, it is noted that for hydrogen and oxygen reaction, the reversible cell efficiency can differ by as much as 14%, depending on whether the product water is liquid or vapor, or whether the HHV or LHV is used for the efficiency calculation under identical operating condition. Since for most of hydrocarbon fuels that contain hydrogen (including hydrogen itself, hydrocarbons, alcohols and to a lesser extent coal), there exist two values for the change in the enthalpy and Gibbs function, e.g.,

$$\text{For natural gas (methane, } CH_4\text{):}\quad \frac{\mathrm{LHV}}{\mathrm{HHV}} = 0.90$$

For coals of typical hydrogen and water content: $\dfrac{\text{LHV}}{\text{HHV}} = 0.95\text{–}0.98$

Therefore, different efficiency values result, depending on which heating value ($-\Delta h$) is used for the efficiency calculation. Typically in fuel cell analysis, the HHV is used unless stated otherwise.

It should be emphasized that from the preceding analysis it is known that for most fuel cell reactions, the reversible efficiency, η_r, decreases as the fuel cell operating temperature is increased. This effect is important in considering high temperature fuel cells, namely the molten carbonate fuel cells (MCFCs) and the solid oxide fuel cells (SOFCs). For example, Figure 1.8 indicates that the reversible cell efficiency is reduced to lower 70% (based on LHV) for hydrogen and oxygen reaction at the typical operating temperature of 1000°C for SOFCs, as opposed to around 95% at 25°C as discussed above. This significant reduction in the reversible cell efficiency seems against high temperature fuel cells. However, the irreversible losses, to be described below, decrease drastically as temperature is increased, so that the practical fuel cell performance (such as efficiency and power output under practical operating condition) increases. Therefore, further analysis should be needed for efficiency under practical operating condition rather than the idealized reversible condition, which is the focus of the following discussion.

1.5.2 Irreversible Energy Losses and Various Forms of Irreversible Energy Efficiency

For fuel cells, the reversible cell potential and the corresponding reversible efficiency are obtained under the thermodynamically reversible condition, implying that there is no rigorous occurrence of continuous reaction or electrical current output. For practical applications, a useful amount of work (electrical energy) is obtained only when a reasonably large current I is drawn from the cells because the electrical energy output is through the electrical power output, which is defined as

$$\text{Power} = EI \quad \text{or} \quad \text{Power density} = EJ \tag{1.80}$$

However, both the cell potential and efficiency decrease from its corresponding (equilibrium) reversible values because of irreversible losses when current is increased. These irreversible losses are often called polarization, overpotential or overvoltage[6] in literature, and they originate primarily from three sources: activation polarization, ohmic polarization and concentration polarization. The actual cell potential as a function of current is the results of these polarizations; therefore, a plot of the cell potential versus current output is conventionally called a polarization curve. It should be noticed that

[6] The term "polarization", "overpotential" (or "overvoltage") has been loosely used in literature to denote cell potential (or voltage) loss. For their subtle differences, Ref. [1].

the magnitude of electrical current output depends largely on the active cell area, therefore, a better measure is the current density, J (A/cm^2), instead of current, I, itself; and the unit A/cm^2 is often used rather than A/m^2 as the unit for the current density because square meter is too large to be used for fuel cell analysis.

A typical polarization curve is illustrated in Figure 1.9 for the cell potential as a function of current density. The ideal cell potential-current relation is independent of the current drawn from the cell, and the cell potential remains equal to the reversible cell potential. The difference between the thermoneutral voltage and the reversible cell potential represents the energy loss under the reversible condition (the reversible loss). However, the actual cell potential is smaller than the reversible cell potential and decreases as the current drawn is increased due to the three mechanisms of irreversible losses: activation polarization, ohmic polarization and concentration polarization. The activation polarization, η_{act}, arises from the slow rate of electrochemical reactions, and a portion of the energy is lost (or spent) on driving up the rate of electrochemical reactions in order to meet the rate required by the current demand. The ohmic polarization, η_{ohm}, arises due to electrical resistance in the cell, including ionic resistance to the flow of ions in the electrolyte and electronic resistance to the flow of electrons in the rest of the cell components. Normally, the ohmic polarization is linearly dependent on the cell current. Concentration polarization, η_{conc}, is caused by the slow rate of mass transfer resulting in the

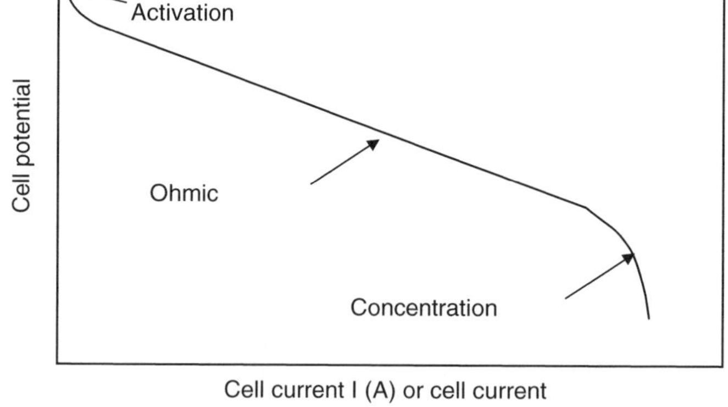

Figure 1.9 Schematic of a typical polarization curve. The cell potential for a fuel cell decreases as the current drawn from the cell is increased due to activation, ohmic and concentration polarizations.

depletion of reactants in the vicinity of active reaction sites and the overaccumulation of reaction products which block the reactants from reaching the reaction sites. It usually becomes significant, or even prohibitive, at high current density when the slow rate of mass transfer is unable to meet the high demand required by the high current output. As shown in Figure 1.9, concentration polarization is often the cause of cell potential decrease rapidly to zero. The current (density) corresponding to the zero cell potential is often called the limiting current (density), and evidently it is controlled by the concentration activation. From Figure 1.9, it is also clear that activation polarization occurs at small current density, while concentration polarization occurs at high current density. The linear drop in the cell potential due to resistance loss occurs at immediate current density, and practical fuel cell operation is almost always located within the ohmic polarization region.

Figure 1.9 also indicates that even at zero current output from the fuel cell, the actual cell potential is smaller than the idealized reversible cell potential. This small difference in cell potential is directly related to the chemical potential difference between the cathode and anode. So that even at zero external load current, there are electrons delivered to the cathode, where oxygen ions are formed, and migrate through the electrolyte to the anode where they deionize to release electrons. The electron released migrates back to the cathode to continue the process or "exchange." The ionization/deionization reactions preceding at a slow rate yield an extremely small current, often called exchange current I_0 or exchange current density J_0, and the cell potential is reduced below the reversible cell potential. Therefore, exchange current arises from the fact that electrons migrating through the electrolyte rather than through the external load, and about 0.1–0.2 V of cell potential loss results from the exchange process. Consequently, the efficiency of a real fuel cell is about (8–16%) lower than the reversible cell efficiency, η_r, even at close to zero current output.

The exchange current density J_0 is very small; it is at least about 10^{-2} A/cm^2 for H_2 oxidation at the anode, and about 10^{-5} times slower for O_2 reduction at the cathode. In comparison, the O_2 reduction process at the cathode is so slow that competing anodic reactions play a significant role, such as oxidation of electrocatalyst, corrosion of electrode materials, oxidation of organic impurities in the anode structure. All these anodic reactions result in the corrosion of electrodes, thereby limiting the cell life unless appropriate counter-measures are taken.

It should be pointed out that the cell potential loss resulting from the exchange current diminishes when the current drawn through the external load is increased beyond a certain critical value. As the external current is increased, the cell potential decreases as shown in Figure 1.9, thus the driving force for the exchange current is reduced, leading to a smaller exchange current – this is the only form of energy losses that decreases when external current is increased.

From the above discussion, it becomes clear that the actual cell potential E is lower than the reversible cell potential E_r, and the difference is due to the potential losses due to the above irreversible loss mechanisms. Therefore:

$$E = E_r - (\eta_{\text{act}} + \eta_{\text{ohm}} + \eta_{\text{conc}}) \tag{1.81}$$

The irreversible energy loss as heat (or waste heat generation) per mole fuel consumed can be easily obtained, because the entropy generation is, according to Eqs. (1.21) and (1.81)

$$\wp_s = \frac{nF(E_r - E)}{T} = \frac{nF(\eta_{\text{act}} + \eta_{\text{ohm}} + \eta_{\text{conc}})}{T} \tag{1.82}$$

Then, Eq. (1.10) becomes for the total heat loss from the fuel cell

$$q = T\Delta s - T\wp_s = \underbrace{T\Delta s}_{\text{Reversible loss}} - \underbrace{nF(\eta_{\text{act}} + \eta_{\text{ohm}} + \eta_{\text{conc}})}_{\text{Irreversible losses}} \tag{1.83}$$

Since the entropy change is negative ($\Delta s < 0$) for most fuel cell reactions, the heat generated is negative as well, implying that energy as heat is lost from the fuel cell shown in Figure 1.1 for both reversible and irreversible losses.

Because $T\Delta s = \Delta h - \Delta g$ by the definition of the Gibbs function change for fuel cell reactions, Eq. (1.83) can be written as

$$\frac{q}{nF} = \frac{T\Delta s}{nF} - (\eta_{\text{act}} + \eta_{\text{ohm}} + \eta_{\text{conc}}) = \frac{\Delta h - \Delta g}{nF} - (\eta_{\text{act}} + \eta_{\text{ohm}} + \eta_{\text{conc}}) \tag{1.84}$$

Considering the definition for the thermoneutral voltage and the reversible cell potential, the above expression becomes

$$\frac{q}{nF} = -E_{tn} + E_r - (\eta_{\text{act}} + \eta_{\text{ohm}} + \eta_{\text{conc}}) \tag{1.85}$$

Combining with Eqs. (1.81) and (1.85) reduces to

$$\frac{q}{nF} = -(E_{tn} - E) \tag{1.86}$$

Hence, the equivalent cell potential loss due to the energy loss from the fuel cell as heat is equal to the difference between the thermoneutral voltage and the actual cell potential.

The rate of heat loss per mole fuel consumed in the fuel cell, Eq. (1.83), can be expressed as an equivalent power loss:

$$P_{\text{heat loss}} = I\left(\frac{q}{nF}\right) = -I(E_{tn} - E) = I\left[\frac{T\Delta s}{nF} - (\eta_{\text{act}} + \eta_{\text{ohm}} + \eta_{\text{conc}})\right] \tag{1.87}$$

This expression is important in determining the cooling requirements of fuel cell stacks.

After the above description of the irreversible energy losses, we can now introduce several forms of energy efficiency that would be useful in the analysis of fuel cell performance.

1.5.2.1 Voltage Efficiency η_E

The voltage efficiency is defined as:

$$\eta_E = \frac{E}{E_r} \tag{1.88}$$

Because the actual cell potential E is compared with the maximum possible cell potential E_r allowed by the second law, the voltage efficiency is really a specific form of the exergy efficiency, representing the degree of departure of the cell operation from the idealized thermodynamically reversible condition. As shown in Eq. (1.81), $E < E_r$, hence $\eta_E < 1$. For example, for H_2 and O_2 reaction at the standard reference state forming liquid water, we know the reversible cell potential is equal to 1.229 V. If such a cell is operating at a cell potential of 0.7 V, the corresponding voltage efficiency can be determined as

$$\eta_E = \frac{E}{E_r} = \frac{0.7}{1.229} = 0.57\%$$

For the same cell, but if the reaction product water is gaseous, then the corresponding voltage efficiency becomes

$$\eta_E = \frac{E}{E_r} = \frac{0.7}{1.185} = 0.59\%$$

1.5.2.2 Current Efficiency η_I

The current efficiency is a measure of how much current is produced from a given amount of fuel consumed in fuel cell reaction; and it is defined as

$$\eta_I = \frac{I}{nF\left(\frac{dN_F}{dt}\right)} \tag{1.89}$$

where $\frac{dN_F}{dt}$ represents the rate of fuel consumption in the fuel cell (mol/s).

The current efficiency would be less than 100% if part of the reactants participates in non-current-productive side reactions, called parasitic reactions, such as reactants cross-over the electrolyte region, incomplete conversion of reactants to desired products, reaction with the cell components, or even reactant leakage from the cell compartment due to sealing problem, etc. For most practical fuel cells, especially at operating conditions where the current output is sufficiently larger than zero (without the effect of exchange current discussed previously), the current efficiency is about 100%. This is because for practical fuel cells, all the parasitic reactions are undesirable and would have been removed by appropriate design.

However, for direct methanol fuel cells about 20% of the liquid methanol can cross-over to the cathode side through the proton-conducting polymer membrane, implying the current efficiency is only about 80% for such cells. Similarly, if the polymer membrane is too thin, say 50 μm or thinner, hydrogen crossover in a proton exchange membrane fuel cell may not be negligible, depending on the operating current density drawn from the cell. On the other hand, hydrogen cross-over is minimal and can be neglected if the membrane used is sufficiently thick, such as Nafion 117 – a commonly used polymer membrane electrolyte for proton exchange membrane fuel cells. Therefore, specific care should be taken for the current efficiency in practical fuel cells.

1.5.2.3 Overall Free Energy Conversion Efficiency, η_{FC}

The overall free energy conversion efficiency is defined as the product of the reversible efficiency, voltage and current efficiency:

$$\eta_{FC} = \eta_r \times \eta_E \times \eta_I \tag{1.90}$$

If the current efficiency is 100% as is often the case for well-designed practical fuel cells, substituting the definitions for the various efficiencies into the above equation lead to

$$\eta_{FC} = \frac{\Delta g}{\Delta h} \times \frac{E}{E_r} \times 1 = \frac{E_r}{E_{tn}} \times \frac{E}{E_r} \times 1 = \frac{E}{E_{tn}}$$

or

$$\eta_{FC} = \frac{E}{E_{tn}} \tag{1.91}$$

Therefore, the overall free energy conversion efficiency is really the overall efficiency for energy conversion process occurring within the fuel cell. Because the thermoneutral voltage is a fixed value for a given fuel and oxidant under a given operating condition of temperature and pressure, the overall energy conversion efficiency for fuel cells is proportional to the actual cell potential

$$\eta_{FC} \sim E \tag{1.92}$$

This is a significantly important result. Once the actual cell potential is determined, the energy conversion efficiency of the fuel cell is known as well. This is the primary reason that in fuel cell literature, it is almost always that the cell polarization curve is given without specifically showing the cell energy efficiency as a function of the current. Further, Eq. (1.92) implies that the fuel cell efficiency will depend on the current output in the same way as the cell potential, i.e., decrease as the current output is increased.

1.5.2.4 Fuel Cell System Efficiency η_s

A fuel cell system is composed of one or multiple fuel cell stacks and a number of auxiliary equipment, which would also have its own energy efficiency of η_{aux}. Therefore, the total fuel cell system efficiency is equal to the product of the efficiency values for the fuel cell and all the auxiliary equipment:

$$\eta_s = \eta_{FC} \times \eta_{aux} \tag{1.93}$$

Figure 1.7 shows a comparison of fuel cell system efficiency with the efficiency of a number of competing technologies as a function of scale. The efficiency is determined based on the LHV.

1.5.3 Efficiency Loss in Operating Fuel Cells: Stoichiometry, Utilization and Nernst Loss

In an operating fuel cell, reactant composition changes between the inlet and outlet of the fuel cell along the flow path over the electrode surface because reactants are consumed to yield current output and reaction products are formed along the way as well. The change in reactant composition results in additional loss of cell potential beyond those losses described in the preceding section. This potential loss arises from the fact that the cell potential E adjusts to the lowest electrode potential given by the Nernst equation, Eq. (1.54), for the various reactant compositions at the exit of the anode and cathode chambers. This is because electrodes are usually made of good electronic conductors and consequently they are iso-potential surfaces. The cell potential E may not exceed the minimum local value set by the Nernst equation. This additional cell potential loss is often also called the Nernst loss, which is equal to the difference between the inlet and exit Nernst potentials determined based on the inlet and exit reactant compositions. According to Eq. (1.54), this additional cell potential loss due to the consumption of reactants in the cell is, when the reactant streams are arranged in a concurrent flow

$$\eta_N = \frac{\Re T}{nF} \ln K_{out} - \frac{\Re T}{nF} \ln K_{in} = \frac{\Re T}{nF} \ln \frac{K_{out}}{K_{in}} \tag{1.94}$$

where K_{in} and K_{out} are the equilibrium constant for partial pressure evaluated at the cell inlet and outlet gas compositions.

In the case of a fuel cell where both fuel and oxidant flow is in the same direction (concurrent), the minimum Nernst potential occurs at the flow outlet. When the reactant flows are in counter-flow, cross-flow, or more complex arrangements, it becomes difficult to determine the location of the minimum potential due to the reactant consumption. Appropriate flow channel design for the anode and cathode side can minimize the Nernst loss.

Equation (1.94) also implies that the Nernst loss will be extremely large, approach infinity if all the reactants are consumed in the in-cell electrochemical reaction leading to zero reactant concentration at the cell outlet. To reduce the Nernst loss to an acceptable level for practical fuel cell operations, reactants are almost always supplied more than the stoichiometric amount required for the desired current production. The actual amount of reactants supplied to a fuel cell is often expressed in terms of a parameter called stoichiometry, S_t:

$$S_t = \frac{\text{Molar flow rate of reactants supplied to a fuel cell}}{\text{Molar flow rate of reactants consumed in the fuel cell}} = \frac{\dot{N}_{in}}{\dot{N}_{consumed}} \tag{1.95}$$

For example, for proton exchange membrane fuel cells, typical operation uses:

$S_t \approx (1.1{\sim}1.2)$ for H_2,
$S_t \approx 2$ for O_2 (pure or in air).

Therefore, stoichiometry really represents the actual flow rate for the reactant delivered to the fuel cell, or how much reactant is consumed in the fuel cell for current production for a given amount of reactant supply. Because there are normally at least two types of reactant for fuel cell, one as fuel and another as oxidant, stoichiometry can be defined for either reactant. For MCFC, there are two reactants in the oxidant stream, namely, CO_2 and O_2, then the stoichiometry for the oxidant is commonly defined for the species in deficiency.

Alternatively, reactant flow rate can be expressed in terms of a parameter called utilization, U_t:

$$U_t = \frac{\text{Molar flow rate of reactants consumed in a fuel cell}}{\text{Molar flow rate of reactants supplied into the fuel cell}} = \frac{\dot{N}_{consumed}}{\dot{N}_{in}} = \frac{1}{S_t} \tag{1.96}$$

Clearly, stoichiometry and utilization are inversely proportional to each other. Although both parameters are used in practice, stoichiometry is more frequently used in literature for proton exchange membrane fuel cells, and utilization is often used for intermediate and high temperature fuel cells (Phosphoric acid fuel cells (PAFCs), MCFCs and SOFCs).

For properly designed practical fuel cells, no reactant cross-over or leakage out of the cell may occur in general, therefore, the rate of reactant consumed within the cell is equal to the difference between the molar flow rate into and exiting the cell. Because there are two reactants for fuel cell reactions: a fuel and an oxidant, the parameter stoichiometry (or utilization) can be defined, respectively, for the fuel and oxidant. For example, the stoichiometry for the fuel may be expressed as

$$S_{t,F} = \frac{\dot{N}_{F,in}}{\dot{N}_{F,consumed}} = \frac{\dot{N}_{F,in}}{\dot{N}_{F,in} - \dot{N}_{F,out}} = \frac{1}{U_{t,F}} \tag{1.97}$$

The stoichiometry or utilization for the oxidant can be written similarly.

Effect of reactant utilization on the reversible cell potential is illustrated in Figure 1.10 with the corresponding reactant composition at the cell outlet, in terms of mole fractions, as a function of utilization given in Table 1.2. It is seen that the reactant composition at the cell outlet decreases, hence the reversible cell potential decreases as well when the utilization factor is increased. The decrease is rapid when utilization goes beyond about 90%. In practical fuel cell operation, 100% utilization (or unity stoichiometry) will result in reactant concentrations vanishing at the cell exit, then the Nernst loss becomes dominant and the cell potential is reduced to zero – this is certainly undesirable situation

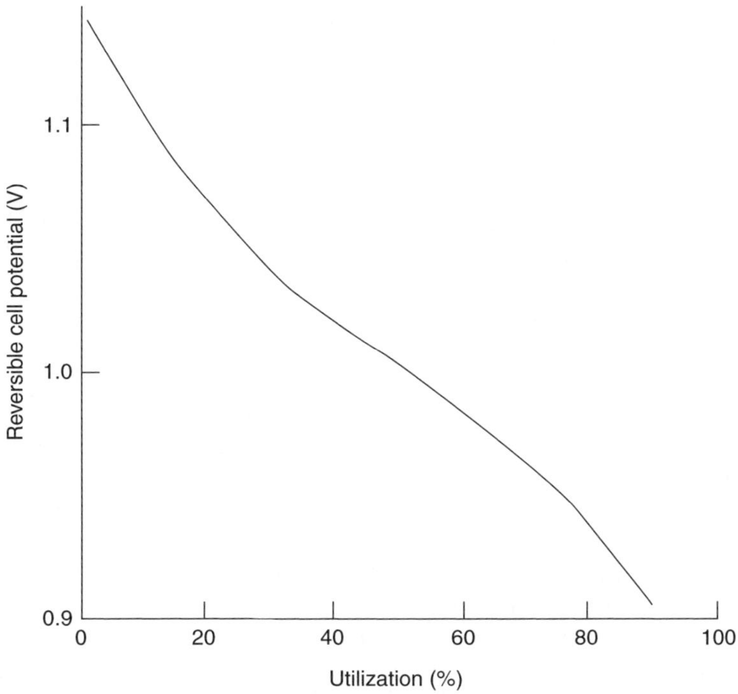

Figure 1.10 Reversible cell potential as a function of reactant utilization (both fuel and oxidant utilizations are set equal) for an MCFC operating at 650°C and 1 atm. Reactant compositions at the cell inlet: 80% H_2/20% CO_2 mixture saturated with $H_2O(g)$ at 25°C for the fuel gas; and 60% CO_2/30% O_2/10% inert gas mixture for the oxidant gas [4].

Table 1.2 Reactant composition[a] at cell outlet as a function of utilization for MCFC operating at 650°C and 1 atm [4]

Reactant	Utilization[b] (%)				
	0	25	50	75	90
Anode[c]					
X_{H_2}	0.645	0.410	0.216	0.089	0.033
X_{CO_2}	0.064	0.139	0.262	0.375	0.436
X_{CO}	0.130	0.078	0.063	0.033	0.013
X_{H_2O}	0.161	0.378	0.458	0.502	0.519
Cathode[d]					
X_{CO_2}	0.600	0.581	0.545	0.461	0.316
X_{O_2}	0.300	0.290	0.273	0.231	0.158

[a] Reactant compositions (H_2 and CO for anode and O_2 and CO_2 for cathode) are given in mole fractions in the table.
[b] Same utilization for fuel and oxidant is used for the calculation.
[c] Fuel gas composition at the anode inlet: 80% H_2/20% CO_2 saturated with H_2O at 25°C; fuel gas composition at the anode outlet: calculated based on compositions for water–gas shift reaction in equilibrium.
[d] Oxidant gas composition at the cathode inlet: 30% O_2/60% CO_2/10% inert gas; oxidant gas composition at the cathode outlet: calculated based on the utilization or amount of consumption in the cell.

that needs to be avoided. Therefore, typical operation requires that the utilization be about (80–90%) for fuel, and 50% for oxidant in order to balance the Nernst loss with the parasitic losses associated with the reactant supply.

As shown in Eq. (1.94), the additional Nernst loss due to the reactant depletion in the cell is directly proportional to the cell operating temperature. For example, for a H_2-air fuel cell, a change in the reactant gas composition from the cell inlet to the outlet that leads to a 60 mV cell potential loss at near room temperature (25°C) would result in a loss of 300 mV at 1200°C. Therefore, the additional Nernst loss arising from the reactant gas composition changes within the cell becomes more significant when the cell operating temperature is increased, and may become a serious loss mechanism for high temperature fuel cells. Figure 1.11 shows the reversible cell potential at the cell inlet and outlet for hydrogen and oxygen reaction forming gaseous water product at 1 atm as a function of temperature. The outlet Nernst potentials (i.e., the reversible cell potential at the cell outlet) are determined for oxygen utilization of 50%, and hydrogen utilizations of 85%, 90% and 95%, respectively; as well as for the utilization of 95% for both hydrogen and oxygen. It is clearly seen that the outlet Nernst potential decreases when either utilization or temperature is increased.

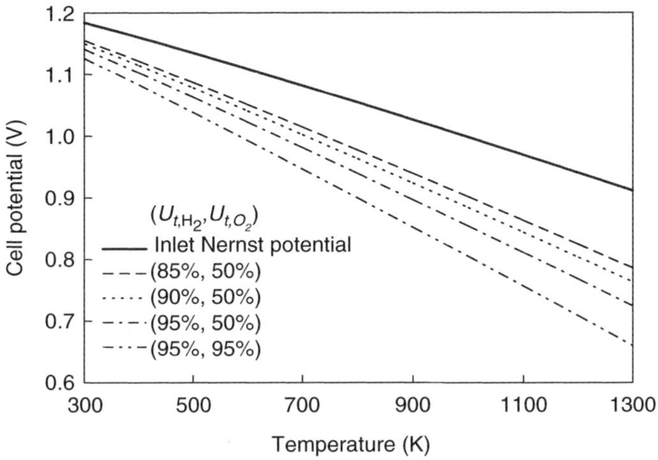

Figure 1.11 Inlet and outlet Nernst potential as a function of temperature and utilization for the reaction of $H_2(g) + \frac{1}{2}O_2(g) \rightarrow H_2O(g)$ at 1 atm.

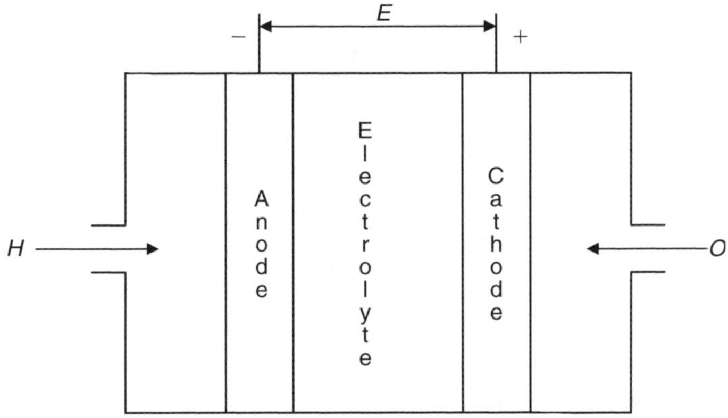

Figure 1.12 Schematic of dead-end design for reactant supply to a fuel cell when pure hydrogen and pure oxygen are used as reactants.

If pure hydrogen is used as fuel, the anode compartment can be designed as a dead-end chamber for hydrogen supply, as shown in Figure 1.12. Similarly, if pure oxygen is used as oxidant, a dead-end cathode compartment can be employed. However, inert impurities in the reactant gas will accumulate at the anode and cathode compartment, and they must be removed either periodically or continuously in order to maintain a good fuel cell performance. Periodic purging or continuous bleeding can be implemented for this purpose, but these results in a small loss of fuel, and hence less than 100% utilization.

From the above discussion, it is evident that 100% utilization for reactants is practically an unwise design. Since in-cell fuel utilization will never be 100% in practice, the determination of in-cell energy conversion efficiency and the cell potential must take utilization factor into consideration. If the fuel exiting the fuel cell is discarded (not recirculated back to the cell or not utilized for other useful purpose such as providing heat for fuel preprocessing), then the overall energy conversion efficiency must be equal to the overall fuel cell efficiency given in Eq. (1.90) multiplied by the utilization to take into account the fact that not all the fuel is being used for electric energy production.

1.6. Concluding Remarks

This chapter is focused on the analysis of fuel cell performance (cell potential and energy conversion efficiency) under the idealized reversible condition, and misconception regarding fuel cell energy conversion efficiency is clarified. The effect of operating conditions, namely, temperature, pressure and reactant concentration, on the reversible cell potential is investigated. It is shown that both fuel cells and heat engines have the same maximum theoretical efficiency, which is equivalent to the Carnot efficiency, when operating on the same fuel and oxidant. However, fuel cells are free from the high temperature limit composed by materials on heat engines and less irreversibilities associated with heat rejection. Hence, fuel cells can have higher practical efficiencies. Further, possibility of over 100% fuel cell efficiency is ruled out from the strict consideration of definition for efficiency and the second law of thermodynamics. Both reversible and irreversible energy loss mechanisms are described for fuel cells, waste heat generation in a fuel cell is provided, and various forms of efficiency are defined. Finally, the Nernst potential loss arising from the reactant consumption in practical cell is considered, and issues related to reactant utilization are outlined.

1.7 Nomenclature

C_P	Specific heat at constant pressure
$\bar{c}_P$	Average specific heat at constant pressure (J/mol K)
E	Cell potential (V)
E_r^o	Standard reversible cell potential (V)
E_r	Reversible cell potential (V)
e	Electric charge per electron (=1.6023 × 10^{-19} Coulomb/electron)
F	Faraday constant (=96,487 Coulomb/mole electron)
G	Gibbs free energy (J)
g	Specific Gibbs free energy (J/mol)
H	Enthalpy (J)
HHV	Higher heating value (J)
h_P, h_R	Specific enthalpy of products and reactants (J/mol)
$h_{f,i}$	Specific enthalpy of formations of species i (J/mol)
$h_{s,i}$	Specific sensible of enthalpy of species i (J/mol)

$h^{\circ}_{f,i}$	Specific enthalpy of formations of species *i* at reference conditions (J/mol)
I_0	Exchange current (A)
J_0	Exchange current density (A/m^2)
LHV	Lower heating value (J)
Mi	Chemical formula for species *i*
N	Number of moles
$\dot{N}$	Rate of mole (mole/s)
n	Number of mole electron
P	Pressure (pas)
$\dot{Q}$	Rate of heat (W)
q	Heat (J/mole)
S	Entropy (J/mol K), stoichiometry
T	Temperature (K)
$\dot{W}$	Power (W)
w	Specific work (J/mol)
X	Mole fraction
ν'_i, ν''_i	Number of moles for species *i* in the reactant and products mixture
ν'_F	Number of moles of fuel

Greek

Δ	A finite difference of a property
η	Efficiency
μ	Chemical potential
ν	Stoichiometric reaction coefficient
$\Re$	Universal gas constant
ℓ	Liquid
$\wp_s$	Entropy production (J/mol)
$\wp_{s,\mathrm{HE}}$	Entropy production in heat engine
$\wp_{s,FC}$	Entropy production in fuel cell
$\wp_{s,\mathrm{TER}}$	Entropy production in thermal energy reservoir
K	Equilibrium constant

Subscripts

act	Activation overpotential
conc	Concentration
C.V.	Control volume
elec	Electrical work
EX	Exhaust
F	Fuel
FC	Fuel cell
g	Gas
H	High temperature
I	Current

i	The chemical species
L	Low temperature
Max	Maximum
ohm	Ohmic resistance
OX	Oxidation
aux	Auxiliary
P	Products of combustion
R	Difference between products and reactants
S	Entropy
tn	Thermoneutral overpotential
0	Standard temperature and pressure condition

REFERENCES

1. X. Li, *Principles of Fuel Cells*, Taylor and Francis, New York, 2006.
2. A. J. Appleby and F. R. Foulkes, *Fuel Cell Handbook,* Van Nostrand Reinhold, New York, 1989.
3. L. J. M. J. Blomen and M. N. Mugerwa (Ed.), *Fuel Cell Systems*, Plenum Press, New York, 1993.
4. J. H. Hirschenhofer, D. B. Stauffer, and R. R. Engleman, *Fuel Cells A Handbook* (Revised 3), U.S. Department of Energy, 1994.
5. K. Kordesch and G. Simader, *Fuel Cells and Their Applications*, VCH, New York, 1996.
6. A. J. Appleby, Fuel cell technology: status and future prospects. *Energy*, 21 (1996) 521–653.
7. W. Vielstich, H. Gasteiger, A. Lamm (Ed.,), *Handbook of Fuel Cells – Fundamentals, Technology, Applications*, Wiley, 2003.
8. A. J. Appleby, Characteristics of fuel cell systems. In *Fuel Cell Systems* (Eds.) L. J. M. J. Blomen and M. N. Mugerwa), Plenum Press, New York, 1993.
9. K. Wark Jr., *Advanced Thermodynamics for Engineers,* McGraw-Hill, New York 1995.
10. C. Haynes, Clarifying reversible efficiency misconceptions of high temperature fuel cells in relation to reversible heat engines. *J. Power Sources*, 92 (2001) 199–203.
11. E. A. Lutz *et al.*, Thermodynamics comparison of fuel cells to Carnot cycle. *Int. J. Hydrogen Energy*, 27 (2002) 1103–1111.
12. S. E. Wright, Comparison of the theoretical performance potential of fuel cells and heat–engines, *J. Renewable Energy*, 29 (2004) 179–195.
13. H. Hassanzadeh and S. H. Mansouri, Efficiency of ideal fuel cell and Carnot cycle from a fundamental perspective, *Proceedings of IMechE* Vol. 219 Part A: *J. Power Energy*, 2005.
14. Xianguo Li, Fuel cells and thermodynamic performance, 8th Annual international Conference ASME, Energy Week, 28–30 January 1997.
15. Xianguo Li, Fuel cells – the environmentally friendly energy converter and power generator, *Int. J. Global Energy Issues*, Vol. 17, Nos. ½, 2002.
16. E. Barendrech, Electrochemistry of fuel cells. In *Fuel Cell Systems* (Eds. L. J. M. J. Blomen and M. N. Mugerwa), Plenum Press, New York, 1989.
17. J. J. Baschuk and Xianguo Li, Carbon monoxide poisoning of proton exchange membrane fuel cells, *Int. J. Energy Res.*, 25 (2001) 695–713.
18. J. J. Baschuk and Xianguo Li, Mathematical model of PEM fuel cell incorporating CO poisoning and O_2(air) bleeding, *Int. J. Global Energy Issues*, 20(3) (2003) 245–276.
19. M. J. Moran and H. N. Shapiro, *Fundamentals of Engineering Thermodynamics*, 3rd edn Wiley, New York, 1995.

CHAPTER TWO

Macroscopic Modeling of Polymer-Electrolyte Membranes

Adam Z. Weber *and* John Newman

Contents

Lawrence Berkeley National Laboratory, Department of Chemical Engineering, University of California, Berkeley, CA 94720-1462, USA

Advances in Fuel Cells
0080453945

Abstracts

In this chapter, the various approaches to the macroscopic modeling of transport phenomena in polymer-electrolyte membranes are discussed. This includes general background and modeling methodologies, as well as exploration of the governing equations and some membrane-related topic of interest.

2.1 INTRODUCTION

The polymer-electrolyte or proton-exchange membrane (PEM) is the heart of the so-called PEFC sandwich (i.e., membrane, catalyst layers, and diffusion media), and hence, the entire fuel cell. It is this electrolyte that makes polymer-electrolyte fuel cells (PEFCs) unique. Correspondingly, the electrolyte must have very specific properties; it needs to conduct protons but not electrons, as well as inhibit gas transport in the separator but allow it in the catalyst layers. Furthermore, the membrane is one of the most important items in the crucial topic of PEFC water management. It is for these reasons as well as others that modeling and experiments of the membrane have been pursued more than any other layer [1].

Although there have been various membranes used, none is more researched or seen as the standard than the Nafion® family by E. I. du Pont de Nemours and Company. Like almost all PEFC membranes, Nafion® is a copolymer with essentially hydrophobic and hydrophilic moieties. Specifically, Nafion® is a copolymer of polytetrafluoroethylene and polysulfonyl fluoride vinyl ether; its formula is given in Figure 2.1. These perfluorinated sulfonic acid (PFSA) ionomers have fixed anions, which are sulfonic acid sites, and consequently, by electroneutrality, the concentration of positive ions remains fixed. They also exhibit many interesting properties such as a high conductivity, prodigious water uptake, and high anion exclusion, to name a few. Nafion® is the main membrane discussed in this review.

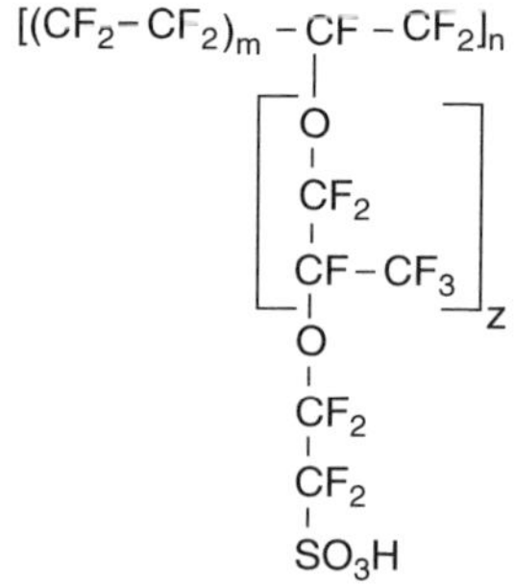

Figure 2.1 Chemical formula of a PFSA membrane, where m is between 6 and 13 and z is 3 for Nafion®.

Although Nafion® is the focus, the models presented can be (and some have been) easily adapted for other membranes such as hydrocarbon ones. Such an adaptation requires altering the various membrane physical and transport properties, but not the governing equations (i.e., the same underlying physics govern transport of the various species). The main reason is that the models presented herein are macroscopic and basically average over the microscale heterogeneities that make the membranes unique. In a similar fashion, although the models reviewed and discussed are primarily for hydrogen-fueled PEFCs, they can be used and appropriately altered for other systems like direct-methanol fuel cells (DMFCs).

The focus of this review is to discuss the different macroscopic PEFC-membrane models with the overall goal of presenting a picture of the various types of transport in the membrane. Although the majority of the relevant literature models has been examined, there are undoubtedly some that were left out. This is especially the case for those models that have a membrane model but it is not the important feature. For such cases, reference is made to a representative example model from the research group, even though there may be multiple papers published from that group using that model but focusing on other PEFC layers and phenomena. In terms of time frame, this review focuses on models that have been published through the middle of 2005.

There are four other recent review articles dealing with membrane modeling that should be noted. First, those of Weber and Newman [1] and Wang [2] examined macroscopic PEFC models through the end of 2004, but only briefly touched on the membrane models. Similarly, that of Fimrite *et al.* [3] did not go into a lot of depth in terms of membrane models, and it focused mainly on those that use concentrated solution theory. Finally, the review of Kreuer *et al.* [4] examined microscopic and atomistic membrane models in great detail, and thus they are not addressed in this chapter.

This review chapter can be broken down into various sections. First, background is presented discussing modeling methodologies and a physically based, qualitative model of the structure of the membrane. Next, the general governing equations are discussed in terms of the various modeling approaches, including proton, water, energy, and gas fluxes. The third main section deals with models aimed at predicting transport properties such as conductivity and especially the membrane water content. In the final main section, some specialized topics in membrane modeling are mentioned including membrane in the catalyst layers, effect of ions in the membrane, durability, and DMFCs.

2.2 Background

At its simplest, a membrane model should predict the water flux and the potential drop through the membrane. The potential drop is necessary in calculating polarization effects, and the water flux directly impacts the water management of the fuel cell, which is perhaps the most significant component of PEFC performance. The models used to predict these two quantities vary greatly in complexity. While some try to understand fundamentally what is occurring in the membrane, others just calculate the values and perhaps only empirically. While the former are useful in optimizing and designing membranes, the latter are favored in complex simulations such as those examining stack or three-dimensional (3-D) effects [1]. Most of the models discussed in this chapter focus on the more complex and macroscopic approaches.

A good model of the membrane must contain certain key elements. Foremost among these is that it must be based on and agree with the physical reality and phenomena that have been observed with these membranes. Furthermore, expressions for the various properties of the membrane should have the relevant dependences such as on temperature and water content. These property expressions can be found by everything from empirical relations to detailed molecular modeling. The water content should also be modeled or accounted for and allowed to vary in a systematic and continuous fashion. Finally, a model should describe the three main fluxes in the membrane, as shown in Figure 2.2.

The three main fluxes through the membrane are a proton flux that goes from anode to cathode, a water electro-osmotic flux that develops along with the proton flux, and a water-gradient flux. This last flux is sometimes known as the water back flux or back-diffusion flux, and, as discussed in Section 2.3.2, has various interpretations including diffusion, convection, and combinations thereof. Furthermore, these interpretations often serve to differentiate membrane models. In addition to the above three fluxes, there are also fluxes due to crossover of oxygen and hydrogen, which are described in Section 2.3.4.

This section is comprised of two main themes that set the stage for the presentation and discussion of the various models in later sections. The first theme is the modeling methodology, and the second is the physical model or representation of membrane structure as a function of water content. Before proceeding to discuss these themes, some mention should be made about empirical models and PEFC performance in general.

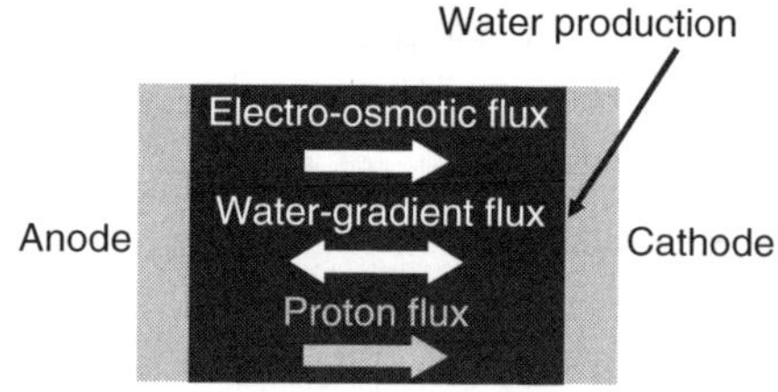

Figure 2.2 Schematic of the membrane showing the various fluxes through it.

2.2.1 PEFC Performance and Empirical Models

As noted above, a membrane model is mainly about predicting performance issues due to transport in the membrane. The performance of a fuel cell is most often reported in the form of a polarization curve, as shown in Figure 2.3. Roughly speaking, the polarization curve can be broken down into three main regions. At low currents, the behavior of a fuel cell is dominated by kinetic losses, which are not directly attributable to the membrane. As the current is increased, ohmic losses become a factor in lowering the overall cell potential. These ohmic losses are mainly from ionic losses in the membrane both in the separator and the electrodes. At high current densities, mass-transport limitations become increasingly important. These losses are due to reactants not being able to reach the reaction sites. Typically, oxygen is the problem due to flooding of the cathode by liquid water, and thus the water management role that the membrane plays is important. Of greater impact by the membrane is that if it becomes dehydrated, a limiting current can be realized due to the protons inability to reach the reactive sites.

To capture the above effects, including those in the membrane, empirical models have been used. These are not rigorous models, *per se*, but instead are educated curve fits using at most semiempirical expression [1,5,6]; an example of such a curve is [1,7,8]

$$V = U^{\theta} - b\log(i) - R_{\Omega}i + b\log\left(1 - \frac{i}{i_{\text{lim}}}\right) \tag{2.1}$$

where R_{Ω} is the constant resistance in the fuel cell, b is the Tafel slope, U^{θ} is the standard cell potential, and V, i, and i_{lim} are the cell potential, current density, and limiting current density, respectively. While such an approach is

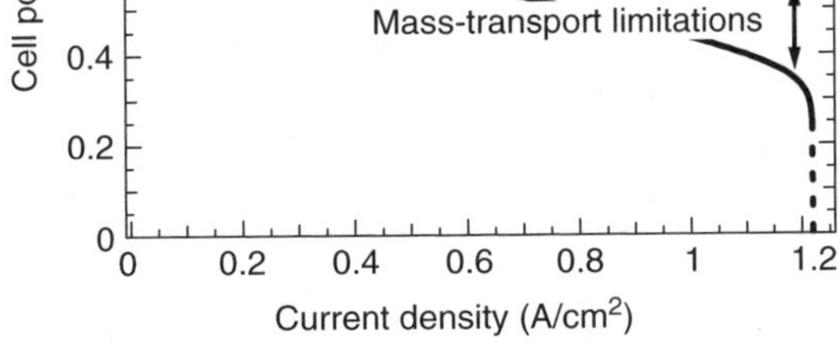

Figure 2.3 PEFC polarization curve showing the various potential losses.

useful in modeling complex geometries and stacks, it is not predictive and not truly a model of the membrane. This is especially apparent in examining Figure 2.3, where one cannot easily distinguish the three regions from one another just by looking at the polarization curve (solid line). For example, the so-called ohmic or linear portion of the curve actually has mass transport and other effects in it, as seen in the breakdown of the curve in the figure. Thus, a fit to a polarization curve is not truly yielding the resistance of the membrane, which may not even have a uniform conductivity. Furthermore, the impact of the role that the membrane plays in water management cannot be quantified using such a simple approach.

To expand on the last point, empirical models also may or may not examine the water balance, and if they do, then a net water flux through the membrane is used. This value is either calculated from experimental water-balance data or just assumed. The latter is typically used when the goal of the model is to examine some other PEFC layer such as the diffusion media or catalyst layers. The assumption of a net water flux greatly aids in the convergence and robustness of a program; however, like using a resistance above, it makes the model less predictive and helpful in understanding membrane phenomena.

2.2.2 Modeling Methodologies

In terms of both quantitative and qualitative modeling, PEMs have been modeled within two extremes, the macroscopic and the microscopic, as discussed in recent review articles [1,2,4]. The microscopic models provide the fundamental understanding of processes like diffusion and conduction in the membrane on a single-pore or even molecular level. They allow for the evaluation of how small perturbations like heterogeneity of pores and electric fields affect transport, as well as the incorporation of small-scale effects. Although the microscopic models may provide more realistic conditions and factors, they require a lot more knowledge of the microstructure and are much more computationally expensive. For these reasons and also to allow modeling of entire PEFC behavior, macroscopic models are more commonly used, although some microscopic details should be incorporated into them. This review focuses on macroscopic models or those that utilize a macrohomogeneous approach.

In a macrohomogeneous approach, the exact geometric details of the modeling domain are neglected. Instead, the domain is treated as a randomly arranged structure that can be described by a small number of variables. Furthermore, transport properties within the domain are averaged over the membrane volume. Thus, all variables are defined at all positions within the domain. Averaging is performed over a region that is small compared to the size of the membrane, but large compared to its microstructure.

The macroscopic membrane models can be placed into three main categories. The differentiation is basically made on how the models treat water movement in the membrane. The first category treats water transport as if by diffusion, and thus is termed diffusive. This type of model implicitly assumes that the membrane is a single phase and was popularized for fuel cells by Fuller and Newman [9] and Springer *et al.* [10] A problem with this approach is that when the membrane is saturated, it does not make sense to have only a diffusive flow since the concentration of water in the membrane is uniform; there is no concentration gradient. Hence, another model should be used in this case (e.g., a hydraulic one).

The second category of membrane models treats the water movement as if by convection, and thus is termed hydraulic. This system explicitly treats the membrane as two phases. The most recognized type of this model was done by Bernardi and Verbrugge [11,12]. A problem with these models arises for the case of a membrane in a low-relative-humidity reservoir. In such a system there is not a continuous liquid pathway across the medium, and the membrane matrix interacts significantly with the water due to the binding and solvating of the sulfonic acid sites. Thus, a concentration gradient and not one of hydraulic pressure of liquid water, which might not even be defined, seems to be the more appropriate driving force; a one-phase model should be used.

The third category treats the water movement as if by a combination of the above two methods, and thus is termed combination. Such models aim to explain transport across the whole range of possible membrane water contents and bridge the gap between the two categories above. Essentially, the two approaches above operate at one limit of water concentration, and are then somehow averaged between those limits [13,14]. The three main model categories are examined in more detail and in terms of equations in Section 2.3.2

2.2.3 Membrane Physical Picture

To model the membrane, one requires a physical basis or picture in order to ensure the correct mathematical treatment and application of the underlying physics and phenomena. The crucial component for such a picture is how the membrane changes and interacts with water as a function of water content.

The water content of a membrane is typically given in terms of λ, a ratio of the moles of water per mole of sulfonic acid sites. This value is normally given in the form of a water-uptake isotherm [15–20], as shown in Figure 2.4. Such an isotherm is typically determined experimentally, but there are some models that aim to predict at least part of the curve as discussed in Section 2.4.1. The dashed line in Figure 2.4 represents what is known as Schröder's paradox [21], a phenomenon which occurs in many different polymers and gels. In essence, the paradox is that the membrane exhibits a difference in water uptake (and therefore other properties) only due to the phase of the reservoir in contact with the membrane (i.e., the chemical potential remains constant). As seen in Figure 2.4, the water content of the membrane in a

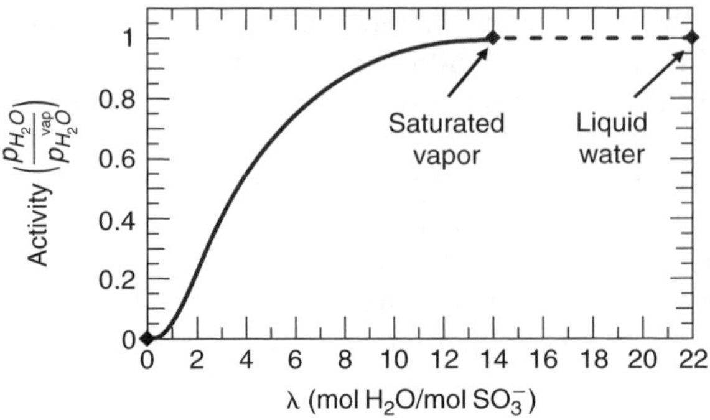

Figure 2.4 Water-uptake isotherm at 25°C showing the effect of Schröder's paradox.

saturated-vapor reservoir is significantly lower than that in a liquid–water reservoir. This is an important issue since fuel cells are often operated with humidified gases, resulting in situations where there is liquid water on the cathodic side of the membrane and only water vapor on the anodic side. With this introduction, one can now dissect the isotherm and relate it to the membrane microstructure.

The general structure of Nafion®, and ionomers in general, as a function of water content has been the source of many studies, as recently reviewed by Mauritz and Moore [22] and Kreuer *et al.* [4]. The experimental data show that a hydrated membrane phase segregates into ionic and matrix or nonionic phases. The ionic phase is associated with the hydrated sulfonic acid groups, and the matrix phase with the polymer backbone. Thus, water is associated with the hydrophilic ionic phase and not the hydrophobic matrix phase. The actual way in which the phases segregate within the polymer depends on the water content and other factors including pretreatment procedures, operating temperature, side-chain length, and equivalent weight, to name a few. Finally, based on various experimental data, all species transported through the membrane move by way of the ionic phase [10,23,24]. Therefore, the nonionic fluorocarbon matrix can be taken as inert. Its roles are to add mechanical strength and hydrophobicity, thus aiding in the membrane microstructure and preventing dissolution.

Starting from the dry-membrane case, water initially hydrates the sulfonic acid sites and allows them to dissociate partially. This first water is very tightly bound and hard to move. The next few water molecules serve to hydrate the sulfonic acid sites further beyond their primary shell. This allows for a reorganization of the membrane microstructure into ionic and nonionic domains. Consequently, the conductivity of the membrane greatly increases when this occurs ($\lambda \approx 2$) [25]. As the membrane continues to take up water, the ionic domains continue to organize and enlarge. The ideal picture of this phase segregation is known as the cluster-network model first proposed by Hsu and Gierke [24]. In this idealized picture, the water is contained in a spherical domain about 4 nm in diameter into which the polymer side-chains infiltrate. The inverse micelle domains form based on a balance between the surface or electrostatic energy due to the coulombic repulsions of the sulfonic acid groups and the elastic or deformation energy due to the work required to deform the polymer matrix. The ionic clusters are connected by interfacial regions or bridging-site pathways about 1 nm in diameter [26]. These were determined by Hsu and Gierke to be transient connections with a stability on the order of ambient thermal fluctuations, in agreement with molecular-dynamics simulations [4,27,28]. While the cluster-network model is an idealized picture, it provides a useful visualization of the polymer phase-separated microstructure.

As the membrane becomes more hydrated, the sulfonic acid sites become associated with more water, allowing for a less bound and more bulk-like water to form. This new water is no longer strongly influence by the dielectric properties of the sulfonic acid groups and is essentially enlarging the ionic domains by filling them in with water. This is why there is a flattening out of the slope above $\lambda = 6$ in the uptake isotherm (see Figure 2.3). The extreme case is when the membrane is placed in a liquid–water reservoir, where the ionic domains swell and a bulk-like liquid–water phase comes into existence throughout the membrane. The way in which this rearrangement and phase-transition-type behavior occurs is currently unknown exactly. However, it is probably due to the interfacial properties of the membrane, such as the fluorocarbon-rich skin on the surface of Nafion® [29,30] or the removal of a liquid-vapor meniscus at the membrane surface [31]. Overall, the final picture of a liquid-equilibrated membrane is a porous structure, with average channel and cluster sizes between 1 and 2 nm and 2 to 4 nm, respectively [32,33].

In summary, Figure 2.5 is a schematic of how water interacts with the membrane. In the first panel, at low water contents, the water is strongly bound and solvates the sulfonic acid groups. Additional water causes the water to become less bound, with some bulk-like water forming, the second panel. With more water uptake, ionic domains grow and form interconnections with each other through a percolation-type phenomenon. The connections, or collapsed channels, are transitory and have low concentrations of sulfonic acid sites (similar to the first two panels). The lower left panel corresponds to a membrane that is in contact with saturated water vapor, where such a cluster-channel network has formed. When there is liquid water at the boundary of the membrane, structural reorganization and a phase transition or capillary condensation occurs, allowing for bulk-like liquid water to exist in the channels, resulting in a pore-like structure, the final panel in Figure 2.5. Because the channels are now filled with liquid, the uptake of the membrane has increased without a change in the chemical potential of the water (i.e., Schröder's paradox). Finally, it was noted above that the cluster-channel representation is very idealized. A more realistic representation based on experimental data is represented in Figure 2.6. In the figure, clusters and

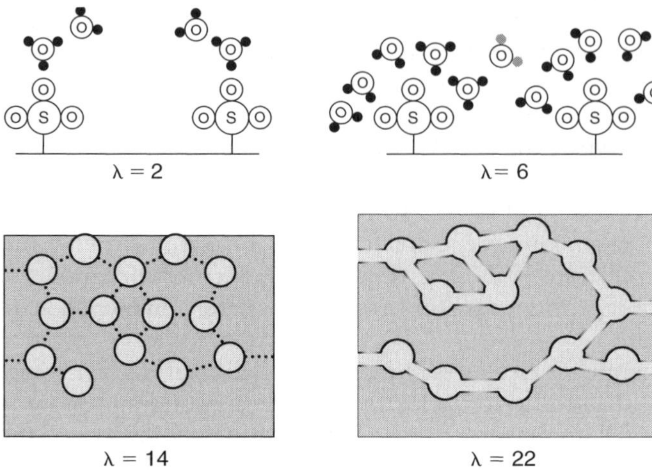

Figure 2.5 Membrane structure and interactions as a function of water content, λ (moles of water per mole of sulfonic acid sites). The top two schematics represent water interacting with the sulfonic acid sites, where the gray waters are more loosely bound. The bottom two schematics are cross-sectional representations of a saturated-vapor- and liquid-equilibrated membrane, respectively, where the gray area is the fluorocarbon matrix, the black is the polymer side chain, the light gray is the water, and the dotted line is a collapsed channel or bridging sulfonic acid site. (Schematics adapted from Refs. [3] and [14] with permission of The Electrochemical Society, Inc.)

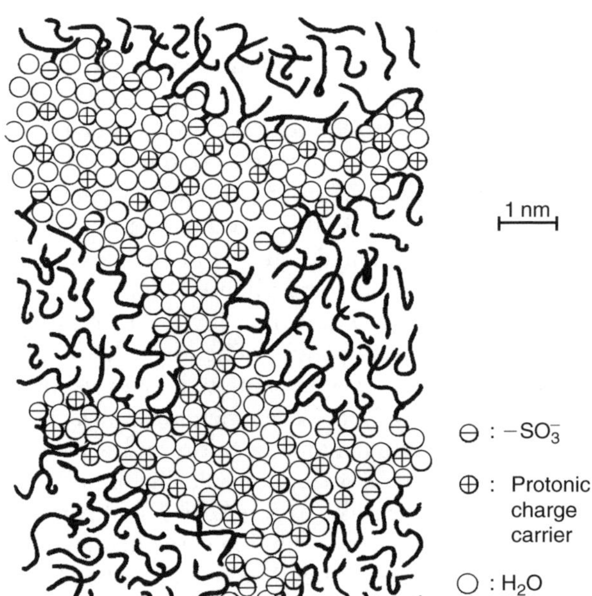

Figure 2.6 More realistic representation of a liquid-equilibrated membrane (reproduced from Ref. [33]).

channels can still be identified, but the clusters are not spherical. Instead, they are more like pore-junction regions where channels cross.

2.3 Governing Equations and Treatments

One needs the same number of equations as unknowns. Typically, this requires four equations for the four unknowns in the membrane of water flux, water chemical potential, electrical potential, and current density. As mentioned above, the main difference between membrane models basically comes down to how they treat transport, and specifically water transport, in the membrane. While some models may use driving forces of pressure or concentration, depending on what is assumed to be dominating, the correct one to use is chemical potential. As will be shown, this driving force is thermodynamically rigorous as it incorporates the other driving forces.

Besides the transport equations, there are also the other governing equations, most of which are the same for the different models. The membrane modeling equations can be broken down into four main types. The first are the conservations equations, the second are the transport relations, the third are the equilibrium relationships, and the fourth are the auxiliary or supporting relations, which include variable definitions, property expressions, and such relations as Faraday's law. In this section, the various equations are examined. The discussion is divided based on the phenomenon that the governing equation addresses. Furthermore, the auxiliary equations and equilibrium relationships may depend on the modeling approach and equations, or may even be separate models (see Section 2.4).

2.3.1 Material Balances

The conservation of material equations are essentially identical for the various membrane models. Furthermore, they are also relatively simple since there is nothing occurring in the membrane in terms of reactions or source terms. It is necessary to write a material balance for each independent component in the membrane. In differential form, this can be expressed as

$$\frac{\partial c_i}{\partial t} = -\nabla \cdot \mathbf{N}_i \tag{2.2}$$

where c_i and $\mathbf{N}_i$ are the concentration and flux density of species i, respectively.

The term on the left side of the equation is the accumulation term, which accounts for the change in the total amount of species i held in the membrane within a differential control volume. This term is assumed to be zero for almost all of the membrane models discussed in this review because they are at steady state (except for those discussed in Section 2.5.3). The term on the right side of the equation keeps track of the material that enters or leaves the control volume by mass transport.

Because protons are the only mobile ionic species in the membrane, the proton flux and current density can be related through Faraday's law,

$$\mathbf{N}_{H^+} = \frac{\mathbf{i}}{F} \tag{2.3}$$

where **i** is the superficial current density in the membrane and F is Faraday's constant. If one assumes electroneutrality,

$$\sum_i z_i c_i = 0 \tag{2.4}$$

where z_i is the charge number of species i, then by conservation of charge and substitution of Eq. (2.3) into (2.2) results in the governing equation for the current (or proton flux) of

$$\nabla \cdot \mathbf{i} = 0 \tag{2.5}$$

The assumption of electroneutrality implies that the diffuse double layer, where there is significant charge separation, is small compared to the volume of the domain, which is normally the case (for exceptions see Section 2.5.2). Also, double-layer charging is ignored in the above equation since double layers are only expected to exist at the membrane interfaces and not inside the membrane.

2.3.2 Proton and Water Transport Phenomena

The major species being transported in the membrane are water and protons, since the anionic sulfonic acid sites are tethered to the membrane backbone. Thus, two transport equations are required. As discussed in Section 2.2.2, there are three main categories of models, the diffusive, the hydraulic, and the combination. Each is discussed below in turn, but before discussing them, some mention should be made about simple approaches.

2.3.2.1 Simple Models

The simplest models of the membrane that treat transport in the membrane in a nontrivial manner (i.e., they consider species transport and not just fit a polarization curve with empirical parameters) make several assumptions and use some simple transport expressions along with the material conservation equations above. These models are aimed at examining effects outside the membrane (e.g., cathode flooding [34–48]) or when only general trends are desired [49–54]. Furthermore, these models assume constant values of transport properties in the membrane and thus are not applicable when the water content of the membrane is expected to vary (e.g., membrane dehydration). Such assumptions also limit the predictability of these models.

Since the membrane is stationary, only the water and protons move in the membrane system. For the proton movement, the simplest treatment is to use Ohm's law (Eq. (2.29) in differential form):

$$\mathbf{i} = -\kappa \nabla \Phi \tag{2.6}$$

where κ is the ionic conductivity of the membrane (which is assumed uniform) and Φ is the electrical potential in the membrane, which is defined with reference to a standard hydrogen electrode at the same temperature and electrolyte (e.g., Nafion®) as the solution of interest (i.e., it carries its own extraneous phases with it). The above expression can easily be integrated to yield a resistance for use in a polarization equation (see Section 2.2.1).

For the movement of water through the membrane, a value of β, the net water flux per proton flux,

$$\beta = \frac{\mathbf{N}_w}{\mathbf{i}/F} \tag{2.7}$$

is often assumed or is calculated based on the water flux in the anode or cathode region. Most models that use this approach also treat the catalyst layer as an interface, which allows for the water flux to become a boundary condition that is directly related to β. While this approach has merit in terms of convergence issues and allowing for analytic expressions to describe the membrane, it has limited usage under most PEFC conditions since β is not known *a priori*.

Before proceeding to discuss more rigorous treatments of transport in the membrane, two mentions should be made about hybrid approaches. In the first approach [55], the transport equations are solved in a more rigorous fashion as discussed below, and a database generated of values of β and $\mathbf{i}$ as a function of the ratio of λ at the anode and cathode and a dimensionless Peclet number for water in the membrane. Such a database is useful for complicated and computationally costly simulations (e.g., 3-D computational-fluid-dynamics ones). In the second hybrid approach [56–58], the conductivity is modeled rigorously, and the transport of water is basically ignored. The reason for this is that the models are more concerned with examining the polarization effects of the cell and not the water management. While this approach is all right for optimization, very complex models, or effects outside of the membrane, ignoring PEFC water management is not generally advisable.

2.3.2.2 Diffusive Models

The diffusive models treat the membrane system as a single, homogeneous phase in which water and protons dissolve and move by concentration and potential gradients. They correspond more-or-less to the vapor-equilibrated membrane (see Figure 2.5), or in other words a membrane at lower water content. Many membrane models, including some of the earliest ones, treat the system in such a manner. The diffusive models allow one to predict such effects as membrane dehydration, as shown in Figure 2.7. As the current density is increased, the water content of the membrane decreases, causing a larger ohmic drop due to its effect on conductivity. Furthermore, the profiles become more curved as the current density is increased because of the interactions between the water-gradient and electro-osmotic fluxes in the membrane (see Eq. (2.11)). There are two main ways in which the diffusive transport can be treated, first by dilute solution theory and second by concentrated solution theory. The former approach is simpler to implement and may yield sufficiently correct answers, especially in terms of current density. However, it lacks the universality and rigor of the latter approach, which can be applied for all water contents.

Dilute solution theory Dilute solution theory starts with the use of the Nernst–Planck equation [59]:

$$\mathbf{N}_i = -z_i\, u_i\, Fc_i \nabla \Phi - D_i \nabla c_i + c_i \mathbf{v} \tag{2.8}$$

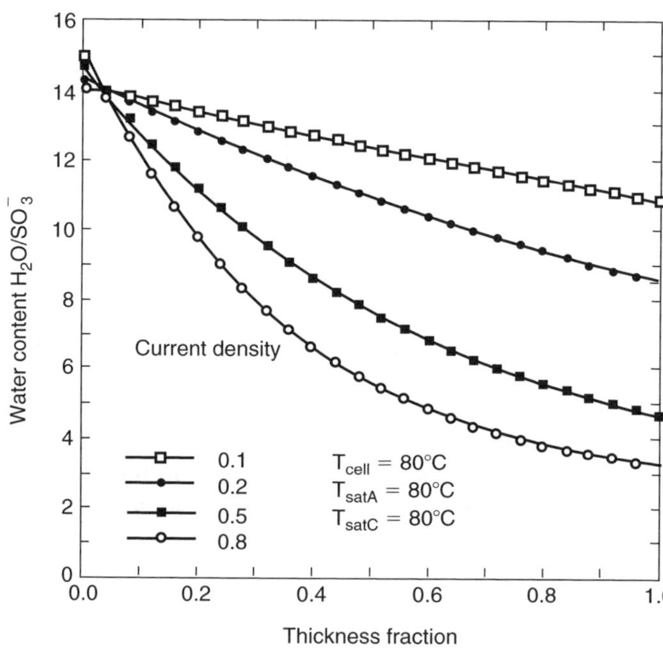

Figure 2.7 Simulation results using a diffusive (one-phase) membrane model showing membrane dehydration: λ as a function of membrane position (cathode on the left) for different current densities (reproduced from Ref. [10] with permission of The Electrochemical Society, Inc).

The first term in the expression is a migration term, representing the motion of charged species that results from a potential gradient. The migration flux is related to the potential gradient ($-\nabla\Phi$) by a charge number, z_i, concentration, c_i, and mobility, u_i. The second term relates the diffusive flux to the concentration gradient using a diffusion coefficient, D_i. The final term is a convective term and represents the motion of the species as the bulk motion of the solvent carries it along. For one-phase treatment, the solvent is the membrane, and thus $\mathbf{v} = 0$.

Dilute solution theory considers only the interactions between each dissolved species and the solvent. The motion of each charged species is described by its transport properties, namely, the mobility and the diffusion coefficient. These transport properties can be related to one another at via the Nernst–Einstein equation [59–61]:

$$D_i = RTu_i \tag{2.9}$$

where R is the ideal-gas constant and T is the absolute temperature. So long as the solute species are sufficiently dilute that the interactions among them can be neglected, material balances can be written based on the above expression for the flux (Eq. (2.8)).

For the protons in the membrane, Eq. (2.8) reduces to Ohm's law, Eq. (2.6). For the movement of water in the membrane, the Nernst–Planck equation reduces to Fick's law,

$$\mathbf{N}_w = -D_w\nabla c_w = -D_\lambda\nabla\lambda \tag{2.10}$$

because water has a zero valance. The second equality comes because the concentration and diffusion coefficient of water can be written as functions of λ if desired, as discussed in Section 2.4.3. The models that utilize the Fick's law treatment are interested mainly in modeling effects outside of full cells [38,41,62–67]. Thus, they typically assume a constant diffusivity, resulting in a linear gradient of λ in the membrane. This allows one to estimate the water flux by just knowing the water content of the membrane at its anode and cathode interfaces.

While Eq. (2.10) stems from dilute solution theory, more rigorous membrane models incorporate the observation that the moving protons induce a flux of water in the same direction. Technically, this electro-osmotic flow is a result of the proton–water interaction, and is not a dilute solution effect since the membrane is taken to be the solvent. As shown in the next section, the electro-osmotic flux is proportional to the current density and can be added to the diffusive flux to get the overall flux of water:

$$\mathbf{N}_w = \xi\frac{\mathbf{i}}{F} - D_w\nabla c_w \tag{2.11}$$

where ξ is the electro-osmotic coefficient, the number of water carried per proton in the absence of a concentration gradient. The above equation and

Ohm's law have been used successfully for most of the models that treat the membrane as a single phase [10,62,63,65,66,68–82]. The deviations and complications in the models arise from what functions are used for the various membrane transport properties and water content (i.e., the constitutive and supporting relations but not the governing-equation framework), as discussed in Section 2.4.

Concentrated solution theory For an electrolyte with three species, it is more rigorous and almost as simple to use concentrated solution theory. Concentrated solution theory takes into account the binary interactions between all of the species. In addition, it uses a more general driving force, namely, that of chemical potential. As discussed later, this allows for the concentrated solution theory equations to be valid for both the diffusive and hydraulic models, the only difference being in the interpretation of the chemical potential. Consequently, if concentrated solution theory is used, the model can easily become a combination one (i.e., valid for all water contents), as discussed in Section 2.3.2.4.

For membranes, concentrated solution theory was initially used by Bennion [83] and Pintauro and Bennion [84]. To do the analysis, one starts with the equation of multicomponent transport:

$$\mathbf{d}_i = c_i \nabla \mu_i = \sum_{j \neq i} K_{i,j} \left(\mathbf{v}_j - \mathbf{v}_i \right) \tag{2.12}$$

where $\mathbf{d}_i$ is the driving force per unit volume acting on species i and can be replaced by a chemical potential gradient of species i [85], $K_{i,j}$ are the frictional interaction parameters between species i and j, and $\mathbf{v}_i$ is the velocity of i relative to a reference velocity (that of the membrane, which is stationary). The treatment of Eq. (2.12) can now be done in one of two ways.

The first treatment is to introduce a concentration scale and relate the frictional coefficients to binary interaction parameters:

$$K_{i,j} = \frac{c_T x_i x_j RT}{D_{i,j}} \tag{2.13}$$

where $D_{i,j}$ is the binary interaction parameter between species i and j, c_T is the total concentration, and x_i is the mole fraction of species i. Doing the above substitution into Eq. (2.12) results in the so-called binary friction model,

$$\nabla_{\mu_i} = \sum_{j \neq i} \frac{RTx_j}{D^e_{i,j}} \left(\frac{\mathbf{N}_j}{c_j} - \frac{\mathbf{N}_i}{c_i} \right) - \frac{RT}{D^e_{i,m}} \left(\frac{\mathbf{N}_i}{c_i} \right) \tag{2.14}$$

where the m denotes the interaction with the membrane and e denotes an effective property of the membrane. As discussed by Fimrite *et al.* [3,86], this treatment is similar to that of the dusty-fluid model applied to the membrane

[87–89], but accounts for the bulk movement of water in a more consistent manner using a different reference frame. This analysis is akin to the analysis of gas movement in porous media presented by Weber and Newman [90]. It should be noted that since the above treatment also implicitly accounts for convection, the model is more of a combination model and not just a diffusion one, as mentioned in the next section. A proton and water form of Eq. (2.14) can then be used to model the transport in the membrane, which results in equations similar to Nernst–Planck (Eq. (2.8)), except that the cross terms are accounted for (see Eqs. (2.16) and (2.17)). The models that use the above formulation are primarily focused on modeling the conductivity of the membrane, and are discussed in Section 2.4.3.

While Eq. (2.14) can be used, it is somewhat complicated and requires predicting or measuring the necessary binary interaction parameters. A simpler and just as general approach is to use Eq. (2.12), but instead of changing the frictional coefficients into binary interaction parameters, the equations are inverted instead:

$$\mathbf{N}_i = -\sum_j L_{i,j} c_i c_j \nabla \mu_j \tag{2.15}$$

where the $L_{i,j}$'s are related directly to the $K_{i,j}$'s [90]. For the three-component membrane system, there are $\frac{1}{2}N(N-1) = 3$ independent transport properties that are needed to characterize the system. The $L_{i,j}$'s can be related to experimentally measured transport properties using a set of three orthogonal experiments [84,90,91]. Doing this results in the proton and water governing transport equations:

$$\mathbf{i} = -\frac{\kappa \xi}{F} \nabla \mu_w - \kappa \nabla \Phi \tag{2.16}$$

and

$$\mathbf{N}_w = \xi \frac{\mathbf{i}}{F} - \alpha_w \nabla \mu_w \tag{2.17}$$

respectively, where α_w is the transport coefficient of water. The above equations have also been arrived at using an irreversible thermodynamics approach [92,93].

Upon comparison of Eq. (2.16) to (2.6), it is seen that the proton–water interaction is now taken into account. This interaction is usually not too significant, but it should be considered when there is a large water gradient (e.g., low-humidity or high-current-density conditions). Upon comparison of Eq. (2.17) to (2.11), it is seen that the equations are basically identical in form where the concentration and diffusion coefficient of water have been substituted for the chemical potential and transport coefficient of water, respectively. Almost all of the models using the above equations make similar substitutions for these variables [3,9,90,94–96].

The exceptions to this are the models of Janssen [97] and Weber and Newman [91], where the chemical potential is used directly. Janssen takes the transport coefficient as a fitting parameter, and Weber and Newman relate it to experimental data. Both models demonstrated good agreement with water-balance data under a variety of conditions. Janssen states that using a chemical-potential driving force does not necessitate making the distinction between diffusive or convective flow in the membrane. However, as Meyers [98] points out, by assuming the membrane system is a single phase, it cannot support a pressure difference inside it. The only way that a single-phase membrane model can have a pressure difference across it is if the chemical potential or water concentration is being altered at the boundaries. This problem is why single-phase membrane models cannot adequately describe transport for fully hydrated membranes where the driving force is the liquid pressure. For this case, one needs to use a two-phase model, although the above concentrated solution theory equations remain the same (only the transport coefficient and chemical-potential interpretation change) as noted below.

2.3.2.3 Hydraulic Models

In opposition to the single-phase treatment above are the models that assume the membrane system is two phases. This type of model corresponds to the liquid-equilibrated membrane shown in Figures 2.5 and 2.6 (i.e., high water contents where there is a bulk-like liquid–water phase in the membrane). In this structure, the membrane is treated as having pores that are filled with liquid water. Thus, the two phases are water and membrane.

The addition of a second-phase allows for the membrane system to sustain a pressure gradient in the water because of a possibly unknown stress relation between the membrane and fluid at every point in the membrane. However, diffusion of water becomes meaningless since the water is assumed to be pure in the models discussed here. Furthermore, unlike the models discussed above, the water content of the membrane is usually assumed to remain constant ($\lambda = 22$) as long as the membrane remains fully liquid equilibrated and has been pretreated appropriately. For the transition case between liquid- and vapor-equilibration, see Sections 2.3.2.4 and 2.4.1.1.

The first model to describe the membrane using a hydraulic approach was that of Bernardi and Verbrugge [11,12], which was based on earlier work by Verbrugge and Hill [99,100]. This model utilizes a dilute solution approach that uses the Nernst–Planck equation (Eq. (2.8)) to describe the movement of protons, except that now $\mathbf{v}$ is not equal to zero because they move in the separate water phase. The velocity of the water is give by Schlögl's equation [100,101]:

$$\mathbf{v} = -\left(\frac{k}{\mu}\right)\nabla p_L - \left(\frac{k_\Phi}{\mu}\right) z_f c_f F \nabla \Phi \tag{2.18}$$

where k and k_Φ are the effective hydraulic and electrokinetic permeability, respectively, p_L is the hydraulic or liquid pressure, μ is the water viscosity, and z_f and c_f refer to the charge and concentration of fixed ionic sites, respectively.

In the above system, the movement of water can be attributed to a potential gradient and a pressure gradient. The movement of water by a pressure gradient is determined primarily by an effective permeability of water moving through the pore network. This approach is quite useful for describing PEFC systems as long as the membrane is well hydrated with a uniform water content. Such a treatment does not necessarily lend itself to describing the flux of water resulting when there is a water-activity gradient across the membrane (i.e., when the membrane is not fully hydrated). Many other models use the same approach and equations as Bernardi and Verbrugge, especially for systems wherein the membrane is expected to be well hydrated (e.g., saturated gas feeds) [102–109].

Instead of the dilute solution approach above, concentrated solution theory can also be used to model liquid-equilibrated membranes. As done by Weber and Newman [91], the equations for concentrated solution theory are the same for both the one- and two-phase cases (Eqs. (2.16) and (2.17)) except that chemical potential is replaced by hydraulic pressure and the transport coefficient is related to the permeability through comparison to Darcy's law. Thus, Eq. (2.17) becomes

$$\mathbf{N}_w = \xi \frac{\mathbf{i}}{F} - \frac{k}{\mu \bar{V}_w} \nabla p_L \tag{2.19}$$

where $\bar{V}_w$ is the molar volume of water. Furthermore, Eq. (2.18) can be arrived at from Eq. (2.12) using a similar analysis of replacing the chemical potential driving force with one of pressure [3], as mentioned above.

An interesting aspect of hydraulic models is that the water-pressure profile inside the membrane is essentially linear (instead of the curved ones seen in Figure 2.7). This is because the water content and consequently the properties are assumed uniform. Also, this assumption means that the net water flux in the membrane should be constant with respect to current density because all of the transport is linearly proportional to the current density. While this is true for the case of a zero pressure difference, it is not the case when there is a pressure differential because this is an extra force driving water from the cathode to the anode, as shown in Figure 2.8. The figure clearly demonstrates that such pressure differences can significant alter PEFC water management, which is one reason why differentials are sometimes used. In Figure 2.8, the value of β goes to negative infinity (net water flux is from cathode to anode) as the current approaches zero by its definition (see Eq. (2.7)). However, under these conditions a hydraulic model may not be valid, which is why the predicted β values are lower than those usually observed [110]. In other words, the pressure effects are over emphasized if one only using a

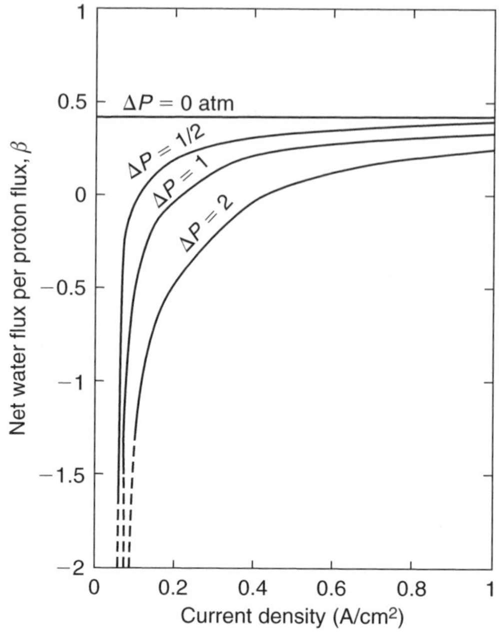

Figure 2.8 Simulation results using a hydraulic (two-phase) membrane model showing the net water flux per proton flux, β, as a function of current density and cathode-to-anode pressure difference (reproduced from Ref. [12] with permission of The Electrochemical Society, Inc).

hydraulic-model approach. Overall, as long as water is on both sides of the membrane, the hydraulic treatment remains valid, and the effect of pressure difference is significant.

2.3.2.4 Combination Models

The two approaches above can be contrasted to one another. In the first approach, Section 2.3.2.2, water moves by diffusion and pressure-driven flow is excluded as a separate driving force. In the second approach, Section 2.3.2.3, only pressure-driven flow is used and there is no diffusive flow. Furthermore, the former approach assumes one phase and the latter two phases. In essence, the diffusive models describe transport in vapor-equilibrated membranes, and the hydraulic models in liquid-equilibrated ones. Thus, to model the full spectrum of water contents, some kind of superposition of the two is required [13,14]. In other words, a model for the transition region between the two modes is necessary.

Perhaps the easiest way to do the superposition is to combine linearly the two driving forces for water, as many models have done [38,55,66,67, 106,111–116]:

$$\mathbf{N}_w = \xi \frac{\mathbf{i}}{F} - D_w \nabla c_w - c_w \frac{k}{\mu} \nabla p_k \tag{2.20}$$

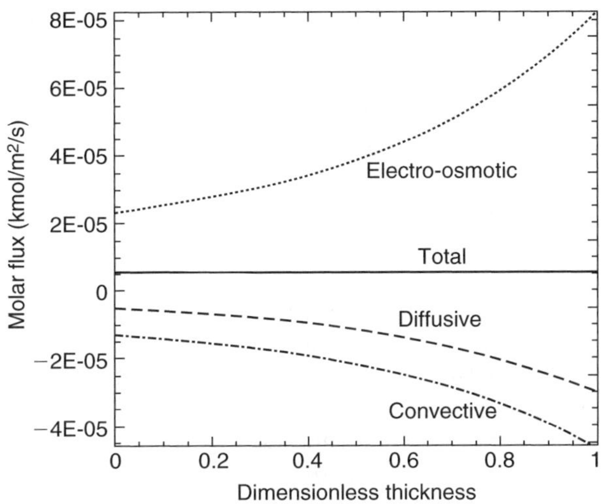

Figure 2.9 Water-flux profiles showing the contributions to the various fluxes, where the anode is much drier than the cathode (reproduced from Ref. [55] with permission of The Electrochemical Society, Inc).

where p_k can be the gas- or liquid-phase pressure. The first problem with the above approach is that it modifies only the water-transport equation and not the proton one. Secondly, and of more importance, the meaning of the two different gradients is somewhat muddled. In essence, this is a dilute solution approach and the two driving forces should not necessarily be valid for all water contents. Furthermore, the addition of both driving forces means that an extra variable is being added to the set of unknowns, and consequently an additional relation is required (e.g., assumption of equilibrium between the concentration and pressure). While this treatment is neither really rigorous nor consistent, it does help to fit and explain data.

Before discussing other approaches, it is of interest to examine the magnitude of the terms in Eq. (2.20). Figure 2.9 shows the various fluxes for a simulation where the anode is much drier than the cathode. As can be seen, the convective flux is much larger than the diffusive one, which is to be expected (e.g., see Figure 2.8). The two fluxes are almost parallel, which means that the driving forces are basically linearly related. While this is more happenstance than design, it does agree with the more physical arguments discussed below. The curvature of the convective and diffusive fluxes is mainly due to the shape of the electro-osmotic flux, since the total water and proton fluxes must remain constant for the steady-state simulation (see Eq. (2.2)). The electro-osmotic flux curves because the electro-osmotic coefficient is a function of water content. This flux is also much larger than the other two, and the three combine to a total flux that is close to zero, which is one reason why PEFCs operate.

Because a single-phase model cannot sustain a pressure gradient, Meyers [98] included pressure-driven flow by allowing for a discontinuity in pressure at the membrane/solution interface, even though the electrochemical potential of all soluble species is continuous. He argues that the additional mechanical stresses compressing the membrane should be indistinguishable from the thermodynamic pressure, and thus the thermodynamic pressure might be discontinuous at the interface. In essence, this approach is similar to using the chemical-potential, concentrated-solution-theory approach described below, although it is not as consistent.

As noted above, concentrated solution theory accounts for both convection and diffusion in the membrane. While multiple dilute solution approaches have been used for the various water contents, the concentrated solution theory approach and equations are independent of the water content. The key is using the correct parameter values and in the interpretation of the single driving force of chemical potential [3,91,97]. This driving force combines those of pressure and activity:

$$\nabla \mu_w = RT \nabla \ln a_w + \bar{V}_w \nabla p_k \tag{2.21}$$

Using this definition in the governing transport Eqs. (2.16) and (2.17), one can describe transport across all water contents.

Weber and Newman [91] and later models [117–120] use the above treatment. They assume that the two transport modes (diffusive for a vapor-equilibrated membrane and hydraulic for a liquid-equilibrated one) operate in parallel with the single driving force above and single net water and proton fluxes across the membrane. Consequently, their governing transport equations remain identical for all water contents. What does change is the value of the property expressions. To model the transition region between the vapor- and liquid-equilibrated modes, they introduce a parameter known as the fraction of expanded channels:

$$S = \frac{\lambda - \lambda_V|_{a_w=1}}{\lambda_L - \lambda_V|_{a_w=1}} \tag{2.22}$$

where $\lambda_V|_{a_w=1}$ and λ_L are the values of λ for the membrane in equilibrium with saturated vapor and liquid at the operating temperature, respectively. They then use this parameter to average the vapor- and liquid-equilibrated transport properties. Although averaging the two equations by this fraction is not necessarily rigorous, it has a physical basis and the correct limiting behavior (i.e., all vapor-equilibrated when there is no bulk-like water, $S = 0$, and all liquid-equilibrated when there is bulk-like water throughout, $S = 1$), with a relatively sharp transition, as expected for a phase transition. The problem with the above approach is that one has to have a method for calculating S. While there is no consensus, most who use this or a similar approach calculate S using a capillary framework, as discussed in Section 2.4.1.1. Another possibility is to a assume a value for S, such as that of the liquid saturation in contact with the membrane [121].

The Weber and Newman approach can be compared to that of assuming separate gradients (Eq. (2.20)). One problem with their approach is that it violates Schröder's paradox in that a small slope in chemical potential exists across the transition region, which is not necessarily the case if the driving forces are taken to be separate. However, the violation amount is small, and, if one makes the transition region infinitesimally thick, there is a zero difference in chemical potential across the transition in harmony with Schröder's paradox. The resulting issue is that it is unknown at what chemical potential this should occur and also whether this is physically accurate. In all, while the two combination treatments are similar and capture many of the same effects, that of Weber and Newman stems from a more consistent and rigorous basis using concentrated solution theory. It also allows one to examine the full range of water contents in the membrane as shown in Figure 2.10.

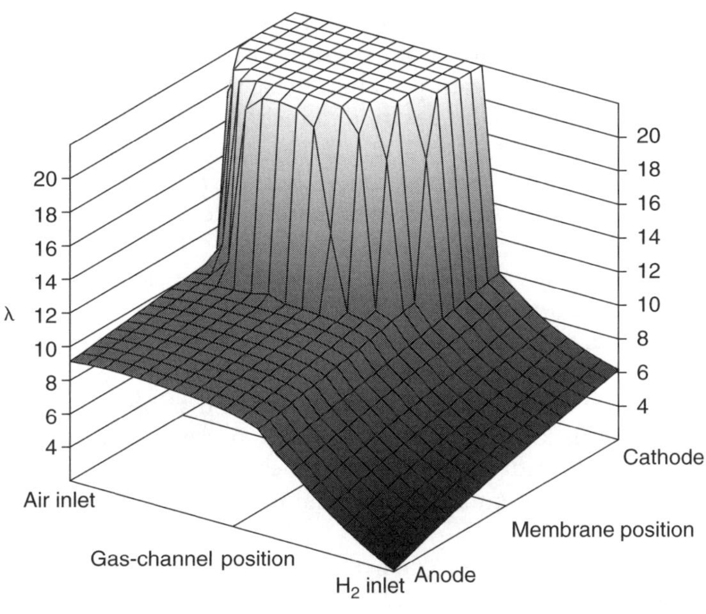

Figure 2.10 Membrane water content as a function of position both along the gas channel and through the thickness of the membrane for the case of countercurrent operation, dry feed gases, and 0.4 A/cm^2 (reproduced from Ref. [118] with permission of The Electrochemical Society, Inc).

The figure demonstrates a maximum penetration depth of liquid water into the membrane near the air inlet. This maximum is caused by the higher liquid pressure at the membrane/cathode interface, which is due to a balance between the hydrogen and air relative humidity. When there is no liquid water in the membrane, λ decreases gradually with both positions in the membrane, showing dehydration effects similar to that in Figure 2.7. The transition between transport modes and the modeling of Schröder's paradox as a continuous change in water content with a very sharp transition region is also clearly shown. Finally, the figure emphasizes the need to have the anode gases humidified. Overall, only through combination models can a physically based description of transport in membranes be accomplished that accounts for all of the experimental findings.

2.3.3 Energy Balance

Consideration of temperature distributions in the membrane is important since the water content and most of the transport properties are functions of temperature. In addition, a temperature change in the membrane can cause an appreciable change in water content if the membrane is close to saturation [88,91,122]. Furthermore, the membrane is somewhat resistive and insulating, resulting in possibly significant temperature gradients. However, some simulations show only very minor changes in temperature through a 1-D membrane slice. This provides some justification for the often used assumption of isothermal behavior. In other words, the energy balance becomes a specification of the temperature. While most models still treat the membrane in this fashion, some have started to include nonisothermal effects and behavior [46,47,65,71,73,79,89,102,105,111,123–125]. Those models that are nonisothermal along the gas channel but assume that the PEFC sandwich remains isothermal are not discussed in this section since the membrane is essentially taken to be isothermal [9,66,67,76]. Typically, those models that include nonisothermal effects in the membrane are of higher dimensionality and contain multiple PEFC-sandwich layers.

Similar to the material balances, the energy balance of the membrane is relatively straightforward, and is essentially the same for the different models. In the membrane, there is heat generation and transfer. Heat is generated due to what is known as ohmic or Joule heating, and it is transferred by convection and conduction. To begin the analysis, the energy conservation equation for the membrane is written in the form [59,126],

$$\rho \hat{C}_p \left(\frac{\partial T}{\partial t} + \mathbf{v} \cdot \nabla T \right) = -\nabla \cdot \mathbf{q} + \sum_i \bar{H}_i \nabla \cdot \mathbf{J}_i \tag{2.23}$$

where various terms of the global energy balance have been ignored due to no reactions in the membrane (see Section 2.3.1) and because it is a single system (i.e., its volume fraction is unity).

In the above expression, the first term represents the accumulation and convection of enthalpy, respectively, where $\hat{C}_p$ and ρ are the heat capacity and density of the membrane system, and may depend on the water content. On the right side of the equation, the first term is due to the heat flux which is related to Fourier's law:

$$\mathbf{q} = -\sum_i \bar{H}_i \mathbf{J}_i - k_T \nabla T \tag{2.24}$$

where $\bar{H}_i$ is the partial molar enthalpy of species i, $\mathbf{J}_i$ is the flux density of species i relative to the mass-average velocity, and k_T is the effective membrane thermal conductivity. The second term on the right side of Eq. (2.23) represents enthalpy changes due to diffusion. Substitution of Eq. (2.24) into (2.23) yields

$$\rho \hat{C}_p \left(\frac{\partial T}{\partial t} + \mathbf{v} \cdot \nabla T \right) = \nabla \cdot (k_T \nabla T) - \sum_i \mathbf{J}_i \cdot \nabla \bar{H}_i \tag{2.25}$$

To get the final modeling equation, the summation on the right side of the above expression must be evaluated. While some models neglect the summation altogether [46,65,71,73,105,124], others rightfully do not. While the enthalpy gradient of the water in the membrane can be taken as insignificant, that of the protons cannot and this results in the phenomenon of Joule heating [59,127,128].

Joule heating is the heat that is generated due to the passage of current. Hence, it is measure of the inefficiency of the movement of current, and since the membrane is somewhat nonconductive, it is important to model it. The expression for Joule heating can be derived from the second term on the right side of Eq. (2.25) using the definition of current density and relating the gradients of enthalpy and electrochemical potential:

$$-\mathbf{J}_{H^+} \cdot \nabla \bar{H}_{H^+} = -\mathbf{i} \cdot \nabla \Phi = \frac{\mathbf{i} \cdot \mathbf{i}}{\kappa} \tag{2.26}$$

The final governing energy-balance equation is determined by substituting in for Joule heating,

$$\rho \hat{C}_p \left(\frac{\partial T}{\partial t} + \mathbf{v} \cdot \nabla T \right) = \nabla \cdot (k_T \nabla T) + \frac{\mathbf{i} \cdot \mathbf{i}}{\kappa} \tag{2.27}$$

This equation is the one that is most often used in modeling nonisothermal behavior in PEFC membranes, although the convective part on the left side is sometimes ignored [47].

The inclusion of nonisothermal behavior can extend further than just adding the above energy balance. For example, one also requires that the various property values including the water content be functions of

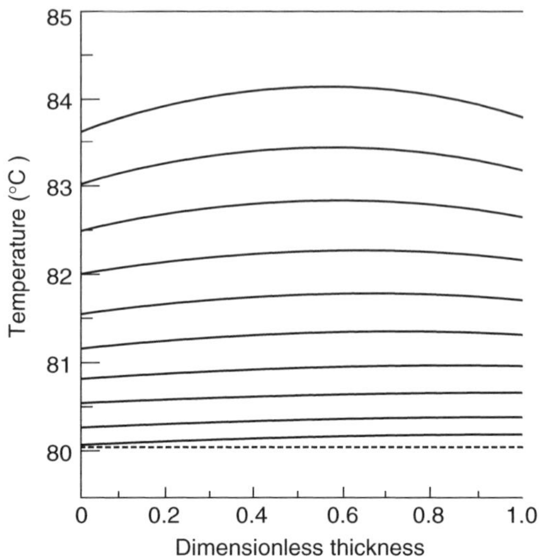

Figure 2.11 Temperature profile in the membrane as a function of current density for humidified feeds (adapted from Ref. [125]).

temperature. Most of the models at least partially ignore this issue and use instead effective values, which limits the analysis of nonisothermal effects. Another consideration is that, to be rigorous, nonisothermal and temperature-gradient effects need to be included in the other governing equations. For instance, because the chemical potential is undefined in a temperature field, a term must be added to account for this. Thus, a thermodynamically rigorous expression for the driving forces for transport must be used:

$$\mathbf{d}_i = c_i[\nabla\mu_i + \overline{S}_i\nabla T] \tag{2.28}$$

where $\overline{S}_i$ is the partial molar entropy of species i. In addition to the above driving force, expressions must be added that result in thermal diffusion [59]. However, because thermal diffusion is small relative to other effects like conduction, and there are small gradients in temperature in the membrane, this type of transport can be neglected. Similarly, the effect of diffusion on heat flux, the Dufour effect, is likewise neglected [59]. In addition, heats of mixing for ideal gases are zero, and for the other components are assumed negligible [129]. Overall, since the above effects can usually be assumed negligible, the governing conservation and equations described in Sections 2.3.1 and 2.3.2 remain valid and do not require alteration.

Figure 2.11 shows temperature profiles from simulations by Ramousse *et al.* [125]. The profiles clearly show that the membrane temperature has a maximum and its magnitude and gradient increase with current density. This occurs because there is more Joule heating and heat production as the current density increases. The magnitude of the temperature gradient is not particularly large, meaning that the membrane is relatively isothermal for this case. If one were to operate with dry or even cool gases, then a larger and more linear temperature profile would exist [125]. Also, the more thermally conductive the cathode and anode, the flatter the membrane temperature profile. It should be noted that the simulation did not consider water evaporation/condensation, which may have significant heat effects.

2.3.4 Gas Crossover

In a PEFC, oxygen and hydrogen crossover is important because of the obvious performance loss, the development of a mixed potential, and durability issues [130–132]. Furthermore, crossover becomes increasingly important as the membranes used become thinner in order to reduce their ohmic drop. Presented in this section are the governing equations and approaches which have been used to model this phenomenon. The issue of methanol crossover in DMFCs is discussed separately (see Section 2.5.4).

Almost all models that deal with multiple PEFC sandwich layers at least implicitly include crossover. They do this by using a constant resistance, which is typically a fitting parameter and used to decrease the cell potential:

$$\Delta\Phi = iR_{\Omega} \tag{2.29}$$

While in theory this resistance is just the resistance due to interlayer contact, crossover can often be incorporated into it since it can be approximated to be linear with current density. This is especially valid for saturated feeds because the membrane is hydrated and the concentration of hydrogen is relatively fixed if there is no reformate. Furthermore, the typical crossover current density is on the order of only 5 mA/cm^2, and, because hydrogen is smaller than oxygen, its crossover flux is typically an order of magnitude higher; consequently, oxygen crossover is mostly ignored. While the above approach is attractive in its simplicity, it has the drawback of not predicting and accounting for the actual hydrogen and oxygen fluxes through the membrane, which cause higher reactant depletion. Since these fluxes are smaller than the fluxes due to reaction, one might consider them to be inconsequential. However, without them, one cannot predict the experimentally measured open-circuit potential, although this is not that important except perhaps in startup and shutdown scenarios. Of more importance is the contemporary concern of PEFC durability. To predict such effects as peroxide generation on the anode due to oxygen reduction, the crossover flux of oxygen must be known.

Before discussing explicit models of crossover, mention should be made concerning the mixed potential. A mixed potential mainly results due to hydrogen crossover and subsequent reactions at the cathode that generate and consume current locally. Some models [42,133] that do not explicitly account for crossover consider the mixed potential by using an experimentally fit function for the open-circuit potential rather than a thermodynamically derived one:

$$U^{\theta} = -\frac{\Delta G}{nF} \tag{2.30}$$

where ΔG is the Gibbs free energy of the reaction and n is the number of electrons transferred in the reaction. While this empirical fix is adequate, it is also not rigorous.

The first model to include crossover explicitly was that of Bernardi and Verbrugge [12]. They assume that there is a gas volume fraction in the membrane that remains constant. Although this does not necessarily agree with the physical picture presented and experimental evidence, it does allow for the calculation of gas crossover through the membrane. The way that it is included is through the convective–diffusion equation [126] (substitute Eq. (2.8) into Eq. (2.2)):

$$D_i \nabla^2 c_i = \mathbf{v} \cdot \nabla c_i \tag{2.31}$$

where the diffusion coefficients of oxygen and hydrogen in Nafion® have been experimentally determined. The use of such a simple relation can be justified since the gases in the membrane are in low concentration and do not interact significantly with each other; a dilute-solution approach is valid. The models that follow the Bernardi and Verbrugge framework treat gas crossover more-or-less the same [102–109,111,114,134], although some allow for a changing gas-phase volume fraction [135,136].

Other models that consider crossover have gotten around the problem of the assumption of a separate gas phase in the membrane by having the gases first dissolve in the membrane and then move by diffusion. Thus, they use both a diffusivity and a solubility. To simplify matters, Weber and Newman [91] combined these two parameters into a permeation coefficient. Thus, their proposed flux equation, which is used with the conservation-of-mass equation (Eq. (2.2)) to yield the governing equation, is of the form:

$$\mathbf{N}_i = -\psi_i \nabla p_i \tag{2.32}$$

where ψ_i and p_i are the permeation coefficient and partial pressure of species i, respectively. They argue that a permeation coefficient is better to use because it allows for a single variable to describe the transport, instead of two, each with their dependences on temperature, etc., which may even offset each other. Furthermore, using the above equation allows for a simple boundary condition of continuous partial pressure of the gas at the membrane interface. Even though the equation uses a partial-pressure driving force, this is due to the use of permeation coefficients. The real driving force is chemical potential, and the proposed physical model is not violated; there is no separate gas phase in the membrane. Finally, to be rigorous, one should add a convective term to Eq. (2.29) as was done in Eq. (2.31). However, this convective term now signifies the amount of gas carried with the net water flux in the membrane, and a back-of-the-envelope calculation shows that such a convective flow is at least an order of magnitude less than the permeation one; it can be disregarded to a first approximation.

The permeation coefficients, like the other transport properties, are expected to depend mainly on the membrane water content and temperature. To put the permeation coefficients into context and to show agreement with the physical model of the membrane, the permeation coefficients for hydrogen for the liquid-equilibrated, dry, and saturated-vapor-equilibrated cases as a function of temperature are given in Figure 2.12. Also included in the graph are the values for hydrogen permeation in water and Teflon®. As a point of interest, the permeation coefficient of oxygen is about 2/3 that of the hydrogen coefficient. As can be seen in the figure, the permeation-coefficient values are basically bounded, with the liquid-equilibrated values higher than the vapor-equilibrated ones. The values are bounded because at higher water contents the gases mainly move through the bulk-like liquid water, and under dry conditions, the membrane is very similar to Teflon®.

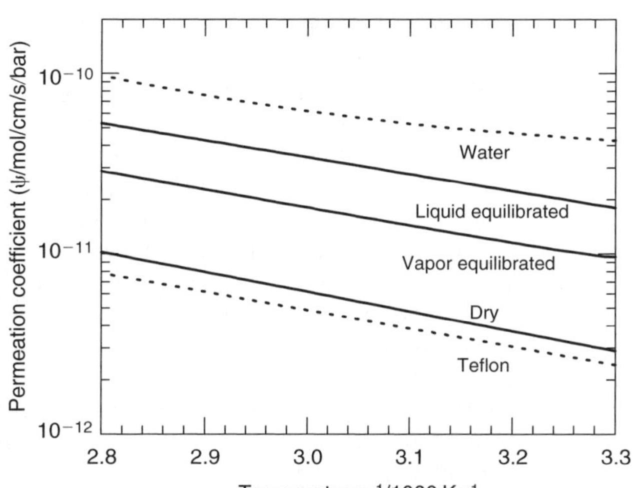

Figure 2.12 Arrhenius plot of the hydrogen permeation coefficient as a function of temperature for a liquid-equilibrated membrane, a vapor-equilibrated membrane, and a dry membrane [14]. Also plotted are the hydrogen permeation coefficients in water [137] and Teflon® [138].

2.3.5 Boundary Conditions and Solution Methods

The above governing equations require boundary conditions in order to be solved. The boundary conditions which are used vary slightly from simulation to simulation. For those simulations which consider effects in the membrane and other layers, internal boundary conditions serve as the necessary ones for the membrane equations. Thus, continuity in superficial flux densities and interstitial concentrations such as liquid pressure and chemical potential are used along with such variables as temperature (i.e., thermal equilibrium is assumed). For membrane-only models, the boundary conditions are typically the operating conditions like the current density and water activity at the two membrane edges. In addition, a reference-potential boundary condition is often used on one side of the membrane. Finally, there are those models that treat catalyst layers as interfaces [74,97,118,139,140]. For these simulations, the membrane boundary conditions can include source terms such as reactant consumption and heat generation.

Due to the complexity and interconnectivity of the governing equations and constitutive relationships, most PEFC models are solved numerically. To do this, a control-volume approach is used. This approach is based on dividing the modeling domain into a mesh. In between mesh points, there are finite elements or boxes. Using Taylor-series expansions, the governing equations are cast in finite-difference form. Next, the flux from the left half-box is set equal to that entering the right half-box; hence, mass is rigorously conserved. This approach requires that all vectors be defined at half-mesh points and all scalars at full-mesh points. The exact details of the numerical methods can be found elsewhere (e.g., see Ref. [141]) and are not the purview of this chapter. The above approach is essentially the same as that used in computational-fluid-dynamics packages (e.g., Fluent® or Comsol®) or discussed in Appendix C of Newman and Thomas-Alyea [59].

While almost all models are solved in the above fashion, analytic solutions are obtainable in certain instances [34,68,104,116,121,135,142–144]. The problem is that such models typically make assumptions like uniform properties, which make the solution of limited significance. Furthermore, the power of digital computers is continually progressing such that the computational cost of running simulations becomes manageable. However, despite the limitations, analytic solutions can provide insight, and, due to their simplicity, are very valuable for complex simulations in which the membrane is only a minor part. Another advantage is that the solutions are in closed form (i.e., the solution is general and the equations need only be solved once), and this is helpful in optimization and parametric studies.

The analytic models use essentially the dilute solution and conservation equations described above, and are isothermal. Furthermore, they normally integrate the equations directly to solve them for such parameters as the membrane resistance. This is accomplished by either assuming a net water flux through the membrane and doing the straightforward integration, or by utilizing boundary conditions such as semi-infinite diffusion. While such a boundary condition is valid only under specific PEFC conditions such as for minor species or at low current densities, it enables the governing equations to be solved analytically. One analytic approach that is noteworthy is to expand all of the transport properties using power series and use these expressions in solving the governing equations [34,68,69,145]. Such an approach allows for variable properties, although the expressions must be fit to data and are not necessarily physically accurate.

Some analytic solutions can be considered 0-D models because they provide single equations that describe the membrane phenomena [1]. Although similar to empirical expressions (see Section 2.2.1), this kind of 0-D models is more rigorous as it stems from solving the governing equations, albeit with simplifying assumptions. A good example is using linear gradients in water content and Ohm's law to get expressions for the net water flux and potential drop through the membrane. Such expressions can then by incorporated into full cell models or matched at the membrane boundaries with analytic solutions of the other PEFC regions.

2.4 Membrane-Property Models

In Section 2.3, the governing equations and models are presented. In order to solve those equations, expressions for the various transport properties and variables are required. There are three main ways that these coefficients are determined, either empirically, semiempirically, or through modeling. Most PEFC models by far use empirical and semiempirical expressions. Such expressions are experimentally obtained, and the difference between the two is that semiempirical expressions use some theory and perhaps other values to guide the form of the expressions. The models for the various transport parameters and *ab-initio* simulations for uptake, etc., are largely microscopic in nature; they have recently been reviewed in the literature [4] and are not discussed in this chapter. There are three main sets of parameters which are required and discussed in this chapter. The first, and perhaps most important, is the determination of the value of the water content of the membrane, λ. Second, is the way in which swelling and dimensional changes are considered in modeling, which is related to the water content. Finally, transport-property models and expressions are discussed.

2.4.1 Water Content

As seen in the modeling equations, the water content is a key variable. Not only can it be a driving force and integral part of PEFC water management, but also it is the crucial parameter that almost all property expressions are related to, such as conductivity, as discussed in Section 2.4.3. Because of the water content's importance, it has received more modeling attention than other membrane properties. Such models span from simple curve fits to more elaborate chemically based ones. All of the models seek to predict and explain at least part of the uptake isotherm as shown in Figure 2.4, including perhaps the higher uptake from liquid water than from saturated vapor (i.e., Schröder's paradox). The various modeling approaches for uptake along with Schröder's paradox are examined below. Before proceeding with the discussion, one should note that all of the models below are trying to predict the water content, which is a property that goes into the governing equations. In essence, these models are speciation ones that compliment the macrohomogeneous approaches and equations discussed in Section 2.3.

The majority of PEFC models use an empirical curve fit for the uptake isotherm. More than anything else, the polynomial fit of Springer *et al.* [10] is used:

$$\lambda = \begin{Bmatrix} 0.043 + 17.81a_w - 39.85a_w^2 + 36a_w^3 & \text{for } 0 < a_w \leq 1 \\ 14 + 1.4(a_w - 1) & \text{for } 1 < a_w \leq 3 \end{Bmatrix} \tag{2.33}$$

The fit is for an isotherm at 30°C. The first part represents the uptake isotherm, and the second term is to handle the experimentally observed higher water contents with liquid water. This latter expression is a linear interpolation from $\lambda = 14$ to $\lambda = 16.8$ at an activity of water of 3. Obviously, such a fit does not make sense due to the use of supersaturated activities. In essence, the curve fit violates Schröder's paradox by having the activity and thus chemical potential of water change. While the polynomial fit is not rigorous, it does allow ease of use in modeling, especially when one might have to take the derivative of the curve (see Section 2.4.3). A problem with the fit is that it does not take into account the observed lower uptake at higher temperatures (see e.g., Refs [146–148]), although this can be done with other curve fits or by extrapolating the curve if some data points are known at higher temperatures [149]. Finally, there are some models that use the same approach except use a fourth-order polynomial [86,133].

Equation (2.33) has a maximum water content of $\lambda = 16.8$ for essentially a liquid-equilibrated membrane. The reason why this is used instead of the often reported value of 22 is that it was what was experimentally observed by Springer *et al.* However, the uptake is known to depend on the pretreatment procedure due to the ability of the polymer to reorient itself as its glass-transition temperature is reached around 100°C. Thus, the value of 16.8 is valid for an "N" or normal-form membrane, and 22 is good for an "E" or expanded-form membrane [17,19]. Most pretreatment procedures today ensure that the membrane is as close to its E form as possible due to its higher conductivity. However, it is unknown how PEFC assembly and compression affect the water uptake and membrane form.

Almost all models assume local equilibrium between the water vapor outside and water in the membrane; this is explicitly accounted for in the models discussed below. However, due to the long hydration times of the membrane and the possible presence of a diffusion layer, mass-transfer resistances can be incorporated if desired [71,82,149,150]. For the hydraulic models, the water content is not as important since the membranes are assumed to be fully liquid equilibrated. An exception to this is the model of Eikerling *et al.* [135,136] as mentioned in Section 2.3.2.4. This model allows for the existence of swollen and nonswollen pores, much like the expanded and collapsed channels of Weber and Newman [91]. Such an approach allows for the full range of water contents to be modeled, as discussed later.

Before proceeding to introduce the more detailed chemically and thermodynamically based uptake models, some mention should be made of the more microscopic cluster models. Typically, these models examine the interactions and forces within a cluster and try to predict cluster size and hence swelling and water uptake. While most of these models are complex and involve *ab-initio* approaches [4], there are some that discuss the phenomena from a macroscopic approach [151–153] that is in line with the focus of this review. These models examine the electrostatic and elastic energies and do a force balance to yield the amount of water in a cluster. This is somewhat akin to an osmotic-pressure analysis. Like most of the more microscopic models, these models provide insight into the membrane morphology and how swelling and water uptake are related and can be modeled, but they are not really useful for full-cell modeling.

The most useful set of uptake models are those that strike a balance between robustness and complexity [31,76,88,91,94,95,122,154,155]. They can explain and predict water uptake as a function of operating conditions and membrane properties, but are macroscopic and wieldy enough to be used in various simulations. They are typically chemically and physically based. The simplest of these models may be that of Thampan *et al.* [88], who use a Brunauer–Emmett–Teller (BET) equation to fit the curve. Such an approach allows one to fit physically meaningful parameters, such as those dealing with multiple water-layer coverages and energies. This approach also allows for consideration of temperature effects on uptake. As a general point, almost all of the chemically based models include temperature effects, which are known to occur but are seldom included in membrane modeling, as mentioned above.

The rest of the chemical models utilize thermodynamics and chemical equilibria explicitly. In the first two, chemical equilibrium is assumed between protons and water with a hydronium ion that is bound or solvated to the sulfonic acid site. This equilibrium is based on the assumption that some water is tightly bound to the protons and sulfonic acid sites in the membrane [4,13,22,156]. In the last one, different ionization and hydration steps are considered, although only the primary one is used.

In the chemical model of Meyers and Newman [18], the equilibrium is used to relate the electrochemical potentials of the species inside the membrane:

$$\mu_{H^+} + \mu_w = \mu_{H_3O^+} \tag{2.34}$$

which are also subject to the boundary condition:

$$\mu_w - \mu_w^{ref} = RT \ln a_w \tag{2.35}$$

where μ_w^{ref} is the reference chemical potential, or the chemical potential of water at unit activity, and the activity is that of the water vapor outside the membrane. Expressing the electrochemical potentials according to the thermodynamic basis of Meyers and Newman [18] for multicomponent transport in a PEM, along with electroneutrality and a mass balance on water in the membrane, leads to two equations that must be solved simultaneously:

$$\begin{aligned} a_w &= K_2(\lambda - \lambda_{H_3O^+})\exp[\phi_2\lambda_{H_3O^+}]\exp[\phi_3\lambda] \\ K_1 &= \frac{\lambda_{H_3O^+}}{(1-\lambda_{H_3O^+})(\lambda - \lambda_{H_3O^+})}\exp[\phi_1\lambda_{H_3O^+}]\exp[\phi_2\lambda] \end{aligned} \tag{2.36}$$

where $\lambda_{H_3O^+}$ is the moles of hydronium ions (or bound water) per mole of sulfonic acid sites, the K's are essentially equilibrium constants, and the other parameters are groupings of standard states and binary interaction parameters, and are used as physically meaningful fitting parameters [18].

To account for temperature and the much stronger solvation and interaction energies at low water contents, Weber and Newman [91] modified the above model. They fit the temperature dependence of the equilibrium constant, K_2, to experimental data. For the low water contents, they use a simple empirical expression to correct the uptake. Although such a correction is not rigorous, the lack of data at low water contents and the unimportance of this region during PEFC operation and simulation justify the approach.

Using the same equilibrium concept as above, Futerko and Hsing [122] and Tsonos [155] use a Flory–Huggins approach [157] instead of one based explicitly on a single Gibbs function [18,158]. Such an analysis results again in two equations that must be solved simultaneously:

$$\begin{aligned} a_w &= [1 - \phi'_m]\exp\left[\left(1 - \frac{\bar{V}_w}{\bar{V}_m}\right)\phi'_m + \chi\phi'^2_m\right] \\ K &= \frac{\lambda_{H_3O^+}}{(1-\lambda_{H_3O^+})a_w} \end{aligned} \tag{2.37}$$

where K is the equilibrium constant, χ is the Flory interaction parameter and is a linear function of ϕ'_m, and ϕ'_m is a modified volume fraction of the membrane that accounts for association of the hydronium ions:

$$\phi'_m = \frac{\bar{V}_m + \lambda_{H_3O^+}\bar{V}_w}{\bar{V}_m + \lambda\bar{V}_w} \tag{2.38}$$

where $\bar{V}_\mathrm{m}$ is the partial molar volume of the dry membrane:

$$\bar{V}_m = \frac{E_W}{\rho_{m,o}} \tag{2.39}$$

where $\rho_{m,o}$ is the density of the dry membrane and E_W is the membrane's equivalent weight. The above definition of ϕ'_m accounts for membrane swelling effects as discussed in Section 2.4.2. Temperature is considered by fitting the values of K and χ to data.

The final approach to be mentioned is that of Datta and coworkers [31, 154]. It is perhaps one of the most complete because it allows one to calculate the entire water-uptake curve including the liquid-equilibrated value. Their model is developed using a similar Flory–Huggins approach and chemical equilibrium between water vapor, bound (i.e., hydronium), and free water molecules. The main difference from those above is that Datta and coworkers also incorporate the energetic effects due to swelling and any interfaces in a membrane pore on the chemical-potential expressions.

For the membrane pressure expression, their first model uses a spring constant as a fitting parameter, and their later model refines this and uses a swelling model [154,159]:

$$\Pi_m = \frac{2}{3} G(\phi_m^{1/3} - \phi_m^{7/3}) \tag{2.40}$$

where Π_m is the pressure value, ϕ_m is the volume fraction of the membrane and is given by Eq. (2.38) with the second (hydronium) term on the numerator neglected, and G is the shear modulus of the membrane and is a function of water content:

$$G = \frac{1}{3} E_0 \exp\left(-2.1753\left(\frac{\lambda\bar{V}_m}{\bar{V}_m + \lambda\bar{V}_w}\right)\right) \tag{2.41}$$

where E_0 is the dry-membrane value of the Young's modulus. For the interfacial pressure relationship, they use the Young–Laplace equation [160]:

$$\Pi_\gamma = -\frac{2\gamma\cos\theta}{r} \tag{2.42}$$

where γ is the surface tension of water, r is the membrane pore radius, and θ is the contact angle of water with the membrane surface. These pressures are then incorporated into the Flory–Huggins approach, and the two equations to be solved simultaneously become

$$\ln\left(\frac{\bar{V}_w(\lambda-\lambda_{H_3O^+})}{\bar{V}_m+\bar{V}_w\lambda}\right)+\left(1-\frac{\bar{V}_w}{\bar{V}_m}\right)\phi'_m+\chi\phi'^2_m-\ln a_w$$

$$=-\frac{\bar{V}_w}{RT}\left[\frac{2}{3}G(\phi_m^{1/3}-\phi_m^{7/3})-\Pi_\gamma\left(1+\frac{\bar{V}_m}{\bar{V}_w\lambda}\right)\right] \quad (2.43)$$

$$\lambda_{H_3O^+}=1800\frac{a_w}{1-a_w}\left(\frac{1-6a_w^5+5a_w^6}{1+999a_w-1000a_w^6}\right)$$

where the second equation deals with the way they have assumed chemical equilibrium. The same set of equations works for both liquid- and vapor-equilibrated membranes, with the difference being that the last term in the top equation, Π_γ, does not exist for the liquid-equilibrated case. The physical reason is that there is no meniscus or phase boundary with liquid water as there is for vapor. This concept and the above approach is discussed in more depth below. Although most of the various parameters in the above set of equations have known temperature functions, they do not explore temperature effects.

It is readily apparent that all of the more detailed approaches of modeling water uptake involve the same phenomena, namely, chemical equilibria of bound, free, and vapor water, as well as a thermodynamic basis. The approaches are mainly just different in how the thermodynamic functions are developed and used. The approach taken, especially the two-phase one of Datta and coworkers, are all on the right track, although none is truly definitive.

As a further point of comparison, one can discuss and compare the various isotherm models including the empirical and semiempirical expressions. Since all of them are fit to experimental data at 30°C, they all do a good job there. However, when one changes the conditions (e.g., membrane equivalent weight, temperature, etc.) to those not where the data was fit, the more rigorous chemical-modeling equations are more accurate and predictive. The sensitivity of the water uptake is also explored in the relevant papers in terms of the physical parameters such as the Flory interaction parameter [122,154], the dissociation constant [18,31,154], and the membrane properties [154]. The main drawback to using the more rigorous models is that they require the solution of two simultaneous equations (i.e., there is no single closed-form equation). This does slow down computation time and necessitates numerical solution and derivatives, but the physical and realistic gains offset these issues.

2.4.1.1 Vapor- to Liquid-Equilibrated Transition (Schröder's Paradox)

As alluded to throughout this chapter, there are physical and property changes depending on the reservoir phase in contact with the membrane, i.e., Schröder's paradox. It is known as a paradox because one has seemingly two different membrane states even though the chemical potential of the water has not changed. However, contemporary thinking is that it is essentially a phase transition that is occurring in the membrane. Thus, the liquid- and a gas phase-equilibrated membranes remain in equilibrium even though they have slightly different internal microstructures. While a gas phase is essentially excluded from the membrane, the two phases are more akin to bulk-like and associated or bound water in the nanopores of the membrane. One can also think of this as collapsed and expanded channels or pores as introduced in Sections 2.2.3 and 2.3.2.4. The collapsed channels can be thought of regions with associated water, and the expanded ones as where there is a bulk-like liquid–water phase. Such a physical picture is helpful in understanding how to account for the paradox.

Schröder's paradox is an observed phenomenon that should be considered in any membrane model where the membrane does not remain fully hydrated or dehydrated. There are various methods to account for the paradox. The easiest way is to ignore it, which a majority of the models do. Next, it can be treated as a discontinuity, which results in both numerical problems and deciding the discontinuity location. Another approach is to assume a functional form of the water content such that λ and a_w continue to increase (see Eq. (2.33)). However, this violates Schröder's paradox to a significant degree. Finally, there are the more phenomenological and rigorous ways to account for it. These methods are essentially all based on using capillary phenomena and equations and methodology from two-phase flow and vapor-liquid equilibrium. There are three main groups who have used such an approach, Weber and Newman [91], Eikerling *et al.* [136], and Datta and coworkers [31,154].

The key way to model the change in water content is to use a parameter that is essentially the fraction of expanded channels (Eq. (2.22)), S, as discussed in Section 2.3.2.4. Basically, this factor linearly interpolates between the water content of the vapor-equilibrated membrane with that of the liquid-equilibrated one. While it is a linear function in terms of water content, it is not necessarily in terms of activity, pressure, or other variables. This approach allows for a continuous transition between vapor- and liquid-equilibrated membrane water contents, although the transition can be relatively sharp, which is not unexpected for a phase-change-type behavior. The key difference between the models is essentially how they determine S.

The most microscopic of the three approaches is that of Eikerling *et al.* They use a random network of collapsed or expanded pores with a pore-size distribution. They assume that impregnation by liquid water is easier than condensation, although the pores may contain vapor. To determine how the water moves and if the pores are expanded, they use capillary equations of the form of Eq. (2.42), where Π_γ is the capillary pressure. Their model is aimed more at percolation and water permeability and proton conductivity than water uptake, although it can easily be modified to yield uptake.

Weber and Newman utilize an approach where Π_γ is calculated using a form of Eq. (2.42) but where it is applied throughout the channels inside the membrane and where the parameters have been given a different meaning. In their treatment, the contact angle and surface tension represent the various interactions between bulk-like water and the membrane-channel surface such as membrane deformation and coulombic repulsion and attraction energies between membrane, bound water, and free or bulk-like water. For this case, Eq. (2.42) is simplified to contain only a single unknown constant, Γ, that accounts for these various energetic interactions:

$$r_c = \frac{\Gamma}{p_L} \tag{2.44}$$

where Π_γ has been replaced by the liquid pressure and r_c is the incipient expanded channel radius such that smaller radii are not expanded (no bulk-like water phase) and larger ones are. Γ is assumed to be a function of temperature in the same way as the elastic modulus and Flory parameter are assumed to be (i.e., inverse relation) [122,161]. To calculate S, a bundle-of-capillaries approach is used that integrates a channel or ionic-domain size distribution from infinity to r_c [119], resulting in

$$S = \frac{1}{2}\left[1 - \operatorname{erf}\left(\frac{\ln r_c - \ln(1.25)}{0.3\sqrt{2}}\right)\right] \tag{2.45}$$

This approach has a problem that Schröder's paradox is violated when a single driving force is used, as mentioned in Section 2.3.2.4. The reason is that S changes over a small range of chemical potential due to the assumption of a channel-size distribution. To correct this requires either knowing which chemical potential the transition occurs at (e.g., that of bulk liquid to vapor water) or using something akin to a Kelvin equation which would thermodynamically support a chemical-potential gradient and coexistence of both expanded and swollen channels [160,162]. More experimental data is required to determine the validity of the approach and refine it.

The final model that considers the transition region is that of Datta and coworkers discussed in the previous section. This treatment is similar to those above except that the capillary phenomena occur at the boundary of the membrane channel and not within its volume like in the other two methods. Their approach also accounts for the swelling pressure explicitly. In a certain sense, they determine Γ more explicitly. In their approach, the last term on the right side of the top expression in Eq. (2.43) is absent when in contact with liquid (this is similar to making $S = 0$ for vapor and $S = 1$ for liquid). Consequently, the liquid-equilibrated uptake is higher than the vapor-equilibrated one. In other words, the meniscus that forms when in contact with water vapor and inerts results in an additional energy compared to the liquid-equilibrated case that the membrane must equilibrate with. While the approach and model development of Datta and coworkers is the most consistent, rigorous and provides good insight into the paradox, it has not been used in a full-cell simulation. In addition, it is unknown what occurs when liquid is on one side and vapor on the other (i.e., where is the phase-transition point?).

The approach of Datta and coworkers and those of the other treatments could be brought together. This would involve making the Π_γ term in Eq. (2.43) a function of some variable like liquid pressure and not just the reservoir phase. In addition, a pore-size distribution should probably be added to allow for a slightly more gradual transition. Experimentation needs to be undertaken to prove such a model. For example, this model would predict a disappearance of the paradox as the system pressure reaches the vapor pressure due to the disappearance of the inert gases and the meniscus, or even different water uptake from helium than from air. Overall, there is still a need to understand the vapor- to liquid-equilibrated phase transition and the nature of Schröder's paradox. There is opportunity to relate it in more depth to polymer phase segregation, ionic and nonionic moieties, membrane elastic properties, and operating conditions.

2.4.2 Membrane Swelling (Thickness)

It is well documented that Nafion® and other PEMs swell to a significant extent, which is related to their water uptake as discussed above. In fact, liquid-equilibrated Nafion® swells around 60 to 70% in volume compared to its dry dimensions. The dimensional changes are proportional to the water content, and, upon examination of the uptake isotherm (Figure 2.4), seem to correspond with the existence of loosely bound and bulk-like water, in agreement with the physical model (see Section 2.2.3). These dimensional changes are important because they directly impact the magnitude of the gradients and the concentration values in the membrane. The thickness is key because many models treat the sandwich and membrane in 1-D, and thus the potential drop and net water flux can change dramatically if the thickness does. The latter will change because the water-gradient flux is affected but the electro-osmotic flux is not since it is proportional to the current density only. In terms of modeling, the detailed models for water uptake described in Section 2.4.1 already basically consider the volume change effect of swelling.

As discussed, both the volume change and the thickness change are important for membrane modeling. These effects have been treated by three main approaches. The simplest treatment is to assume a volume and thickness of the swollen membrane, normally based on experimental data. This assumption is fine as long as there is not a large change in water content across the membrane and the value was chosen correctly. The first models to use this approach were by Verbrugge and coworkers [11,12,163], who examined a fully liquid-equilibrated membrane. They used factors of 1.29 and 1.72 for the membrane thickness and volume, respectively. Many others have used this approach, typically with the same values for the thickness and volume [62,64,65,82,97,103,104,108, 112,164,165].

The next approach accounts implicitly for swelling by folding it into the governing equations and properties using a coordinate transformation. The concept is to transform the actual thickness coordinate, z, into one that is pegged to the dry thickness of the membrane, z'. Doing this allows for swelling to be considered but not actually tracked. Hence, all of the properties like diffusion coefficients as well as concentrations are cast in this new form, and the dry membrane conditions used. This approach is by far the most often used [38–40,66,67,70,77–79,81,111,120,164,166,167], especially in those models that assume a linear gradient in the membrane and are just solving the dilute solution equations, as it strikes a balance between complexity and ease of use.

The transformation can be done in various ways. Springer *et al.* [10] popularized using an expansion factor of

$$z' = \frac{z}{1 + s\lambda} \tag{2.46}$$

where s is the swelling factor and has a value of $s = 0.0126$. Using their transformation, water concentrations are written as

$$c_w = \frac{1}{\bar{V}_m}\frac{\lambda}{1 + s\lambda} \tag{2.47}$$

and the diffusion coefficients also have to be recalculated [10]. Other transformations include using either the change in density [90,96] or the total volume with water content, where one can assume constant and additive partial molar volumes [29,168–171],

$$V = \bar{V}_m + \lambda\bar{V}_w \tag{2.48}$$

Incidentally, such an assumption is used in many of the more rigorous uptake models discussed in Section 2.4.1 (e.g., see Eq. (2.38)).

While the above approach does a good job in accounting for swelling, it is somewhat complicated and requires careful implementation to ensure that all terms and parameters are transformed. It also allows for different volumes and thicknesses when there is a gradient of water content in the membrane. Such a "local" swelling effect is somewhat unphysical since the membrane length is a uniform, albeit unknown *a priori*, value. However, applied correctly, the approach is numerically convenient because the thickness of the membrane remains uniform in the modified coordinate system.

A more straightforward and rigorous, albeit more numerically intensive, approach is to account for swelling explicitly [88,91,117,135,149,161,172]. Essentially, this is done by doing a mass balance on the membrane and using the above wet-volume expression (Eq. (2.48)) with the average membrane water content. That expression is used to alter all of the variables that are a function of the water volume fraction or concentration. To account for the thickness change of the membrane, the average water content is again used:

$$l = l_o\left(1 + 0.36\,\frac{\hat{\lambda}\bar{V}_w}{\bar{V}_m}\right) \tag{2.49}$$

where l and l_o are the swollen and dry membrane thicknesses, respectively, and $\hat{\lambda}$ is the average value of λ in the membrane. In the above expression, a value of 0.36 is used instead of 0.29 due to the slight anisotropy in membrane swelling [29,91,169,173]. Since $\hat{\lambda}$ is not known *a priori*, one must iterate over the entire simulation until the value converges [91,118].

To get around this obstacle, Newman and coworkers [172,174] developed the following set of two governing differential equations. The first equation is an expression of the average water content as an integral,

$$\hat{\lambda} = \frac{1}{l}\int_0^{z=l} \lambda(z)dz = \int_0^1 \lambda(\varsigma)d\varsigma \tag{2.50}$$

where the equation has been nondimensionalized. While this provides the first equation, the second arises from the thickness being a scalar quantity that is uniform:

$$\frac{dl}{d\varsigma} = 0 \tag{2.51}$$

These two equations are solved with the swelling boundary condition given by Eq. (2.49) and applied at $\varsigma = 1$. Treatment of swelling in the above fashion ensure conservation of the membrane mass, as discussed by Meyers and Newman [174].

One question that may arise is does the membrane swell inside of the fuel cell where it is under compression. To answer this question, Weber and Newman [161] used a simple stress analysis along with their membrane model to predict membrane expansion. They determined that the membrane essentially swells freely within a PEFC, i.e., the membrane pushes against the other sandwich layers and expands. This is because of its prodigious water uptake and the large force required to compress it. If desired, a fractional or degree-of-constraint term can be added to the equations above in determining the membrane thickness and volume to account for partial swelling. Such a parameter would then be determined through a stress balance.

2.4.3 Transport Properties

The transport properties used are different depending on which modeling approach is taken. Normally they involve a proton diffusion coefficient or conductivity, an electro-osmotic coefficient, a permeability, a water diffusion coefficient, and/or a transport coefficient. The macroscopic property models typically apply the underlying governing-equation framework (as introduced in Section 2.3) and apply more constitutive relations like speciation in order to examine the specifics of a property. Often, the analysis is done using concentrated solution theory since it is the most appropriate choice, especially for looking on a detailed level. While the outcome of most of these models is to yield parameter values or simple expressions for use in the governing macrohomogeneous equations, they can be useful in membrane design and optimization analysis. These models are also typically much more complex than the water-uptake ones describe above. The discussion below is loosely separated based on the parameter being discussed, although many are interrelated. Due to the large numbers of expressions used for the various parameters by the various models, only general comments are made about them and the discourse is mainly about the macroscopic-modeling issues of them. While in principle these issues could extend to the experimental techniques that are used to measure the properties since many require models and detailed theoretical analysis, this is beyond the scope of this review and not discussed. Also not discussed are models and property expressions for minor components such as crossover gases, since these either have not been modeled or are mentioned elsewhere in this chapter.

Before moving on, mention should be made about the relationship between these property expressions and Schröder's paradox. Since it is known that the different water contents and membrane structures yield different properties, this has to be considered. Whether one is averaging the property values for a single equation or using multiple driving forces or a single modeling treatment, the experimentally observed effects should be included in the macroscopic model. Most frequently, these effects are incorporated by having different property expressions for different water contents. An example is the electro-osmotic coefficient, which displays a large increase above vapor-equilibrated water contents, or the water diffusion coefficient, which is often separated in up to four different expressions depending on the water content of the vapor-equilibrated membrane. Once again, due to the number of expressions, the above issue is only briefly touched upon below, and for a more detailed discussion of Schröder's paradox.

The ionic conductivity of the membrane is essentially a measure of the proton movement in the membrane. The conductivity can be defined in a certain fashion using the Nernst–Einstein equation (Eq. (2.9)) for the mobility,

$$\kappa = \frac{F^2 D_{H^+} c_{H^+}}{RT} \tag{2.52}$$

This equation implies that protons are the only mobile charge carries in the membrane, which is typically the case, although the presence of anions will affect the conductivity (see Section 2.5.2 and Eq. (2.76)). In the membrane, protons move both by the vehicle (i.e., molecular diffusion) and Grotthuss (i.e., through proton complexes and reorientations) mechanisms as mentioned in Section 2.2.3 and reviewed in Ref. [4].

The value of the conductivity has been measured by many techniques and in many laboratories (e.g., see Ref. [175]). The resulting empirical expressions are often used in modeling. In addition, semiempirical expressions have been used. Such models explain the temperature dependence of the conductivity by the change in the equilibrium constant for the dissociation of the sulfonic acid sites and the activation energies for the Grotthuss and vehicle mechanisms [88,91,176]. They also ascribe the experimentally observed almost linear dependence on water content to the formation of less bound water and percolation effects. The latter was first used for Nafion® by Hsu and coworkers [25], where the volume fraction of water in the membrane is used to determine the conductivity. While this gives some meaning to the conductivity and agrees with the experimental data by showing a several order-of-magnitude change in the conductivity at very low water contents, it has shortcomings such as the fact that it cannot predict the flattening and decrease of the conductivity at higher water contents approaching infinite dilution (i.e., the polymer is dissolved in an infinite amount of water). Furthermore, the conductivity change is not necessarily due to network percolation but perhaps just to the existence of more mobile protons (i.e, those outside the inner hydration shells of the sulfonic acid sites, see Figure 2.5). A more rigorous (compared to just using some general expressions fit to data) percolation-type model for conductivity is presented by Eikerling *et al.* [135,136,156], who use statistical relations, effective medium approximations, and different pore structures. They also summarize various proton-conduction mechanisms and examine single-pore or microscopic models of conduction.

While most of the conductivity models fall in the realm of microscopic models, which examine detailed potential and ion distributions as mentioned later and in Section 2.5.2, there are some that are more macrohomogeneous.

The main approach of these models is using concentrated solution theory, and in terms of a binary-friction model (Eq. (2.14)) [3,86,88,89]. For visualization purposes of the various species, one can use Figure 2.6. To apply this model to conductivity, a water-uptake model as describe in Section 2.4.1 and a dissociation model are required.

For the dissociation model, protons are considered both from the sulfonic acid site and from water complexes such as hydronium ions. An expression for the degree of dissociation, α, can easily be found through chemical equilibria [88], resulting in

$$\alpha = \frac{(\lambda + 1) - \sqrt{(\lambda + 1)^2 - 4\lambda(1 - 1/K_{A,C})}}{2(1 - 1/K_{A,C})} \tag{2.53}$$

where $K_{A,C}$ is the equilibrium dissociation constant and is a function of temperature. This equation can be combined with the transport equations to yield an expression for the conductivity in the form:

$$\kappa = \frac{F^2}{RT\lambda\bar{V}_w}(f - f_0)^{1.5} f(\alpha, D_{i,j}, \lambda) \tag{2.54}$$

where the second part is a Bruggeman and percolation expression, as discussed above, where f is the volume fraction of water in the membrane, and the last part is some function of the dissociation constant and the three binary interaction parameters between water, membrane, and protons. Equation (2.54) along with an isotherm model is then fit to experimental data to determine the necessary parameters. The resultant equation and parameter values provide insight into conduction such as the limiting step and energy required, effect of equivalent weight, etc.

Almost all of the macroscopic membrane models utilize the electro-osmotic coefficient as one of the key experimentally measured parameters. This coefficient, sometimes termed the electro-osmotic drag coefficient, is the transport number of water in the membrane. It is a measure of the number of water molecules that are carried with each proton in the absence of a concentration gradient. The electro-osmotic coefficient arises mainly due to the complexes that solvate the protons and the vehicle mechanism, and depends solely on temperature and water content, at least macroscopically. Most models use empirical (e.g., depends linearly on water content at high water content) or semiempirical expression for the coefficient, although the latter is much preferred.

An example of a semiempirical argument is as follows [91]. For a vapor-equilibrated membrane, the electro-osmotic coefficient has a value close to unity since basically only a hydronium ion is moving through the membrane.

However, for a liquid-equilibrated membrane, the coefficient has a much larger value due to the ability to form extended complexes like Zundel, $H_5O_2^+$, and Eigen, $H_9O_4^+$, ions with the bulk-like-water molecules. Furthermore, it has a dependence on temperature with an activation energy that describes the data and stems from the amount of energy needed to break a hydrogen bond in the bulk-like water in Nafion® [91], which is the limiting step in the Grotthuss mechanism [28,156,177].

There are also some models that try to predict the electro-osmotic coefficient or use the electro-osmotic flux to determine membrane properties and structural aspects [4,178–181]. Since electro-osmotic flow is intricately related to proton conduction, the conductivity models mentioned above can in principle determine the electro-osmotic coefficient with some minor adjustments. Of the models that only examine electro-osmotic flow, all of them except that of Breslau and Miller [178] use a traditional electro-osmotic approach of examining the fluid flow due to that in the double layer next to the membrane surface inside of a pore (e.g., see Ref. [59]). Such an approach calculates the potential distribution and proton concentration in a membrane pore with a charged wall. This is accomplished by solving Poisson's equation (Eq. (2.73)) with a distribution function. Once solved, hydrodynamic equations are used to determine the velocity of the water due to the presence of the charged particles and an electric field. This approach is discussed in some more detail in Section 2.5.2, and the reader is referred to a review on the subject [4] and Chapter 9 of Ref. [59]. As noted, Breslau and Miller use a different approach. They treat the water in a more macroscopic way in terms of hydrodynamic forces and drag. Their model and approach is relatively simple, and is notable in that it was one of the first to treat this subject in PEFC membranes.

The other form of water movement in the membrane is due to the gradient flux. For convective or hydraulic flow, an experimentally measured value for the permeability is used, which can be made into a transport coefficient by substitution of pressure with chemical potential [91]. The reason for just using such an empirical approach is that the membrane is typically assumed to remain fully hydrated for this case. An exception is the model of Eikerling *et al.* [135], which essentially treats the membrane as being a porous medium in which gas and liquid can flow. They then use statistical and percolation arguments along with capillary and two-phase flow equations to determine an effective permeability that varies with water content. In this fashion, they treat the whole water-content range with a single transport coefficient, although they do have a separate gas-phase that exists within the membrane.

Similar to the permeability, those models that utilize a water diffusion coefficient normally use one or more empirical expressions. There are no real macroscopic models aimed at predicting water diffusion, mainly due to its microscopic nature. The only models used are those which are required for analyzing the experimental data. While there are no models to examine, some discussion is required on the different diffusion coefficients which are used. As mentioned in section 2.3.2.2, there are various driving forces that can be used for the gradient-flux equation, each with their own diffusion coefficient.

In the most general case, one uses the chemical potential as the driving force [91]:

$$\mathbf{N}_w = -\frac{c_w D_{\mu_w}}{RT(1 - x_w)} \nabla \mu_w = -\alpha_w \nabla \mu_w \tag{2.55}$$

where D_{μ_w} is the water diffusion coefficient in the membrane related to a chemical-potential driving force or the thermodynamic diffusion coefficient. The expression may or may not contain the mole fraction term in the denominator depending on how the experimental data were interpreted. While this equation and coefficient can be used as is, if one wants to use something like Fick's law (Eq. (2.10)) or the dilute solution theory water equation (Eq. (2.11)), the chemical diffusion coefficient can be related to the water-content driving force:

$$\mathbf{N}_w = -c_w D_{\mu_w} \frac{d(\ln a_w)}{d\lambda} \nabla \lambda = -D_\lambda \nabla \lambda \tag{2.56}$$

where the derivative of activity term is a thermodynamic factor or correction due to nonunity activity coefficients [126]. To determine the factor, differentiation is performed using either water-uptake expressions or models (see Section 2.4.1). As pointed out in the literature [10,182], at low water contents ($\lambda \approx 3$) there is peak in the value of D_λ (due to an inflection in the isotherm; see e.g., Figure 2.4) which depends highly on the way in which the thermodynamic factor is calculated. This is one reason why using D_{μ_w} and a chemical-potential driving force may be preferable. Finally, one can use a concentration of water driving force, resulting in

$$\mathbf{N}_w = -\frac{\lambda \bar{V}_m}{(\bar{V}_m + \lambda \bar{V}_w)^2} D_\lambda \nabla c_w = -D_w \nabla c_w \tag{2.57}$$

where swelling is considered (i.e., an expression akin to Eq. (2.47) was used for the concentration). The above equations define the various diffusion coefficients in terms of each other in order to help the reader see their interchangeability and relation to the various modeling driving forces. Only a single diffusion coefficient is necessary for modeling water movement due to its gradient flux.

In terms of semiempirical justification and values, the diffusion coefficient inside the membrane is typically smaller than that of pure water and with a slightly higher activation energy attributed to the confined geometry and strong interactions between water and the membrane. Furthermore, the diffusion coefficient, D_{μ_w}, is typically an almost a linear function of water content [90,91,183], in agreement with the membrane physical picture.

2.5 Special Topics

This section discusses some special aspects, topics, and models related to the membrane. These issues are worth mentioning in terms of macroscopic membrane modeling but have not been discussed above. The first topic examines the issues related to membrane in the catalyst layer with a focus on the reaction site. The second topic relates modeling efforts to model electrolytes and impurity ions in the membrane. The final two topics focus on membrane durability, including transient models, and DMFCs, respectively.

2.5.1 Membrane in the Catalyst Layers

Up until now the focus of the chapter has been on the membrane as a separator. In this section, some comments are made about the treatment of the membrane in the catalyst layer; for a more detailed discussion on catalyst-layer modeling the reader is referred to another review [1]. Ionomer is placed in catalyst layers in order to provide a conduction path for the protons and extend the reaction zone. In essence, it is used to make the layer into a porous electrode. These porous electrodes are the most complicated of all the PEFC-sandwich regions since they contain all of the various phases as well as supported electrocatalyst. The resistance of the membrane in the catalyst layers can be appreciable and has an effect on the reaction-rate distribution. This section is divided into two main themes based on the use of an embedded macrohomogeneous model [1] for the catalyst layer. The first focuses on how to modify the governing equations for the membrane in a catalyst layer, and the second examines the nature of the membrane and the reaction site in terms of modeling. The former can be considered changing the membrane equations on a macroscopic length scale (i.e., the catalyst layer), and the latter on a more microscopic, local one (i.e., the reaction site).

2.5.1.1 Governing Equations and Properties (Layer Length Scale)

Since membrane exists throughout the catalyst layer, the modeling treatments and equations discussed in the previous sections of this chapter remain valid, although they may have to be altered slightly. All of the porous-electrode-type models incorporate the following changes [1].

In terms of properties, the only difference is that since the membrane does not occupy the entire volume of the layer, something like a Bruggeman expression is required to alter the transport-property values [184–187]:

$$Y = Y_o \frac{\varepsilon_m}{\tau_m} = Y_o \varepsilon_m^{1.5} \tag{2.58}$$

where Y stands for a property such as conductivity, the O denotes a bulk property value, and ε_m and τ_m are the membrane-phase volume fraction and tortuousity, respectively.

In terms of the governing equations, the appropriate source terms must be added to the conservation equations. The transport equations and constitutive relations remain the same except for the use of modified property expressions. For the current balance (Eq. (2.5)), the transfer current between the membrane and electronically conducting solid must be included:

$$\nabla \cdot \mathbf{i}_m = -\nabla \cdot \mathbf{i}_s = a_0 i_h \tag{2.59}$$

where $\nabla \cdot \mathbf{i}_m$ represents the total anodic rate of the electrochemical reaction per unit volume of electrode, m and s denote the membrane and solid phase, respectively, a_0 is the specific interfacial reaction area, and i_h is the transfer current for reaction h. For the hydrogen oxidation reaction (HOR) and the oxygen reduction reaction (ORR), the transfer current expressions are [1] as follows:

$$i_{HOR} = i_{0_{HOR}} \left[\frac{p_{H_2}}{p_{H_2}^{ref}} \exp\left(\frac{\alpha_a F}{RT} (\eta_{HOR}) \right) - \exp\left(\frac{-\alpha_c F}{RT} (\eta_{HOR}) \right) \right] \tag{2.60}$$

and

$$i_{ORR} = -i_{0_{ORR}} \left(\frac{p_{O_2}}{p_{O_2}^{ref}} \right) \exp\left(\frac{-\alpha_c F}{RT} (\eta_{ORR}) \right) \tag{2.61}$$

respectively, where α_a and α_c are the anodic and cathodic transfer coefficients, respectively, *ref* denotes the value at its reference conditions, and η_h is the overpotential for reaction h.

For the mass balance (Eq. (2.2)), the appropriate source terms must be included. Because it is assumed that the electrochemical reactions occur in the membrane (see Section 2.5.1.2), the mass-balance expression for species i in the membrane becomes

$$\frac{\partial c_i}{\partial t} = -\nabla \cdot \mathbf{N}_i + S_i + R_i \tag{2.62}$$

where R_i is the transfer of species i into or out of the membrane through mass-transfer or equilibrium expressions as discussed in Sections 2.4.1 and 2.5.1.2, and S_i is the generation/consumption term of species i that is given by Faraday's law:

$$S_i = \sum_h s_{i,h} \frac{i_h}{n_h F} \tag{2.63}$$

where $s_{i,h}$ is the stoichiometric coefficient of species i participating in reaction h (−1 for hydrogen in the HOR, and −1 and 2 for oxygen and water in the ORR, respectively), and n_h is the number of electrons transferred in reaction h (2 for HOR and 4 for ORR).

For the energy balance (Eq. (2.23)), the appropriate source terms must be added, yielding [1]:

$$\rho \hat{C}_p \left(\frac{\partial T}{\partial t} + \mathbf{v} \cdot \nabla T \right) = -\nabla \cdot \mathbf{q} + \sum_i \bar{H}_i \nabla \cdot \mathbf{J}_i + \sum_h a_0 i_h (\eta_{s_h} + \Pi_h) + \sum_g \Delta H_g R_g \tag{2.64}$$

where the last term on the right includes enthalpy generation/consumption due to chemical reactions (*e.g.*, evaporation) and the second to last term that due to the electrochemical reactions. The latter contains expressions for both the reversible and irreversible heat generation, as derived by Newman and coworkers [59,128,129,188]. The irreversible heat generation is represented by the surface overpotential, and the reversible by the Peltier coefficient [189]. While almost all of the catalyst-layer models use the mass- and current-balance equations, only a few [37,44,47,71,89,105,111,190] use the above energy equation.

2.5.1.2 Reaction-Site Models (Local Length Scale)

While the previous section discusses the changes to the governing equations on the layer length scale, it did not discuss the local length scale or where the reaction site is. As mentioned, it is assumed that the reactions occur within the membrane phase of the catalyst layers. However, there are various modeling methodologies and pictures for how this occurs [1]. To determine the correct modeling equations and how the membrane is involved, an analysis is done where three models are chosen and fit to experimental data. The three models are 1-D PEFC sandwich models that are identical except for their rate expressions [172,191], which are given in terms of the ORR below for brevity, and can easily be adapted for the HOR.

In the first model [11,12,37,53,89,102,111,172,192,193], denoted PE, just the porous-electrode equations are used (as given in the previous section), and the reaction site is assumed to be just the catalyst interface.

In the second model [46,50,124], denoted PEF, the porous-electrode equations are used, and the reaction site is assumed to be an interface covered by a thin membrane film. Hence, the current balance for the ORR can be expressed as

$$\nabla \cdot \mathbf{i}_m = 4Fc_{O_2}^{bulk} \left[\frac{1}{\dfrac{\delta_{film}}{AD_{O_2,film}} + \dfrac{1}{k'}} \right] \tag{2.65}$$

where *bulk* signifies the concentration outside the *film*, the reference concentration is that in the membrane in equilibrium with the reference pressure, δ_{film} and A are the thickness and specific external surface area of the *film*, respectively, and

$$k' = \frac{a_0 i_{0_{ORR}}}{4Fc_{O_2}^{ref}} \exp\left(-\frac{\alpha_c F}{RT}(\eta_{ORR}) \right) \tag{2.66}$$

In the third model [56,57,76,103,194–199], denoted PEA, the porous-electrode equations are used and the reaction site is assumed to be a spherical agglomerate composed of supported catalyst, membrane, and possible gas micropores. For this model, the current balance for the ORR is given by

$$\nabla \cdot \mathbf{i}_m = a_0 i_{ORR} E \tag{2.67}$$

where E is the effectiveness factor, which for a first-order reaction can be written as [126,200]:

$$E = \frac{1}{3\phi^2}(3\phi \coth(3\phi) - 1) \tag{2.68}$$

where ϕ is the Thiele modulus for the system [201]:

$$\phi = \frac{R_{agg}}{3}\sqrt{\frac{k'}{D_{O_2,agg}^{e}}} \tag{2.69}$$

where R_{agg} is the radius of the agglomerate. In the above derivation, the agglomerate was assumed to be equipotential. If this assumption is not used or if the reaction is not first order, than an analytic expression is not obtainable; a separate numerical model must be used.

The three models were fit to various experimental polarization curves, where the PE model had two fitting parameters and the PEF and PEA models each had three, all of which were similar in their effect. The fits are shown in Figure 2.13. As is readily apparent, the PEA model fits the experimental data much better than the others, in agreement with other literature comparisons that used simpler PEFC and membrane models [204–206]. For the three cases, the PEF model mainly follows the PE model in terms of fitting the curve, with the possible exception of case (*c*) where the PEF model results in a better fit. Overall however, the PE and PEF models result in a pronounced "knee" in the mass-transfer portion of the curve. Thus, one can surmise that the added fitting parameter and diffusion resistance provided by the PEF model has only a minimal impact. Therefore, a single barrier to oxygen transport is not limiting, and one needs combined reaction and diffusion and significant interaction with the membrane. Thus, the agglomerate model is the best to use for the reaction site of membrane, electrocatalyst, and gas (either dissolved or free).

The reason why the PEA model fits best and an effectiveness factor is required is that it allows for a more gradual decrease in the reaction rate or transfer current as a function of decreasing potential, as shown in Figure 2.14. This is because the oxygen partial pressure decreases gradually, due to the

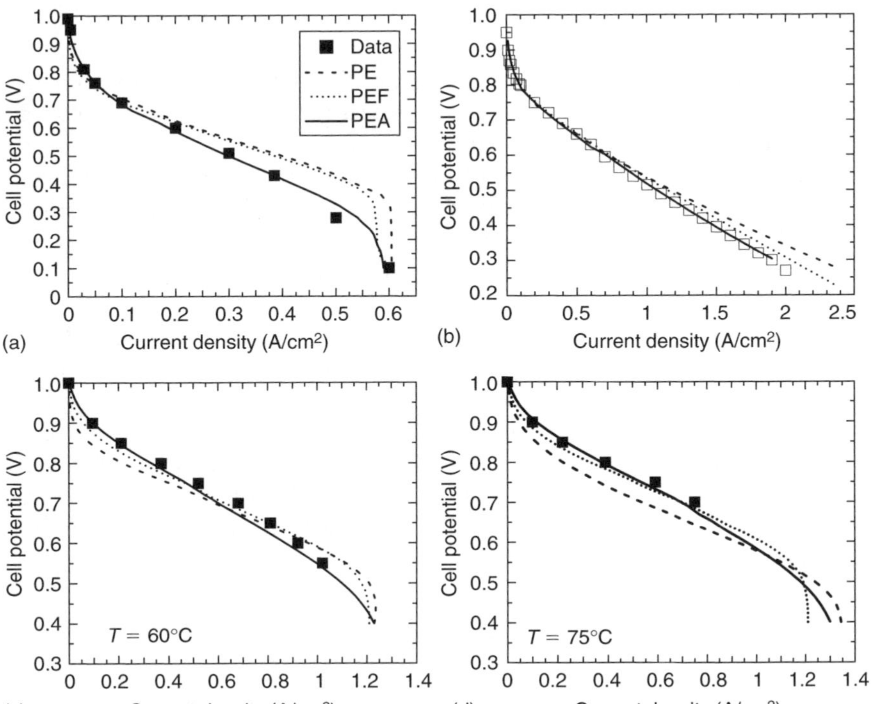

Figure 2.13 Comparisons of the three test models, PE (dashed), PEF (dotted), and PEA (solid), to experimental data (squares) from (a) Ticianelli *et al.* [202], (b) Gloaguen *et al.* [204], and Qi and Kaufman [203] at (c) 60°C and (d) 75°C.

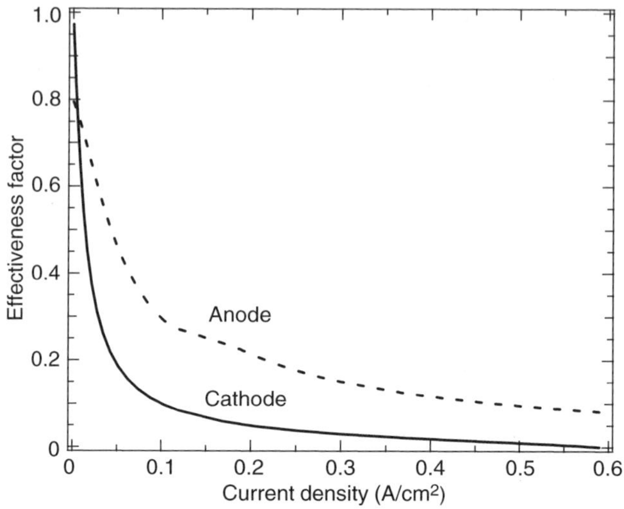

Figure 2.14 Anode and cathode effectiveness factors as a function of current density for case (a) from Figure 2.13 using the PEA model.

variable reaction penetration depth into the agglomerate because of the consideration of simultaneous reaction and diffusion. Further examination of the figure shows that, at low current densities, the effectiveness factor for the anode is lower than for the cathode because of the facile kinetics of the HOR, but is higher at higher current densities due to the higher hydrogen concentration in the anode catalyst layer and faster hydrogen diffusion in the agglomerate. However, it is the cathode catalyst layer and its effectiveness factor that are limiting in all the cases examined.

In summary, the membrane model for the layer length scale is basically the same as for the membrane itself, except that the properties need to be adjusted and source terms have to be added to the conservation equations. For the local length scale, the reaction site is best treated using an agglomerate model because it allows for an effectiveness factor that changes gradually with potential versus that of a constant for a bare or film-covered reaction site. Finally, a membrane film can be added on top of agglomerate in the PEA model to stimulate a diffusion barrier without reaction, however this also adds another unknown.

2.5.2 Impurity Ion and Electrolyte Effects

The existence of other ions in the membrane besides protons and sulfonic acid sites (e.g., platinum, iron, sodium, etc.) complicates the transport and modeling of the membrane. In fact, the other ions often act as poisons which decrease PEFC performance, although additional protons from acids may increase the conductivity [163]. The ions can come from a variety of sources including end or bipolar plates, corrosion reactions, catalyst dissolution

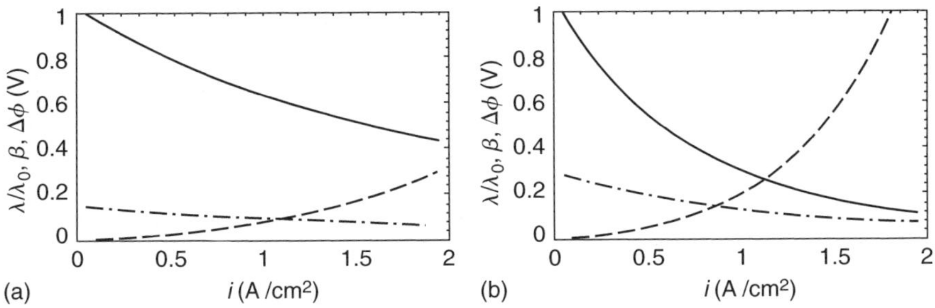

Figure 2.15 Comparison of no contamination (a) and 10% Na^+ contamination on the cathode side of the membrane (b) on dimensionless water content (solid line), net water flux (dot-dash line), and membrane potential drop (dashed line) as a function of current density (reproduced from Ref. [209]).

(see Section 2.5.3), and contaminants left over from manufacturing. To model such effects requires the incorporation of additional phenomena into the governing equations. Because these effects are mainly of secondary importance in fuel cells, the developed models are aimed at other applications of the membranes such as in separation processes and the chlor-alkali industry. For these reasons and because they are more microscopic in approach, the model treatments are discussed only briefly below.

There are three main electrolyte issues for the membrane, which can be categorized as effects due to cation exchange, the presence of electrolyte in the reservoir next to the membrane, and the presence of electrolyte inside the membrane pores. PFSA membranes readily undergo cation exchange, e.g., sodium ions will replace the protons associated with the sulfonic acid sites. Such a replacement has obvious implications in terms of changing the transport properties of the membrane [207], with the decrease in proton conduction the most expected and significant due to the inability to form hydrogen bonds and utilize the Grotthuss mechanism. More than any other group, Okada and coworkers have modeled such effects [143,208–210]. They analytically solve the dilute-solution equations described in Section 2.3, but incorporate the changing transport properties, usually through some kind of linear average. An example of their simulation results is shown in Figure 2.15.

The figure demonstrates how having 10% contamination of sodium ions on the cathode side of the membrane greatly influences the water management and ohmic resistance through the membrane. These effects are especially significant at high current densities. One can see that the water content decreases and the net water flux is higher, resulting in significant dehydration of the membrane, which is apparent in the much higher resistance values. The net water flux is higher due to a much larger electro-osmotic flux through the membrane in its sodium rather than hydrogen or acid form. This effect and the changes in the other transport properties cause the resistance of the membrane and the net water flux to be nonlinear functions of the current density, as one might have expected. Overall, the results clearly demonstrate the significant problems and changes associated with having cation impurities and part of the membrane cation exchanged.

While cations may exchange with the protons in the membrane, they can also exist in the hydrophilic phase within the membrane as neutral combinations with anions such as chloride, even though one expects some anion exclusions and some cation exchange due to the presence of the anionic sulfonic acid sites. Thus, something like platinum in the membrane should decrease the membrane's conductivity in the same fashion as described above. These impurity ions may also cause other physicochemical changes such as different microstructure due to crosslinking, ion-pair formation, etc. [208]. They will also alter the uptake properties of the membranes.

To calculate the distribution of ions in the membrane in contact with a reservoir, one needs to consider chemical equilibrium among the various species in the membrane and their counterparts in the external reservoir. To do this, one needs to calculate the concentration distribution of each ion inside of the pore. This distribution can be written as a modified Boltzmann distribution [4,211–214]:

$$c_i(r) = c_i^{ext}\exp\left[-\frac{z_i F\Phi(r)}{RT} - \frac{A_i}{RT}\left(\frac{\varepsilon_0}{\varepsilon_r(r)} - \frac{\varepsilon_0}{\varepsilon_r^{ext}}\right)\right] \tag{2.70}$$

where r is the radial position of the pore, *ext* denotes the external reservoir, A_i is an ion hydration constant of species i (see Ref. [213] for values), and $\varepsilon_r/\varepsilon_0$ is the dielectric constant of the medium, where ε_0 is the permittivity of free space. In the above expression, the first term represents electrostatic attraction/repulsions of ions, and the second relates the effects of a changing dielectric medium. If the dielectric constant is uniform and the same as the reservoir, then the expression becomes a normal Boltzmann distribution.

In order to calculate the distribution, the dielectric-constant and potential distributions must be known. The dielectric-constant distribution can be determined using Booth's equation [215]:

$$\frac{\varepsilon(r)}{\varepsilon_0} = n^2 + \frac{3\left(\dfrac{\varepsilon_r^{ext}}{\varepsilon_0} - n^2\right)}{\zeta\nabla\Phi(r)}\left[\coth(\zeta\nabla\Phi(r)) - \frac{1}{\zeta\nabla\Phi(r)}\right] \tag{2.71}$$

where n is the refractive index of the solution and ζ is

$$\zeta = \left(\frac{5\eta}{2kT}\right)(n^2 + 2) \tag{2.72}$$

where k is Boltzmann's constant and η is the dipole moment of the solvent molecule. Finally, the potential distribution can be calculated by solving Poisson's equation [59]:

$$\nabla \cdot (\varepsilon(r)\nabla\Phi(r)) = -\rho_e = -F\sum_{i=1}^{m} z_i c_i(r) \tag{2.73}$$

where ρ_e is the charge density and the summation is over all charged species. Solving Eqs. (2.70), (2.71), and (2.73) simultaneously yields the concentration distributions of the various ions and the potential distribution within the pore.

The potential and dielectric-constant distributions for various pore sizes as calculated by Pintauro and Verbrugge [214] are shown in Figure 2.16. The

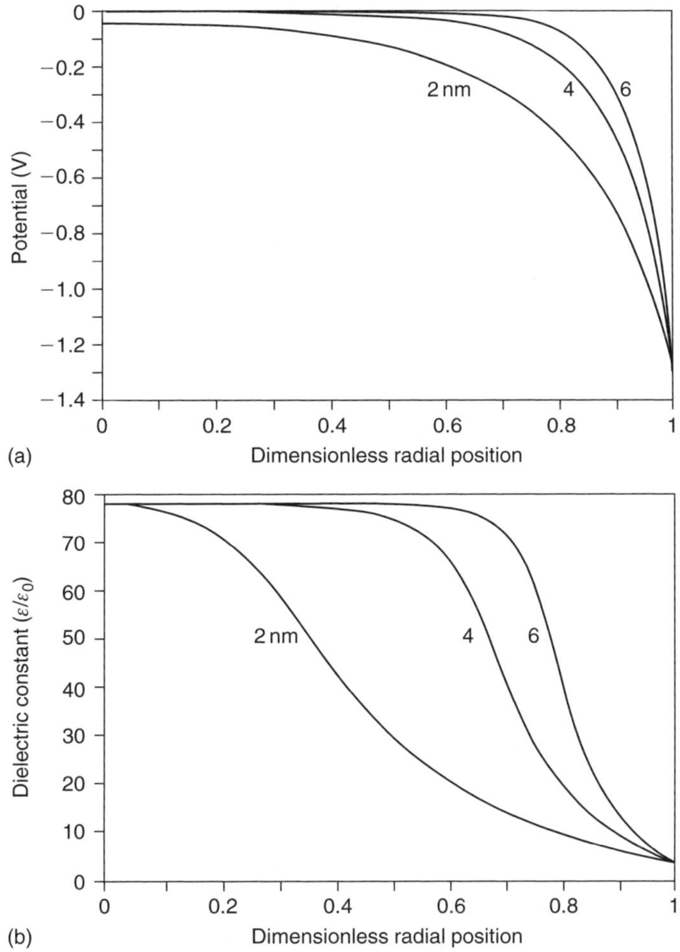

Figure 2.16 Potential (a) and dielectric constant (b) profiles as a function of pore size (reproduced from Ref. [214]).

figure clearly demonstrates that the dielectric constant and potential are not uniform within a significant distance of the pore wall, in agreement with other analyses [4,213]. The potential deviates due to both dielectric-constant variation and especially hydration effects as also shown by Pintauro and Verbrugge. As can be seen in the figure, pore sizes around 2 nm and below do not have any regions of uniform potential or dielectric constant. This size correlates with partially hydrated pore sizes (i.e., when $\lambda < 7$), and one can draw the inference that bulk-like water forms when the potential and dielectric-constant distributions flatten and become equal to that of water, as mentioned in Section 2.2.3. In terms of ion distributions and uptake, the overall partitioning of salts is shown to be dependent on the monovalent rather than the divalent cation, with the larger monovalent cations being more selective than the smaller ones [212,213]. In all, the above approach allows one to predict ion partitioning by these membranes, which is of special interest in separation applications. Furthermore, the potential-distribution calculation is also a key component of most microscopic models [4], including the ones for conductivity and electro-osmotic flow.

The last topic of discussion in this section is that of transport of electrolytes through the membrane. This transport is seen in dissolution studies where one finds platinum and alloy catalyst movement through the membrane after operation [130,216–219]. To model such effects, the same basic approaches and equations as described in Section 2.3 have been used [92, 216,220–222].

The major differences are that not all of the current is carried by protons and the system is no longer only three components. The major modification of the equations involves a more general definition of the current density:

$$\mathbf{i} = F\sum_i z_i \mathbf{N}_i \tag{2.74}$$

This also means that a more general expression can be derived for the potential gradient,

$$\nabla\Phi = -\frac{\mathbf{i}}{\kappa} - \frac{F}{\kappa}\sum_i z_i D_i \nabla c_i \tag{2.75}$$

where the conductivity can be defined as

$$\kappa = F^2 \sum_i z_i^2 u_i c_i \tag{2.76}$$

In addition to the above equations, material balances need to be applied for each species (Eq. (2.2)), the current and energy balances remain the same (Eqs. (2.5) and (2.23), respectively), and either a Nernst–Planck equation (Eq. (2.8)) or a multicomponent transport one (e.g., Eq. (2.15)) is required depending on the approach taken. Finally, either electroneutrality

(Eq. (2.4)) can be assumed or, for the more microscopic models, Poisson's equation (Eq. (2.73)) along with Eqs. (2.70) and (2.71) can be incorporated into the governing set of equations.

Due to the presence of more species, more properties are required. These can take the form of mobilities, diffusion coefficients, conductivity, and transference numbers [59]:

$$t_j = \frac{z_j^2 u_j c_j}{\sum_i z_i^2 u_i c_i} \tag{2.77}$$

which is the fraction of current carried by the ion j in the absence of concentration gradients. Without other ions, this value is 1 for protons, and it can be used as an indication of the current transport efficiency. As a side note, the electro-osmotic coefficient is basically the transference number of water. For the multicomponent systems, concentrated solution theory, which is given in terms of binary interaction parameters that can be related to the above experimentally measurable properties, predicts the correct number of transport properties whereas dilution solution does not [59]. In essence, all of the interactions are treated explicitly. Furthermore, to be rigorous, one must also consider activity coefficients for neutral combinations and thermodynamic factors (see Section 2.4.3 and Ref. [59]).

2.5.3 Membrane Durability and Transient Modeling

The most important issue facing PEFCs today is probably that of lifetime and durability. As the heart of the fuel cell, the membrane is a key part in understanding failure. This is especially true in those cases where the membrane is either chemically or physically altered, resulting in pinhole formation and catalyst layer delamination [130,219], or where the cause of failure can be traced to such issues as gas crossover through the membrane [223]. In addition, the poisoning effects described above in terms of cations in the membrane (including platinum) can lead to performance loss [223–225]. In terms of modeling, there has been some work on durability and failure mechanisms, but it is relatively new; understanding failure is one of the main goals of current and future modeling activities.

Any model of failure and durability is inherently a transient one. This is because these effects occur due to fluctuations in operating conditions like potential or during transient operation (e.g., startup and shutdown). In terms of transient membrane modeling, various groups have examined the behavior of the water content of the membrane [40,72,82,89,108,120,226,227]. The governing equations are essentially the same as those in Section 2.3, although with the time dependent terms considered. They clearly show that operation with dry gases and high stoichiometries leads to cases of unstable operation

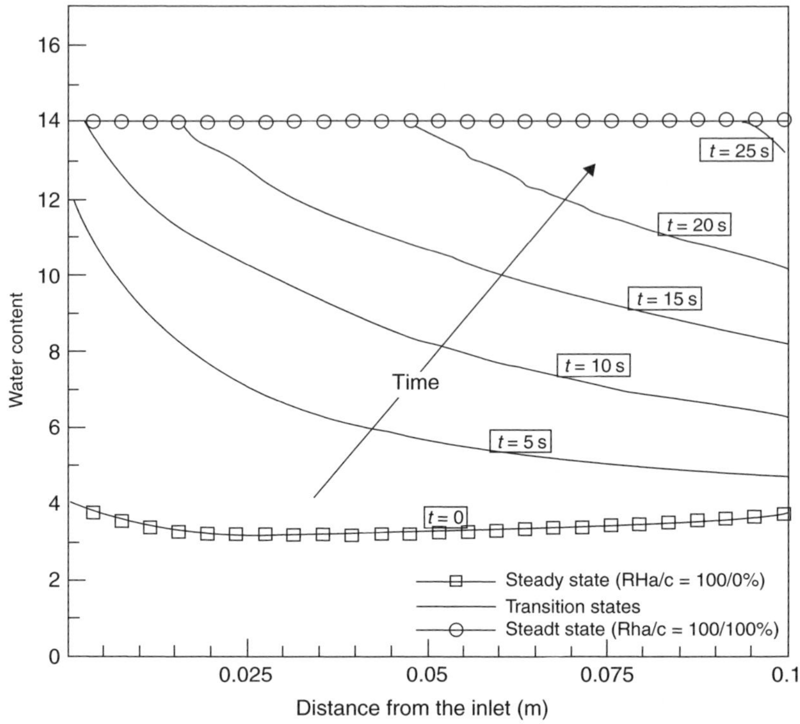

Figure 2.17 Evolution of water-content profiles when the cathode inlet changes from dry to fully humidified (reproduced from Ref. [226]).

where the membrane dries out and the fuel cell fails. They have also shown much more instability in operation due to the transients in water management. This last is shown in Figure 2.17 in terms of the time lag that it takes for a membrane to become fully hydrated after a step change in relative humidity. This lag causes a very nonuniform current distribution along the channel not to mention water buildup at the cathode, both of which can lead to instabilities and possible durability issues. It is worth noting that the overall-system-response time constant is the same as that for membrane hydration.

In terms of membrane failure mechanisms, the most modeled is pinhole formation due to membrane degradation by chemical attack of hydroxyl radicals [132,223,228]. The hydroxyl radicals are generated due to chemical and electrochemical reactions, and may come from the formation of peroxide in the ORR or through oxygen crossover and a two-electron reaction at the anode [131,229]. It is also generated during potential cycling on the cathode where oxide and hydroxide layers form on the platinum catalysts [217]. The generated radicals attacks the membrane end groups and essentially unzips the polymer [132]. While the qualitative picture is known and improvements have been made in terms of durability, there has been no quantitative model of this effect yet.

Other membrane failure mechanisms include mechanical effects during cycling (especially with humidity and temperature) [130,219], carbon corrosion [230–232], and cation impurities and movement in the membrane [130,216–219,224]. For the first mechanism, there has been no substantial modeling effort as of yet, although there is ongoing work on this topic in the community. For the second topic, Reiser *et al.* [232] put forth a reverse-current decay mechanism in which carbon corrosion occurs due to inadequate fuel quantities and possible crossover during startup and shutdown. For the third topic, it has been suggested that these ions come from platinum dissolution [216–218] at the cathode or other sources of impurities such as iron from the end plates [224].

In terms of modeling, some effort has been done on the dissolution mechanism of platinum [217,233] and on platinum migration using dilute solution equations as mentioned in the preceding section [216]. Also mentioned there are issues associated with cation exchanged membranes causing water management problems and changes in transport properties and perhaps membrane microstructure [208]. Of note is the model of Chen *et al.* [144], which shows the time and required current density for stable operation is increased with impurity ions in the membrane. Finally, in terms of incorporating failure, Jiang and Chu [234] use an empirical membrane deterioration term that limits the maximum water content of the membrane in their model.

2.5.4 Direct-Methanol Fuel Cells

The membranes used for DMFCs are typically the same as those for hydrogen fuel cells (i.e., Nafion®), although they may be thicker and have slightly smaller and more tortuous pores. This is because performance loss due to methanol crossover is more important than membrane conductivity, especially since the membrane remains fully hydrated.

Since the membranes are basically identical, the same can be said of the membrane models [121,174,235–243]. A benefit of DMFCs is that liquid water exists on both sides of the membrane, and thus it remains liquid-equilibrated throughout. This means that the membrane model can be simplified, especially with respect to conductivity. However, water flow becomes more complicated due to methanol crossover. Unlike gas crossover in hydrogen fuel cells, methanol can exist in appreciable quantities due to its miscibility with water. This means that it cannot be taken as a minor component, and the appropriate transport equations must be used.

The most common approach for modeling water and methanol movement in the membrane is by using the convective diffusion equation (Eq. (2.31)) for both water and methanol, where the bulk velocity is given by the Schlögl equation (Eq. (2.18)). A slightly more complicated approach is to use a convective diffusion equation for methanol that also includes an electro-osmotic term with a different coefficient than that of water. In essence, both of these are basically dilute-solution approaches. While they will yield general trends and answers, it is problematic in terms of determining the correct property expressions like diffusion and electro-osmotic coefficients that should be used in the equations. Finally, the uniform membrane hydration along with a simple methanol diffusion equation allow for analytic models and for the equations to be solved exactly for specific cases, such as that of constant and equimolar feeds [121,241,242].

A more realistic and complicated approach is to use a concentrated solution theory analog [174]. The equations used are derived the same as in section 2.3.2.2, except that there is the additional methanol equation along with its coefficients and cross coefficients, all of which have to be determined. The concentrated solution theory approach also accounts for the fact that uptake of the membrane both in terms of water and methanol depends on the concentration of the outside reservoir. While three-phase diagrams exist for this system (e.g., see Ref. [148]), there is a minimal amount of modeling to predict the uptake. The best and most rigorous treatment for water uptake is that of Meyers and Newman [18, 174] using their chemical model. Starting from a single Gibbs function for the membrane [158], they derive expressions for the chemical potential of water and methanol in the membrane, and equate those to the chemical potentials of water and methanol in the reservoir. This is essentially the same treatment as their chemical model for uptake from water vapor (see Section 2.4.1), except that now it is from a liquid water/methanol reservoir. Such a modification requires careful consideration of the appropriate reference states and determination of activity coefficients. Overall, modeling a DMFC membrane is very similar to modeling a PEFC membrane (the underlying physics are the same), except now there is liquid water throughout although it is not pure (the methanol must be accounted for with its own governing equation and interactions).

2.6 Summary

In this chapter, we have examined the different models for polymer-electrolyte membranes used in fuel cells operating with hydrogen. The major focus has been on the governing equations and approaches for transport of the various species within the membrane. This included a discussion of the various governing equations and property values such as membrane water content. Concentrated solution theory was shown to be the most rigorous approach, and the one that can best address the full range of membrane water contents and conditions. Some related special topics of interest including membrane treatment in catalyst layers and durability issues were also discussed.

Full membrane models were not directly compared to each other; instead they were broken down into their constitutive parts. The reason for this is that validation of the models is usually accomplished by comparison of simulation to experimental polarization data (e.g., Figure 2.3), which necessitates at least a 1-D sandwich model of the entire cell. Furthermore, in fitting this data, the models not only vary in their complexity and treatments, but also in their number and kind of fitting parameters. This is one reason why it is hard to justify one approach over another by just looking at the modeling results. In general, it seems reasonable that the more complex models, which are based on physical arguments and do not contain many fitting parameters, are perhaps closest to reality. Of course, this assumes that they fit the experimental data and observations under a variety of operating conditions and parameter values.

This review has highlighted the important effects that should be modeled. These include a robust water-uptake model, a combination model for transport that remains valid for both vapor- and liquid-equilibrated membranes, and inclusion of membrane swelling, among others. For any model, a balance must be struck between the complexity required to describe the physical reality and the additional costs of such complexity. In other words, while more complex models more accurately describe the physics of the transport processes (e.g., detailed pore models of potential distributions for proton conduction), they are more computationally costly and may contain many fitting parameters. Overall, membrane modeling and understanding has progressed substantially in the last few years, and the research community is now closing in on good descriptions of the underlying transport phenomena, microstructure, etc. Such treatments mean that simulations can now begin to optimize and perhaps even aid in designing PEFC membranes. Hopefully, this review has shown and broken down for the reader the complexities and approaches undertaken in modeling polymer-electrolyte membranes.

2.7 Notation

Roman

a_i = Activity of species i

a_0 = Interfacial area of reaction between the solid and membrane phases (cm^{-1})

A = Specific external surface area (cm^{-1})

A_i = Ion hydration constant of species i (J/mol)

b = Tafel slope (V)

c_i = Interstitial concentration of species i (mol/cm^3)
c_T = Total solution concentration (mol/cm^3)
C_p = Membrane system heat capacity (J/g/K)
$\mathbf{d}_i$ = Driving force per unit volume acting on species i (J/cm^4)
D_i = Fickian diffusion coefficient of species i (cm^2/s)
$D_{i,j}$ = Diffusion coefficient of i in j (cm^2/s)
D_{μ_w} = Diffusion coefficient of water related to a chemical potential driving force (cm^2/s)
D_λ = Diffusion coefficient of water related to a water content driving force (cm^2/s)
E = Effectiveness factor, defined by Eq. (2.68)
E_W = Membrane equivalent weight (g/equiv)
E_0 = Young's modulus reference value (N/cm^2)
f = Water volume fraction in the membrane
F = Faraday's constant (96487 C/equiv)
G = Membrane shear modulus (N/cm^2)
ΔG = Gibbs free energy of reaction (J/mol)
$\bar{H}_i$ = Partial molar enthalpy of species i (J/mol)
ΔH_l = Heat or enthalpy of reaction l (J/mol)
i = Superficial current density through the membrane (A/cm^2)
$\mathbf{i}$ = Superficial current density (A/cm^2)
i_h = Transfer current density of reaction h per interfacial area (A/cm^2)
i_{o_h} = Exchange current density for reaction h (A/cm^2)
i_{lim} = Limiting current density (A/cm^2)
$\mathbf{J_i}$ = Flux density of species i relative the mass-average velocity (mol/cm^2/s)
k = Boltzmann's constant (1.38065×10^{-23}/J/K)
k = Effective permeability (cm^2)
k'_h = Reaction-rate portion of the Thiele modulus for reaction h (mol/bar/cm^3/s^1)
k_T = Effective thermal conductivity (J/cm^2/K)
k_ϕ = Electrokinetic permeability (cm^2)
$K_{i,j}$ = Frictional coefficient of interaction between species i and j (J/s/cm^5)
K_l = Equilibrium coefficient for reaction l
l = Membrane thickness (cm)
$L_{i,j}$ = Inverted frictional coefficient of interaction between species i and j (cm^5/J/s)
M_i = Molecular weight of species i (g/mol)
n = Solution refractive index
n = Number of electrons transferred in a reaction
N = Number of species
$\mathbf{N}_i$ = Superficial flux density of species i (mol/cm^2/s)
p = Total thermodynamic pressure (N/cm^2)
p_i = Partial pressure of species i (N/cm^2)

p_k = Total pressure of phase k (N/cm^2)
p_w^{vap} = Vapor pressure of water (N/cm^2)
$\mathbf{q}$ = Superficial heat flux ($J/cm^2/s$)
r = Membrane pore radius (cm)
r_c = Critical pore radius (μm)
R = Ideal-gas constant (8.3143 J/mol/K)
R_{agg} = Agglomerate radius (cm)
R_i = Rate of transfer of species i to/from the membrane ($mol/cm^3/s$)
R_Ω = Total constant ohmic resistance ($\Omega\ cm^2$)
$s_{i,h}$ = Stoichiometric coefficient of species i participating in reaction h
S = Fraction of expanded channels
S_i = Rate of generation/consumption of species i ($mol/cm^3/s$)
$\bar{S}_i$ = Molar entropy of species i (J/mol/K)
t = Time (s)
t_j = Transference number of species j
T = Absolute temperature (K)
u_i = Mobility of species i ($cm^2/mol/J/s$)
U^θ = Standard cell potential (V)
$\mathbf{v}$ = Superficial solution velocity (cm/s)
$\mathbf{v}_i$ = Superficial velocity of species i (cm/s)
V = Cell potential (V)
$\bar{V}_i$ = (partial) Molar volume of species i (cm^3/mol)
x_i = Mole fraction of species i
Y = Generic property such as conductivity or transport coefficient
z = Through-plane (thickness) distance of the membrane (cm)
z' = Modified membrane thickness coordinate to account for swelling (cm)
z_i = Valence or charge number of species i

Greek

α = Transport coefficient ($mol^2/J/cm/s$)
α_a = Anodic transfer coefficient
α_c = Cathodic transfer coefficient
β = Net water flux per proton flux through the membrane
γ = Surface tension (N/cm)
δ_{film} = Membrane film thickness (cm)
ε_k = Volume fraction of phase k
ε_r = Dielectric constant of the medium
ε_0 = Permittivity of free space (8.8542×10^{-14} F/cm)
Γ = Energetic interaction parameter of the membrane with bulk-like water (N/cm)
ς = Dimensionless membrane thickness
ξ = Electro-osmotic coefficient

Π_m = Pressure relation for membrane expansion (N/cm^2)
Π_g = Pressure relation due to interfacial meniscus (N/cm^2)
Π_γ = Peltier coefficient for reaction h (V)
ρ = System density (g/cm^3)
ρ_e = Electric charge density (C/cm^3)
η = Dipole moment of the solvent (C/cm)
η_h = Electrode overpotential of reaction h (V)
η_s = Surface overpotential of reaction h (V)
θ = Contact angle (degrees)
κ = Ionic conductivity (S/cm)
λ = Moles of water per mole of sulfonic acid sites
$\lambda_V|_{a_w=1}$ = Maximum value of λ for a membrane in contact with water vapor at unit activity and the operating temperature
λ_L = Moles of water per mole of sulfonic acid sites for liquid-equilibrated membrane
λ = Average membrane water content
$\lambda_{H_3O^+}$ = Moles of hydronium ions (bound water) per mole of sulfonic acid sites
μ = Viscosity (Pa/s)
μ_i = Electrochemical potential of species i (J/mol)
τ_k = Tortuosity of phase k
ϕ = Thiele modulus, defined by Eq. (2.69)
ϕ_l = Weber and Newman chemical model parameter
ϕ_m = Membrane volume fraction (prime indicates slightly modified)
Φ = Electrical potential (V)
χ = Flory interaction parameter
ψ_i = Permeation coefficient of species i (mol/bar/cm/s)

Subscripts/superscripts

e = Effective transport property
ext = External reservoir conditions
f = Fixed ionic site
H^+ = Proton
HOR = Hydrogen oxidation reaction
m = Membrane
o = Initial or reference value
ORR = Oxygen reduction reaction
ref = Parameter evaluated at the reference conditions
s = Solid phase
w = Water

ACKNOWLEDGMENTS

We would first like to thank the financial support of the the Assistant Secretary for Energy Efficiency and Renewable Energy, Office of Hydrogen, Fuel Cell, and Infrastructure Technologies, of the US Department of Energy under contract number DE-AC02-05CH11231. We would also like to thank Trung V. Nguyen and Tim Zhao for inviting us to write this chapter.

REFERENCES

1. A. Z. Weber, and J. Newman, *Chem. Rev.*, 104 (2004) 4679.
2. C. Y. Wang *Chem. Rev.*, 104 (2004) 4727.
3. J. Fimrite, H. Struchtrup, and N. Djilali, *J. Electrochem. Soc.*, 152 (2005) A1804.
4. K. D. Kreuer, S. J. Paddison, E. Spohr and M. Schuster, *Chem. Rev.*, 104 (2004) 4637.
5. B. Carnes and N. Djilali, *J. Power Sources*, 144 (2005) 83.
6. M. V. Williams, H. R. Kunz, and J. M. Fenton, *J. Electrochem. Soc.*, 152 (2005) A635.
7. A. A. Kulikovsky, *Electrochem. Commun.*, 4 (2002) 845.
8. H. A. Liebhafsky, E. J. Cairns, W. T. Grubb, L. W. Niedrach, in G. J. Young, and H. R. Linden (Eds.), *Fuel Cell Systems*, Vol. 47, American Chemical Society, Washington, DC, 1965, p. 116.
9. T. F. Fuller, and J. Newman, *J. Electrochem. Soc.*, 140 (1993) 1218.
10. T. E. Springer, T. A. Zawodzinski, and S. Gottesfeld, *J. Electrochem. Soc.*, 138 (1991) 2334.
11. D. M. Bernardi, and M. W. Verbrugge, *AIChE J.*, 37 (1991) 1151.
12. D. M. Bernardi, and M. W. Verbrugge, *J. Electrochem. Soc.*, 139 (1992) 2477.
13. K. D. Kreuer, *Solid State Ionics*, 97 (1997) 1.
14. A. Z. Weber, and J. Newman, *J. Electrochem. Soc.*, 150 (2003) A1008.
15. N. Cornet, G. Gebel, and A. de Geyer, *J. De Physique Iv*, 8 (1998) 63.
16. F. Meier, and G. Eigenberger, *Electrochim. Acta*, 49 (2004) 1731.
17. J. T. Hinatsu, M. Mizuhata, and H. Takenaka, *J. Electrochem. Soc.*, 141 (1994) 1493.
18. J. P. Meyers, and J. Newman, *J. Electrochem. Soc.*, 149 (2002) A710.
19. T. A. Zawodzinski, C. Derouin, S. Radzinski, R. J. Sherman, V. T. Smith, T. E. Springer, and S. Gottesfeld, *J. Electrochem. Soc.*, 140 (1993) 1041.
20. T. A. Zawodzinski, T. E. Springer, J. Davey, R. Jestel, C. Lopez, J. Valerio, and S. Gottesfeld, *J. Electrochem. Soc.*, 140 (1993) 1981.
21. P. Schröder, *Z. Phys. Chem.*, 45 (1903) 75.
22. K. A. Mauritz, and R. B. Moore, *Chem. Rev.*, 104 (2004) 4535.
23. M. Falk, *Can. J. Chem.*, 58 (1980) 1495.
24. W. Y. Hsu, and T. D. Gierke, *J. Membr. Sci.*, 13 (1983) 307.
25. W. Y. Hsu, J. R. Barkley, and P. Meakin, *Macromolecules*, 13 (1980) 198.
26. F. P. Orfino, and S. Holdcroft, *J. New Mater. Electrochem. Syst.*, 3 (2000) 285.
27. A. Vishnyakov, and A. V. Neimark, *J. Phys. Chem. B*, 105 (2001) 9586.
28. E. Spohr, P. Commer, and A. A. Kornyshev, *J. Phys. Chem. B*, 106 (2002) 10560.
29. G. Gebel, P. Aldebert, and M. Pineri, *Polymer*, 34 (1993) 333.
30. R. S. McLean, M. Doyle, and B. B. Sauer, *Macromolecules*, 33 (2000) 6541.
31. P. Choi, and R. Datta, *J. Electrochem. Soc.*, 150 (2003) E601.
32. J. Divisek, M. Eikerling, V. Mazin, H. Schmitz, U. Stimming, and Y. M. Volfkovich, *J. Electrochem. Soc.*, 145 (1998) 2677.

33. K. D. Kreuer, *J. Membr. Sci.*, 185 (2001) 29.
34. V. Gurau, F. Barbir, and H. Liu, *J. Electrochem. Soc.*, 147 (2000) 2468.
35. T. E. Springer, M. S. Wilson, and S. Gottesfeld, *J. Electrochem. Soc.*, 140 (1993) 3513.
36. A. A. Kulikovsky, *Fuel Cells*, 1 (2001) 162.
37. D. Bevers, M. Wohr, K. Yasuda, and K. Oguro, *J. Appl. Electrochem.*, 27 (1997) 1254.
38. W. He, J. S. Yi, and T. V. Nguyen, *AIChE J.*, 46 (2000) 2053.
39. D. Natarajan, and T. V. Nguyen, *J. Power Sources*, 115 (2003) 66.
40. D. Natarajan, and T. V. Nguyen, *J. Electrochem. Soc.*, 148 (2001) A1324.
41. J. S. Yi, and T. V. Nguyen, *J. Electrochem. Soc.*, 146 (1999) 38.
42. Z. H. Wang, C. Y. Wang, and K. S. Chen, *J. Power Sources*, 94 (2001) 40.
43. J. H. Nam, and M. Kaviany, *Int. J. Heat Mass Transfer*, 46 (2003) 4595.
44. T. Berning, and N. Djilali, *J. Electrochem. Soc.*, 150 (2003) A1598.
45. S. J. Ridge, R. E. White, Y. Tsou, R. N. Beaver, and G. A. Eisman, *J. Electrochem. Soc.*, 136 (1989) 1902.
46. S. Mazumder, and J. V. Cole, *J. Electrochem. Soc.*, 150 (2003) A1510.
47. E. Birgersson, M. Noponen, and M. Vynnycky, *J. Electrochem. Soc.*, 152 (2005) A1021.
48. J. L. Yuan, B. Sunden, M. Hou, and H. M. Zhang, *Num. Heat Transf. A*, 46 (2004) 669.
49. C. Marr, and X. G. Li, *J. Power Sources*, 77 (1999) 17.
50. Y. W. Rho, S. Srinivasan, and Y. T. Kho, *J. Electrochem. Soc.*, 141 (1994) 2089.
51. R. F. Mann, J. C. Amphlett, M. A. I. Hooper, H. M. Jensen, B. A. Peppley, and P. R. Roberge, *J. Power Sources*, 86 (2000) 173.
52. H.-S. Chu, C. Yeh, and F. Chen, *J. Power Sources*, 123 (2002) 1.
53. J. Leddy, A. E. Iverson, N. E. Vanderborgh, in J. R. Selman, and H. C. Maru (Eds.), *Electrochemical and Thermal Modeling of Battery, Fuel Cell, and Photoenergy Conversion Systems*, PV 86–12, The Electrochemical Society Proceeding Series, Pennington, Nj, 1986, pp. 310.
54. J. C. Amphlett, R. M. Baumert, R. F. Mann, B. A. Peppley, P. R. Roberge, and T. J. Harris, *J. Electrochem. Soc.*, 142 (1995) 1.
55. S. Mazumder, *J. Electrochem. Soc.*, 152 (2005) A1633.
56. D. T. Song, Q. P. Wang, Z. S. Liu, M. Eikerling, Z. Xie, T. Navessin, and S. Holdcroft, *Electrochim. Acta*, 50 (2005) 3347.
57. W. Sun, B. A. Peppley, and K. Karan, *Electrochim. Acta*, 50 (2005) 3359.
58. L. Wang, and H. Liu, *J. Power Sources*, 134 (2004) 185.
59. J. Newman, and K. E. Thomas-Alyea, *Electrochemical Systems*, 3rd edn., John Wiley & Sons, New York, 2004.
60. A. Einstein, *Ann. Physik.*, 17 (1905) 549.
61. W. Nernst, *Z. Phys. Chem.*, 2 (1888) 613.
62. S. Dutta, S. Shimpalee, and J. W. Van Zee, *J. Appl. Electrochem.*, 30 (2000) 135.
63. S. Dutta, S. Shimpalee, and J. W. Van Zee, *Int. J. Heat Mass Transfer*, 44 (2001) 2029.
64. S. Shimpalee, S. Greenway, D. Spuckler, and J. W. Van Zee, *J. Power Sources*, 135 (2004) 79.
65. W. K. Lee, S. Shimpalee, and J. W. Van Zee, *J. Electrochem. Soc.*, 150 (2003) A341.
66. T. V. Nguyen, and R. E. White, *J. Electrochem. Soc.*, 140 (1993) 2178.
67. J. S. Yi, and T. V. Nguyen, *J. Electrochem. Soc.*, 145 (1998) 1149.
68. T. Okada, G. Xie, and Y. Tanabe, *J. Electroanal. Chem.*, 413 (1996) 49.
69. T. Okada, G. Xie, and M. Meeg, *Electrochim. Acta*, 43 (1998) 2141.
70. J.-T. Wang, and R. F. Savinell, *Electrochim. Acta*, 37 (1992) 2737.
71. N. P. Siegel, M. W. Ellis, D. J. Nelson, and M. R. von Spakovsky, *J. Power Sources*, 115 (2003) 81.
72. H. van Bussel, F. G. H. Koene, and R. Mallant, *J. Power Sources*, 71 (1998) 218.
73. S. Shimpalee, and S. Dutta, *Num. Heat Transf. A*, 38 (2000) 111.
74. A. A. Kulikovsky, *J. Electrochem. Soc.*, 150 (2003) A1432.

75. D. R. Sena, E. A. Ticianelli, V. A. Paganin, and E. R. Gonzalez, *J. Electroanal. Chem.*, 477 (1999) 164.
76. K. Dannenberg, P. Ekdunge, and G. Lindbergh, *J. Appl. Electrochem.*, 30 (2000) 1377.
77. H. Meng, and C. Y. Wang, *J. Electrochem. Soc.*, 151 (2004) A358.
78. H. Meng, and C. Y. Wang, *J. Electrochem. Soc.*, 152 (2005) A1733.
79. H. Ju, H. Meng, and C. Y. Wang, *Int. J. Heat Mass Transfer*, 48 (2005) 1303.
80. U. Pasaogullari, and C. Y. Wang, *J. Electrochem. Soc.*, 152 (2005) A380.
81. U. Pasaogullari, C. Y. Wang, and K. S. Chen, *J. Electrochem. Soc.*, 152 (2005) A1574.
82. P. Berg, K. Promislow, J. St Pierre, J. Stumper, and B. Wetton, *J. Electrochem. Soc.*, 151 (2004) A341.
83. D. N. Bennion, *Mass Transport of Binary Electrolyte Solutions in Membranes, Water Resources Center Desalination Report No. 4. 1966*, Department of Engineering, University of California: Los Angeles, CA.
84. P. N. Pintauro, and D. N. Bennion, *Ind. Eng. Chem. Fundam.*, 23 (1984) 230.
85. J. O. Hirschfelder, C. F. Curtiss, and R. B. Bird, *Molecular Theory of Gases and Liquids*, John Wiley & Sons, Inc., New York 1954.
86. J. Fimrite, B. Carnes, H. Struchtrup, and N. Djilali, *J. Electrochem. Soc.*, 152 (2005) A1815.
87. E. A. Mason, and A. P. Malinauskas, *Gas Transport in Porous Media: The Dusty-Gas Model*, Elsevier, Amsterdam 1983.
88. T. Thampan, S. Malhotra, H. Tang, and R. Datta, *J. Electrochem. Soc.*, 147 (2000) 3242.
89. M. Wohr, K. Bolwin, W. Schnurnberger, M. Fischer, W. Neubrand, and G. Eigenberger, *Int. J. Hydrog. Energy* 23 (1998) 213.
90. T. F. Fuller, Ph.D. Dissertation, University of California, Berkeley, CA, 1992.
91. A. Z. Weber and J. Newman, *J. Electrochem. Soc.*, 151 (2004) A311.
92. B. Auclair, V. Nikonenko, C. Larchet, M. Metayer, and L. Dammak, *J. Membr. Sci.*, 195 (2002) 89.
93. S. Koter, and C. H. Hamann, *J. Non-Equilib. Therm.*, 15 (1990) 315.
94. P. Futerko, and I. M. Hsing, *Electrochim. Acta*, 45 (2000) 1741.
95. I. M. Hsing, and P. Futerko, *Chem. Eng. Sci.*, 55 (2000) 4209.
96. A. C. West, and T. F. Fuller, *J. Appl. Electrochem.*, 26 (1996) 557.
97. G. J. M. Janssen, *J. Electrochem. Soc.*, 148 (2001) A1313.
98. J. P. Meyers, Ph.D. Dissertation, University of California, Berkeley, 1998.
99. M. W. Verbrugge, and R. F. Hill, *J. Electrochem. Soc.*, 137 (1990) 1131.
100. M. W. Verbrugge, and R. F. Hill, *J. Electrochem. Soc.*, 137 (1990) 886.
101. R. Schlögl, *Z. Phys. Chem.*, 3 (1955) 73.
102. V. Gurau, H. Liu, and S. Kakac, *AIChE J.*, 44 (1998) 2414.
103. L. Pisani, G. Murgia, M. Valentini, and B. D'Aguanno, *J. Electrochem. Soc.*, 149 (2002) A898.
104. G. Murgia, L. Pisani, M. Valentini, and B. D'Aguanno, *J. Electrochem. Soc.*, 149 (2002) A31.
105. T. Berning, D. M. Lu, and N. Djilali, *J. Power Sources*, 106 (2002) 284.
106. D. Singh, D. M. Lu, and N. Djilali, *Int. J. Eng. Sci.*, 37 (1999) 431.
107. J. J. Baschuk, and X. Li, *J. Power Sources*, 86 (2000) 181.
108. S. Um, C.- Y. Wang, and K. S. Chen, *J. Electrochem. Soc.*, 147 (2000) 4485.
109. B. R. Sivertsen, and N. Djilali, *J. Power Sources*, 141 (2005) 65.
110. G. J. M. Janssen, and M. L. J. Overvelde, *J. Power Sources*, 101 (2001) 117.
111. A. Rowe, and X. G. Li, *J. Power Sources*, 102 (2001) 82.
112. P. Costamagna, *Chem. Eng. Sci.*, 56 (2001) 323.
113. K. Hertwig, L. Martens, and R. Karwoth, *Fuel Cells*, 2 (2002) 61.
114. L. X. You, and H. T. Liu, *Int. J. Heat Mass Transfer*, 45 (2002) 2277.
115. P. W. Li, L. Schaefer, Q. M. Wang, T. Zhang, and M. K. Chyu, *J. Power Sources*, 115 (2003) 90.

116. M. De Francesco, E. Arato, and P. Costa, *J. Power Sources*, 132 (2004) 127.
117. A. Z. Weber, and J. Newman, *J. Electrochem. Soc.*, 152 (2005) A677.
118. A. Z. Weber, and J. Newman, *J. Electrochem. Soc.*, 151 (2004) A326.
119. A. Z. Weber, R. M. Darling, and J. Newman, *J. Electrochem. Soc.*, 151 (2004) A1715.
120. C. Ziegler, H. M. Yu, and J. O. Schumacher, *J. Electrochem. Soc.*, 152 (2005) A1555.
121. A. A. Kulikovsky, T. Wuster, A. Egmen, and D. Stolten, *J. Electrochem. Soc.*, 152 (2005) A1290.
122. P. Futerko, and I. M. Hsing, *J. Electrochem. Soc.*, 146 (1999) 2049.
123. Y. Y. Shan, and S. Y. Choe, *J. Power Sources*, 145 (2005) 30.
124. S. Mazumder, and J. V. Cole, *J. Electrochem. Soc.*, 150 (2003) A1503.
125. J. Ramousse, J. Deseure, O. Lottin, S. Didierjean, and D. Maillet, *J. Power Sources*, 145 (2005) 416.
126. R. B. Bird, W. E. Stewart, and E. N. Lightfoot, *Transport Phenomena*, 2nd edn, John Wiley & Sons, Inc., New York, 2002.
127. J. Newman, *Ind. Eng. Chem. Res.*, 34 (1995) 3208.
128. L. Rao, and J. Newman, *J. Electrochem. Soc.*, 144 (1997) 2697.
129. K. E. Thomas, and J. Newman, *J. Electrochem. Soc.*, 150 (2003) A176.
130. J. Xie, D. L. Wood, D. M. Wayne, T. A. Zawodzinski, P. Atanassov, and R. L. Borup, *J. Electrochem. Soc.*, 152 (2005) A104.
131. V. Stamenkovic, B. N. Grgur, P. N. Ross, and N. M. Markovic, *J. Electrochem. Soc.*, 152 (2005) A277.
132. D. E. Curtin, R. D. Lousenberg, T. J. Henry, P. C. Tangeman, and M. E. Tisack, *J. Power Sources*, 131 (2004) 41.
133. N. P. Siegel, M. W. Ellis, D. J. Nelson, and M. R. von Spakovsky, *J. Power Sources*, 128 (2004) 173.
134. T. Zhou, and H. Liu, *Int. J. Trans. Phen.*, 3 (2001) 177.
135. M. Eikerling, Y. I. Kharkats, A. A. Kornyshev, and Y. M. Volfkovich, *J. Electrochem. Soc.*, 145 (1998) 2684.
136. M. Eikerling, A. A. Kornyshev, and U. Stimming, *J. Phys. Chem. B*, 101 (1997) 10807.
137. R. H. Perry, and D. W. Green, J. O. Maloney (Ed.) *Perry's Chemical Engineers' Handbook*, 7th ed., McGraw-Hill, New York, 1997.
138. T. Sakai, H. Takenaka, N. Wakabayashi, Y. Kawami, and E. Torikai, *J. Electrochem. Soc.*, 132 (1985) 1328.
139. D. M. Bernardi, *J. Electrochem. Soc.*, 137 (1990) 3344.
140. R. Bradean, K. Promislow, and B. Wetton, *Num. Heat Transf. A*, 42 (2002) 121.
141. S. Patankar, *Numerical Heat Transfer and Fluid Flow*, Hemisphere Publishing Corporation 1980.
142. A. A. Kulikovsky, *Electrochem. Commun.*, 4 (2002) 527.
143. T. Okada, *J. Electroanal. Chem.*, 465 (1999) 1.
144. F. L. Chen, Y. G. Su, C. Y. Soong, W. M. Yan, and H. S. Chu, *J. Electroanal. Chem.*, 566 (2004) 85.
145. W. M. Yan, F. L. Chen, H. Y. Wu, C. Y. Soong, and H. S. Chu, *J. Power Sources*, 129 (2004) 127.
146. D. R. Morris, and X. D. Sun, *J. Appl. Polym. Sci.*, 50 (1993) 1445.
147. K. Broka, and P. Ekdunge, *J. Appl. Electrochem.*, 27 (1997) 117.
148. C. M. Gates, and J. Newman, *AIChE J.*, 46 (2000) 2076.
149. S. H. Ge, X. G. Li, B. L. Yi, and I. M. Hsing, *J. Electrochem. Soc.*, 152 (2005) A1149.
150. T. V. Nguyen, and N. Vanderborgh, *J. Membr. Sci.*, 143 (1998) 235.
151. V. K. Datye, P. L. Taylor, and A. J. Hopfinger, *Macromolecules*, 17 (1984) 1704.
152. B. Dreyfus, *J. Polym. Sci., Part B: Polym. Phys.*, 21 (1983) 2337.
153. K. A. Mauritz, and C. E. Rogers, *Macromolecules*, 18 (1985) 483.
154. P. Choi, N. H. Jalani, and R. Datta, *J. Electrochem. Soc.*, 152 (2005) E84.

155. C. Tsonos, L. Apekis, and P. Pissis, *J. Mater. Sci.*, 35 (2000) 5957.
156. M. Eikerling, A. A. Kornyshev, A. M. Kuznetsov, J. Ulstrup, and S. Walbran, *J. Phys. Chem. B*, 105 (2001) 3646.
157. J. M. Prausnitz, R. N. Lichtenthaler, and E. G. Azevedo, *Molecular Thermodynamics of Fluid-Phase Equilibria*, 3rd edn, Prentice-Hall, Inc., Upper Saddle River, New Jersey, 1999.
158. T. J. Edwards, M.S. Thesis, University of California, Berkeley, 1974.
159. V. Freger, *Polymer*, 43 (2002) 71.
160. F. A. L. Dullien, *Porous Media: Fluid Transport and Pore Structure*, 2nd edn, Academic Press, Inc., New York, 1992.
161. A. Z. Weber, and J. Newman, *AIChE J.*, 50 (2004) 3215.
162. J. Bear, *Dynamics of Fluids in Porous Media*, Dover Publications, Inc., New York, 1988.
163. M. W. Verbrugge, and R. F. Hill, *J. Phys. Chem.*, 92 (1988) 6778.
164. Y. Wang, and C. Y. Wang, *J. Electrochem. Soc.*, 152 (2005) A445.
165. S. Um, and C. Y. Wang, *J. Power Sources*, 125 (2004) 40.
166. K. R. Weisbrod, S. A. Grot, N. E. Vanderborgh, in A. R. Landgrebe, S. Gottesfeld and G. Halpert (Eds.), *First International Symposium on Proton Conducting Membrane Fuel Cells*, PV 95–23, The Electrochemical Society Proceeding Series, Pennington, NJ, 1995, pp. 153.
167. P. N. Pintauro, and D. N. Bennion, *Ind. Eng. Chem. Fundam.*, 23 (1984) 234.
168. G. Gebel, *Polymer*, 41 (2000) 5829.
169. C. H. Hsu, and C. C. Wan, *J. Power Sources*, 115 (2003) 268.
170. R. S. Yeo, and C. H. Cheng, *J. Appl. Polym. Sci.*, 32 (1986) 5733.
171. J. A. Elliott, S. Hanna, A. M. S. Elliott, and G. E. Cooley, *Polymer*, 42 (2001) 2251.
172. A. Z. Weber, Ph.D. Dissertation, University of California, Berkeley, 2004.
173. G. Gebel, P. Aldebert, and M. Pineri, *Macromolecules*, 20 (1997) 1425.
174. J. P. Meyers, and J. Newman, *J. Electrochem. Soc.*, 149 (2002) A718.
175. S. Slade, S. A. Campbell, T. R. Ralph, and F. C. Walsh, *J. Electrochem. Soc.*, 149 (2002) A1556.
176. P. Choi, N. H. Jalani, and R. Datta, *J. Electrochem. Soc.*, 152 (2005) E123.
177. N. Agmon, *Chem. Phys. Lett.*, 244 (1995) 456.
178. B. R. Breslau, and I. F. Miller, *Ind. Eng. Chem. Fundam.*, 10 (1971) 554.
179. G. Karimi, and X. Li, *J. Power Sources*, 140 (2005) 1.
180. S. Koter, *J. Membr. Sci.*, 166 (2000) 127.
181. A. Narebska, and S. Koter, *Electrochim. Acta*, 32 (1987) 449.
182. S. Motupally, A. J. Becker, and J. W. Weidner, *J. Electrochem. Soc.*, 147 (2000) 3171.
183. T. A. Zawodzinski, M. Neeman, L. O. Sillerud, and S. Gottesfeld, *J. Phys. Chem.*, 95 (1991) 6040.
184. R. E. De La Rue, and C. W. Tobias, *J. Electrochem. Soc.*, 106 (1959) 827.
185. D. A. G. Bruggeman, *Ann. Physik.*, 24 (1935) 636.
186. G. Li, and P. P. Pickup, *J. Electrochem. Soc.*, 150 (2003) C745.
187. D. M. Bernardi, Ph.D. Dissertation, University of California, Berkeley, 1986.
188. D. M. Bernardi, E. Pawlikowski, and J. Newman, *J. Electrochem. Soc.*, 132 (1985) 5.
189. J. N. Agar, in P. Delahay (Ed.), *Advances in Electrochemistry and Electrochemical Engineering*, 3, Vol. John Wiley & Sons, New York, 1963, p. 32.
190. H. C. Ju, C. Y. Wang, S. Cleghorn, and U. Beuscher, *J. Electrochem. Soc.*, 152 (2005) A1645.
191. A. Z. Weber, J. Newman, in J. W. Van Zee, T. F. Fuller, S. Gottesfeld, and M. Murthy (Eds.), *Proton Conducting Membrane Fuel Cells Iv*, The Electrochemical Society Proceeding Series, Pennington, NJ, 2004.
192. M. Eikerling, and A. A. Kornyshev, *J. Electroanal. Chem.*, 453 (1998) 89.
193. L. X. You, and H. T. Liu, *Int. J. Hydrog. Energy*, 26 (2001) 991.
194. L. Pisani, M. Valentini, and G. Murgia, *J. Electrochem. Soc.*, 150 (2003) A1558.
195. M. L. Perry, J. Newman, and E. J. Cairns, *J. Electrochem. Soc.*, 145 (1998) 5.

196. M. Maja, P. Tosco, and M. Vanni, *J. Electrochem. Soc.*, 148 (2003) A1368.
197. F. Jaouen, G. Lindbergh, and G. Sundholm, *J. Electrochem. Soc.*, 149 (2002) A437.
198. R. P. Iczkowski, and M. B. Cutlip, *J. Electrochem. Soc.*, 127 (1980) 1433.
199. Z. N. Farhat, *J. Power Sources*, 138 (2004) 68.
200. H. S. Fogler, *Elements of Chemical Reaction Engineering*, 2nd edn, Prentice-Hall, Upper Saddle River, NJ, 1992.
201. E. W. Thiele, *Ind. Eng. Chem.*, 31 (1939) 916.
202. E. A. Ticianelli, C. R. Derouin, A. Redondo, and S. Srinivasan, *J. Electrochem. Soc.*, 135 (1988) 2209.
203. Z. G. Qi, and A. Kaufman, *J. Power Sources*, 109 (2002) 38.
204. F. Gloaguen, P. Convert, S. Gamburzev, O. A. Velev, and S. Srinivasan, *Electrochim. Acta*, 43 (1998) 3767.
205. K. Broka, and P. Ekdunge, *J. Appl. Electrochem.*, 27 (1997) 281.
206. S. H. Chan, and W. A. Tun, *Chem. Eng. Tech.*, 24 (2001) 51.
207. T. Okada, G. Xie, O. Gorseth, S. Kjelstrup, N. Nakamura, and T. Arimura, *Electrochim. Acta*, 43 (1998) 3741.
208. T. Okada, *J. New Mater. Electrochem. Syst.*, 4 (2001) 209.
209. T. Okada, *J. Electroanal. Chem.*, 465 (1999) 18.
210. T. Okada, N. Nakamura, M. Yuasa, and I. Sekine, *J. Electrochem. Soc.*, 142 (1995) 4143.
211. S. W. Capeci, P. N. Pintauro, and D. N. Bennion, *J. Electrochem. Soc.*, 136 (1989) 2876.
212. R. Tandon, and P. N. Pintauro, *J. Membr. Sci.*, 136 (1997) 207.
213. J. R. Bontha, and P. N. Pintauro, *Chem. Eng. Sci.*, 49 (1994) 3835.
214. P. N. Pintauro, and M. W. Verbrugge, *J. Membr. Sci.*, 44 (1989) 197.
215. F. J. Booth, *J. Chem. Phys.*, 19 (1951) 391.
216. R. M. Darling, and J. P. Meyers, *J. Electrochem. Soc.*, 152 (2005) A242.
217. R. M. Darling, and J. P. Meyers, *J. Electrochem. Soc.*, 150 (2003) A1523.
218. M. Schulze, A. Schneider, and E. Gulzow, *J. Power Sources*, 127 (2004) 213.
219. J. Xie, D. L. Wood, K. L. More, P. Atanassov, and R. L. Borup, *J. Electrochem. Soc.*, 152 (2005) A1011.
220. S. Koter, *J. Membr. Sci.*, 206 (2002) 201.
221. Y. H. Yang, and P. N. Pintauro, *AIChE J.*, 46 (2000) 1177.
222. A. G. Guzman-Garcia, P. N. Pintauro, M. W. Verbrugge, and R. F. Hill, *AIChE J.*, 36 (1990) 1061.
223. W. Liu, K. Ruth, and G. Rusch, *J. New Mater. Electrochem. Syst.*, 4 (2001) 227.
224. A. Pozio, R. F. Silva, M. De Francesco, and L. Giorgi, *Electrochim. Acta*, 48 (2003) 1543.
225. J. Kiwi, N. Denisov, Y. Gak, N. Ovanesyan, P. A. Buffat, E. Suvorova, F. Gostev, A. Titov, O. Sarkison, P. Albers, and V. Nadtochenko, *Langmuir*, 18 (2002) 9054.
226. Y. Wang, and C. Y. Wang, *Electrochim. Acta*, 50 (2005) 1307.
227. S. Kim, S. Shimpalee, and J. W. Van Zee, *J. Power Sources*, 135 (2004) 110.
228. M. Pianca, E. Barchiesi, G. Esposto, and S. Radice, *J. Fluorine Chem.*, 95 (1999) 71.
229. U. A. Paulus, T. J. Schmidt, H. A. Gasteiger, and R. J. Behm, *J. Electroanal. Chem.*, 495 (2001) 134.
230. K. H. Kangasniemi, D. A. Condit, and T. D. Jarvi, *J. Electrochem. Soc.*, 151 (2004) E125.
231. L. M. Roen, C. H. Paik, and T. D. Jarvi, *Electrochem. Solid-State Lett.*, 7 (2004) A19.
232. C. A. Reiser, L. Bregoli, T. W. Patterson, J. S. Yi, J. D. L. Yang, M. L. Perry, and T. D. Jarvi, *Electrochem. Solid-State Lett.*, 8 (2005) A273.
233. H. Wendt, T. Brenscheidt, A. Fischer, Philos. Trans. Royal Soc. Lond. Seri. Mathe. Phys. Eng. Sci. 354 (1996) 1627.
234. R. Z. Jiang, and D. Chu, *J. Power Sources*, 92 (2001) 193.
235. K. Scott, P. Argyropoulos, and K. Sundmacher, *J. Electroanal. Chem.*, 477 (1999) 97.
236. K. Scott, W. M. Taama, P. Argyropoulos, and K. Sundmacher, *J. Power Sources*, 83 (1999) 204.

237. J. R. Fan, G. L. Hu, J. Yao, and K. F. Cen, *Energy Fuels*, 16 (2002) 1591.
238. P. Argyropoulos, K. Scott, and W. M. Taama, *J. Power Sources*, 79 (1999) 169.
239. E. Birgersson, J. Nordlund, H. Ekstrom, M. Vynnycky, and G. Lindbergh, *J. Electrochem. Soc.*, 150 (2003) A1368.
240. G. Murgia, L. Pisani, A. K. Shukla, and K. Scott, *J. Electrochem. Soc.*, 150 (2003) A1231.
241. A. A. Kulikovsky, *J. Electrochem. Soc.*, 152 (2005) A1121.
242. A. A. Kulikovsky, *Electrochem. Commun.*, 6 (2004) 1259.
243. S. S. Sandhu, R. O. Crowther, and J. P. Fellner, *Electrochim. Acta*, 50 (2005) 3985.

CHAPTER THREE

On the Prospects for Phosphonated Polymers as Proton-Exchange Fuel Cell Membranes

Benoît Lafitte *and* Patric Jannasch

Contents

Division of Polymer & Materials Chemistry, Lund University, SE-221 00 Lund, Sweden

Advances in Fuel Cells
0080453945

Abstract

One of the major challenges in advancing polymer electrolyte membrane fuel cell technology lies currently in the development of new durable polymeric membranes that will allow fuel cell operation at high temperatures without extensive humidification requirements. Access to these membranes promises important benefits concerning the complexity, cost and performance of the fuel cell system. In this context, membranes functionalized with covalently linked phosphonic acid may potentially show some crucial advantages in relation to the commonly employed sulfonated membranes. Because of the hydrogen bonding and amphoteric properties of the phosphonic acids, the former membranes may transport protons through structure diffusion under low-humidity conditions. At high water contents the protons may instead be transported through the dynamics of the water, much in the same way as in conventional sulfonated membranes. Furthermore, phosphonated polymers generally show a high hydrolytic and thermal stability due to the strength of the carbon–phosphorus bond, which is especially critical under high-temperature operation. However, it is clear that the molecular architecture of the phosphonated polymers will require a very careful design in order to reach these advantageous membrane properties. In addition, phosphonated polymers are in general more complicated to prepare than the corresponding sulfonated ones. The present treatise describes the nature of the phosphonic acids and critically surveys the current literature on synthetic approaches to useful phosphonated polymers and the properties of phosphonated membranes. The synthetic approaches include chemical modification of polymers by attaching phosphonic acid via alkyl, aryl and perfluoroalkyl groups, as well as direct polymerization using phosphonated monomers. In addition, some potential research directions for the development of new efficient fuel cell materials are outlined.

3.1 INTRODUCTION

The limitations of the fuel cell are today very much determined by the performance and durability of the proton conducting membrane. The function of the membrane is basically to efficiently separate the anode and the cathode sides and to conduct protons within a wide range of operation conditions. Some of the membrane materials available today possess a number of the properties required, but no single material seem to satisfy the complete range of requirements set by the fuel cell system. The proton conducting membranes employed in the fuel cells are typically based on hydrated sulfonated polymers in their protonated form [1–7]. For example, the current state-of-the-art membranes (e.g., Nafion®) are based on perfluorosulfonic acid polymers. These membranes are characteristically phase separated into a percolating network of nanopores containing water, and a hydrophobic polymer-rich phase domain. The water in the nanopores dissociates the acid units and functions as the proton solvent which facilitates the transport of the acidic protons. Thus, membranes based on sulfonated polymers rely critically on water for the proton conductivity. Consequently, these membranes typically offer quite satisfactory fuel cell performance below 90°C, as long as the degree of humidification is sufficient [2].

There is currently a strong demand from fuel cell system developers to increase the operational temperature and reduce the requirement for humidification. The background is the pressing need to reduce the complexity and the cost, and to increase the efficiency of systems for both automotive and stationary applications [8,9]. Consequently, there is a strong call for new types of membranes that allow operation in a wide temperature window, typically from sub-zero degrees up to approximately 120°C, at low humidification with water vapor partial pressures below 0.5 bar, or with no humidification at all [9]. These conditions are expected to lead to, e.g., a greatly simplified water and heat management, a higher catalytic CO-tolerance and a reduced need for heat rejection capacity. The important advantages reached on the system level may thus facilitate a broad commercialization of fuel cells, perhaps especially for automotive applications. As mentioned above, the proton conductivity of polymer membranes such as Nafion® characteristically suffers greatly under such conditions due to the loss of the water. Despite many years of efforts to modify the properties of commercially available perfluorosulfonic acid membranes of the Nafion® type, the demands have not yet been met, although significant improvements have been made, e.g., by the inclusion of inorganic nanoparticles [10, 11]. Perfluorosulfonic acid membranes are, however, presently developed further [12,13].

The current insufficient properties of the perfluorosulfonic acid membranes have generated an intensive research activity to develop proton conducting polymers and membranes based on hydrocarbon polymers [2, 4–7, 14–17]. This approach has a larger flexibility in the design of the polymer architecture, and thus also when it comes to manipulating and optimizing membrane properties. Especially, the preparation and investigation of membranes based on durable sulfonated main-chain aromatic polymers, such as polysulfones (PSUs), polyketones and polyimides, have shown that these ionomers may provide improved properties in some respects, as compared with the perfluorosulfonic acid membranes. However, although a very large number of different ionomers have been investigated so far, the existence of a membrane that can satisfactory meet all the demands required has not yet been reported.

The present situation has widened the focus of the research area, and increased the interest of investigating alternative "acidic" or "protogenic" groups which have the ability to facilitate proton conductivity under low-humidity conditions. These groups include e.g., the phosphonic acid and different heterocycles such as imidazole and benzimidazole [18–25], and are perhaps best described as amphoteric. Thus, polymers carrying these groups have quite different properties than their sulfonated analogs. For example, phosphonated polymers generally show a higher degree of hydrogen bonding and lower water uptake than their sulfonated analogs at comparable ion-exchange capacities. Moreover, phosphonated model compounds have recently been shown to possess an attractive combination of properties that further motivates investigations into the properties of phosphonated polymers as proton conductors under low-humidity conditions [24]. However, the area of phosphonated membranes for fuel cells is still an emerging field and the vast majority of the phosphonated polymers reported in the literature up until now have not been molecularly designed or studied as proton conducting materials.

The aim of the present chapter is to review the current status of the area of phosphonated polymers and membranes, and to assess the potential of these materials for use in fuel cells. Because this is a relatively new topic for most researchers in the field, the perspective on the properties and synthesis of phosphonated polymers is kept quite broad. First, both general and specific characteristics of the phosphonic acid group will be discussed. In the next chapter, the synthetic approaches to polymers containing phosphonic acid groups will be reviewed by first assessing the different synthetic routes available for the organic polymer chemist for the synthesis of the carbon–phosphorus (C—P) bond, and then providing an overview of the different strategies employed so far to obtain phosphonated polymers. This chapter is followed by a summary and critical discussion of the properties of proton conducting membranes based on phosphonated polymers, including the performance in fuel cells. Next, the prospects of phosphonated polymers for proton conduction under low-humidity conditions are discussed on the basis of a number of recent reports in the literature. Finally, a brief outlook is given in an attempt to identify how to advance this emerging field.

3.2 General Characteristics of Phosphonic Acids

When it comes to the design of novel ionomers based on phosphonic acid units, it is important to consider the specific characteristics of this acidic group in relation to those of the sulfonic acid group. These characteristics have then to be taken into account in the preparation of efficient proton conducting membranes for fuel cells. In general terms, the phosphorus atom is a multivalent, non-metallic element of the nitrogen group in the periodic system and is commonly found in inorganic phosphate rocks and in all living cells. Unfortunately, a widespread confusion is seen in the literature concerning the nomenclature of organophosphorus compounds. First, the correct spelling of the element is phosphorus. The word phosphorous does exist, but it is the adjectival form for the lower valency. Consequently, just as sulfur forms sulfurous and sulfuric compounds, phosphorus forms phosphorous and phosphoric compounds. The nomenclature based on the oxidation state of the phosphorus center eliminates the confusion and helps to promote the understanding of the chemistry. Scheme 1 presents the molecular structures of the different tricoordinate and tetracoordinate oxyacids

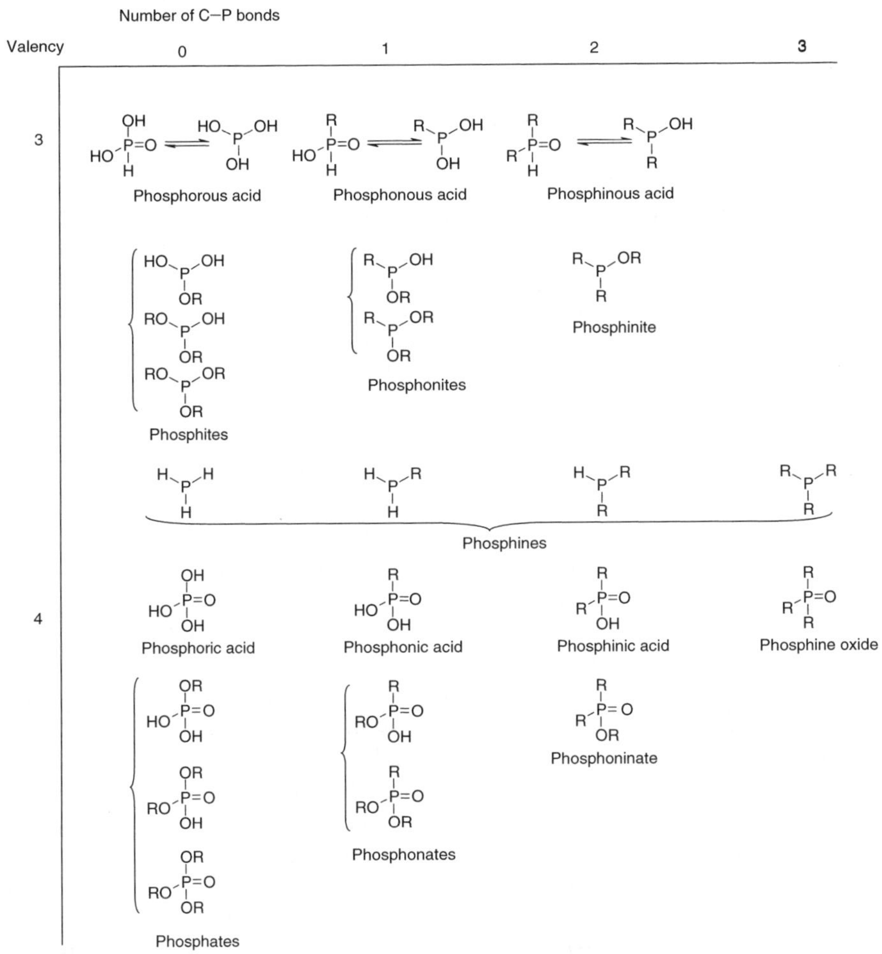

Scheme 1 Nomenclature of organophosphorus compounds.

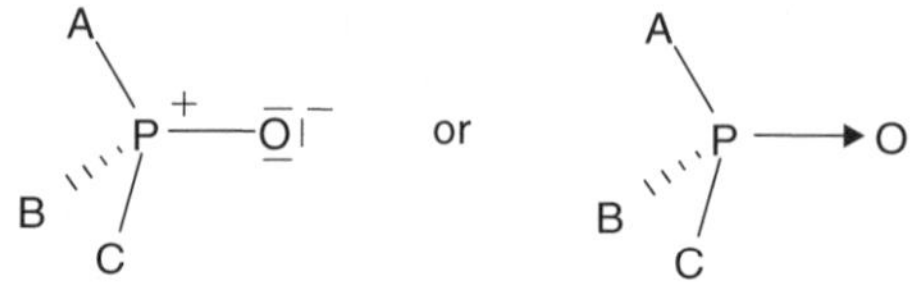

Scheme 2 Alternative representations of the phosphoryl linkage.

arranged in accordance with their general names based on the oxidation state and the number of the C—P bonds. It is worth noting that throughout the present chapter, the phosphoryl group of phosphonates and phosphonic acids will be denoted by P═O or P(O), keeping in mind the polar nature of the linkage. Indeed, this bond is highly polar and is often represented in other ways (see Scheme 2).

The development of organophosphorus chemistry has classically been driven by the use of phosphorus-based compounds as insecticides, agricultural chemicals, flame retardants, drugs, etc. [26]. For instance, phosphonates often act as inhibitors of enzymes due, in part, to the high stability of the C—P bond and to their structural similarity to phosphates [27]. Novel application areas have emerged including the use of chiral diphosphorus reagents associated with metals for stereospecific syntheses [26]. In general, phosphonic acids occupy an important place among the organic derivatives of phosphorus. Together with their amides and esters, they probably outnumber all other compounds containing C—P bonds. All this emphasizes the specific properties of phosphonate and phosphonic acid compounds, and polymers derived thereof, over other common functionalized molecules based on sulfonates and sulfonic acid for instance. An example of this specificity is found in the area of ion-exchange resins where the lack of selectivity of sulfonic-acid-based resins for specific ions can limit their practical use. Indeed, Boyd *et al.* have shown that the ion-exchange reaction alone does not have a wide enough range in free energy of reaction to allow for selectivity [28]. The change in enthalpy of reaction for sulfonic-acid-based resins with alkali metal ions is actually in the order of only 6.3 kJ/mol [29]. As a result, much effort has been devoted to the development of novel ion-exchange resins based on phosphonic acid units. The synthesis of such resins was first reported in the 1950s [30,31]. Compared to the relatively weak binding ability of sulfonic groups, phosphonic-acid-based polymers have shown a strong interaction with a variety of substrates such as metals [32,33] and heavy metals [34,35], as well as radioactive actinides [36]. The binding ability of the phosphonic acid groups is a result of either an ion exchange or a co-ordination process. Indeed, the polarity of the phosphoryl linkage P═O provides a Lewis base character to the oxygen, resulting in the ability to chelate metal cations and adhere strongly to other substrates. This adhesive property has been exploited in different applications such as improved adhesion of polyethylene to metallic surfaces [37], and dentine adhesives for which different phosphonic-acid-based monomers have also been prepared and used as primers [38–40]. It has also been found that the presence of a phosphonic acid functionality in a polymer increases the biocompatibility and the adhesion to the tooth due to chelation with calcium ions in the tooth surface [41]. Monomers bearing phosphonic acid groups are also of interest because of their wet adhesion resistance when applied on galvanized steel substrates [42,43] and their good anticorrosive surface passivation properties for steel and iron [43–45]. Sundell *et al.* have shown that polymers carrying bis(phosphonic acid) units bond effectively to noble metals such as Palladium (Pd) [46]. Such interaction is of prime concern for fuel cell applications as noble metals, and especially Platinum (Pt), are used as catalysts in the preparation of fuel cell electrodes. This interaction may influence the properties of the membrane-electrode assembly (MEA) as it will most probably have an effect on the interaction between the polymer and the catalyst surface. However, Schuster *et al.* have reported that the reaction rates for hydrogen oxidation and oxygen reduction on Pt surfaces were similar for systems based on phosphonic acid as compared those based on slightly hydrated sulfonic acid [24]. Flame retardancy is another typical characteristic of phosphorus-containing polymers. In the solid state, these polymers form poly(phosphoric acid) when heated, which catalyzes the formation of an intumescent char that protects the surface from further burning [47,48]. This effect is also observed for polymers containing phosphonate groups [49,50].

When it comes to the preparation of novel ionomers for fuel cells, one major difference between polymers based on phosphonic acid units and conventional sulfonated polymers is the important chemical transformations that the former acidic units may undergo (see Scheme 3). In addition, phosphonic acids are usually thermally stable compounds because of the high bond energy of the C—P bond, calculated to be 259 kJ, which may be compared with a typical bond energy of 268 kJ for the carbon–carbon bond [51]. Although more hydrolytically stable than sulfonic acids, some observations summarized by Freedman and Doak [51] are of interest when it comes to the design of hydrolysis-resistant phosphonic acid ionomers for fuel cells. First, aryl phosphonic acids tend to be more prone to hydrolysis in the presence of strong electron-donating groups in the *ortho* or *para* position. In this situation the C—P bond is weakened, which may lead to the formation of free phosphoric acid. In contrast to this, alkyl phosphonic acids appear to be weakened by certain electron-withdrawing groups in close proximity. In particular, compounds in which a carbonyl group is situated directly next to the phosphonic acid are rather unstable. However, it should be

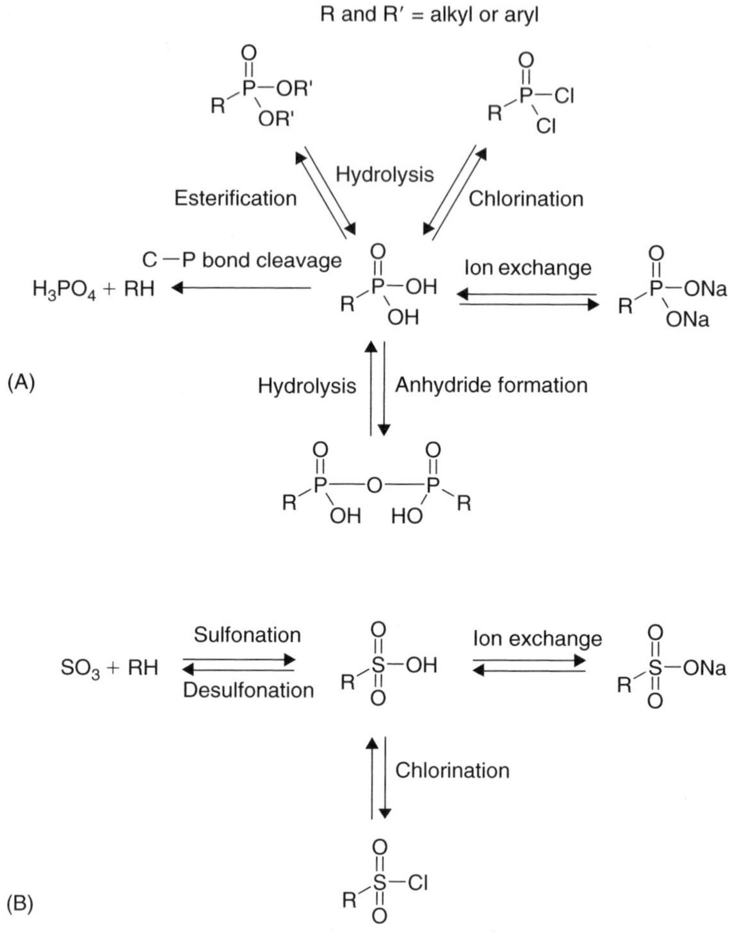

Scheme 3 Comparison of important transformations of phosphonic acid units (A) and sulfonic acid units (B).

noted that trifluoromethylphosphonic acid has been reported to be a very stable compound toward hydrolysis [52].

Phosphonic acids possess two ionizable acid moieties. For aryl and alkyl phosphonic acid the p*K*a values are typically between 2–3 and 7–8 for the first and second proton, respectively [51]. The first acidic moiety has an ionization potential intermediate to those of sulfonic and carboxylic acids. Because of the intermediate acid strength of the first acid moiety and the higher ion-exchange capacity (*IEC*) of the phosphonic acid, as compared with sulfonic and carboxylic acids at the same acid concentration, many of the properties of ionomers based on phosphonic acid may be found

Table 3.1 Acid dissociation constants in water for selected phosphonic acid compounds [51]

Acid	pKa$_1$	pKa$_2$
$CF_3P(O)(OH)_2$	1.2	3.9
$C_6H_5CH_2P(O)(OH)_2$	2.3	7.6
$CH_3P(O)(OH)_2$	2.4	7.5
$(HO)_2(O)PCH_2P(O)(OH)_2$	<2 and 2.6	6.9 and 10.3
$C_6H_5P(O)(OH)_2$	1.8	7.1
$C_6H_5SO_3H$	<1	–

intermediate to the properties of sulfonated and carboxylated ionomers or, perhaps, even comparable to sulfonated ionomers. Alkyl phosphonic acids are in general weaker than phosphoric acid (pKa_1 = 1.97 and pKa_2 = 6.82) and its monoalkyl esters [53]. The pKa values of the phosphonic acids generally increases with an increase in the number of carbon in the alkyl chain, or with the degree of branching of the carbon chain as a direct consequence of the inductive effect of saturated carbons [51]. As far as aryl phosphonic acids are concerned, the effects of the substituents on the acidity of phenylphosphonic acid have been found to be quite similar to those in the case of benzoic acid [51]. Special cases arise when the substituents are located *ortho* to the phosphonic groups. In most of these cases, the fact that the inductive effect of the *ortho* position is larger than the one of the *para* position cannot account for the differences observed. Freedman and Doak have suggested that these differences are most probably due to intramolecular hydrogen bonding between the substituents in the *ortho* position and the phosphonic acid [54]. The acidic strengths of a few selected generic molecular structures based on phosphonic acid and other acidic groups are summarized in Table 3.1 for comparison.

Because of their amphoteric character in the Brønsted sense, phosphonic acids are usually found to be associated by hydrogen bond formation in both the solid state and in certain organic solvents. Despite a considerable protonic dissociation in aqueous solutions of these acids, Kosolapoff and co-workers have argued that hydrogen bonded aggregates of phosphonic acid molecules are present even under these conditions, implying that the hydrogen bonds of phosphonic acids are exceptionally strong [55–57]. Schuster *et al.* have recently investigated functionalized model compounds bearing sulfonic acid, imidazole or phosphonic acid with respect to their transport, stability and reactivity properties at different levels of hydration [24]. In their study, it was clearly shown that 1-heptylphosphonic acid had a much lower molecular diffusion than 1-heptylsulfonic acid and 2-heptylimidazole, indicating that the former specie most probably existed in the form of strong molecular aggregates. The pronounced amphoteric character of phosphonic acid allows structure diffusion of protons through fast hydrogen bond breaking and forming processes in the hydrogen bond network through a so-called Grotthuss mechanism [58]. This results in a high proton conductivity in a low-humidity atmosphere, whereas sulfonic acid strongly depends on a minimum level of hydration (see Scheme 4). Phosphonic acids were also found to retain water molecules up to 180°C, thus avoiding the condensation of phosphonic acid through the formation of P—O—P linkages and the release of water at high temperatures [24]. This is a very important characteristic if the proton conducting ionomer is to be used at 100–180°C. Finally, the amphoteric character of phosphonic acids also accounts for low values of water uptake as compared to sulfonic acids. Indeed, Schuster *et al.* found that 1-heptylsulfonic acid takes up 10 water molecules per sulfonic acid group at 100°C and a relative humidity close to 100%, whereas 1-heptylphosphonic acid was found to take up only 1.5 water molecules per phosphonic acid group under the same conditions [24]. The water uptake in sulfonic acid compounds is mainly driven by the very large gain in entropy resulting from the solvation of the highly dissociated protons by water molecules. In the case of phosphonic acids, the acidic protons are preferentially stabilized by autoprotolysis. The low levels of water uptake imply that high levels of phosphonation are necessary in the design

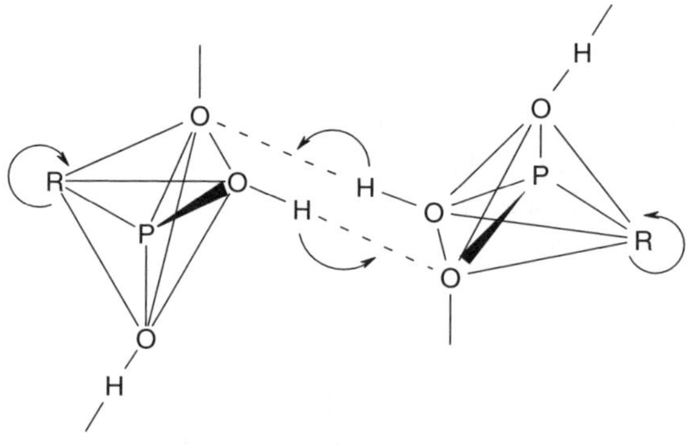

Scheme 4 Hydrogen bonding and cooperativity of proton transfer in phosphonated compounds.

of efficient ionomers. Consequently, there is a need for novel approaches in the design and synthesis of ionomers based on phosphonic acid, that are different from those concerning sulfonic-acid-based ionomers.

3.3 SYNTHETIC APPROACHES TO POLYMERS CONTAINING PHOSPHONIC ACID GROUPS

3.3.1 Preparation of Phosphonic Acid Derivatives

Sulfonic acid units have by far been the acidic group of choice when it comes to preparing ionomer membranes for fuel cells. The primary reason can be ascribed to the high acidity of the sulfonic acid and that this group can be introduced into polymeric structures by a number of convenient methods [15–17,59]. In order to develop proton conducting polymers suitable for fuel cell applications, it is important to use hydrolytically stable acidic groups. Consequently, the synthesis of ionomers owing their acidic character to hydrolytically unstable monoalkyl or monoaryl phosphate, which are linked to the polymer via C—O—P bonds (see Scheme 5), will not be discussed in this review, although they have been mentioned as candidates for fuel cell membranes [60]. These polymers may be of interest for the preparation of membranes doped with phosphoric acid. The development of these membranes doped with phosphoric acid has been reviewed previously [23,61]. The preparation of phosphonic acid ionomers requires the use of a specific chemistry, selected to yield the desired polymer structure. It certainly is beyond the scope of the present review to provide a comprehensive survey on the synthesis of C—P bonds. The reader is referred to several specific reviews on this topic for more detailed information [26,62–68]. We will instead focus on selected synthetic pathways that are of particular interest for the preparation of ionomers suitable for application in fuel cells.

Prior to 1950, the synthesis of C—P bonds was essentially achieved via Friedel–Craft and Michaelis–Arbuzov reactions [51]. Although these reactions remain of great importance in the preparation of organophosphorus compounds today, progress in the area has lead to new efficient synthetic pathways [26]. Most of these pathways are now frequently employed in the preparation of phosphonated polymers by chemical modification, irradiation grafting or polymerization of phosphonated monomers. With regards to the specific reactions required for their synthesis, the preparation of C—P bonds can be divided into the formation of alkyl–phosphorus (C_{Al}—P) and aryl–phosphorus (C_{Ar}—P) bonds. Moreover, a special case arises in the synthesis of perfluoroalkyl–phosphorus (C_F—P) bonds, as further discussed below.

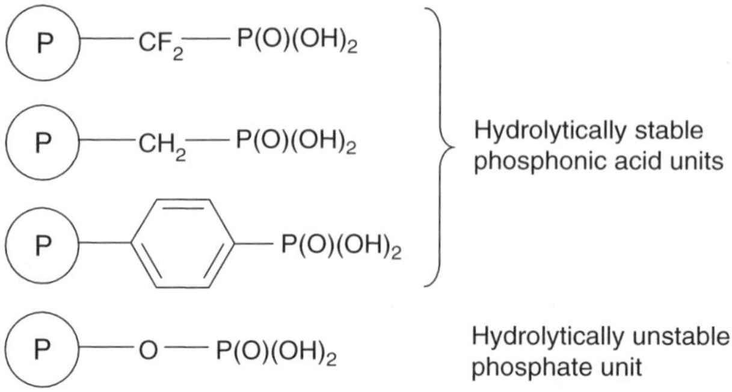

Scheme 5 Schematic representation of different ionomers based on phosphorus–oxyacids (P denotes a polymer scaffold).

3.3.1.1 Formation of Alkyl–Phosphorus Bonds

The Michaelis–Arbuzov rearrangement is historically the most important approach when it comes to the synthesis of C_{Al}—P bonds. In addition, this is a highly versatile reaction. It proceeds via the reaction of an alkyl halide with trialkyl phosphite, yielding the valence-expanded dialkyl alkylphosphonate and an alkyl halide by-product (see pathway I(a) in Scheme 6) [69–71]. Any halide capable of reacting by a S_N2 mechanism is suitable for such a reaction. Their reactivity decreases in the following order: RCO—X > RCH_2—X > R_2CH—X >> R_3C—X, where X is a halogen atom and R an alkyl or aryl substituents. Furthermore, alkyl iodides are more reactive than alkyl bromides, which are in turn more reactive than alkyl chlorides [26]. Since the reaction proceeds by nucleophilic attack of the unshared electron pair of the phosphorus on the alkyl halide, it will be promoted if the phosphorus atom is substituted with electron-donating groups. Consequently, electron-withdrawing substituents will retard the same reaction. In general, the Michaelis–Arbuzov rearrangement proceeds in good yields without the aid of a catalyst. However, the use of catalysts may be required for the substitution of aryl halides, as further discussed in Section 3.3.1.2 [69]. Complications generally arise due to the self-alkylating ability of trialkyl phosphites, especially in the presence of catalytic amounts of alcohol, alkyl halides, aluminum alkyls or inorganic iodides [72]. It has been found that, in general, the synthesis of alkyl phosphonates with long alkyl chains requires the use of more reactive conditions, which leads to the formation of by-products [73]. This implies that the use of such reactions might be limited in connection with aliphatic polymers. The use of the salt of trivalent phosphorus-centered oxyacids was found to improve the results. This reaction, the Michaelis–Becker rearrangement, involves the same alkylhalide as in the Michaelis–Arbuzov rearrangement [74]. The same considerations regarding the reactivity of the alkyl halide apply to the former rearrangement

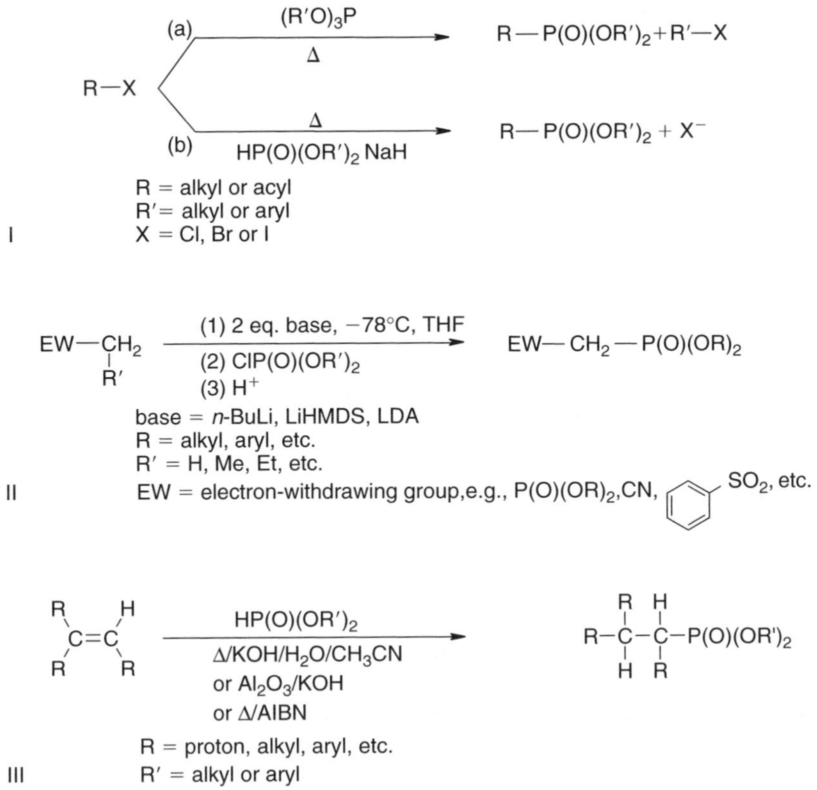

Scheme 6 Important synthetic pathways for the formation of alkyl–phosphorus bonds.

(see pathway I(b) in Scheme 6). In Michaelis–Becker rearrangements, dibutylphosphite is often used as the reagent because of the high solubility of its salt in the most common organic solvents. A notable aspect of the Michaelis–Becker rearrangement is that milder conditions can be used as compared to the Michaelis–Arbuzov rearrangement.

Another important synthetic route to the formation of C—P bonds is based on the use of phosphorus halogen compounds. First, P(III) halogen compounds may act as nucleophilic reagents because of their unshared electron pair. However, these compounds are essentially acid halides with highly reactive phosphorus–halogen linkages, and are therefore susceptible to facile displacement by nucleophilic reagents including carbanionic compounds [26]. The use of organometallics, like Grignard reagents, to displace halogens from phosphorus is not novel [75]. The most common reactants used either as a direct reagent, or as an intermediate for the formation of C—P bonds, are PCl_3 for the P(III) system and $P(O)Cl_3$ for the P(V) system.

However, the high reactivity may present difficulties for a given synthesis, and this is true for phosphorus–polyhalogen species which can undergo multiple reactions with nucleophilic compounds. Organolead and organomercury compounds have been found to provide a controlled reaction. Unfortunately, the difficulties involved in their preparation and their extreme toxicity make them unsuitable [76,77]. Organolithium compounds, on the other hand, provide very strong and efficient alkylating and arylating reagents for the generation of C—P bonds. It is generally found that lithium reagents readily add to quinquevalent phosphorus compounds at low temperatures in a selective manner, thus avoiding the problem of side reactions leading to low yields. An extensive literature survey has recently been published by Savignac and co-workers in which the synthesis of phosphonates by nucleophilic substitutions at phosphorus, so-called $S_NP(V)$ reactions are reviewed [78]. Although the Michaelis–Becker and Michaelis–Arbuzov reactions remain the two most important methods for the formation of C—P bonds, nucleophilic substitution reactions at phosphorus appear to constitute a more versatile method which is under continuous development. Consequently, the nucleophilic substitution and functionalization of dialkyl chlorophosphates in the presence of a strong base is a powerful method for the preparation of alkyl phosphonates (see pathway II in Scheme 6) [78]. In general, any molecules possessing strong electron-withdrawing groups which allow the displacement of an acidic hydrogen by a strong base can be used in such a reaction. One example is the reaction between a lithiated methylsulfone, produced by metalation with lithium diisopropylamide (LDA), and diethylchlorophosphate in tetrahydrofuran at −78°C in the presence of *N,N,N′,N′*-tetramethylethylenediamine [79]. A special aspect of this approach is that it may be used to prepare diphosphonates. Indeed, dialkyl alkylphosphonates can be metalated with LDA at low temperatures to yield α-phosphorylated carbanions since phosphonate units are strongly electron withdrawing (see pathway II in Scheme 6). The carbanion reacts readily at the same temperature with dialkyl chlorophosphate to produce the thermally stable tetraalkyl lithiomethylenediphosphonate in quantitative yields. This procedure allows the preparation of elaborated diphosphonates substituted with alkyl, aryl or halogen groups [78,80]. The incorporation of diphosphonated sites is interesting for the preparation of densely phosphonated polymers, which may prove advantageous when it comes to the preparation of phosphonated polymers for fuel cell membranes.

Yet another convenient approach to the formation of alkyl phosphonates is the Pudovik reaction in which compounds containing labile P—H bonds, typically dialkyl hydrogenophosphonate, react with unsaturated functions such as alkenes, alkynes, carbonyls and imines (see pathway III in Scheme 6) [81]. The Pudovik reaction may proceed either via an ionic or a radical mechanism depending on the structure of the unsaturated function and the phosphorus reagent [82]. Notably, the free-radical addition of diethyl hydrogenophosphonate on alkenes was found to be the preferred route for the synthesis of phosphonated alkanes having long alkyl chains, in competition with the conventional Michaelis–Arbuzov and Michaelis–Becker reactions [73]. Such an approach is appealing since a wide variety of monomers and polymers containing unsaturated functions are commercially available.

3.3.1.2 Formation of Aryl–Phosphorus Bonds

The preparation of compounds in which the phosphonic acid group is directly linked to an aromatic ring is of great interest due to the generally high chemical and thermal stability of aromatic polymers as compared to aliphatic polymers. A number of well-established procedures exist to tether phosphonic acid units onto aromatic rings. However, all these reactions are not necessarily appropriate for polymeric materials. For example, systematic studies of typical Friedel–Crafts reactions between phosphorus trichloride and polystyrene (PS) showed that the trifunctionality of phosphorus trichloride leads to crosslinked products regardless of the reaction conditions [83, 84]. Modifications of the classical approach are necessary to control the reactivity and to minimize side reactions. It should be noted that the use of stannic chloride, which is soluble in organic media, was found to limit the number of electrophilic substitution reactions occurring on the multiple halogen–phosphorus center to only one, even in the case of highly activated aromatic rings (see pathway IV in Scheme 7) [85]. The main drawback of this reaction is the necessity of a subsequent oxidation step to yield the phosphonic acid groups.

Classical Michaelis–Arbuzov or Michaelis–Becker approaches to the formation of aromatic C—P are not generally feasible (see pathway I.a in Scheme 7). However, nucleophilic aromatic substitutions occur under particular circumstances, often with mechanisms quite different from those involving the formation of alkyl C—P bonds, provided that suitable substituents are present. One approach involves the photoinitiated reaction of aryl iodides with trialkyl phosphites that gives products corresponding to those of a simple Michaelis–Arbuzov reaction (see pathway III in Scheme 7) [86–88]. This represents a convenient pathway that can usually be carried at low temperatures. However, the access to the appropriate aryl iodides is often limited. In order to facilitate the formation of aromatic C—P bonds, a variety of transition metal salts have been considered as catalysts. In one approach, aryl halides were reacted with trialkyl phosphites in the presence of Ni(II) halide salts, such as $NiCl_2$, at elevated temperature to yield the corresponding aryl phosphonate (see pathway I.b in Scheme 7) [89, 90]. Similar reactions have been performed to substitute phosphorus for the halide in vinylic chlorides [91]. However, the use of these reactions is limited because of the harsh conditions, including heating at around 160–190°C, and the fact that the reaction often results in low yields. Recently, the use

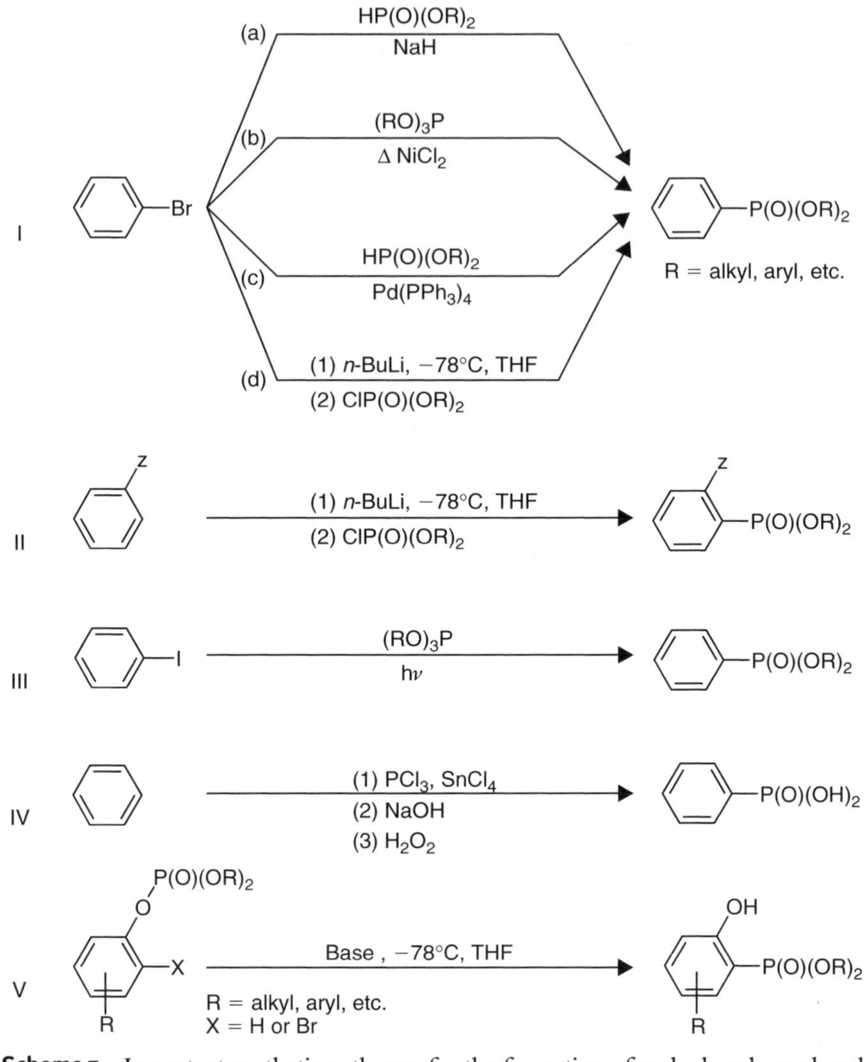

Scheme 7 Important synthetic pathways for the formation of aryl–phosphorus bonds (R represents an alkyl or an aryl group, X a hydrogen or a bromine atom and Z an *ortho*-directing group).

of Pd(0) catalyst in the presence of a tertiary amine has proven to be very efficient for the substitution on aromatic halides by a variety of P(O)H-type compounds (see pathway I(c) in Scheme 7) [92]. The latest development of this coupling reaction has been driven by its use in the preparation of phosphine ligands for Pd species. A pioneering work in the area has been performed by Hirao *et al.* [93] who reported the Pd-catalyzed coupling of aryl and vinyl bromides with dialkyl phosphites, giving rise to dialkyl arylphosphonates and dialkyl vinylphosphonates, repectively. For these early experiments, the preferred reagents were $Pd(PPh_3)_4$ and Et_3N, either neat or in toluene solution. Although resembling a Michaelis–Becker type of reaction, the mechanism is quite different. For instance, the role of the tertiary amine is to regenerate the catalytic Pd(0) species rather than to generate the phosphorus anion.

As mentioned in Section 3.3.1.1, the displacement of halogen atoms at phosphorus by organolithium is a convenient pathway for the formation of C—P bonds. Consequently, the reaction of aryllithium compounds with chlorophosphonic acid esters will produce the corresponding aryl phosphonates. It is of particular interest that aryllithium reagents may be prepared by transmetalation of the corresponding bromoderivatives (see pathway I(d) in Scheme 7). The first reported investigation in this area was carried out by Dawson and Burger [75,94]. A typical reaction involves quenching the aryllithium compound with chlorophosphonic acid ester in THF at low temperature to yield the corresponding aryl phosphonate in good yields [78]. Furthermore, the presence of strong electron-withdrawing groups will make neighboring aromatic hydrogens acidic enough to be metalated by organolithium reagents, which allows the preparation of aryllithium compounds from non-halogenated species as well (see pathway II in Scheme 7). These different routes to aryllithiums give a high degree of freedom in the synthesis of different aryl phosphonates. However, the use of organolithium chemistry is not confined to displacement reactions at phosphorus. Indeed, an important development of this chemistry is the phosphate–phosphonate rearrangement, characterized by a base-induced migration of a dialkoxyphosphoryl group from oxygen to carbon (see pathway V in Scheme 7). This rearrangement, which was simultaneously reported by two laboratories [95, 96], has recently been reviewed in detail by Taylor [97] and is applicable for the formation of a wide variety of different phosphonated phenols. The rearrangement is also referred to as an anionic phospho-fries rearrangement. The driving force for the reaction is that the Li—O bond is stronger than the C—Li bond, which outweighs the energy loss in the transition from a P—O to a P—C bond [78,97]. The application of such a rearrangement to dialkyl arylphosphates to yield dialkyl arylphosphonates is a useful tool for the preparation of phosphonated monomers. Phosphate esters of substituted phenols, i.e., dialkyl arylphosphates, are readily obtained from phenols and dialkyl phosphites in the presence of triethylamine, in a so-called Atherton–Todd reaction [98], or by the reaction of phenolate anions with dialkyl chlorophosphates. Different phosphate esters might be employed although sterical hindrance can impede the rearrangement in some cases [99]. By far the most widely used base is LDA which, due to its bulkiness, is not nucleophilic. However, *n*-butyllithium (*n*-BuLi) is often employed despite the risk of a competitive nucleophilic attack at phosphorus. Although phosphate groups have an *ortho*-directing effect in the metalation process, a metal-halogen exchange is preferred since it decreases reaction times and results in higher yields and fewer by-products. It has been shown that the presence of electron-donating substituents on the aromatic ring generally facilitates the reaction, while electron-withdrawing substituents have a detrimental effect. Notably, the use of aromatic diols leads to potentially useful diphosphonated monomers can be prepared. Such monomers have been synthesized [97,100], and may allow the preparation of diphosphonated poly(arylene ether ketone)s and poly(arylene ether sulfone)s by direct polycondensations.

3.3.1.3 Formation of Perfluoroalkyl–Phosphorus Bonds

In addition to the C_{Al}—P and C_{Ar}—P bonds, the C_F—P bond is also of great interest when it comes to the synthesis of proton conducting polymers. As mentioned in Chapter 3.2, the C_F—P bond is generally quite stable. Furthermore, the proximity of the electron-withdrawing fluorine atoms enhances the acidity and hygroscopicity of the phosphonic acid unit. As discussed in Section 3.3.1.1, the formation of C_{Al}—P bonds is readily achieved by the use of several reactions such as Michaelis–Arbuzov and Michaelis–Becker. Unfortunately, these approaches cannot generally be applied to perfluoroalkyl halides. The strongly electron-withdrawing perfluoroalkyl groups cause an electron deficiency on the halogen atom, preventing any nucleophilic attack in most cases. However, some fluoroalkyl halides have been reported to participate in these types of reactions [101,102]. Other pathways consist either of the electrophilic fluorination of phosphonate carbanions, the nucleophilic fluorination of functionalized phosphonate substrates, the syntheses via fluorinated phosphonate carbanions, the use of transition metal-catalyzed addition reactions, or radical approaches [68]. For instance, perfluoroalkyl iodides can react with tetraethyl pyrophosphite via a thermally induced radical reaction in the presence of di-*tert*-butyl peroxide in an autoclave [103]. Notably, Burton *et al.* found that a milder reaction, where a mixture of tetraethyl pyrophosphite and perfluoroalkyl iodide was subjected to UV irradiation at room temperature, allows for the preparation of a wide variety of perfluoroalkyl phosphonates in good yields (see Scheme 8A) [104]. This type of reaction was also used for the preparation of perfluoroalkyl compounds functionalized with both sulfonic acid and phosphonic acid for evaluation as potential fuel cell electrolytes [105,106]. In order to avoid the use of the expensive tetraethyl pyrophosphite, DesMarteau and co-workers developed another approach involving the reaction between diethyl chlorophosphate and the corresponding perfluoroalkyl magnesium halide (see Scheme 8B) [107]. The latter reactant can be prepared by treating perfluoroalkyl iodide with conventional Grignard reagents in ether at low temperature. These multi-step reactions allow the preparation of useful monomers which may be copolymerized with tetrafluoroethylene (TFE), as will be discussed in Section 3.3.3 [107–109]. It is worth noting that organolithium reagents have also been used to perform $S_NP(V)$ reactions in the preparation of perfluoroalkyl phosphonates. Indeed, Obayashi *et al.* lithiated difluoromethylphosphonate with LDA in THF at −78°C, resulting in the formation of the corresponding organolithium reagent. Several difluoromethylphosphonates were produced in good yields by treating various electrophiles, including diethyl chlorophosphate, with this organolithium reagent (see Scheme 8C) [110]. Yet another approach reported in the literature is telomerizations with the phosphonic group present in the telogen [111,112]. Fluorination of non-fluorinated compounds is often a convenient approach as the latter compounds are more readily available (see Sections 3.3.1.1 and 3.3.1.2). Alkyl phosphonic compounds may be fluorinated using $FClO_3$ or Et_2NSF_3 [113–115]. A quite common approach is the fluorination of acyl phosphonates using diethylaminosulfur trifluoride (DAST) (see Scheme 9a) [116,117]. Acyl phosphonates are conveniently synthesized by several different procedures including the Michaelis–Arbuzov reaction [118]. The fluorination of acyl phosphonates using DAST allows the synthesis of *tert*-butyl-protected difluoro(aryl)- or (naphthalenyl)methylphosphonates, which under mildly acidic deprotection conditions yield the desirable phosphonic acids. However, the practical problems involved in performing such reactions on a multigram scale, as well as the use of large amounts of the expensive DAST, and the risk of uncontrolled exothermic reactions, limit its applicability [119]. Therefore, an alternative procedure has been developed which consists of an electrophilic fluorination of benzylic phosphonates. In this procedure, α-carbanions of benzylic phosphonates are first prepared in situ using a strong base and are then reacted with *N*-fluorobenzenesulfonimide (NFBS) (see Scheme 9b) [120].

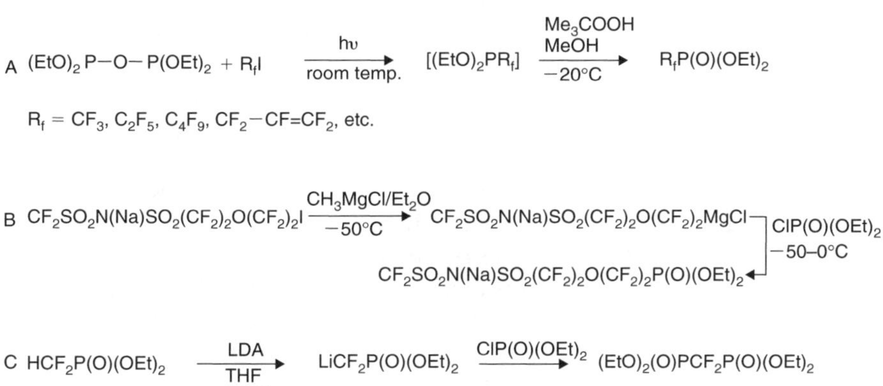

Scheme 8 Synthetic routes to perfluoroalkyl phosphonates via (A) an UV-induced radical reaction between tetraethyl pyrophosphite and fluorinated iodides, (B) a nucleophilic attack at phosphorus using a Grignard reagent, and (C) a nucleophilic attack using an organolithium reagent.

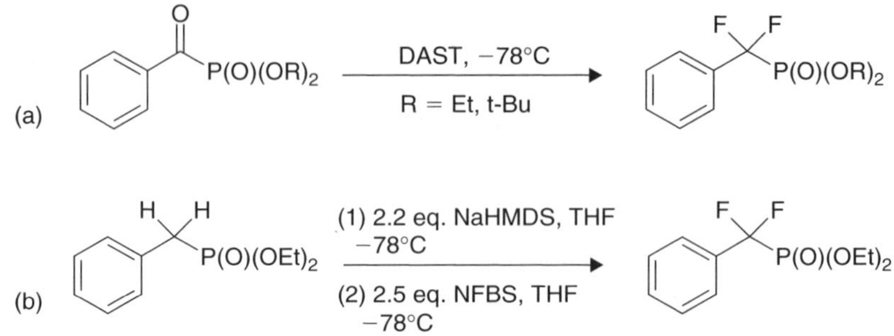

Scheme 9 Preparation of aryl(difluoromethylenephosphonates) through fluorination [116,117,120].

3.3.1.4 The Hydrolysis of Phosphonate Esters to Form Phosphonic Acids

In most cases, the preparation of phosphonic acids, as described in Sections 3.3.1.1–3.3.1.3, requires a second step to convert phosphonate esters or phosphonic acid chlorides into the acids. Phosphonic acid chlorides are readily converted into phosphonic acid when brought in contact with water. However, the hydrolysis, or dealkylation, of phosphonic acid esters is more difficult. In the literature, different approaches have been used leading either to the complete transformation to phosphonic acid units, or to monoacid/monoester units (see Scheme 10). Alkaline hydrolysis using aqueous sodium hydroxide usually results in the formation of monoacid/monoester-substituted compounds [121]. The hydrolysis of phosphonate esters proceeds via a trigonal bipyramidal intermediate (phosphorane) which is an electron-rich state [122]. Consequently, the presence of electron-withdrawing units attached to phosphorus generally increases the rate of the hydrolysis. However, it is worth noting that basic hydrolysis of perfluoroalkyl phosphonate esters often results in the cleavage of the C_F—P bond to yield —CF_2H, rendering such a procedure unsuitable in some cases [108]. Another selective method toward mono dealkylation is the treatment of the phosphonate esters with halogenide ions [123]. A typical procedure involves the use of solutions of NaI in ketones (usually acetone or 2-butanone) under reflux. This method is very selective for the formation of monoesters/monosalts, but it has been found that only the methyl esters of phosphonates are reactive enough [124]. The method of choice to produce phosphonic acids from phosphonate esters is usually reflux in concentrated HCl or HBr solutions. This protocol has been successfully used to obtain almost quantitative conversions to the diacid [51]. The treatment in these acidic solutions is usually suitable since polymers for use in fuel cell applications should be chemically resistant in acidic environment. Acid hydrolysis is usually conducted under reflux over long periods of time, especially in the case of esters of perfluoroalkyl phosphonic acids for

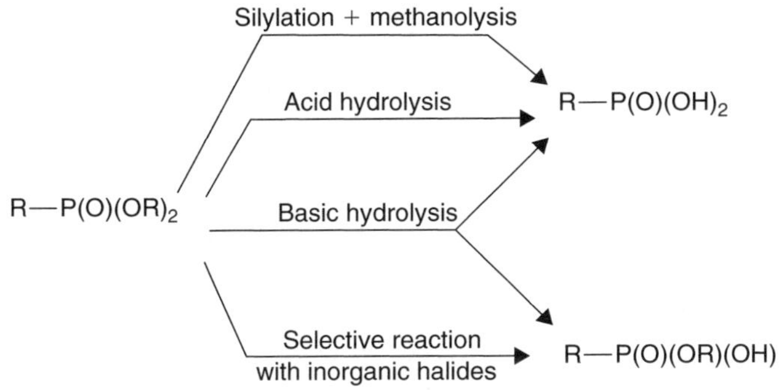

Scheme 10 Schematic representations of different pathways yielding monoacid monoester phosphonate and/or phosphonic acid groups from phosphonate esters.

which reaction times up to 10 days are required [109]. These harsh conditions might in some cases be detrimental for the stability of certain C—P bonds, as discussed in Chapter 3.2. This probably explains the results reported by Jaffres *et al.*, where the attempt to hydrolyze 6-hydroxy-2-napthylphosphonate ester yielded 2-naphtol and free phosphoric acid [125]. Consequently, hydrolysis reactions and their conditions have to be employed with caution. Notably, it has been found that *tert*-butyl esters can be cleaved under much milder acidic conditions by using 85% trifluoroacetic acid in organic solvents [126,127]. Rabinowitz was first to report the successful dealkylation of phosphonate esters using trimethylchlorosilane in the 1960s [128]. This reaction proceeds via the formation of the corresponding bis(trimethylsilyl)phosphonates. The latter compound readily hydrolyzes into phosphonic acids in contact with water and alcohols at room temperature. This approach is particularly advantageous for polyfunctional molecules which can be selectively converted into phosphonic acid derivatives [129–131]. The reaction has been successfully used on a wide variety of different phosphonates, including methyl, ethyl and isopropyl derivatives.

3.3.2 Preparation of Phosphonated Ionomers by Chemical Modification

As mentioned in the introduction, an extensive worldwide research is currently ongoing to develop new ionomers for fuel cells based on durable aromatic main-chain polymers. One attractive way to achieve such ionomers is by chemical modification of commercially available polymers. Chemical modification offers a versatile route to introduce acidic units either directly on the polymer main chain, or on side chains to the polymer main chain, and has recently been reviewed [17,59]. In general, polymers undergo the same reactions as their low-molecular-weight analogs, and it is generally assumed that the reactivity of a functional group is in the same range regardless if it is placed on a small molecule or a polymer, as long as the reaction is carried out under homogeneous conditions under the appropriate conditions. Several considerations have to be taken into account in the choice of an efficient chemical reaction to modify a polymer: (1) high yields are required in order to achieve high levels of modification, (2) the concentration of the reagent within the polymer coils in solution is important for the overall rate of reaction, (3) the polymer may precipitate from the reaction media as the reaction proceeds, thereby limiting the extent of modification, (4) side reactions may result in crosslinked polymers, (5) sterical hindrance will affect the polymer reactivity, especially if the functional group is close to the polymer main chain, (6) neighboring-group effects of already reacted segments on the remaining reactive sites might be detrimental [132]. The phosphonation of a number of common polymers has been studied so far, usually with the aim to enhance solubility in different solvents or to improve polymer–polymer miscibility [133–138]. However, the current demands for polymer membranes for fuel cells have revitalized the development of new ionomers based on phosphonated polymers, as reflected in the number of patents filed in the past 5 years [139–160].

The use of chloro- or bromomethylated polymers as precursors for the preparation of phosphonated polymers has attracted a lot of attention. Indeed, as discussed in Section 3.3.1.1, such groups may readily be converted to alkyl phosphonates by using the well-established Michaelis–Arbuzov and Michaelis–Becker rearrangements (see pathway I.a,b in Scheme 6). In an early work by Cabasso and co-workers, the Michaelis–Arbuzov reaction was employed to prepare phosphonated PS and poly(phenylene oxide) (PPO) to be used in desalination processes (see Scheme 11) [133,161,162]. The two polymers are of particular interest since aromatic polymers have the basic stability to serve as backbone polymers for alternative ionomers to the commonly used Nafion® ionomer [2]. First, Cabasso *et al.* investigated the bromination of PPO and the chloromethylation of PS in order to introduce the halogenated sites needed for the subsequent modification via reaction with triethyl- or trimethylphosphite. In the case of PS, the desired modification was difficult to achieve due to gelation and formation of insoluble products. However, by choosing a polydentate etheral solvent (such as diethyl carbitol) the gelation was avoided. Concerning the phosphonation of PPO the authors reported no problems with crosslinked products and the introduction of up to two phosphonate ester groups per repeating unit of the PPO was achieved. Xu and Cabasso also investigated the use of phosphonated PPO as a proton conducting material for fuel cells [163,164]. These results will be further discussed in Chapter 3.4. Allcock *et al.* have studied the phosphonation of poly(aryloxyphosphazene) via bromomethylated intermediates using the Michaelis–Becker reaction. They reported complete conversion of the bromides by treatment of the polymer with sodium dimethyl phosphite in THF at room temperature for 16 h [165]. Moreover, problems with crosslinking

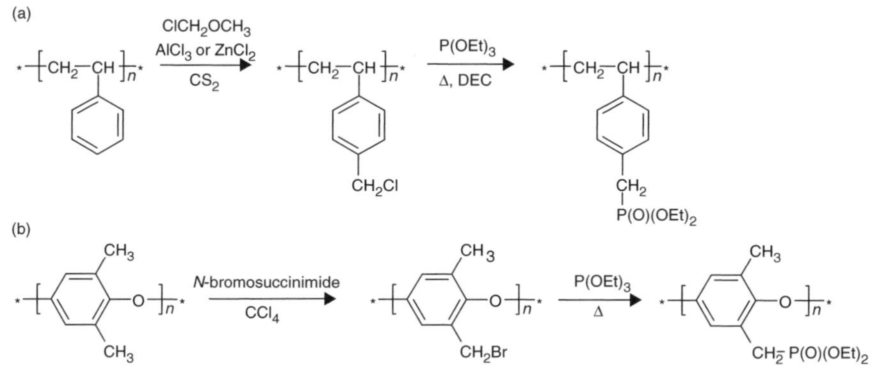

Scheme 11 Phosphonation of poly(*p*-chloromethylstyrene) (a) and poly(2,6-dimethyl-1,4-phenylene oxide) (b) [133].

reactions were not encountered. In another piece of work, Okamoto *et al.* used a Michaelis–Becker reaction to prepare polyimides with pendent phosphonate ester groups (see Scheme 12) [166]. A nearly quantitative replacement of the bromine atoms of the brominated polyimide was achieved by reaction with 5–10 equiv. of triethyl- or trimethylphosphite in *N*-methyl-2-pyrrolidone under reflux. The sensitivity of the five-membered imide rings toward hydrolysis might limit the potential use of the polyimides prepared in this work [167]. In contrast, the six-membered imide rings in napthalenic polyimides have been found to be much more stable toward hydrolysis, especially under fuel cell conditions [3,167].

The chemical modification of polymer substrates in the form of different resins or films is another convenient approach that has been studied extensively. For instance, phosphonic acid and phosphonate ester resins have been synthesized by treating chloromethylated styrene-based polymers with PCl_3 in the presence of $AlCl_3$ [29,168]. By using this reaction, ion-exchange resins having *IEC* up to 6.51 meq/g have been prepared. Different mixtures of ester/monoester/diacid groups were prepared by the careful addition of ethanol/water in various molar ratios. Slow additions were required to avoid any increase in the temperature due to the highly exothermic reaction. Another example is the grafting of monomers directly onto pre-irradiated polymer films. The method is critical to the choice of a suitable polymer substrate with the appropriate morphology and physical properties, and the introduction of reactive free radicals into the polymer chain by irradiation with, e.g., γ-rays. The free radicals can react to initiate graft polymerization of vinyl monomers. This procedure has been used extensively to prepare sulfonated membranes [169]. Vinylbenzyl chloride (VBC) is a commonly used monomer in this context since it provides grafted polymers containing reactive chloromethylene functions. One example of the

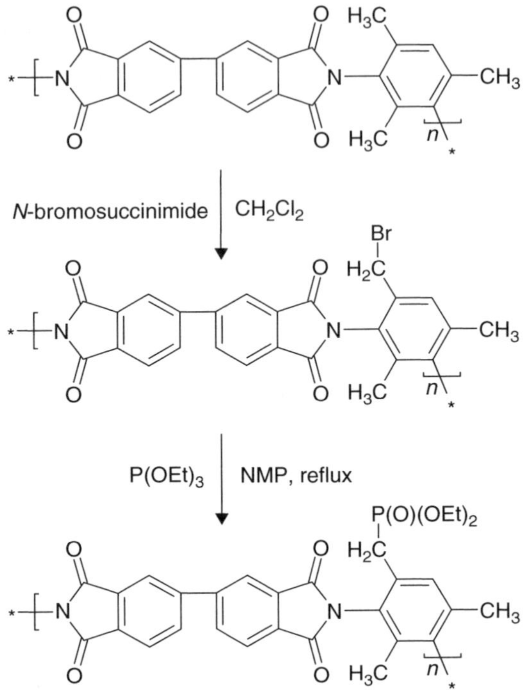

Scheme 12 Synthesis of polyimides with pendent phosphonate ester groups [166].

application of this pathway is the use of the Michaelis– Arbuzov reaction on films prepared by irradiation grafting of VBC onto TFE-ethylene (ETFE) copolymer films or on fluorinated ethylene–propylene (FEP) copolymer films (see Scheme 13) [170]. Temperatures between 100°C and 140°C reaction times up to 18 h were required to achieve high levels of phosphonation. The hydrolysis of the polymers was achieved in hot concentrated aqueous HCl which resulted in a conversion of about 80 mol.% to phosphonic acid groups.

As already mentioned in Section 3.3.1.1, diphosphonate compounds may be synthesized through various pathways, giving rise to elaborated diphosphonates bearing substituents such as alkyl, aryl or halogen groups on the methylene group. Any protons located on the methylene group are acidic due to the electron-withdrawing power of the phosphonate units. Consequently, these methylene groups may be metalated by the use of a strong base such as NaH. The carbanions formed may then be used to displace the chlorine atom of chloromethylene units on polymers. Sundell *et al.* have investigated the formation of poly[ethylene-g-VBC] through irradiation grafting and its subsequent functionalization with bis(phosphonic acid) (see Scheme 14) [171]. By using the sodium salt of tetraisopropyl

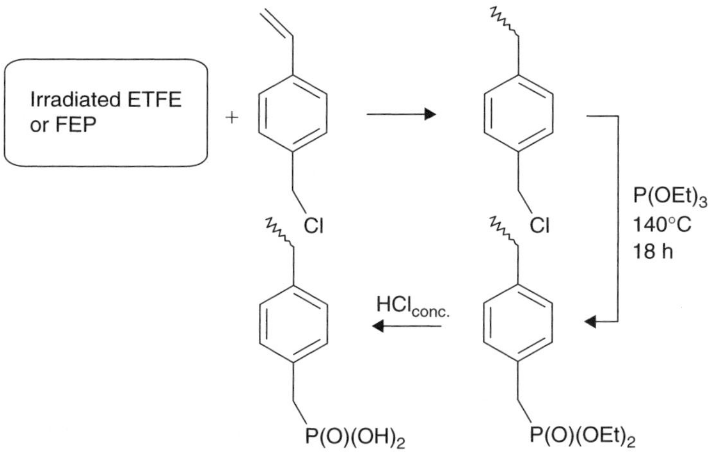

Scheme 13 Synthesis of proton conducting membranes with pendent phosphonic acid groups by irradiation grafting of VBC [170].

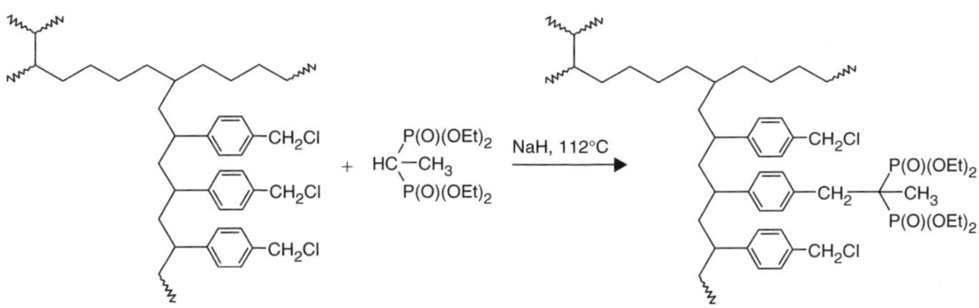

Scheme 14 Synthesis of polyethylene-supported bisphosphonate derivatives [171].

methylenebisphosphonate in a 10-fold excess in refluxing toluene for 24 h, 46% of the chlorine atoms were substituted. By using tetraethyl ethane-1, 1-bisphosphonate, a conversion of 50% was achieved. In both cases, the occurrence of side reactions accounted for the limited yields. In the case of tetraisopropyl methylenebisphosphonate, the remaining acidic hydrogen may also be metalated, resulting in coupling reactions of already grafted bisphosphonates and remaining chloromethylene groups. Obviously, such reactions may limit the application of this method for the preparation of soluble bisphosphonate-functionalized polymers. In the case of tetraethyl ethane-1, 1-bisphosphonate, these coupling reactions are suppressed. However, Sundell *et al.* found that alkylated compounds were formed through the reaction of the bisphosphonate anion with the ethylester groups through a S_N2 mechanism [171]. Finally, the tetraalkyl bisphosphonate groups were hydrolyzed to the free acids by refluxing the films in aqueous HCl. Nearly quantitative conversion of the phosphonate ester to phosphonic acid was observed. Other types of polyfunctional resins have been prepared and investigated using a similar approach [172].

The main drawback of the polymers described just above is the instability of the benzylic methylene groups in an oxidizing environment, which can be expected to be a severe shortcoming of these materials for use in fuel cell applications. Consequently, besides attempting to fluorinate the methylene groups, as described in Section 3.3.1.3, the chemical modification of polymers to introduce C_{Ar}—P bonds is interesting. As discussed in Section 3.3.1.2, a useful approach for the formation of C—P bonds is the reaction between aryllithium and chlorophosphonic acid esters. Allcock *et al.* have recently reported on the successful phosphonation of poly(aryloxyphosphazene)s via lithiation and reaction with diethylchlorophosphate [165]. In an initial study, the reaction was found to be difficult to control as crosslinked materials were obtained already at limited degrees of phosphonation [173]. Consequently, the target degree of phosphonation was reduced from 100% to 20% to obtain non-crosslinked products. Under the proper conditions, the phosphonation via the lithiophenoxy intermediate and reaction with diethylchlorophosphate proceeded with 80–85% conversion of the bromophenoxy side groups to diethylphosphonate side groups, demonstrating the potential of such an approach for the functionalization of poly(aryloxyphosphazene)s [165]. Allcock and co-workers also showed that the limiting factor for the conversion was not the lithiation step, which was quantitative, but rather side reactions involving the highly reactive lithiophenoxy intermediate [165]. The use of *n*-BuLi to lithiate polyphosphazenes does, however, result in partial backbone cleavages and the molecular weight of the modified polymer was found to be about half of that of the parent polymer [165]. This work was further extended to include the use of diphenylchlorophosphate in order to decrease the risk of crosslinking (see Scheme 15) [174]. Furthermore, it was found that by using *tert*-BuLi, only 2 molar equiv. of the reactant was enough for an efficient metal-halogen exchange, as compared with 4–8 molar equiv. for *n*-BuLi. The decrease in the amount of BuLi used for the reaction afforded a lower tendency for backbone cleavage. A rapid quenching of the lithiated polymer was necessary to avoid intermolecular crosslinking reactions although intramolecular coupling always occurred. The overall conversion of the bromophenoxy side groups to diphenyl phosphonates was found to be 50%, i.e., lower than that found with diethyl chlorophosphate. This is most probably a consequence of the bulkiness of diphenylchlorophosphate. It is worth noting, however, that about 70% of the reactive sites had actually reacted with the electrophile. The remaining 20% accounted for intramolecular couplings. Allcock *et al.* made attempts to hydrolyze the diphenylester side groups by treating the polymer with an alkaline solution. This gave mainly the monoester/monoacid derivative at a level of 40%, which is in accordance with the discussion in Section 3.3.1.4.

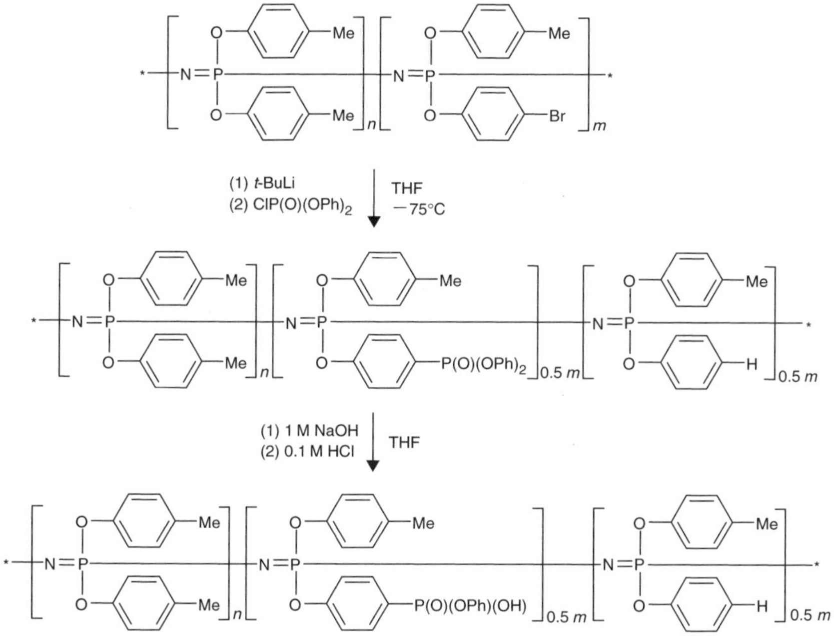

Scheme 15 Synthesis of phenylphosphonic-acid-functionalized poly(aryloxyphosphazene)s [174].

In line with the work of Allcock *et al.* and Lafitte and Jannasch have attempted the phosphonation of PSUs via lithiation and reaction with chlorophosphonic acid esters (see Scheme 16) [175]. PSUs are high-performance thermoplastic polymers with high mechanical, chemical and thermal stability. The PSU backbone is composed of electron-rich arylene ether segments, which are unpolar and flexible, and of electron-poor arylene sulfone segments, which are polar and rigid. The presence of these two segments makes it possible to modify PSU by either electrophilic substitution, such as chloromethylation or halogenation, or by lithiation chemistry. Guiver *et al.* have shown that the PSU can be quantitatively and selectively lithiated with a high degree of control without any chain cleavages [176,177]. As reported by Allcock *et al.*, a large excess (approximately 200%) of chlorophosphonic acid ester was required in order to suppress crosslinking reactions. In the case of PSUs, Lafitte and Jannasch showed that the overall conversion of the lithiated sites was limited to 40% at degrees of lithiation up to one lithiated site per repeating unit. Regardless of the reaction conditions, all attempts to introduce two lithiated sites per repeating units led to crosslinking after addition of diphenyl or diethyl chlorophosphate. The temperature had to be kept at −70°C in order to avoid the reaction of unreacted lithiated sites with already grafted phosphonate units. The limited reactivity of *ortho*-to-sulfone lithiated

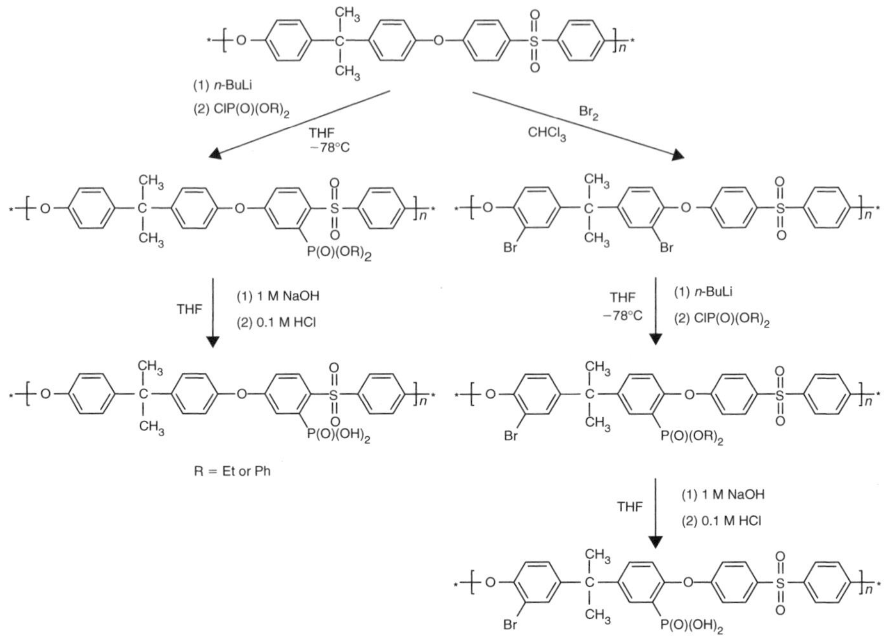

Scheme 16 Phosphonation of PSU via lithiation and reaction with chlorophosphonic acid esters [175].

sites with chlorophosphonic acid ester can partially be explained by sterical hindrance and the electron-withdrawing power of the sulfone group which may to some extent deactivate the lithiated sites. In an attempt to increase the reactivity of the lithiated sites, the PSU main chain was brominated and selectively lithiated by transmetalation with *n*-BuLi to achieve activated *ortho*-to-ether lithiated sites as reported by Guiver *et al.* [178]. In this way, the degree of phosphonation was slightly enhanced to reach 50%. Notably, a complete conversion of diphenyl ester to phosphonic acid occurred during basic hydrolysis, although such a hydrolysis procedure normally yields the monoacid/monoester. As mentioned in Section 3.3.1.4, the hydrolysis proceeds via a trigonal bipyramidal intermediate (phosphorane) which is electron rich. In the hydrolysis of the *ortho*-to-sulfone phosphonated PSUs, the electron-withdrawing power of the phenoxy units, combined with the one of the sulfone group, might have contributed to the stabilization of such the phosphorane intermediate. This may explain the complete hydrolysis in this case. Phosphonation of PSUs via $S_NP(V)$ type reaction provides a non-catalytic reaction that is quite simple to accomplish, although its efficiency for converting lithiated sites to phosphonate units is limited. PSUs having degrees of phosphonation up to 0.4–0.5 with the phosphonic acid placed either on the bisphenol-A or the biphenyl sulfone segments were conveniently synthesized.

In a recent study, Lafitte and Jannasch have also prepared PSUs carrying benzoyl(difluoromethylenephosphonic acid) side chains (see Scheme 17) [179]. In the first step, PSUs were lithiated and reacted with methyl iodobenzoates to prepare *p*- and *o*-iodobenzoyl PSUs. Next, the phosphonated PSUs were prepared via CuBr-mediated cross-coupling reactions between the iodinated polymer and the [(diethoxyphosphinyl)difluoromethyl]zinc bromide. Finally, dealkylation using bromotrimethylsilane afforded the highly acidic $—CF_2—P(O)(OH)_2$ derivatives. The replacement of the iodine atoms by $—CF_2—P(O)(OEt)_2$ units was found to be almost quantitative in the case of *o*-iodobenzoyl PSU. Polymers having degrees of phosphonation up to 0.5 were synthesized and studied.

The first study on the phosphonation of PSU was reported by Ziaja *et al.* [180]. In this work, PSU was phosphonated using a Friedel–Craft reaction. PSU was reacted with PCl_3 in the presence of $SnCl_4$ in dichloroethane to yield PSU–PCl_2. As discussed in Section 3.3.1.2, the use of $SnCl_4$ afforded non-crosslinked materials since the substitution of PCl_3 is limited to one. In a second step, PSU–PCl_2 was reacted with aqueous NaOH to yield PSU–$P(OH)_2$ which was further oxidized to yield phosphonic-acid-functionalized PSU. Degrees of phosphonation up to one phosphonic acid unit per repeating unit of the polymer were achieved. Unfortunately, no detailed information concerning the characterization of these polymers was reported.

Recently, Nuñes *et al.* reported on a Pd-catalyzed phosphonation of polyphenylsulfone (see Scheme 18A) [145,181]. In this study, polyphenylsulfone was first brominated. Then, the use of Michaelis–Arbuzov reaction assisted by $NiCl_2$ was compared with the Pd-catalyzed phosphonation in order to achieve phosphonated polymers having the phosphonic acid units located *ortho* to the ether linkages. The use of the catalyzed Michaelis–Arbuzov approach directly on the brominated polymer was ineffective and slow. This might result from the sterical hindrance of the polymer main chain or from the dilution of the components. Indeed, a solvent had to be employed since triethyl phosphite is a non-solvent for PSU. The slow reaction resulted in the reduction of the metal catalyst by triethylphosphite, which is a common side reaction in such a procedure [90]. The Pd(0)-catalyzed P—C bond formation proved to be much more effective. The reaction was typically carried out in a mixture of diethylphosphite and diphenyl ether or *o*-dichlorobenzene in the presence of Et_3N to take care of the HBr evolution. Temperatures of around 90–120°C were required to achieve a degree of phosphonation of 0.9. The use of higher temperatures led to degradation, and the limitation in the degree of phosphonation is most probably due to sterical hindrance. A very large excess of diethylphosphite had to be used (around 20 equiv.) and 2.5 mol.% of catalyst were required for an efficient phosphonation to occur. One of the main drawbacks of this approach is the limited control of the degree of modification. The complete hydrolysis of the phosphonate ester derivatives was

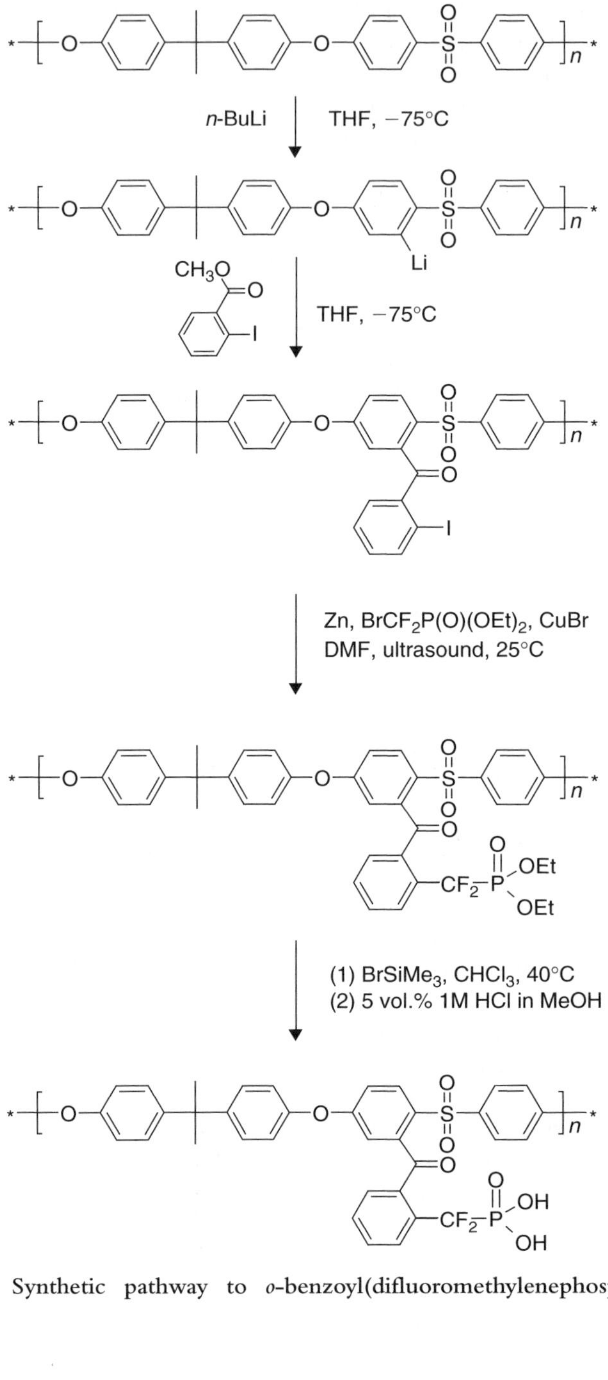

Scheme 17 Synthetic pathway to *o*-benzoyl(difluoromethylenephosphonic acid) PSU [179].

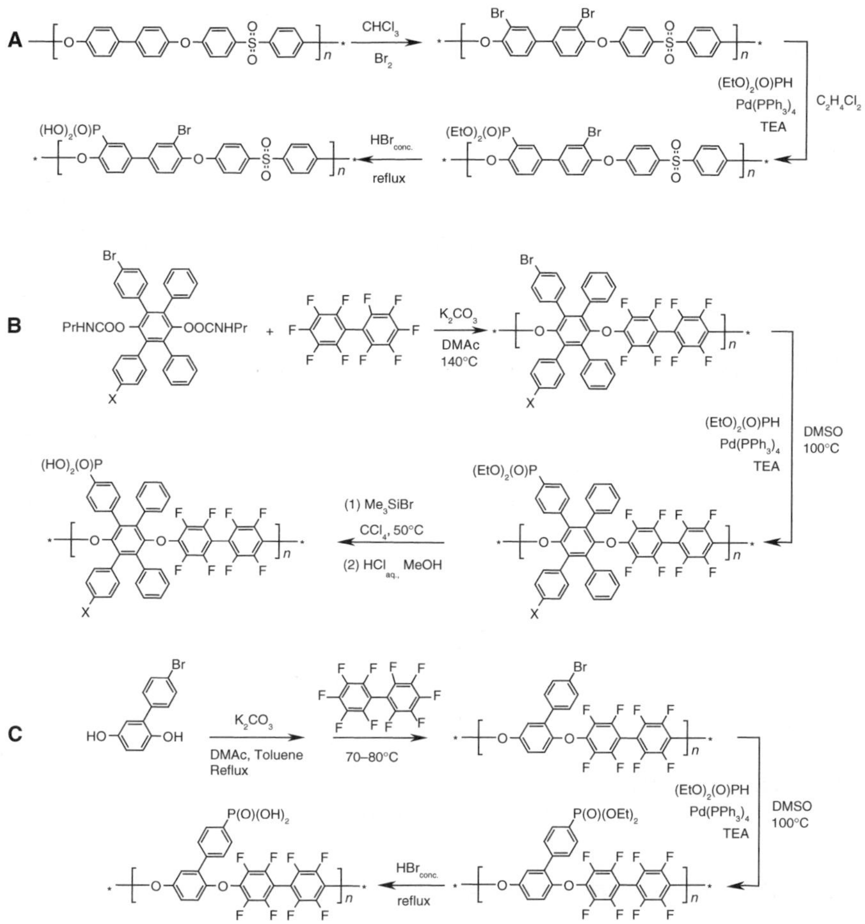

Scheme 18 Synthesis of poly(arylene ether)s with pendent phosphonic acid groups using Pd(0)-catalyzed reactions (X represents a hydrogen or a bromine atom) [182–184].

achieved by keeping the polymer in concentrated HBr under reflux for 48 h. This indicated that acid-catalyzed hydrolysis is more effective that the basic approach. More recently, Nuñes *et al.* have also sulfonated already phosphonated polyphenylsulfones using trimethylsilylchlorosulfonic [182]. The sulfonation reaction was not very effective because of deactivation of the aromatic rings of the polymer by the phosphonate groups. For example, only a degree of sulfonation of 0.09 was reached in the post-sulfonation of a polymer having a degree of phosphonation of 0.6, and could not be increased by changing the reaction conditions.

Substitution reactions of aryl bromides located on a poly(arylene ether) backbone has also been reported by Miyatake and Hay (see Scheme 18B) [183].

The synthesis of phosphonated poly(arylene ether)s having one or two phosphonic acid groups per repeating unit of the polymer was achieved. However, a multi-step monomer preparation was required to yield the desired brominated polymers. In contrast to the work reported by Nuñes *et al.*, a high amount of catalyst was needed (50 mol.%) to achieve 100% conversion of the bromides. The reaction was typically carried out in polar solvents such as DMSO at 100°C for about 70 h. The dealkylation of the phosphonated polymer was performed using bromotrimethylsilane, and resulted in a 92% conversion to phosphonic acid units on the basis of NMR results.

In a very recent study, Guiver and co-workers prepared a semi-fluorinated poly(arylene ether) containing pendant 4-bromophenyl groups that were subsequently converted into phenylphosphonic acid groups using bromotrimethylsilane (see Scheme 18C) [184]. Notably, Guiver and co-workers found that the substitution reaction proceeded very rapidly, i.e., reaching 95% conversion after only 6 min, using a lower loading of the palladium catalyst as compared to the level reported by Miyatake and Hay [183].

Rikukawa and co-workers have reported on the synthesis of phosphonated poly(4-phenoxybenzoyl-1,4-phenylene) (P-PPBP) by a Michaelis–Arbuzov reaction catalyzed with $NiCl_2$ to substitute bromine atoms on P-PPBP (see Scheme 19) [156,185]. Brominated PPBP was prepared by reaction of the neat polymer with *N*-bromosuccinimide and azobis(isobutyronitrile) in DMF. Unfortunately, no detailed information concerning the synthesis was reported, and phosphonations up to 40% were reached. The dealkylation was carried out by a rather uncommon procedure involving the use of dimethyl sulfide and methanesulfonic acid.

3.3.3 Polymerization of Phosphonated Monomers

Chain polymerization of vinyl monomers is a convenient technique for the preparation of a wide variety of functional polymers. Anionic polymerization of diethyl vinylphosphonate in tetrahydrofuran, initiated by either butyllithium or methylmagnesium bromide, has been reported to result in high yields and in high-molecular-weight polymers [186]. On the other hand, free-radical homopolymerization of vinyl phosphorus monomers is in general difficult due to the sterical effects acting on the propagation reactions. The reduction of the propagation rate by sterical interference makes the termination and chain-transfer processes correspondingly more dominant and the polymerization generally results in only low-molecular-weight polymers [187,188]. Chain-transfer reactions usually occur at the alkoxy group linked to the phosphorus atom of the monomer or the polymer [188]. Overberger and co-workers have shown that the introduction of polarizing groups to the double bond, such as aryl groups or esters, decrease the sterical and electron-withdrawing effects of the phosphorus

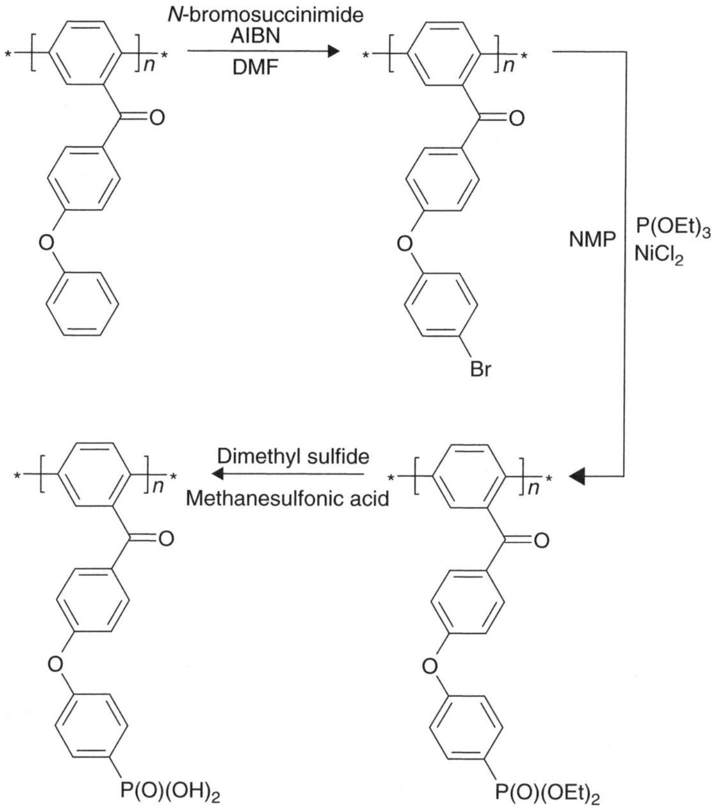

Scheme 19 Phosphonation of P-PPBP [185].

atom leading to a higher reactivity of the double bond [189]. Consequently, a number of phosphonated polymers have been prepared by free-radical polymerization of phosphonated acrylates. Most of these polymers (prepared as adhesion-promoters to different substrates such as dentine [38,40, 190,191] and steel [42,43]) have the phosphonate groups located on hydrolysable ester side chains. Phosphonated acrylates have also been grafted onto low-density polyethylene by free-radical graft polymerization [37]. Efforts have been made to enhance the hydrolytic stability of acrylate polymers by having the polymerizable acrylate group connected to the phosphonic acid group via a hydrolytically stable ether [39] or a carbon–carbon [192] bond. Acvi and Albayrak have reported detrimental sterical effects when the phosphonate groups are close to the double bond, which was consistent with the earlier observations made by Pike and Cohen and Rabinowitz *et al.* [187,191]. Acvi and co-workers reported that bulk polymerization of the phosphonic acid-functionalized vinyl monomers was unsuccessful [190,191]. In a recent paper, Salman *et al.* reported that the reactivities of phosphonic acid monomers in copolymerizations increase with decreasing sterical hindrance and/or increasing hydrogen bonding [192]. The latter effect is most probably due to the rapid increase of the viscosity of the polymerization medium which auto-accelerates the propagation reaction according to the so-called gel effect, as reported by Moszner *et al.* [39]. Boutevin *et al.* have studied the polymerization of phosphonated styrenic monomers [193]. The homopolymerization of diethyl (*p*-vinylbenzyl) phosphonate proceeded to 50% yield. In order to investigate the polymerization of the phosphonic acid derivative, the diethyl phosphonate ester was hydrolyzed using the silylation method. Notably, Boutevin *et al.* reported the formation of P—O—P bridges upon methanolysis of the silylated compound. This was avoided by performing the hydrolysis in the presence of dicyclohexyl amine. The homopolymerization of the phosphonic acid derivative was found to proceed to 85% yield. However, these polymers suffer from the poor oxidative stability of the benzylic groups. Free-radical copolymerizations of styrene and vinyl phosphonates have been carried out by several research groups, and in some cases the polymers obtained were hydrolyzed to produce poly(styrene-co-vinyl phosphonic acid) [194–197]. These copolymerizations are rather difficult because of the large difference in the reactivity ratios between styrene and diethyl vinylphosphonate, 3.25 and ~0, respectively [194]. Wu and Weiss have reported on the properties of these ionomers which will be further discussed in Chapter 3.4 [198]. Sundell *et al.* have also investigated the polymerization of bis(phosphonic acid)-functionalized monomers. In their study, porous poly[styrene-DVB-1-(vinylphenyl)propane-2,2-bis(phosphonic acid)] were prepared by microemulsion polymerization [46].

Stone and co-workers have studied the potential of phosphonic-acid-based polymers for fuel cells by taking advantage of the base chemistry already developed by Ballard for the preparation of sulfonated polymers [199, 200]. In their study, copolymers based on phosphonic-acid-functionalized monomers of α,β,β-trifluorostyrene (TFS) were synthesized by free-radical polymerization (see Scheme 20) [201]. Using a Pd-catalyzed cross-coupling reaction, they synthesized dimethylphosphonate-substituted TFS monomer from 4-iodobenzene phosphonic acid dimethyl ester and trifluoroethenylzinc bromide [202]. Homopolymerizations of dimethylphosphonate-substituted TFS were carried out in the bulk using azobis(isobutyronitrile) as initiator. Both oligomeric and polymeric materials were produced as shown by chromatography and NMR analysis. In this context, it should be mentioned that phosphonated TFS monomer has previously been reported not to homopolymerize via a free-radical mechanism [12]. Attempts by Stone *et al.* to reach higher molecular weights were successful by copolymerizing the monomer with neat TFS in an emulsion process. A reasonably high-molecular-weight copolymer (e.g., $\overline{M_n} = 38{,}100$) of quite low polydispersity (e.g., 2.78) was

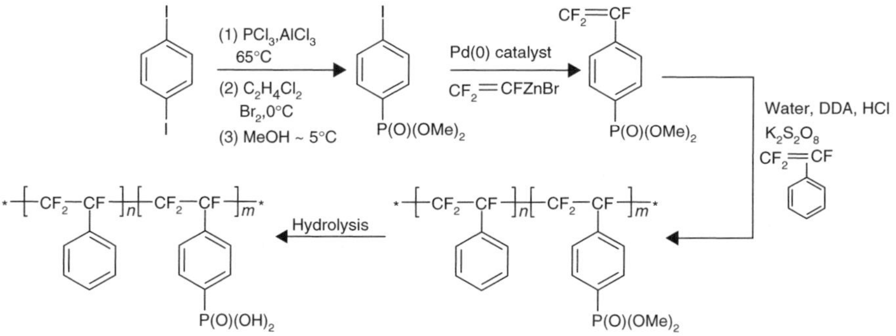

Scheme 20 Preparation of phosphonated TFS monomer and its emulsion polymerization to yield phosphonic acid functionalized polymers for fuel cells [201].

achieved, although the yield remained limited to about 21%. A typical alkaline hydrolysis procedure was used to dealkylate the copolymers where DMSO solutions of the polymers were kept at 80°C for 64 h in the presence of aqueous KOH. The extent of the hydrolysis was determined to be between 90% and 100%, depending on the equivalent weight of the copolymer. Membranes of the different copolymers (either in their ester form or acidic form) were prepared by impregnating a microporous support material with a 5 wt.% solution of the copolymers in DMF. An alternative hydrolysis procedure where the supported membranes were immersed in concentrated HCl and glacial acetic acid at 50–90°C for 48 h was used to produce fully hydrolyzed membranes [201].

Recently, Souzy *et al.* prepared a new 4-[(α,β,β-trifluorovinyl)oxy] benzene phosphonic acid monomer through four different parallel pathways: the Michaelis–Arbuzov, the Michaelis–Becker, the Pd-catalyzed cross-coupling and lithiation and reaction with chlorophosphonic acid esters [203]. They found that the reaction involving Pd triphenylphosphine as the catalyst led to the highest yield (see Scheme 21). This is in line with the discussion in Section 3.3.1.2. The dealkylation of the monomer was performed by silylation and methanolysis to reach a yield of 85%. These monomers have a potential interest for the preparation of fuel cell membranes and their copolymerizations are reported to be under further investigation.

Polycondensation is another attractive type of polymerization that has been successfully used for the preparation of phosphonated polysiloxanes [204, 205]. However, when it comes to the preparation of poly(arylene ether)s from phosphonate ester functionalized monomers and activated aromatic dihalides, the potential displacement of ester units by phenolates may cause both reactivity and solubility problems. Interestingly, the polymerization of potassium phosphonate functionalized monomers, via the nucleophilic route mentioned above, has been carried out by Meng *et al.* with the

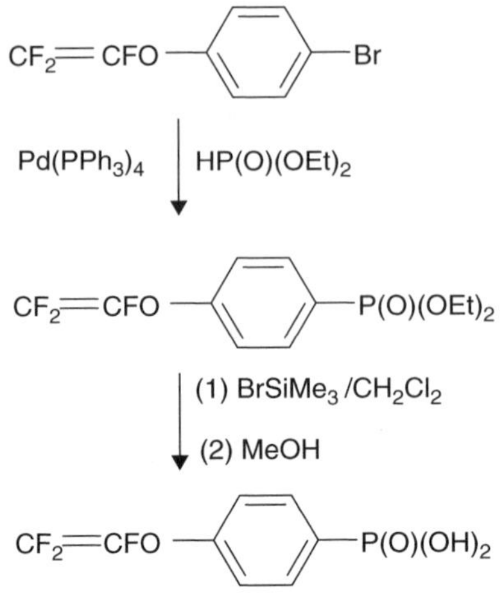

Scheme 21 Synthesis of 4-[(α,β,β-trifluorovinyl)oxy)]benzene phosphonic acid [203].

aim of studying the properties of the resulting polymers as membranes for fuel cells (see Scheme 22) [206,207]. First, the phosphonic acid bisphenol was synthesized through a multi-step chemical pathway to yield a diol to which phenylphosphonic acid was attached via a five-membered imide ring. The synthesis of the necessary precursor to the phosphonated diol has been described separately [208,209]. As mentioned already, one serious drawback of these polymers is the potential instability of the five-membered imide ring under acidic conditions [167]. No apparent side reactions were encountered during the polymerization and high molecular weights were reportedly achieved. Notably, the polymerization was carried out with the monomer in the salt form and not in the ester form. These results are encouraging when it comes to future attempts to prepare phosphonated poly(arylene ether)s by direct polycondensation. As mentioned above, useful diols may be prepared by phosphor-fries rearrangement yielding diphosphonated monomers (see Scheme 23) [100]. Similar monomers have also been prepared and further modified with acrylic groups in order to be used in free-radical polymerizations [40, 41]. In addition, useful phosphonated diols may also be prepared by direct Pd-assisted phosphonation of 6,6′-dibromo-1,1′-binaphthalene-2,2′-diol [125]. Such diols may be used in order to prepare diphosphonated equivalents to the disulfonated polymers developed by McGrath *et al.* [15,210–214].

Kotov *et al.* were the first to report the preparation of Nafion-like polymers-carrying phosphonic acid units instead of sulfonic acid units (see Scheme 24) [109]. In their study, redox initiated emulsion polymerization of phosphonated perfluoro monomers was employed. The hydrolysis of the polymers proved to be difficult and up to 10 days of reflux in HCl/acetic

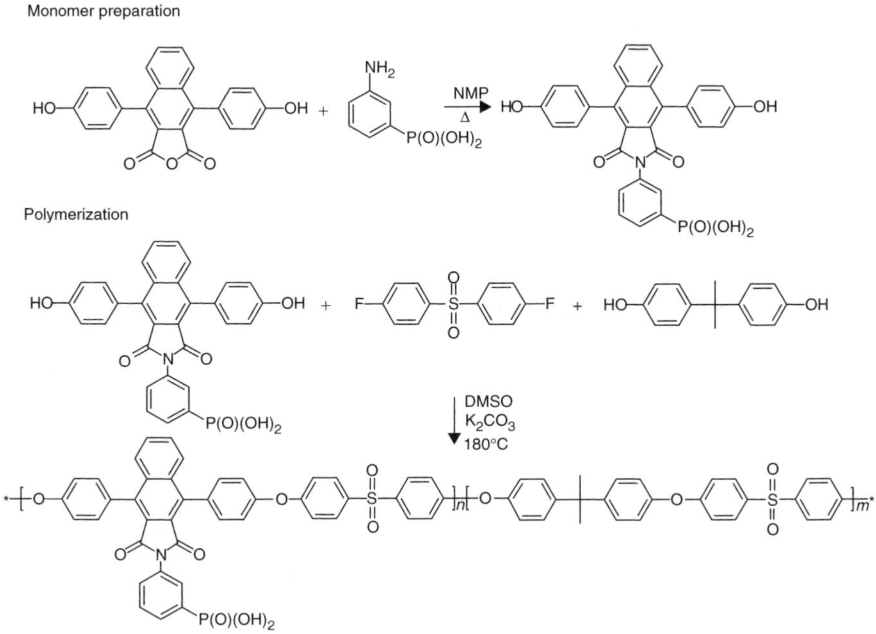

Scheme 22 Synthesis and polycondensation of 4,9-bis-(4-hydroxyphenyl) benzoisobenzofuran-1,3-dione [206,207].

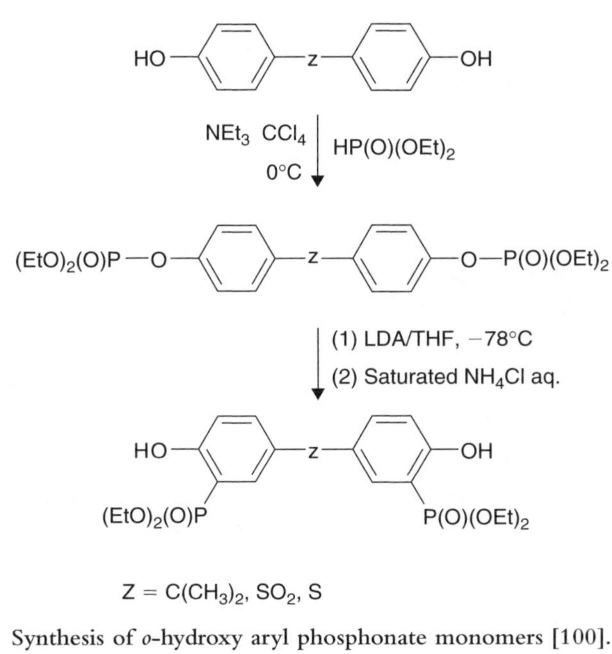

Scheme 23 Synthesis of *o*-hydroxy aryl phosphonate monomers [100].

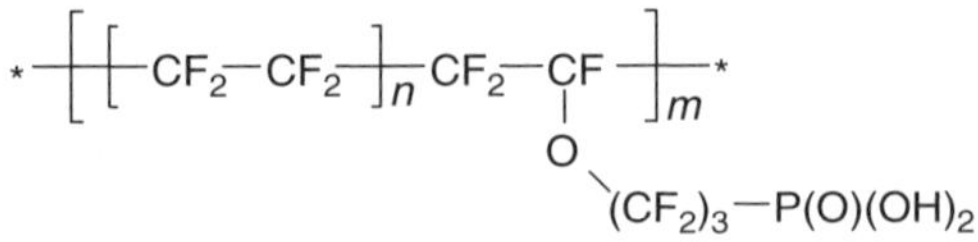

Scheme 24 Molecular structure of the perfluorophosphonic acid polymer prepared by Yamabe *et al.* and Kotov *et al.*, respectively [108,109].

acid mixtures were required. This might explain the unsuccessful hydrolysis via HCl treatment of phosphonic acid esters placed on perfluorinated chains reported by Brondino *et al.* [112]. The silylation route may prove to be more advantageous in this case. However, Gaboyard *et al.* have reported that the hydrolysis of phosphonic acid esters attached to CF_2 groups still requires harsher conditions than usual since heating at 65°C for 48 h was required to complete the dealkylation of the phosphonate ester groups, while 3 h at room temperature is usually sufficient for alkylphosphonate esters [111]. Yamabe *et al.* have investigated the preparation of perfluorophosphonic acid ionomers by first preparing fluorinated monomers through a multi-step pathway (see Scheme 8) [108]. They found that the copolymerization of the phosphonated monomers with TFE proceeded only to a limited conversion of 25%. These copolymers were then successfully converted to the phosphonic acid form by acidic treatment.

3.4 Properties of Membranes Based on Phosphonated Polymers

Comparatively little work has been devoted to the study of phosphonated ionomer membranes for fuel cells. Consequently, it is difficult to present a comprehensive survey of the structure–property relationships for these membranes. However, some general characteristics may be outlined to facilitate the development of new and more optimized materials. In order to compare the different properties of these ionomers it is often practical to do this on the basis of their *IEC*. However, a special complication arises with phosphonic acid units because it carries two acidic protons. In general, only one acidic proton should be considered per phosphonic acid group. Yet, both protons might be consumed during titrations with strong bases. Indeed, Yamabe *et al.* showed that the values of the *IECs* evaluated by titration corresponded to approximately twice the number of phosphonic acid groups [108]. Although no details concerning the titration was given, this result might be explained by the lower p*K*a value of the second acidic proton in perfluorophosphonic acid compounds as compared to non-fluorinated phosphonic acids (see Table 3.1).

3.4.1 Membrane Formation and Microphase Separation

The solubility behavior of phosphonated ionomers is often of major concern when these ionomers are synthesized and handled. Nuñes *et al.* have reported that phosphonated PSUs having more than 0.75 phosphonic acid unit per repeating unit of the PSU were not completely soluble in DMAc or DMSO [145,181]. Instead, highly swollen gels of the densely phosphonated polymers were obtained. These gels were readily dissolved after the addition of 2 vol.% of concentrated HBr. This indicated that strong hydrogen bonded clusters were formed and acted as physical crosslinks between the polymer chains. In turn, this probably explains why the phosphonated poly(arylene ether)s having two phosphonic acid unit per repeating unit, synthesized by Miyatake and Hay were found to only swell in DMSO even though the temperature was raised to 100°C [183]. Notably, Liu *et al.* prepared a phosphonated derivative of fluorinated poly(arylene ether) having one phosphonic acid unit per repeating unit that was readily soluble in DMAc, DMF, DMSO and NMP [184]. Kotov *et al.* were unable to cast films of their perfluorinated ionomers due to insolubility in common low-boiling organic solvents [109]. Films of the different polymers in their phosphonate ester form were, however, successfully prepared by compression molding at 170°C, provided that the polymers were amorphous, i.e., contained a sufficiently high content of phosphonate ester groups. Attempts to hydrolyze these polymer films resulted in membranes of poor quality. Instead, the polymers were hydrolyzed in their powder form and were then hot pressed at 215–240°C, slightly above the glass transition temperature (T_g) of the ionomers. By this method, useful membranes with thicknesses between 100 and 140 μm were obtained.

PS-based ionomers may serve as suitable model compounds for the understanding of the general behavior of phosphonated ionomers and membranes. The reason is that their synthesis is rather straightforward. A quite large number of sulfonated PS ionomers have thus been prepared and studied, which offers a good basis for comparison. Wu and Weiss have prepared poly(styrene-co-vinyl phosphonate) ionomers and studied the structure and the thermal and viscoelastic properties of these copolymers [198]. They reported that the T_g decreased with increasing phosphonate ester concentration, while the T_g's of the phosphonic acid derivative and their salts increased. The phosphonic ester is a large, bulky, non-mesogenic group. When introduced into a polymer structures, it usually increases the free volume, resulting in a decrease in T_g. A similar effect was found in a study of diphenylphosphonate ester-substituted PSUs, for which the T_g decreased by 20–30°C as compared to the neat PSU [175]. The decrease in T_g might allow the preparation of membranes by hot pressing at temperatures in between the T_g and the decomposition temperature. Subsequent hydrolysis of the films would then yield the phosphonic acid polymer membranes. Such a procedure may be of particular interest for high-T_g

aromatic main-chain polymers for which solvent casting is usually the only alternative to obtain membranes. In contrast to the ester derivatives, Wu and Weiss reported that the phosphonic acid copolymers exhibited intermolecular hydrogen bonding and strong dipole–dipole associations that acted as physical crosslinks to restrict the mobility of the backbone resulting in an increase in T_g [198]. These observations have been reported for a number of different systems (see Table 3.2) [174,175,179,184,206,207]. For instance, Meng *et al.* showed that the T_g's of poly(arylene ether)s increased with the phosphonic acid content, and that no T_g was detected for a polymer having one phosphonic acid groups per repeating unit [206,207]. This was in line with the increase in the T_g observed for the phosphonic acid PSUs, as compared to neat PSU [175], and the high T_g's observed for the oligomers based having repeating units of phosphonated metaphenylene synthesized by Rager *et al.* [215].

Wu and Weiss confirmed the presence of physical crosslinking in the ionomers based on phosphonic acid and its salts by the appearance of a rubbery plateau in the storage modulus, measured as a function of temperature by dynamic mechanical analysis (DMA) [198]. This rubbery plateau was found to be broader for the ionomers in the salt form than for the corresponding ionomers in the acidic form (see Figure 3.1). It indicates that the physical crosslinks due to dipole–dipole interactions are stronger and more temperature resistant than those arising from hydrogen bonding. In addition, Wu and Weiss showed that two $\tan\delta$ peaks emerged during the DMA measurements, one arising from the T_g of the polymer hydrocarbon matrix, and one at a higher temperatures due to a transition of an ion-rich phase to indicate microphase separation. From their study, the driving force for microphase separation seemed greater with phosphonic-acid-based ionomers than for corresponding ionomers based on sulfonic acid. However, a more comprehensive study is needed to confirm this observation. The microphase separation and the formation of clusters is of prime importance when it comes to the preparation of polymer membranes for fuel cells since the level of proton conductivity can often be explained on the basis of cluster formation of the ionomer leading to water-filled channels after water uptake. As mentioned in Section 3.3.3, Kotov *et al.* have prepared several co- and terpolymers based on TFE, perfluro(propyl vinyl) ether (PPVE) and phosphonated perfluorovinylethers with different *IECs* [109]. In their study, it was found that the proton conductivity of the copolymer with the highest *IEC* was much lower than that of the copolymers with lower values of *IECs*. Investigations by calorimetry implied that this was a result of poor cluster formation during the compression molding.

3.4.2 Water Uptake and Methanol Crossover

With the exception of anhydrous systems, which will be discussed in Chapter 3.5, high levels of hydration are necessary to form a percolating

Table 3.2 Thermal properties of selected ionomers

Polymer	*IEC* (meq/g)	T_g ester (°C)	T_d ester[b] (°C)	T_g acid (°C)	T_d acid[c] (°C)	Observations concerning the degradation of the ionomers in the acidic form	
Perfluorinated phosphonic acid membranes[a]	2.5–3.0	–	250	–	320	A small weight loss of 2–2.5 wt.% was noted between 20°C and 330°C due to the loss of water	[109]
Phosphonated membrane based on irradiation grafted FEP	–	–	290	–	450	A small weight loss of a few wt.% occurred between 200°C and 450°C	[170]
Phosphonated PSU[a]	0.84	168	260	212	375	A small weight loss of 4 wt.% was noted between 200°C and 320°C attributed to the formation of P—O—P linkages resulting in the loss of water	[175]
Phosphonated PPSU[a]	0.81	–	225	–	330	A small weight loss of 4 wt.% was noted between 20°C and 330°C	[145,181]
P-PPBP[d]	1.31	–	–	–	350	–	[185]
Phosphonated poly(arylene ether)s[e]	1.3–1.9	–	–	254 n/d	502–476	A weight loss of 5 wt.% occurred between 200°C and 480°C	[206,207]

(Continued)

Table 3.2 (*Continued*)

Polymer	*IEC* (meq/g)	T_g ester (°C)	T_d ester[b] (°C)	T_g acid (°C)	T_d acid[c] (°C)	Observations concerning the degradation of the ionomers in the acidic form	
Phosphonated poly(arylene ether)s[e]	1.3–2.3	244 n/d	284–257	n/d	263–257	A weight loss of about 10–12 wt.% occurred between 260°C and 480°C	[183]
Benzoyl(difluoromethylene phosphonic acid) PSU[a]	1.79	169	251	n/d	227	The initial weight loss occurring at 230°C was attributed to the loss of the phosphonic acid units	[179]
Phosphonated derivative of fluorinated poly(arylene ether)[a]	1.8	161	307	n/d	420	The weight loss at 461°C was 5 wt.%	[184]

[a]TGA or DSC experiments carried out under an inert atmosphere (nitrogen, argon or helium) at a heating rate of 10°C/min.
[b]Onset temperature accounting for the loss of ester groups (alkyl or phenyl esters).
[c]Temperature at which a major decomposition occurs.
[d]TGA or DSC experiments carried out under an inert atmosphere (nitrogen, argon or helium) at a heating rate of 5°C/min.
[e]TGA or DSC experiments carried out under an inert atmosphere (nitrogen, argon or helium) at a heating rate of 20°C/min.
n/d: not detected.

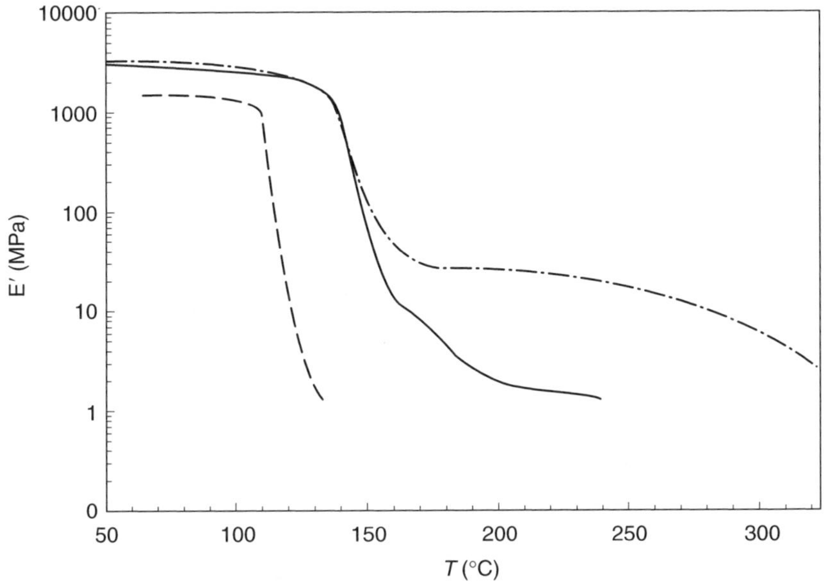

Figure 3.1 Storage modulus as measured by dynamic mechanical analysis of copolymers based on styrene and vinyl phosphonate, containing 6.4 mol.% of diethyl vinylphosphonate in its ester form (dashed line), acid form (solid line) and sodium salt form (dash-dotted line) [198].

network of water-filled channels in which the transport of the protons can be facilitated. However, as seen in Figure 3.2, the water uptake of phosphonated ionomers is in general very low in comparison to sulfonated ionomers having the same *IEC*. As discussed in Chapter 3.2, the water uptake of sulfonated ionomers is mainly driven by the large gain in entropy resulting from the solvation of the dissociated protons. In the case of phosphonic acid ionomers, the acidic units are preferentially stabilized by autoprotolysis, which accounts for the low levels of water uptake. In addition, extensive hydrogen bonding between the polymer chains results in high T_g's which restrict the water uptake of the polymer membranes. Actually, the self-hydrogen bonding in materials containing phosphonic acid can be so effective that only very few phosphonic acid sites are available for hydrogen bonding with water. Indeed, poly(styrene-co-vinylphosphonic acid)s with values of *IECs* up to 1.5 meq/g exhibited a water uptake of only a few weight percent and no conductivities could be detected at 25°C [198]. Phosphonated aromatic main-chain polymers also show low levels of water uptake. For example, the phosphonated poly(arylene ether)s synthesized by Meng *et al.* (*IEC* = 1.93 meq/g) and the phosphonated PSU synthesized by Lafitte *et al.* (*IEC* = 0.84 meq/g) were found to take up only 7.5 wt.% and <2 wt.% water, respectively [175,206,207]. The phosphonated

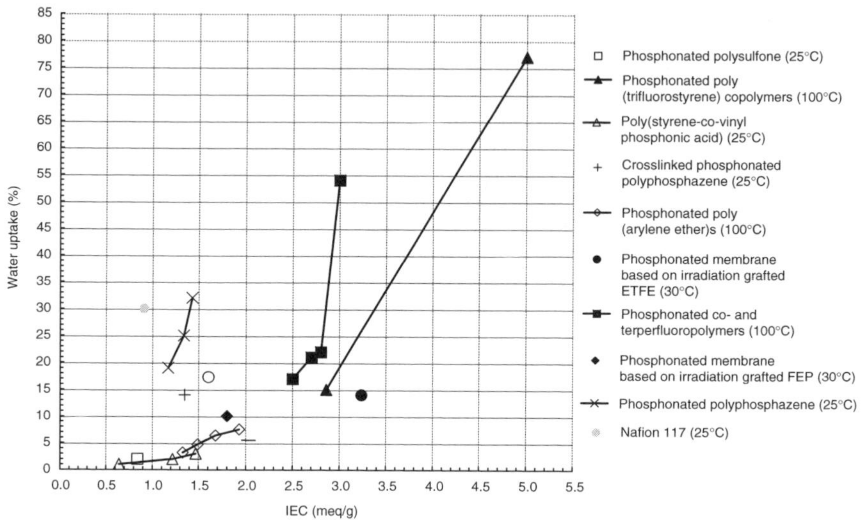

Figure 3.2 Water uptake of phosphonated ionomers as a function of their ion-exchange capacities. Please note that the water uptake of the different membranes was measured at different temperatures (indicated within parenthesis).

poly(arylene ether)s synthesized by Miyatake and Hay can be expected to exhibit low levels of water uptake. Unfortunately, no values were reported in their study [183]. Finally, the phosphonated derivative of fluorinated poly(arylene ether) prepared by Liu *et al.* ($IEC = 1.8$ meq/g) and benzoyl(difluoromethylenephosphonic acid) PSU ($IEC = 1.8$ meq/g) prepared by Lafitte and Jannasch were found to take up 3 and 6 wt.% water, respectively, at room temperature [179,184]. Several authors have reported that casting the ionomers in their lithium salt form, followed by an ion exchange to the protonated form, will increase the water uptake [175,206,207]. It is clear that further investigations of the mechanism of water uptake in phosphonated ionomers has to be carried out in order to identify the most suitable procedure for membrane preparation. As mentioned above, the water uptake is related to the T_g, and phosphonated aromatic main-chain polymers usually possess high T_g's, often above the decomposition temperature (~250–350°C). This may at least partly explain why the water uptake does not necessarily increase significantly after immersing these membranes in hot water. The strong interactions between phosphonic acid units imply that the T_g of the ionomer remains high even in the swollen state. Cabasso *et al.* have demonstrated the relationship between the T_g and the water uptake by carrying out measurements on a phosphonated PPO containing a hydrophobic plasticizer. While such plasticizers normally cause a decrease in the water absorption when incorporated in a hydrophilic polymer, the addition lead to a decrease in the T_g from 120°C to 18°C, which resulted in an increase in the water uptake from 38% to 86% [133]. Allcock and co-workers have been active in the investigation of phosphonated polyphosphazenes as candidates for fuel cell membrane applications. As discussed in Section 3.3.2, Allcock *et al.* synthesized polyphosphazenes bearing phenoxy phosphonic acid side groups [174]. In comparison to their brominated precursors, the phosphonated polyphosphazenes showed an increase in the T_g's of approximately 30°C, which was attributed to sterical hindrance and hydrogen bonding. Membranes having *IEC*-values ranging from 1.17 to 1.43 meq/g were prepared and were subsequently found to take up between 19 and 32 wt.% water at room temperature [216]. However, when the membranes contained poly[(3-methylphenoxy)(phenoxy)phosphazene] instead of poly[(4-methylphenoxy)(phenoxy)phosphazene], a water uptake of only 12 wt.% was measured. This indicated that the water uptake was not only dependant on the phosphonic acid groups, but also to the character of the backbone. Indeed, 3-methylphenoxy side groups are thought to shield the hydrophilic polyphosphazene backbone [216,217]. In general, the water uptake of the phosphonated polyphosphazene was much lower than for the corresponding sulfonated polyphosphazene membranes [216]. Nonetheless, as seen in Figure 3.2, the phosphonated polyphosphazene ionomers absorbed more water at a given *IEC*-value than any of the other ionomers, which is most probably a result of the low T_g's of these polymers, close to room temperature (~40°C). One obvious strategy to increase the water uptake of ionomers is to increase their *IEC*. Schmidt-Naake *et al.* have prepared ionomer membranes with *IEC*-values ranging between 1.4 and 3.2 meq/g by irradiation grafting of VBC onto FEP or ETFE films (see Scheme 13) [170]. These membranes were found to take up between 10 and 14 wt.% water when immersed at room temperature. Measurements at higher temperatures showed no significant change in the water uptake, and an increase of only a few wt.% was noted at temperatures up to 90°C [170]. Stone *et al.* have also prepared ionomers with high values of *IEC* by preparing phosphonated poly(trifluorostyrene)-based ionomers (see Scheme 20). Copolymers having *IEC* of 2.86 and 5.00 meq/g were found to take up 15 and 77 wt.% water, respectively, at 100°C. As discussed in Chapter 3.2, phosphonic acid groups linked to fluorinated carbons are more acidic due to the electron-withdrawing power of the fluorine atoms. This increase in the acidity is most probably beneficial for the level water uptake of the ionomer. Kotov *et al.* have prepared different co- and terpolymers of TFE, PPVE and phosphonated perfluorovinylethers (see Scheme 24) [109]. The copolymers synthesized from the former monomer and perfluorovinylether, having *IEC* of 2.8 and 3.0 meq/g, took up 22 and 54 wt.% water, respectively, after boiling for 4 h.

When it comes to polymer membranes for direct methanol fuel cells (DMFC), the methanol barrier properties are of prime importance. Methanol crossover is a severe drawback of the Nafion® membrane and is extensive at

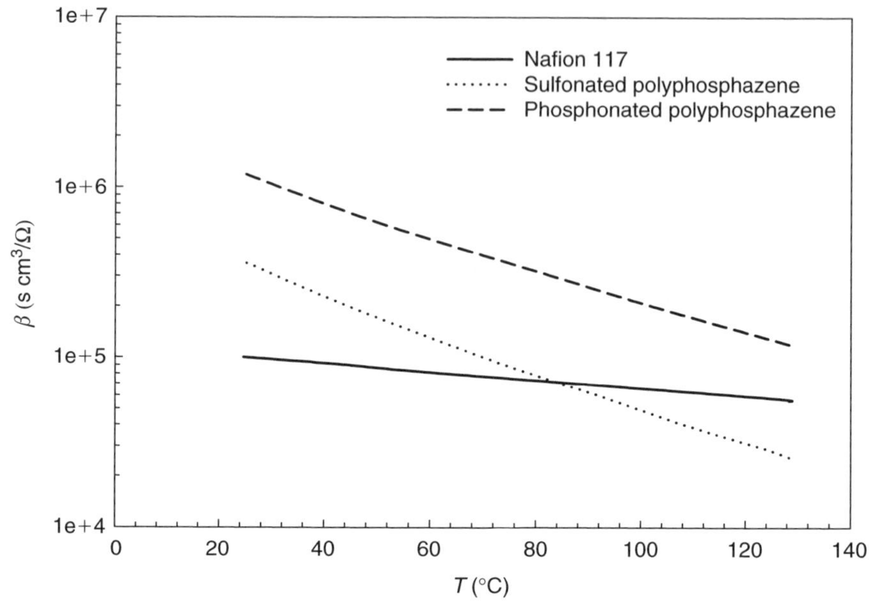

Figure 3.3 Selectivity of Nafion® (solid line), sulfonated polyphosphazene and phosphonated polyphosphazene toward methanol permeation [223].

high temperatures (above 100°C) [218–220]. Allcock and co-workers have investigated the use of their phosphonated polyphosphazenes for use as DMFC membranes [216,221–223]. They showed that the diffusion coefficient of methanol was 40 times lower in the phosphonated polyphosphazene membrane than in the Nafion® 117 membrane, and 10–20 times lower than for a crosslinked sulfonated polyphosphazene membrane at room temperature [221]. The advantage of phosphonated polyphosphazene is seemingly even more pronounced at elevated temperatures. Phosphonated polyphosphazene retained a low methanol permeability at 120°C while the sulfonated polyphosphazene took up excessive amount of water resulting in methanol permeabilities similar to Nafion® [222,223]. In addition, the selectivity, calculated as the ratio between the proton conductivity and the permeability, of the phosphonated polyphosphazene membrane was found to be superior to Nafion® over a wide temperature range between 22°C and 125°C (see Figure 3.3). Thus, the use of phosphonated membranes might allow DMFCs to run at higher temperatures with an increased methanol concentration in the feed.

3.4.3 Proton Conductivity

For membranes which rely on the presence of water, the conductivity depends mainly on the level of water uptake and the formation of an efficient

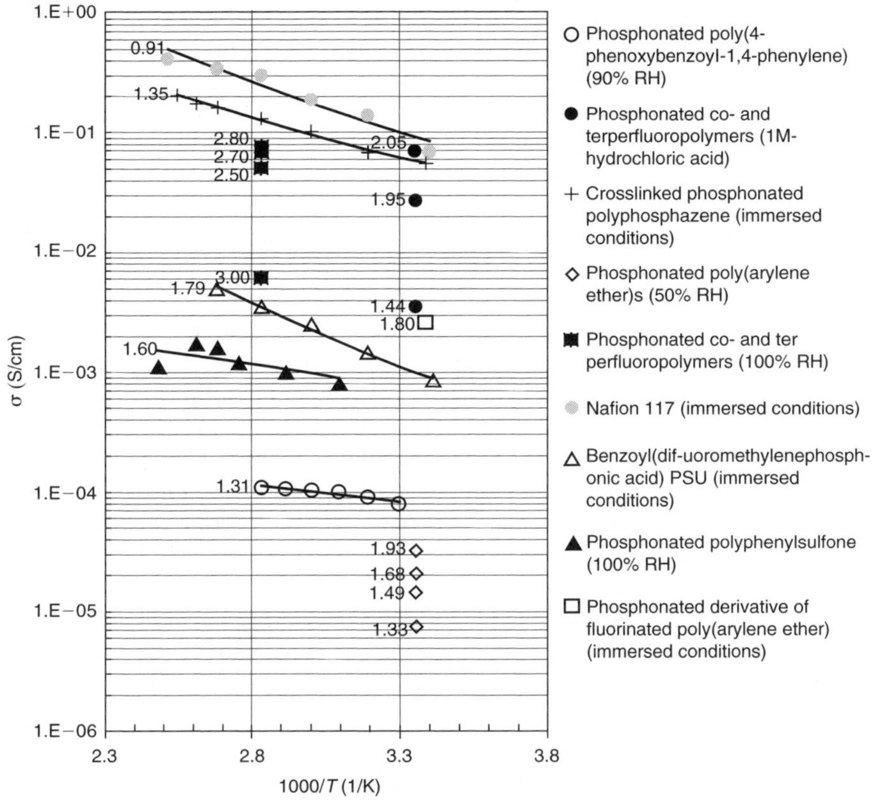

Figure 3.4 Arrhenius conductivity plots of different phosphonated ionomers. Please note that the conductivities have been measured under different experimental conditions, as indicated within parenthesis. The *IEC* (in meq/g) of each ionomer is indicated next to the corresponding set of data.

percolating network through which the protons can be conducted. Conductivity data of different phosphonated ionomers found in the literature are summarized in Figure 3.4. Yamabe *et al.* have found a linear relationship between the *IEC*-value and the proton conductivity [108]. The conductivity data of the perfluorophosphonic acid membranes were measured in 1 M HCl at 25°C using a cell with Pt/Pt electrodes. Conductivities up to 7.0×10^{-2} S/cm for membranes having *IEC*-values up to 2.05 meq/g were measured. These values were close to the conductivity of the Nafion® (*IEC* = 0.91 meq/g) membrane which was equal to 9.0×10^{-2} S/cm, measured under the same conditions. Kotov *et al.* have also reported conductivities of perfluorophosphonic acid membranes analogs to the sulfonated commercial Nafion® membrane [109]. Interestingly, besides having similar conductivities in the range of $7-8 \times 10^{-2}$ S/cm, as measured under saturated water vapor at 80°C, the perfluorophosphonic acid membranes took up less water than the corresponding Nafion® membrane (22 and 30 wt.%, respectively). In this context, it should be mentioned that it is unfortunately often difficult to directly compare the few results reported in the literature because of the differences in the experimental procedures to measure the proton conductivity. Moreover, proton conductivity data are still lacking for a number of ionomers reported as membrane materials for fuel cells [145,170,175,181, 183]. Meng *et al.* have measured the proton conductivity of their phosphonated poly(arylene ether)s at 25°C and 50% relative humidity. They observed a maximum in the conductivity at 0.5 phosphonic acid units per repeating unit of the backbone. A lower conductivity was recorded for ionomers with a higher phosphonic acid content, which was attributed to an increase in the T_g of the ionomer membranes [206,207]. Their considerably lower conductivity, as compared to the perfluorophosphonic acid membranes, is probably a result of the low water content of the former membranes. As mentioned in Section 3.4.2, phosphonated aromatic main-chain polymers require high *IEC*-values in order to take up a sufficient amount of water. For example, P-PPBP membranes having an *IEC*-value of 1.31 had a proton conductivity of only 10^{-4} S/cm at 60°C and 90% relative humidity [185]. Allcock and co-workers have investigated the proton conductivity of water-equilibrated phosphonated polyphosphazene membranes using a four-electrode electrochemical impedance spectroscopy technique. Phosphonated polyphosphazene membranes having an *IEC*-value of 1.43 meq/g were found to have a proton conductivity of 5.4×10^{-2} S/cm at room temperature [216]. Notably, the water uptake could be reduced from 32 to 14 wt.% by crosslinking the polyphosphazene membranes without recording any significant loss in proton conductivity. Moreover, these phosphonated polyphosphazene membranes exhibited proton conductivities five times higher than the corresponding sulfonated polyphosphazene membrane having an *IEC*-value of 1.07 meq/g and a water uptake equal to 38 wt.% [216]. This was attributed to the higher concentration of acidic sites in the phosphonated membranes as compared to the sulfonated membranes. This outlines one of the main differences between the phosphonated and the sulfonated membranes, namely that phosphonated membranes with high values of *IEC* may be employed without extensive water uptake. Zhou *et al.* have reported conductivities of a crosslinked phosphonated polyphosphazene membrane (having an *IEC*-value equal to 1.35 meq/g) that were close to those of Nafion® over the temperature range 20–120°C [223]. Lafitte and Jannasch have showed that membranes based on ionomers having benzoyl(difluoromethylenephosphonic acid) side chains, which are more acidic as compared to arylphosphonic acids, displayed a level of conductivity comparable to that reached by a membrane based on a sulfonated PSU having approximately the same number of acidic units per gram of dry polymer [179]. This indicate that increasing the overall acidity of the phosphonic acid unit improves the water-assisted proton conductivity. Consequently, despite the comparatively low degree of acidity, phosphonic-acid-based membranes can exhibit high proton conductivity coupled with good water uptake characteristics which is a key requirement for their successful use in fuel cells.

3.4.4 Thermal Stability

Much emphasis is currently placed on the development of fuel cell systems capable of sustained operation at temperatures exceeding 100°C [11]. The requirements connected with these conditions place quite severe demands on the membrane materials, especially since the membrane should be able to function, e.g., currently for more than 5000 h for automotive applications. This underlines the need for ionomers with a high thermal, mechanical and chemical stability. The thermal decomposition temperatures of different phosphonated ionomers reported in the literature are collected in Table 3.2 together with the experimental conditions used during the measurements. In general, phosphonated polymers in their acidic form are usually found to lose weight in two separated steps during a thermogravimetrical analysis (see Table 3.2). First, at around 200°C, a small weight loss of only 2–5 wt.% is generally observed. This initial weight loss has been attributed to the loss of water in connection with the formation of anhydrides through the creation of P—O—P linkages [175,224]. These anhydrides may be formed both inter- or intramolecularly. If heat-treated samples are placed in water, these bonds are likely to be hydrolyzed, resulting in the formation of phosphonic acid units (see Scheme 3) [224]. The cleavage of the C—P bond is usually reported to occur at temperatures above 320–350°C, indicating the high stability of this bond [109,170,175,181,185]. Meng *et al.* have reported on an onset of decomposition at around 200°C, while the temperatures at a 5 wt.%-loss was recorded were in the range of 470–500°C [206, 207]. Consequently, the reported onset of decomposition might have been due to anhydride formation, and the irreversible breakdown of the ionomer occurred at a much higher temperature (~480°C). In contrast, Miyatake and Hay reported a loss of phosphonic acid units already at temperatures slightly above 260°C [183]. In addition, Lafitte and Jannasch. have reported the loss of phosphonic acid units above 230°C in benzoyl(difluoromethylenephosphonic acid) PSU [179]. The latter two considerably lower decomposition temperatures of the phosphonic acid units may be explained by their location on electron-rich aromatic rings in the polymers synthesized by Miyatake *et al.*, and the presence of strongly electron-withdrawing —CF_2— groups in those synthesized by Lafitte and Jannasch. As discussed in Chapter 3.2, electron-withdrawing groups substituted on the α-carbon will weaken the C—P bond. In any case, the collected results show a higher thermal stability of phosphonated ionomers in relation to the corresponding sulfonated ionomers [6]. However, one must keep in mind that the thermogravimetrical analyses referred to above were carried out under inert atmospheres. Therefore, further investigations are needed to evaluate the important stability of these novel ionomers under fuel cell conditions. It is worth noting that Meng *et al.* have reported that no changes in the *IEC* were detected after treating the phosphonated poly(arylene ether)s in a 1% aqueous solution of H_2O_2 under reflux for 100 h, indicating the high chemical stability of phosphonated aromatic main-chain polymers toward oxidative attack [206].

3.4.5 Performance of Membranes in Fuel Cells

Schuster *et al.* have investigated the electrochemical behavior of model compounds bearing phosphonic acid, sulfonic acid and imidazole units by cyclic voltammetry at a temperature of 120°C and a potential ranging from −0.05 to +1.5 V [24]. The phosphonated compound showed, in the dry state, similar electrochemical state to the sulfonated compound in the hydrated state. It should be noted that the sulfonated compound suffered from a pronounced electrochemical instability (oxidation and reduction) in the dry state (see Figure 3.5). Notably, the exchange current densities for hydrogen oxidation and oxygen reduction on Pt were similar for both the dry phosphonated compound and the slightly hydrated sulfonated compound. In addition to these model compounds, polymers based on phosphonic acid have been investigated as anhydrous proton conducting polymer electrolytes [225–229]. These results will be further discussed in Chapter 3.5.

As expected, only very few phosphonated ionomer membranes have been studied under fuel cell conditions. Cabasso *et al.* have investigated the use of phosphonated PPO as membranes in H_2/O_2 fuel cells. Unfortunately, difficulties to prepare efficient electrodes using phosphonated PPO/cellulose acetate blends led to serious problems with interfacial resistances accounting for a considerable ohmic loss. The open cell voltage was found to depend largely on the *IEC* of the membranes employed, and the voltage was found to increase progressively with the *IEC*-value. In an attempt to compare with the performance of Nafion® 117, the phosphonated PPO membranes were thermally sandwiched between Nafion®-impregnated electrodes with a Pt loading of 0.5 mg/cm^2. Polarization curves were then recorded for both the Nafion® 117 membrane and the phosphonated PPO membrane at gas pressures of 1 bar (see Figure 3.6a) [163,164]. These data show a reasonable performance of the phosphonated polymer membranes as compared to Nafion® 117 at 25°C. Notably, the performance of the cell employing the phosphonated membrane was found to increase steadily between 100°C and 150°C. However, specific data were not reported due to extensive scattering attributed to problems with the interface between the electrodes and the membranes. Stone *et al.* have evaluated the phosphonated membranes based on trifluorostyrene monomers

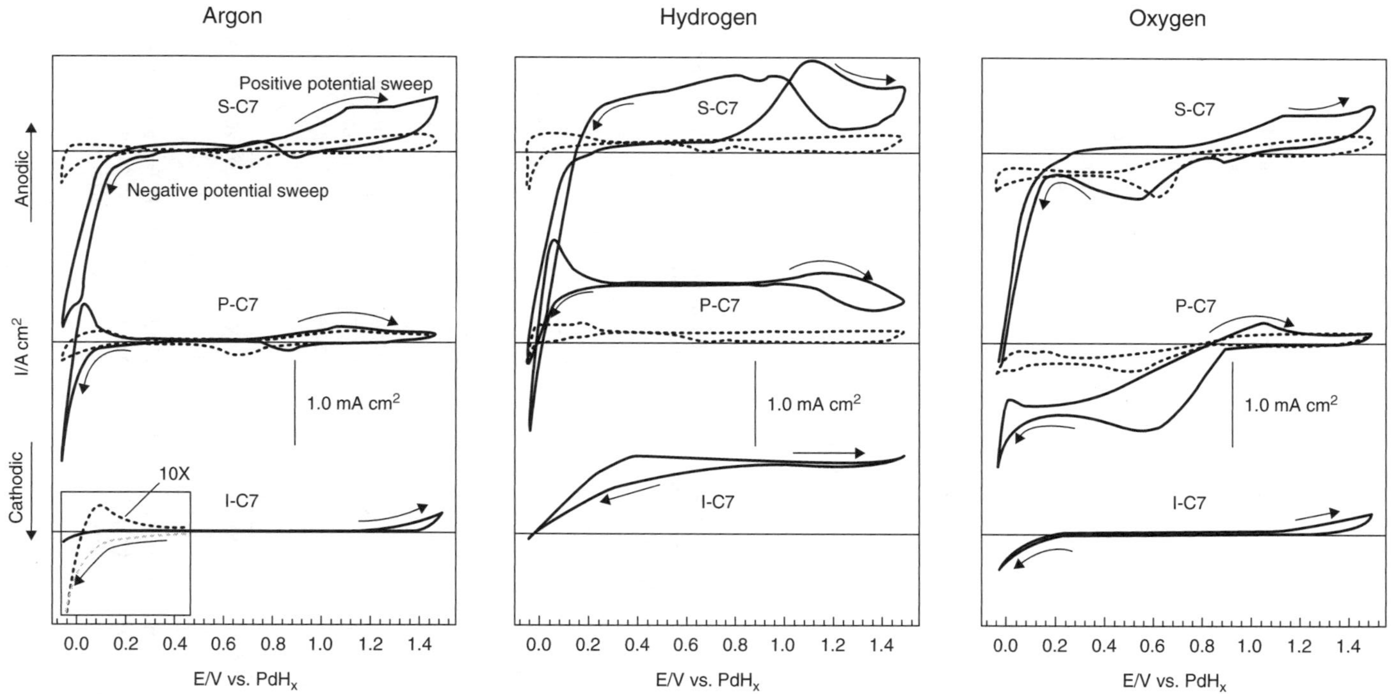

Figure 3.5 Cyclic voltammograms of 2-heptyl-imidazole (I-C7), 1-heptylphosphonic acid (P-C7) and 1-heptylsulfonic acid (S-C7) under argon, hydrogen and oxygen atmospheres at $T = 120°C$ (solid lines) and at $T = 30°C$ for 0.5 M P-C7 and S-C7 aqueous solutions (dotted lines) [24].

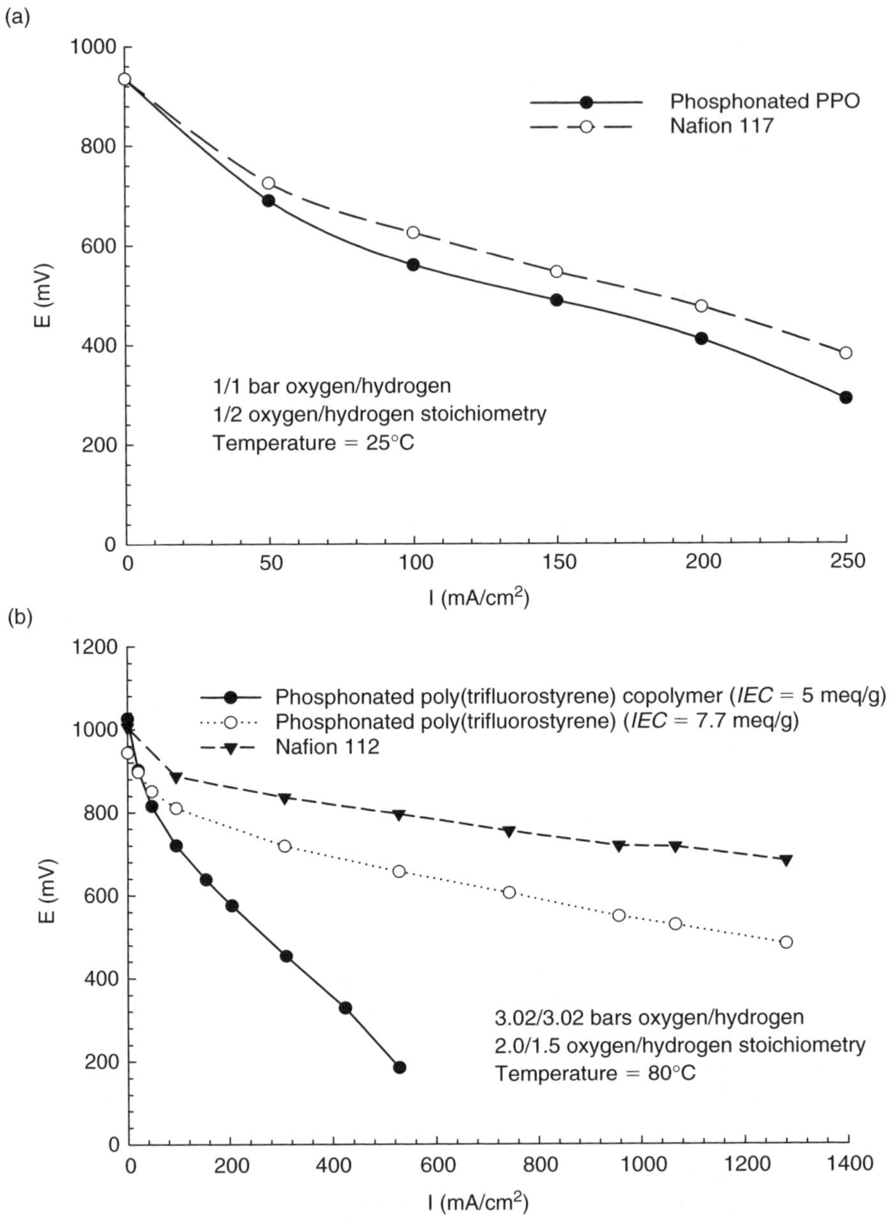

Figure 3.6 Polarization curves of phosphonated PPO [163,164] (a) and phosphonated poly(trifluorostyrene) [201] (b). The polarization curves of Nafion® 117 and 112 are also plotted as a basis for comparison. Please note the different scales in the plots.

using a Pt loading of 1.0 mg/cm^2 and gas pressures of 3 bars in a Ballard MK4B single cell. These data, together with those obtained with Nafion® 112, are shown in Figure 3.6b. The cell containing the supported membrane prepared using phosphonated poly(trifluorostyrene) having an *IEC*-value equal to 7.7 meq/g showed a reasonable performance, only slightly lower than that of the Nafion® 112 cell at 80°C. However, a severe loss in the fuel cell performance was encountered when employing a copolymer having an *IEC*-value equal to 5 meq/g in the preparation of a supported membrane. No efforts were made to evaluate the performance of the ionomers at temperatures above 100°C. As expected, it is difficult to directly compare the fuel cell performance of the materials reported in the literature because of the differences in the experimental setup and the conditions used.

Low-molecular-weight perfluoro-ethylene-1,2-bis-phosphonic acid has been studied as an alternative electrolyte to concentrated phosphoric acid in acid fuel cells [230]. Although slightly outside the primary scope of the present review, results from such an investigation may bring useful information for the understanding of the general characteristics of phosphonated polymers in fuel cells. By using bis(phosphonic acid) the kinetics for oxygen reduction on Pt was enhanced since the overpotential for O_2 reduction on the microfuel cell cathode was smaller in 85% bis-phosphonic acid than in 85% phosphoric acid for current densities up to 200 mA/cm^2 at 100°C. This alternative electrolyte was successfully tested under realistic fuel cell conditions for several hundred hours in the temperature range 150–200°C without deterioration of the performance. Eventually, however, the cathode performance reverted to that typically found with phosphoric acid as electrolyte, indicating the gradual hydrolysis of phosphonic acid affording phosphoric acid under these conditions [230].

3.5 Phosphonated Polymers for Proton Conduction Under Low-Humidity Conditions

3.5.1 Proton Conducting Anhydrous Systems Based on Phosphonic Acid–Base Complexes

In general, anhydrous proton conducting polymer materials conduct via the Grotthuss mechanism (structure diffusion), i.e., the protons are transferred between the charge-carrying solvent groups through the making and breaking of hydrogen bonds in a complex manner [231]. This is in contrast to the vehicle mechanism where the protons are typically transported by charge carriers which are diffusing freely in a liquid. The most obvious way to achieve membrane materials able to conduct protons above 100°C under anhydrous conditions is to replace the water with another proton solvent which is considerably less volatile. Perhaps the most prominent example in this respect are the proton conducting membranes based on strong oxyacids, e.g., phosphoric acid and sulfuric acid, complexed by basic polymers such as polybenzimidazole [11,23]. These membranes are currently under development by several research groups [11,61]. Moreover, weakly basic nitrogen-containing heterocycles such as imidazole, benzimidazole and pyrazole show an amphoteric behavior with an ability to form intermolecular hydrogen bonds, and show proton conduction mechanisms similar to that in water [18]. However, the potential risk of depleting the conducting phase still exists in these materials which still contain low-molecular-weight species.

A handful of studies on anhydrous acid–base complexes involving low-molecular-weight phosphonic acids, as well as phosphonated polymers, have been reported in the literature. In all of these studies, IR spectroscopy has been used to confirm the acid–base complex formation by proton transfer from the phosphonic acid to the basic component, as exemplified in Scheme 25. Yamada and Honma have studied diphosphonic acid complexes with chitosan, a biodegradable polymer [232]. In this case, the diacid partly protonated the amino groups of the chitosan to form acid–base complexes. The conductivity, claimed to be the result of structure diffusion, was measured to be 5×10^{-3} S/cm at 150°C for an electrolyte containing 200 wt% of the diacid. Sevil and Bozkurt have investigated electrolytes based on poly (vinylphosphonic acid) containing 0.5, 1 and 2 mol imidazole per acid unit [229]. The T_g increased with the fraction of imidazole concentration at low imidazole contents, most probably an effect of a decreased chain mobility resulting from the ionic interactions of the complexes. However, at 2 mol imidazole per acid unit, the T_g decreased to −36°C due to plasticization by the excess imidazole. This may be compared to the T_g of −23°C measured for the neat poly(vinylphosphonic acid). The conductivity of thin sample films increased with increasing fractions of imidazole to reach a maximum value of 5×10^{-3} S/cm at 130°C. Unfortunately, the complexes were found by thermogravimetry to decrease in weight already at about 150°C. In a similar study, Yamada and Honma have investigated poly(vinylphosphonic acid) containing various concentrations of imidazole, pyrazole and 1-methylimidazole, respectively [225]. The authors reported that the conductivity measured at 150°C initially increased with increasing concentrations of imidazole, to reach a maximum conductivity of 7×10^{-3} S/cm at 89 mol% imidazole. Increasing concentrations above this value gave decreasing conductivities. Similar maxima in the conductivity were noted for the pyrazole (8×10^{-4} S/cm at 93 mol.%) and 1-methylimidazole (1×10^{-3} S/cm at 91 mol%) complexes. Notably, a fuel cell evaluation of the electrolyte containing 89 mol.% imidazole was performed using dry H_2/O_2 feed gases at 80°C. Employing a MEA based on E-TEK Pt/C electrodes, the experiment showed a power density of approximately 10 mW/cm^2 at a current density of 35 mA/cm^2 and a cell voltage of 0.3 V. Unfortunately, no information

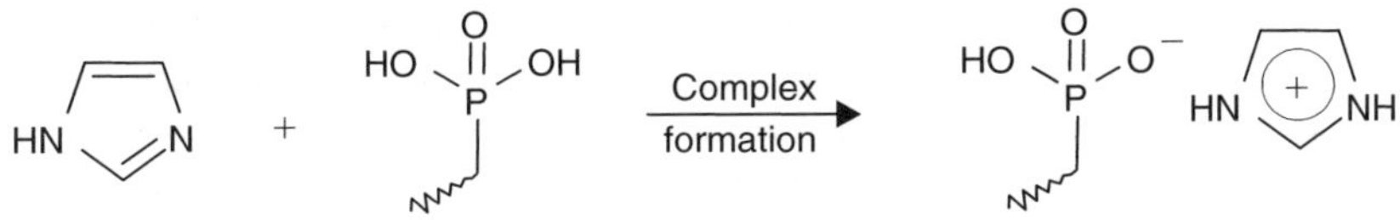

Scheme 25 Illustration of the acid–base complex formation between phosphonic acid and imidazole via proton transfer and the formation of the imidazolium ion.

regarding the duration of the experiment was reported. Bozkurt *et al.* have also copolymerized vinylphosphonic acid and 4-vinylimidazole in order to immobilize both the acidic and basic units in fully polymeric structures [228]. The statistical copolymers rich in the former monomer were found to be insoluble in water. This observation was ascribed to interpolymeric acid–base complexation. Thermogravimetry of the amorphous copolymers under both N_2 and air showed a slight weight loss already at 100–220°C. The conductivity was found to vary in the range 10^{-12}–10^{-6} S/cm at temperatures between 0°C and 160°C, and increased gradually with increasing fraction of the vinylimidazole monomer in the copolymers. The latter finding may be a consequence of an increasing fraction of "free" proton accepting nitrogen sites, which are necessary for efficient proton conduction. The low conductivity of the copolymers may be ascribed to the low segmental mobility of the polymer chains due to the dense physical (ionic) crosslinking. Notably, no T_g was detected below the onset of degradation close to 200°C. Copolymers based on vinylphosphonic acid and vinylpyrrolidone have been reported to show similar levels of conductivities [227].

All the electrolytes containing low-molecular-weight species described in this section reached quite high conductivities in the range of 1×10^{-3} to 5×10^{-3} S/cm at 130–150°C. This is practically in the same range as the conductivity reported for neat imidazole at 130°C, 5×10^{-3} S/cm [7]. However, even though the boiling point of imidazole is 255°C, it shows a significant evaporation rate at temperatures above 100°C. This, in combination with the expected low inherent chemically stability of the studied polymers, raises questions regarding the use of the complexes as fuel cell membrane materials under sustained operation at elevated temperatures.

3.5.2 All-Polymeric Systems for Proton Conduction Under Anhydrous Conditions

Kreuer has pointed out the attractive possibility to develop fully polymeric proton conducting membranes based on immobilized proton solvents [18]. This concept prevents the loss of the proton solvents through evaporation or leakage, and also implies that the proton transport has to take place only via structure diffusion. The immobilized proton solvents should ideally have pronounced and balanced proton donor and acceptor properties to favor a high degree of self-dissociation, giving rise to a high concentration of intrinsic charge carriers. The proton solvent should also possess a high dielectric constant to promote separation of the charge carriers in the intermolecularly hydrogen bonded networks. Furthermore, convenient synthetic pathways must exist so that the proton solvent molecules can be immobilized in the polymer structure in such a way that the proton donor and acceptor groups remain chemically equivalent, and so that they easily can restructure in connection with the proton transport process. The immobilized proton solvent should not become hydrolyzed or form condensates, and should finally also resist reduction and oxidation under the fuel cell operating conditions. A number of fully polymeric proton conductors consisting of imidazole [25] and benzimidazole [19,25] tethered to different backbones have recently been prepared and evaluated. Although the proton conductivities are encouraging, the low intrinsic charge carrier concentration in the phases formed by these heterocycles (resulting from a low degree of self-dissociation) give rise to only moderate values, typically below $10^{-4}-10^{-3}$ S/cm.

In a quite recent paper, Kreuer and co-workers have reported on a critical study of imidazole, sulfonic acid and phosphonic acid under dry and low-humidity conditions [24]. Based on proton conductivity data and self-diffusion coefficients of low-molecular-weight heptyl derivatives of the respective protogenic groups, the authors concluded that the phosphonated compound showed the highest conductivity under anhydrous ("quasi water-free") conditions (e.g. 10^{-2} S/cm at 200°C, seen in Figure 3.7), but the lowest molecular diffusion. This finding strongly indicated structure diffusion to be the dominant proton transport mechanism. Moreover, it was consistent with the presence of a high number of intrinsic charge carriers, which is expected from the distinct amphoteric nature of the phosphonic acid group.

The degree of structure diffusion was found to be much lower for the sulfonic acid and the imidazole compounds. Also, only the conductivity of the phosphonic acid compound could be measured reversibly up to nearly 200°C, indicating a high thermal stability. Cyclic voltammetry experiments revealed that the imidazole compound had the largest electrochemical stability window. However, the study also showed that the phosphonated compound, both in the dry and hydrated state, showed a stability similar to that of the hydrated sulfonated compound. Importantly, the phosphonated derivative was found to retain some water up to 180°C, and no condensation of the acid groups was detected under these conditions. In conclusion, the phosphonic acid compound showed a high proton conductivity with a high degree of structure diffusion, in combination with a high level of thermal and electrochemical stability.

An investigation of the mass and charge transport in molten anhydrous phosphonic acid (phosphorous acid, H_3PO_3) has shown predominantly

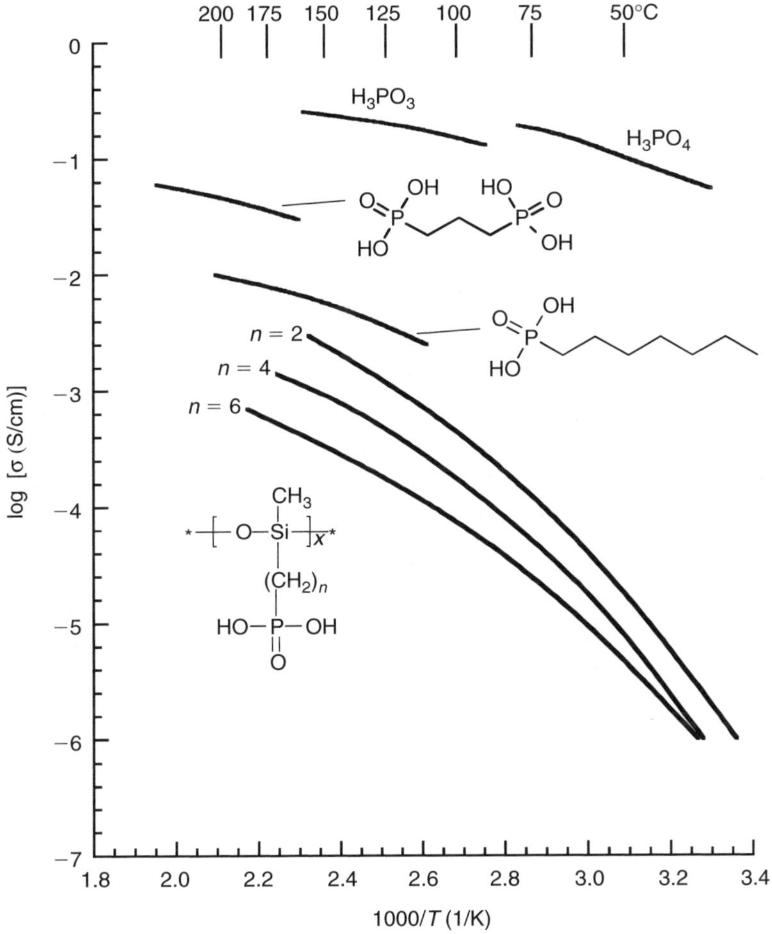

Figure 3.7 Comparative plot containing conductivity data of different phosphonic acid derivatives measured under anhydrous condition [24,226,233]. Also shown is data of phosphoric acid [233].

structure diffusion of protons, and a conductivity of just above 0.10 S/cm at 90°C was measured (see Figure 3.7) [233]. This value may be compared with the conductivity of phosphoric acid (H_3PO_4) which was found to reach approximately 0.11 S/cm under the same conditions. Moreover, the data of the phosphonic acid were consistent with a system having a pronounced amphoteric feature with a high degree of self-dissociation to facilitate a high concentration of protonic charge carriers and a high carrier mobility, quite similar to the characteristics found for the more commonly studied phosphoric acid [231].

3.5.3 Phosphonated Polymers with Intrinsic Proton Conduction Under Low-Humidity Conditions

The comparative study by Kreuer and co-workers suggested that phosphonic acid tethered to polymers may offer a possibility to reach high-proton conductivities in a wide temperature range without the presence of a separate water phase [24]. At low temperatures, the conductivity may be promoted by absorbed water, while at high temperatures, the conductivity may occur within hydrogen bonded networks through structure diffusion. Just very recently, Kreuer and co-workers have reported on the properties of fully immobilized phosphonic acid tethered to polysiloxanes in high concentrations via flexible alkyl spacers [226]. Conductivities of up to 1×10^{-3} S/cm were measured at 130°C and a relative humidity of 37% (see Figure 3.7). This was only slightly higher than the conductivity measured for similar imidazole tethered polymers, despite the higher conductivity of neat phosphonic acid compared to neat imidazole. Thus, the conductivity of the phosphonic acid suffered more from the immobilization than the imidazole. In addition, the conductivity of the immobilized phosphonic acid was found to decrease with increasing length of the spacer unit. This important finding was the opposite of the situation found in the study of immobilized imidazole described above [226]. The observation implies that the hydrogen bonded networks formed by phosphonic acid is more

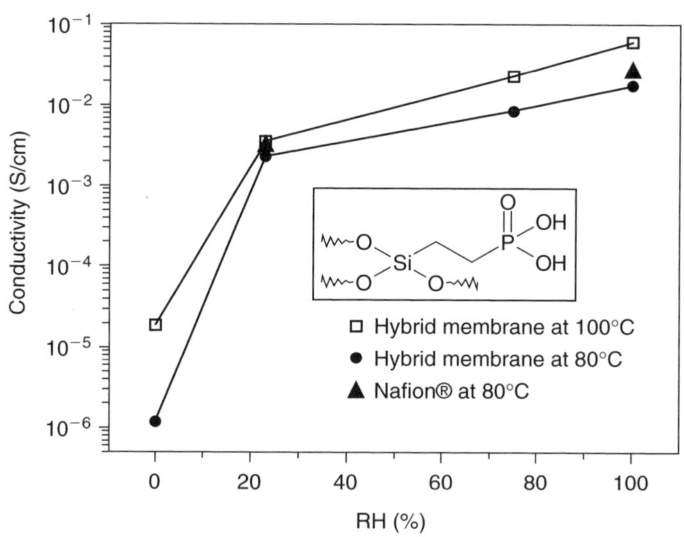

Figure 3.8 The variation of the proton conductivity with the humidity of hybrid inorganic–organic polymer membranes containing phosphonic acid [234]. The inset shows the structural unit carrying the acid units. Also shown is data of Nafion® at 20% and 100% RH [234].

sensitive toward the changes imposed by the immobilization on the chemical and structural environments and the local acid concentration. One possible reason behind this observation may be that larger hydrogen bonded domains, containing a larger number of phosphonic acid units, are required in comparison to the immobilized imidazole. The study of the fully immobilized phosphonic acid provided valuable insights on the structure properties of phosphonated polymers, but also showed that considerable challenges remain. Just very recently, Li *et al.* have reported on the properties of hybrid inorganic–organic polymer membranes containing phosphonic acid prepared by using a sol–gel process [234]. As seen in Figure 3.8, the membranes reached conductivities comparable or even higher than that of Nafion® at 80–100°C and a relative humidity between 20% and 100%. After drying the membranes and measuring the conductivity under dry argon, the value decreased to 2×10^{-5} S/cm at 100°C.

3.6 Summary and Outlook

The published data on fuel cell membranes based on phosphonated polymers is still very limited. However, a careful examination of the results obtained up to now indicate that these membranes may potentially show some advantages over sulfonated ones. This is perhaps especially true when it comes to operation under low-humidity conditions at high temperatures. At high water contents, the protons may be transported through the dynamics of the water, much in the same way as in conventional membranes based on sulfonated membranes. In addition, because of the amphoteric nature of the phosphonic group, there seems to be a possibility to transport protons through structure diffusion within the hydrogen bonded phosphonic acid networks at low water contents. Moreover, phosphonated polymers generally show a higher hydrolytic and thermal stability due to the strength of the C—P bond, which is especially critical under high-temperature operation.

It is clear that the molecular architecture of the phosphonated polymers requires a careful design in order to reach these advantageous properties. For example, the concentration of phosphonic acid groups in the membranes needs to be sufficiently high to enable the formation of percolating hydrogen bonded networks for proton transport under low-humidity conditions, or at least high enough to form a percolating network of water-filled channels when the membranes are kept under high-humidity conditions. A special challenge arises when it comes to the immobilization of phosphonic acid units in polymeric structures without inhibiting the conduction process occurring in "free" phosphonic acid. The very limited results reported up until now indicate that domains with very high local concentrations of phosphonic acid units are necessary, and that the conductivity suffers greatly if this concentration is decreased. However, the conditions necessary for efficient proton transfer may be achieved by a careful design and optimization of the way the phosphonic acid units are attached to the polymer. Concerning this issue, the synthetic polymer chemists may gain important information from molecular simulations and theoretical considerations, perhaps especially regarding the character of hydrogen bonded networks of phosphonic acids. The water uptake of phosphonated polymers is generally much more moderate than for corresponding sulfonated polymers, which swell excessively after reaching a critical temperature or a critical degree of sulfonation. In this respect, most of the polymers studied up until now have an insufficient degree of phosphonation for use as fuel cell membranes. An alternative way to increase the water uptake may be to enhance the acidity by attaching the phosphonic acid via C_F—P bonds. This will obviously change the acid–base character in relation to phosphonic acid units attached through C_{Ar}—P or C_{Al}—P bonds.

A survey of the literature indicates that phosphonation of polymers in general is more difficult than the corresponding sulfonation. The difference is perhaps most striking when it comes to attaching acid groups on aromatic rings. However, a number of pathways to phosphonated polymers exist, either through modification of polymers or by copolymerization using phosphonated monomers. The preparation of phosphonated polymers through the latter method seems to be attractive since high yields and molecular weights have been reported. This method can also be used to prepare block copolymers where densely phosphonated blocks are mechanically stabilized through phase separation from non-phosphonated blocks. Also, given that densely phosphonated polymers are required, it may prove fruitful to investigate the conductivity and behavior of polymers functionalized with diphosphonic acid units. Moreover, phosphates are components of many inorganic compounds, some of which are proton conducting. This opens up possibility to prepare tailored proton conducting inorganic–organic hybrid materials where phosphonated polymers are covalently attached to the inorganic phase in a nanocomposite.

In order to critically assess the possibility of using phosphonated membranes in fuel cells, it is of course of paramount importance to evaluate them *in situ* to study their performance under different operational conditions and to investigate their degradation and lifetime under realistic situations. Unfortunately, there are only a handful of fuel cell studies of phosphonated membranes reported in the literature. The limited data indicates some advantages, e.g., that the MeOH permeability is lower than for corresponding sulfonated membranes. Finally, the development of efficient and durable membrane-electrode assemblies (MEA) will be critical to the development of functional fuel cell components. To our knowledge, the preparation of an MEA using phosphonated polymers has not yet been reported. In summary, much work is still needed to increase the knowledge and the amount of data on phosphonated membrane materials for fuel cells. These novel membranes show a number of distinctly attractive and quite unique properties that may meet the demands from the commercial fuel cell developers, and thus accelerate a broad introduction of fuel cell technology.

ACKNOWLEDGMENTS

This work has been financially supported in part by the Swedish Foundation for Strategic Environmental Research (Mistra).

REFERENCES

1. N. P. Brandon, S. Skinner, and B. C. H. Steele, *Annu. Rev. Mater. Res.*, 33 (2003) 183.
2. J. Roziere and D. J. Jones, *Annu. Rev. Mater. Res.*, 33 (2003) 503.
3. O. Savadogo, *J. New Mater. Elect. Syst.*, 1 (1998) 47.
4. J. A. Kerres, *J. Membr. Sci.*, 185 (2001) 3.
5. M. Rikukawa and K. Sanui, *Prog. Polym. Sci.*, 25 (2000) 1463.
6. A. L. Rusanov, D. Y. Likhatchev, and K. Mullen, *Russ. Chem. Rev.*, 71 (2002) 761.
7. K. D. Kreuer, *J. Membr. Sci.*, 185 (2001) 29.
8. B. C. H. Steele and A. Heinzel, *Nature*, 414 (2001) 345.
9. C. Wieser, *Fuel Cells*, 4 (2004) 245.
10. G. Alberti and M. Casciola, *Annu. Rev. Mater. Res.*, 33 (2003) 129.
11. Q. F. Li, R. H. He, J. O. Jensen, and N. J. Bjerrum, *Chem. Mater.*, 15 (2003) 4896.
12. R. Souzy and B. Ameduri, *Prog. Polym. Sci.*, 30 (2005) 644.
13. K. A. Mauritz and R. B. Moore, *Chem. Rev.*, 104 (2004) 4535.
14. M. A. Hickner, H. Ghassemi, Y. S. Kim, B. R. Einsla, and J. E. McGrath, *Chem. Rev.*, 104 (2004) 4587.
15. W. L. Harrison, M. A. Hickner, Y. S. Kim, and J. E. McGrath, *Fuel Cells*, 5 (2005) 201.
16. J. A. Kerres, *Fuel Cells*, 5 (2005) 230.
17. P. Jannasch, *Fuel Cells*, 5 (2005) 248.
18. K. D. Kreuer, A. Fuchs, M. Ise, M. Spaeth, and J. Maier, *Electrochim. Acta*, 43 (1998) 1281.
19. J. C. Persson and P. Jannasch, *Macromolecules*, 38 (2005) 3283.
20. J. C. Persson and P. Jannasch, *Chem. Mater.*, 15 (2003) 3044.
21. M. F. H. Schuster, W. H. Meyer, M. Schuster, and K. D. Kreuer, *Chem. Mater.* 16 (2004) 329.
22. M. Schuster, W. H. Meyer, G. Wegner, H. G. Herz, M. Ise, M. Schuster, K. D. Kreuer, and J. Maier, *Solid State Ionics*, 145 (2001) 85.
23. M. E. Schuster and W. H. Meyer, *Annu. Rev. Mater. Res.*, 33 (2003) 233.
24. M. Schuster, T. Rager, A. Noda, K. D. Kreuer, and J. Maier, *Fuel Cells*, 5 (2005) 355.
25. H. G. Herz, K. D. Kreuer, J. Maier, G. Scharfenberger, M. F. Schuster, and W. H. Meyer, *Electrochim. Acta*, 48 (2003) 2165.
26. R. Engel, C. Iolani, and JaimeLee, *Synthesis of Carbon-Phosphorus Bonds*, 2nd ed, CRC Press LLC, Boca Raton, Fla., 2004.
27. J. M. Huang and R. Y. Chen, *Heteroat. Chem.*, 11 (2000) 480.
28. G. E. Boyd, F. Vaslow, and Lindenba.S, *J. Phys. Chem.*, 71 (1967) 2214.
29. S. D. Alexandratos, D. R. Quillen, and M. E. Bates, *Macromolecules*, 20 (1987) 1191.
30. E. N. Walsh, T. M. Beck, and A. D. F. Toy, *J. Am. Chem. Soc.*, 78 (1956) 4455.
31. J. I. Bregman and Y. Murata, *J. Am. Chem. Soc.*, 74 (1952) 1867.
32. I. Cabasso, J. Smid, and S. K. Sahni, *J. Appl. Polym. Sci.*, 41 (1990) 3025.
33. K. Riedelsberger, W. Jaeger, and A. Friedrich, *Des. Monomers Polym.*, 3 (2000) 35.

34. E. P. Horwitz, R. Chiarizia, H. Diamond, R. C. Gatrone, S. D. Alexandratos, A. Q. Trochimczuk, and D. W. Crick, *Solvent Extr. Ion Exch.*, 11 (1993) 943.
35. R. Chiarizia, E. P. Horwitz, R. C. Gatrone, S. D. Alexandratos, A. Q. Trochimczuk, and D. W. Crick, *Solvent Extr. Ion Exch.*, 11 (1993) 967.
36. A. W. Trochimczuk and S. D. Alexandratos, *J. Appl. Polym. Sci.*, 52 (1994) 1273.
37. M. Gaboyard, J. J. Robin, Y. Hervaud, and B. Boutevin, *J. Appl. Polym. Sci.*, 86 (2002) 2011.
38. N. Moszner, F. Zeuner, U. K. Fischer, and V. Rheinberger, *Macromol. Chem. Phys.*, 200 (1999) 1062.
39. N. Moszner, F. Zeuner, S. Pfeiffer, I. Schurte, V. Rheinberger, and M. Drache, *Macromol. Mater. Eng.*, 286 (2001) 225.
40. N. Sibold, P. J. Madec, S. Masson, and T. N. Pham, *Polymer*, 43 (2002) 7257.
41. L. Y. Mou, G. Singh, and J. W. Nicholson, *Chem. Commun.* (2000) 345.
42. C. Brondino, B. Boutevin, J. P. Parisi, and J. Schrynemackers, *J. Appl. Polym. Sci.*, 72 (1999) 611.
43. C. Bressy-Brondino, B. Boutevin, Y. Hervaud, and M. Gaboyard, *J. Appl. Polym. Sci.*, 83 (2002) 2277.
44. J. L. Fang, Y. Li, X. R. Ye, Z. W. Wang, and Q. Liu, *Corrosion*, 49 (1993) 266.
45. M. Duprat, F. Moran, and F. Dabosi, *Corros. Sci.*, 23 (1983) 1047.
46. M. J. Sundell, E. O. Pajunen, O. E. O. Hormi, and J. H. Nasman, *Chem. Mater.*, 5 (1993) 372.
47. C. I. Lindsay, S. B. Hill, M. Hearn, G. Manton, N. Everall, A. Bunn, J. Heron, and I. Fletcher, *Polym. Int.*, 49 (2000) 1183.
48. A. F. Grand and C. A. Wilkie, Fire Retardancy of Polymeric Materials, Marcel Dekker Inc., New York, 2000.
49. M. Spirckel, N. Regnier, B. Mortaigne, B. Youssef, and C. Bunel, *Polym. Degrad. Stab.*, 78 (2002) 211.
50. M. Cochez, M. Ferriol, J. V. Weber, P. Chaudron, N. Oget, and J. L. Mieloszynski, *Polym. Degrad. Stab.*, 70 (2000) 455.
51. L. D. Freedman and G. O. Doak, *Chem. Rev.*, 57 (1957) 479.
52. F. W. Bennett, H. J. Emeleus, and R. N. Haszeldine, *J. Chem. Soc.*, Abstracts (1954) 3896.
53. W. D. Kumler and J. J. Eiler, *J. Am. Chem. Soc.*, 65 (1943) 2355.
54. H. H. Jaffe, L. D. Freedman, and G. O. Doak, *J. Am. Chem. Soc.*, 76 (1954) 1548.
55. G. M. Kosolapoff, *J. Am. Chem. Soc.*, 74 (1952) 3427.
56. G. M. Kosolapoff and J. S. Powell, *J. Chem. Soc.* (1950) 3535.
57. E. C. Ashby and G. M. Kosolapoff, *J. Am. Chem. Soc.*, 75 (1953) 4903.
58. N. Agmon, *Abstr. Pap. Am. Chem. Soc.*, 210 (1995) 328.
59. C. Iojoiu, M. Maréchal, F. Chabert, and J.-Y. Sanchez, *Fuel Cells*, 5 (2005) 344.
60. M. Rikukawa, D. Inagaki, K. Kaneko, Y. Takeoka, I. Ito, Y. Kanzaki, and K. Sanui, *J. Mol. Struct.*, 739 (2005) 153.
61. J. A. Asensio and P. Gómez-Romero, *Fuel Cells*, 5 (2005) 336.
62. J.-P. Majoral and Editor, *New Aspects in Phosphorus Chemistry, I* (In: *Top. Curr. Chem.*, 2002; 220), Springer GmbH, Berlin, 2002.
63. J.-P. Majoral and Editor, *New Aspects in Phosphorus Chemistry V* (In: *Top. Curr. Chem.*; 2005, 250), Springer GmbH, Berlin, 2005.
64. J.-P. Majoral and Editor, *New Aspects in Phosphorus Chemistry IV* (In: *Top. Curr. Chem.*, 2004; 232), Springer GmbH, Berlin, 2004.
65. J.-P. Majoral and Editor, *New Aspects in Phosphorus Chemistry II* (In: *Top. Curr. Chem.*, 2003; 223), Springer GmbH, Berlin, 2003.
66. J.-P. Majoral and Editor, *New Aspects in Phosphorus Chemistry III* (In: *Top. Curr. Chem.*, 2003; 229), Springer GmbH, Berlin, 2003.
67. P. Savignac and B. Iorga, *Modern Phosphonate Chemistry*, CRC Press LLC, Boca Raton, Fla., 2003.

68. V. D. Romanenko and V. P. Kukhar, *Chem. Rev.*, 106 (2006) 3868.
69. A. K. Bhattacharya and G. Thyagarajan, *Chem. Rev.*, 81 (1981) 415.
70. A. E. Arbuzov, *J. Russ. Phys. Chem. Soc.*, 38 (1906) 687.
71. A. Michaelis and R. Kaehne, *Chem. Ber.*, 31 (1898) 1048.
72. K. H. Worms and M. Schmidt-Dunker (Eds.) 1976; *Org. Phosphorus Compd.*, 7 (1976) 1.
73. N. Pelaprat, C. Brondino, R. Rigal, and B. Boutevin, *Eur. Polym. J.*, 32 (1996) 761.
74. A. Michaelis and T. Becker, *Chem. Ber.*, 30 (1897) 1003.
75. A. Burger and N. D. Dawson, *J. Org. Chem.*, 16 (1951) 1250.
76. R. C. Larock, *Tetrahedron*, 38 (1982) 1713.
77. M. S. Kharasch, E. V. Jensen, and S. Weinhouse, *J. Org. Chem.*, 14 (1949) 429.
78. F. Eymery, B. Iorga, and P. Savignac, *Tetrahedron*, 55 (1999) 13109.
79. D. Craig, C. J. Etheridge, and A. M. Smith, *Tetrahedron*, 52 (1996) 15267.
80. M. P. Teulade, P. Savignac, E. E. Aboujaoude, S. Lietge, and N. Collignon, *J. Org. Chem.*, 304 (1986) 283.
81. A. N. Pudovik and I. V. Konovalova, *Synthesis* (1979) 81.
82. D. Semenzin, G. EtemadMoghadam, D. Albouy, O. Diallo, and M. Koenig, *J. Org. Chem.*, 62 (1997) 2414.
83. S. D. Alexandratos, M. A. Strand, D. R. Quillen, and A. J. Walder, *Macromolecules*, 18 (1985) 829.
84. F. Popescu, *Rev. Roum. Chim.*, 14 (1969) 1525.
85. J. A. Miles, M. T. Benny, and K. W. Ratts, *J. Org. Chem.*, 40 (1975) 343.
86. H. Ohmori, S. Nakai, and M. Masui, *J. Chem. Soc., Perkin Trans. 1: Org. Bio-Org. Chem.* (1979) 2023.
87. R. Obrycki, and C. E. Griffin, *J. Org. Chem.*, 33 (1968) 632.
88. J. B. Plumb, R. Obrycki, and C. E. Griffin, *J. Org. Chem.*, 31 (1966) 2455.
89. T. M. Balthazor and R. C. Grabiak, *J. Org. Chem.*, 45 (1980) 5425.
90. P. Tavs, *Chem. Ber.-Rec.*, 103 (1970) 2428.
91. G. Sturtz, B. Damin, and J. C. Clement, *J. Chem. Res.*, Synopses (1978) 89.
92. A. L. Schwan, *Chem. Soc. Rev.*, 33 (2004) 218.
93. T. Hirao, T. Masunaga, N. Yamada, Y. Ohshiro, and T. Agawa, *B. Chem. Soc. Japan*, 55 (1982) 909.
94. N. D. Dawson and A. Burger, *J. Org. Chem.*, 18 (1953) 207.
95. R. C. Cambie and B. D. Palmer, *Aust. J. Chem.*, 35 (1982) 827.
96. L. S. Melvin, *Tetrahedron Lett.*, 22 (1981) 3375.
97. C. M. Taylor and A. J. Watson, *Curr. Org. Chem.*, 8 (2004) 623.
98. G. W. Kenner and N. R. Williams, *J. Chem. Soc.*, (1955) 522.
99. D. A. Casteel and S. P. Peri, *Synthesis* (1991) 691.
100. G. A. Consiglio, S. Failla, P. Finocchiaro, and V. Siracusa, *Phosphorus, Sulfur Silicon Relat. Elem.*, 134 (1998) 413.
101. D. J. Burton and R. M. Flynn, *Synthesis* (1979) 615.
102. D. J. Burton and R. M. Flynn, *J. Fluorine Chem.*, 10 (1977) 329.
103. M. Kato and M. Yamabe, *J. Chem. Soc. Chem. Commun.* (1981) 1173.
104. H. K. Nair and D. J. Burton, *J. Am. Chem. Soc.*, 119 (1997) 9137.
105. D. Su, W. Cen, R. L. Kirchmeier, and J. M. Shreeve, *Can. J. Chem.*, 67 (1989) 1795.
106. D. J. Burton, A. S. Modak, R. Guneratne, D. Su, W. Cen, R. L. Kirchmeier, and J. N. M. Shreeve, *J. Am. Chem. Soc.*, 111 (1989) 1773.
107. S. Z. Zhu, G. F. Jin, and D. D. DesMarteau, *Chin. J. Chem.*, 20 (2002) 1268.
108. M. Yamabe, K. Akiyama, Y. Akatsuka, and M. *Kato, Eur. Polym. J.*, 36 (2000) 1035.
109. S. V. Kotov, S. D. Pedersen, W. M. Qiu, Z. M. Qiu, and D. J. Burton, *J. Fluorine Chem.*, 82 (1997) 13.
110. M. Obayashi, E. Ito, K. Matsui, and K. Kondo, *Tetrahedron Lett.*, 23 (1982) 2323.
111. M. Gaboyard, Y. Hervaud, and B. Boutevin, *J. Fluorine Chem.*, 107 (2001) 5.

112. C. Brondino, B. Boutevin, Y. Hervaud, N. Pelaprat, and A. Manseri, *J. Fluorine Chem.*, 76 (1996) 193.
113. G. M. Blackburn and D. E. Kent, *J. Chem. Soc., Perkin Trans. 1: Org. Bio-Org. Chem.*, (1986) 913.
114. G. M. Blackburn and D. E. Kent, *J. Chem. Soc. Chem. Commun.* (1981) 511.
115. C. R. Hall, T. D. Inch, and N. E. Williams, *J. Chem. Soc., Perkin Trans. 1: Org. Bio-Org. Chem.* (1985) 233.
116. B. Ye, and T. R. Burke, *Tetrahedron*, 52 (1996) 9963.
117. M. S. Smyth, H. Ford, and T. R. Burke, *Tetrahedron Lett.*, 33 (1992) 4137.
118. C. E. McKenna and B. A. Kashemirov, *Top. Curr. Chem.*, 220 (2002) 201.
119. T. Yokomatsu, T. Murano, K. Suemune, and S. Shibuya, *Tetrahedron*, 53 (1997) 815.
120. S. D. Taylor, C. C. Kotoris, A. N. Dinaut, and M. J. Chen, *Tetrahedron*, 54 (1998) 1691.
121. R. Rabinowitz, *J. Am. Chem. Soc.*, 82 (1960) 4564.
122. D. W. Allen, B. G. Hutley, and M. T. J. Mellor, *J. Chem. Soc., Perkin Trans. 2: Phys. Org. Chem.* (1977) 789.
123. I. Petnehazy, G. Szakal, and L. Toke, *Synthesis* (1983) 453.
124. B. Boutevin, Y. Hervaud, T. Jeanmaire, A. Boulahna, and M. Elasri, *Phosphorus, Sulfur Silicon Relat. Elem.*, 174 (2001) 1.
125. P. A. Jaffres, N. Bar, and D. Villemin, *J. Chem. Soc., Perkin Trans. 1: Org. Bio-Org. Chem.* (1998) 2083.
126. B. Dhawan and D. Redmore, *Synth. Commun.*, 15 (1985) 411.
127. A. J. Ganzhorn, J. Hoflack, P. D. Pelton, F. Strasser, M. C. Chanal, and S. R. Piettre, *Bioorg. Med. Chem.*, 6 (1998) 1865.
128. R. Rabinowitz, *J. Org. Chem.*, 28 (1963) 2975.
129. C. J. Salomon and E. Breuer, *Tetrahedron Lett.*, 36 (1995) 6759.
130. A. R. Glabe, K. L. Sturgeon, S. B. Ghizzoni, W. K. Musker, and J. N. Takahashi, *J. Org. Chem.*, 61 (1996) 7212.
131. R. Chen and E. Breuer, *J. Org. Chem.*, 63 (1998) 5107.
132. G. Odian, *Principles of Polymerization*, 4th ed, Wiley, New York, NY, 2004.
133. I. Cabasso, J. Jagurgro, and D. Vofsi, *J. Appl. Polym. Sci.*, 18 (1974) 1969.
134. J. Sun and I. Cabasso, *Macromolecules*, 24 (1991) 3603.
135. S. K. Sahni and I. Cabasso, *J. Polym. Sci., Part A: Polym. Chem.*, 26 (1988) 3251.
136. D. Vofsi, S. Yanai, and I. Cabasso, *J. Polym. Sci., Part A: Polym. Chem.*, 22 (1984) 3335.
137. I. Cabasso, S. K. Sahni, and D. Vofsi, *J. Polym. Sci., Part A: Polym. Chem.*, 26 (1988) 2997.
138. J. P. Schroeder, and W. P. Sopchak, *J. Polym. Sci.*, 47 (1960) 417.
139. M. Abe (Nitto Denko Corp., Japan). JP2003257238 (2003).
140. J. Belack, I. Kundler, T. Schmidt, O. Uensal, J. Kiefer, C. Padberg, and M. Weber (Pemeas G.m.b.H., Germany). WO2005023914 (2005).
141. I. Cabasso, Y. Yuan, and F. E. Johnson (USA). US2004028976 (2004).
142. G. Calundann, M. Sansone, B. Benicewicz, E. W. Choe, O. Uensal, and J. Kiefer (Celanese Ventures GmbH, Germany). DE10246459 (2004).
143. S. Fujita, and M. Abe (Nitto Denko Corp., Japan). JP2002083514 (2002).
144. I. Honma (National Institute of Advanced Industrial Science and Technology, Japan). JP2004185891 (2004).
145. K. Jakoby, S. P. Nunes, and K.-V. Peinemann (GKSS-Forschungszentrum Geesthacht G.m.b.H., Germany). DE10148131 (2003).
146. J. Kiefer and O. Uensal (Pemeas G.m.b.H., Germany). WO2005024988 (2005).
147. J. Kiefer and O. Uensal (Celanese Ventures G.m.b.H., Germany; Pemeas GmbH). WO2003096464 (2003).
148. A. Kondo, M. Nakamura, and N. Ito (Toray Industries, Inc., Japan). JP2004319353 (2004).
149. S. Li, M. Lui, K. Hase, M. Nakanishi, W. Li, and J. Ukai (Toyota Technical Center Usa, Inc., USA; Georgia Tech Research Corporation). WO2005027240 (2005).

150. J. Nakao, Y. Sakaguchi, K. Kitamura, S. Takase, S. Hamamoto, and H. Tatemori (Toyobo Co., Ltd., Japan). JP2002146015 (2002).
151. K. Omi, S. Fujii, T. Kuroki, A. Fujiyama, J. Ishikawa, K. Takeda, and M. Tamai (Mitsui Chemicals Inc., Japan). JP2004311302 (2004).
152. Y. Sakaguchi, K. Kitamura, J. Nakao, S. Hamamoto, H. Tatemori, and S. Takase (Toyobo Co., Ltd., Japan). JP2002146022 (2002).
153. S. Sasaki, A. Yashiro, and Y. Hidaka (Sumitomo Chemical Company, Limited, Japan). EP1238998 (2002).
154. T. Suzuki, T. Taniguchi, Y. Morimoto, M. Kawasumi, N. Hasegawa, and A. Kamiya (Kabushiki Kaisha Toyota Chuo Kenkyusho, Japan). EP1110992 (2001).
155. K. Muellen, M. Klapper, V. Sinigersky, W. Wu, O. Uensal, and J. Baurmeister (Celanese Ventures GmbH, Germany). DE10230477 (2004).
156. T. Taniguchi, M. Takami, M. Rikukawa, and Y. Takeoka (Toyota Jidosha K. K., Japan). DE10355619 (2004).
157. O. Uensal, J. Kiefer, and G. Christ (Celanese Ventures G.m.b.H., Germany). WO2003075389 (2003).
158. M. Umeda and A. Yamada (Nagaoka University of Technology, Japan). JP2004335231 (2004).
159. S. Sasaki, A. Yashiro, Y. Hidaka, and Y. Taniguchi (Sumitomo Chemical Company Limited, Japan). FR2834716 (2003).
160. J. Nakao, Y. Sakaguchi, K. Kitamura, S. Takase, S. Hamamoto, and H. Tatemori (Toyobo Co., Ltd., Japan). JP2002146013 (2002).
161. I. Cabasso and T. Chinh Nogh, Gulf South Res. Inst., New Orleans , LA, USA. (1977) 64 pp.
162. I. C. Cabasso, Gulf South Res. Inst., New Orleans, LA, USA. (1979) 72 pp.
163. X. Xu and I. Cabasso, *Polym. Mater. Sci. Eng.*, 68 (1993) 120.
164. X. Xu and I. Cabasso, *Abstr. Pap. Am. Chem. Soci.*, 205 (1993) 78.
165. H. R. Allcock, M. A. Hofmann, and R. M. Wood, *Macromolecules*, 34 (2001) 6915.
166. K. Okamoto, T. Ijyuin, S. Fujiwara, H. Y. Wang, K. Tanaka, and H. Kita, *Polym. J.*, 30 (1998) 492.
167. C. Genies, R. Mercier, B. Sillion, R. Petiaud, N. Cornet, G. Gebel, and M. Pineri, *Polymer*, 42 (2001) 5097.
168. T. Sata, T. Yoshida, and K. Matsusaki, *J. Membr. Sci.*, 120 (1996) 101.
169. L. Gubler, S. A. Gürsel, and G. G. Scherer, *Fuel Cells*, 5 (2005) 317.
170. G. Schmidt-Naake, M. Bohme, and A. Cabrera, *Chem. Eng. Technol.*, 28 (2005) 720.
171. M. J. Sundell, K. B. Ekman, B. L. Svarfvar, and J. H. Nasman, *React. Polym.*, 25 (1995) 1.
172. W. A. Trochimczuk, *Eur. Polym. J.*, 35 (1999) 1457.
173. H. R. Allcock and J. P. Taylor, *Polym. Eng. Sci.*, 40 (2000) 1177.
174. H. R. Allcock, M. A. Hofmann, C. M. Ambler, and R. V. Morford, *Macromolecules*, 35 (2002) 3484.
175. B. Lafitte, and P. Jannasch, *J. Polym. Sci., Part A: Polym. Chem.*, 43 (2005) 273.
176. M. D. Guiver, G. P. Robertson, M. Yoshikawa, and C. M. Tam, ACS *Symp. Ser.*, 744 (2000) 137.
177. M. D. Guiver, J. W. ApSimon, and O. Kutowy, *J. Polym. Sci., Part C: Polym. Lett.*, 26 (1988) 123.
178. M. D. Guiver, O. Kutowy, and J. W. ApSimon, *Polymer*, 30 (1989) 1137.
179. B. Lafitte and P. Jannasch, *J. Polym. Sci., Part A: Polym. Chem.* (in press).
180. J. Ziaja, L. Balogh, and W. M. Trochimczuk, *Prace Naukowe Instytutu Podstaw Elektrotechniki i Elektrotechnologii Politechniki Wroclawskiej*, 34 (1998) 153.
181. K. Jakoby, K. V. Peinemann, and S. P. Nunes, *Macromol. Chem. Phys.*, 204 (2003) 61.
182. E. Parcero, R. Herrera, and S. P. Nunes, *J. Membr. Sci.*, 285 (2006) 206.
183. K. Miyatake and A. S. Hay, *J. Polym. Sci., Part A: Polym. Chem.*, 39 (2001) 3770.

184. B. Liu, G. P. Robertson, M. D. Guiver, Z. Shi, T. Navessin, and S. Holdcroft, *Macromol. Rap. Commun.*, 27 (2006) 1411.
185. S. Yanagimachi, K. Kaneko, Y. Takeoka, and M. Rikukawa, *Synth. Met.*, 135 (2003) 69.
186. S. Gopalkrishnan, in 1988 Rutgers the State University of New Jersey, New Brunswick.
187. R. Rabinowitz, J. Pellon, and R. Marcus, *J. Polym. Sci., Part A*, 2 (1964) 1241.
188. R. M. Pike and R. A. Cohen, *J. Polym. Sci.*, 44 (1960) 531.
189. C. G. Overberger, and E. Sarlo, *J. Polym. Sci., Part A*, 2 (1964) 1017.
190. D. Avci and L. J. Mathias, *J. Polym. Sci., Part A: Polym. Chem.*, 40 (2002) 3221.
191. D. Avci and A. Z. Albayrak, *J. Polym. Sci., Part A: Polym. Chem.* 41 (2003) 2207.
192. S. Salman, A. Z. Albayrak, D. Avci, and V. Aviyente, *J. Polym. Sci., Part A: Polym. Chem.*, 43 (2005) 2574.
193. B. Boutevin, B. Hamoui, J.-P. Parisi, and B. Ameduri, *Eur. Polym. J.*, 32 (1996) 159.
194. C. L. Arcus and R. J. S. Matthews, *J. Chem. Soc.* (1956) 4607.
195. C. L. Arcus and R. J. S. Matthews, *Chem. Ind.* (1958) 890.
196. B. J. Muray, *J. Polym. Sci., Part C: Polym. Sym.* (1967) 1869.
197. M. Banks, J. R. Ebdon, and M. Johnson, *Polymer*, 35 (1994) 3470.
198. Q. Wu and R. A. Weiss, *J. Polym. Sci., Part B: Polym. Phys.*, 42 (2004) 3628.
199. J. Wei, C. Stone, and A. E. Steck (Ballard Power Systems Inc., Can.). WO9508581 (1995).
200. A. E. Steck and C. Stone, *International Symposium on New Materials for Fuel Cell and Modern Battery Systems*, 2nd edn, Montreal, July 6–10, 1997; *New Materials for Fuel Cell and Modern Battery Systems* (1997) 792.
201. C. Stone, T. S. Daynard, L. Q. Hu, C. Mah, and A. E. Steck, *J. New Mat. Elect. Syst.*, 3 (2000) 43.
202. P. L. Heinze and D. J. Burton, *J. Org. Chem.*, 53 (1988) 2714.
203. R. Souzy, B. Ameduri, B. Boutevin, and D. Virieux, *ChemInform*, 36 (2005).
204. S. Lin and I. Cabasso, *J. Polym. Sci., Part A: Polym. Chem.*, 37 (1999) 4043.
205. S. Gallagher, *J. Polym. Sci., Part A: Polym. Chem.*, 41 (2003) 48.
206. Y. Z. Meng, S. C. Tjong, A. S. Hay, and S. J. Wang, *J. Polym. Sci., Part A: Polym. Chem.*, 39 (2001) 3218.
207. Y. Z. Meng, S. C. Tjong, A. S. Hay, and S. J. Wang, *Eur. Polym. J.* 39 (2003) 627.
208. Y. Z. Meng, A. R. Hill, and A. S. Hay, *Polym. Adv. Technol.*, 12 (2001) 206.
209. Y. Z. Meng, I. A. Abu-Yousef, A. R. Hlil, and A. S. Hay, *Macromolecules*, 33 (2000) 9185.
210. W. Harrison, F. Wang, Y. S. Kim, M. Hickner, and J. E. McGrath, *Polym. Prepr. (Am. Chem. Soc., Div. Polym. Chem.)*, 43 (2002) 700.
211. W. L. Harrison, F. Wang, J. B. Mecham, T. E. Glass, M. A. Hickner, and J. E. McGrath, *Abstr. Pap. Am. Chem. Soc.*, 221 (2001) 376.
212. F. Wang, M. Hickner, Q. Ji, W. Harrison, J. Mecham, T. A. Zawodzinski, and J. E. McGrath, *Macromol Symp.*, 175 (2001) 387.
213. F. Wang, T. E. Glass, X. Li, M. Hickner, Y. Kim, and J. E. McGrath (eds.) 2002; Abstracts of Papers, *223rd ACS National Meeting*, Orlando, FL, United States, April 7–11, 2002 (2002) 113.
214. F. Wang, M. Hickner, Y. S. Kim, T. A. Zawodzinski, and J. E. McGrath, *J. Membr. Sci.*, 197 (2002) 231.
215. T. Rager, M. Schuster, K.-D. Kreuer, and J. Maier, *Draft* (2005).
216. H. R. Allcock, M. A. Hofmann, C. M. Ambler, S. N. Lvov, X. Y. Zhou, E. Chalkova, and J. Weston, *J. Membr. Sci.*, 201 (2002) 47.
217. R. Wycisk and P. N. Pintauro, *J. Membr. Sci.*, 119 (1996) 155.
218. V. Tricoli, N. Carretta, and M. Bartolozzi, *J. Electrochem. Soc.*, 147 (2000) 1286.
219. D. Rivin, C. E. Kendrick, P. W. Gibson, and N. S. Schneider, *Polymer*, 42 (2001) 623.
220. S. Hietala, S. L. Maunu, and F. Sundholm, *J. Polym. Sci., Part B: Polym. Phys.*, 38 (2000) 3277.

221. M. V. Fedkin, X. Y. Zhou, M. A. Hofmann, E. Chalkova, J. A. Weston, H. R. Allcock, and S. N. Lvov, Mater. Lett., 52 (2002) 192.
222. S. Lvov, X. Zhou, E. Chalkova, J. A. Weston, C. M. Ambler, A. E. Maher, R. M. Wood, and H. R. Allcock, *Prepr. Symp. Am. Chem. Soc., Div. Fuel Chem.*, 48 (2003) 480.
223. X. Y. Zhou, J. Weston, E. Chalkova, M. A. Hofmann, C. M. Ambler, H. R. Allcock, and S. N. Lvov, *Electrochim. Acta*, 48 (2003) 2173.
224. D. D. Jiang, Q. Yao, M. A. McKinney, and C. A. Wilkie, *Polym. Degrad. Stab.*, 63 (1999) 423.
225. M. Yamada and I. Honma, *Polymer*, 46 (2005) 2986.
226. H. Steininger, M. Schuster, K. D. Kreuer, and J. Maier, *Solid State Ionics*, 177 (2006) 2457.
227. H. Erdemi and A. Bozkurt, *Eur. Polym. J.*, 40 (2004) 1925.
228. A. Bozkurt, W. H. Meyer, J. Gutmann, and G. Wegner, *Solid State Ionics*, 164 (2003) 169.
229. F. Sevil and A. Bozkurt, *J. Phys. Chem. Solids*, 65 (2004) 1659.
230. K. Kanamura, A. Tanaka, D. Gervasio, V. Kennedy, R. Adzic, and E. B. Yeager, *J. Electrochem. Soc.*, 143 (1996) 2765.
231. K. D. Kreuer, S. J. Paddison, E. Spohr, and M. Schuster, *Chem. Rev.*, 104 (2004) 4637.
232. M. Yamada and I. Honma, *Electrochim. Acta*, 50 (2005) 2837.
233. M. Schuster, K. D. Kreuer, H. Rössler, and J. Maier, *Draft* (2005).
234. S. W. Li, Z. Zhou, H. Abernathy, M. L. Liu, W. Li, J. Ukai, K. Hase, and M. Nakanishi, *J. Mater. Chem.*, 16 (2006) 858.

CHAPTER FOUR

Polymer Electrolyte Membranes for Direct Methanol Fuel Cells

Yu Seung Kim *and* Bryan S. Pivovar

Contents

Abstract

This review addresses polymer electrolytes that have been demonstrated in direct methanol fuel cells. This work reviews the material requirements of polymer electrolytes and gives an exhaustive overview of the polymer electrolyte membranes (PEMs) investigated specifically for direct methanol fuel cells (DMFCs). This chapter then provides a framework for interpreting the reported performance and potential of PEMs in DMFC applications. Finally, this review presents and compares currently reported performance of DMFCs, particularly those using alternative membranes with some discussion of future research directions.

4.1 Introduction

Direct methanol fuel cells (DMFCs) are being pushed toward the brink of commercialization because they offer the potential of longer operating

MPA-11, Sensors and Electrochemical Devices, Los Alamos National Laboratory, Los Alamos, NM 87545, USA

Advances in Fuel Cells
0080453945

lifetimes and the ability to refuel versus recharge compared to the batteries typically used in portable power applications. Significant research effort has been focused on the development on new membranes for improved DMFC performance, but there has not been a comprehensive review published specifically for DMFCs or DMFC electrolytes. The goals of this chapter are to: (1) give an overview of the polymer electrolyte membranes (PEMs) investigated specifically for DMFCs; (2) provide a framework for interpreting the performance and potential of membranes in DMFC applications; and (3) present currently reported performance of DMFCs, particularly those using alternative membranes.

In order to achieve these goals, an introduction into polymer electrolyte fuel cells (PEFCs) and comparisons between hydrogen and methanol fuel cells is presented. This is followed by an overview of PEMs, starting with basic characterization techniques of PEMs for DMFC applications. A review of measurement techniques used for determining proton conductivity and methanol permeability is presented as these are the properties most studied for DMFC electrolytes. Water related measurement such as water uptake, diffusion and electro-osmotic drag, and other DMFC relevant measurements are also discussed. Finally, DMFC performance is addressed.

The various types of PEMs developed for DMFCs over the last 10 years are then reviewed. There has been great effort in developing new materials for either cheaper or better performing fuel cell membranes. A number of comprehensive reviews of polymers for PEFCs exist [1–4], and there are others focusing on the overall efforts of specific research groups [5–9], but there has not yet been a comprehensive review on polymer electrolytes for DMFCs or a discussion about the specific properties, approaches and needs necessary for developing and using high performance DMFC PEMs. In light of the current state of DMFC technology and the development of new electrolytes, it is one of the goals of this review to address this area. Nafion and other perfluorinated ionomers will be discussed first. Specific alternative polymer systems will be discussed next, starting with sulfonated polyarylenes. Other membranes based on polyimide (PI), polyphosphazene (PPZ), grafted polystyrene sulfonic acid (PS), polyvinyl alcohol (PVA), and styrene block copolymers will be presented in some detail. Finally, other alternative approaches such as layered or composite membranes will be discussed.

A significant issue in comparing different membranes for DMFC applications is determining proper evaluation criteria. Evaluation criteria of alternative membranes are reviewed in the Section 4.5 of this chapter. Selectivity, or the ratio of proton conductivity to methanol permeability, will be presented as a qualitative basis for the evaluation of DMFC electrolytes. The limitations of using selectivity as a basis such for determining DMFC performance under fuel cell operation will be addressed. Reported performances of a number of polymers will be compared to Nafion in terms of proton conductivity, methanol permeability, selectivity, water uptake, and

DMFC performance. The importance and difficulty of making high performance membrane– electrode assemblies (MEAs) using alternative PEMs will be presented with special emphasis placed on water uptake and membrane–electrode interfacial effects. The need to use optimized DMFC performance as a relevant comparison between polymer electrolytes will be discussed, using a case study of Nafion and a poly(arylene ether sulfone) copolymer as an illustrative example.

Finally, future research directions and the outlook for DMFCs will be speculated upon, including passive DMFCs which are largely ignored in the bulk of this chapter.

4.2 BACKGROUND

PEFCs have become the most popular fuel cell technology for low temperature (<100°C) operation, in the areas of automotive, stationary, and portable power because they offer the promise of highly efficient, long lasting, inexpensive pollution-free power sources that have the ability to be operated on potentially renewable resources. The PEFC is differentiated from other fuel cells by the use of a proton conducting, solid, polymer electrolyte that connects the fuel cell anode and cathode ionically, while separating them physically and electronically. Originally, this fuel cell was referred to as the ion exchange membrane fuel cell (IEMFC). Later it was referred to as the solid polymer fuel cell (SPFC) or the solid polymer electrolyte fuel cell (SPEFC). Recently, the name proton exchange membrane fuel cell (PEMFC) has also become popular [10]. DMFCs are a subset of PEFCs, in which an aqueous solution of methanol is provided as the fuel directly to the fuel cell usually at concentrations below 2 M (6.4 wt%).

While fuel cells were first reported as early as 1839 by Grove and Schonbein, they received little attention before the mid-1960s, when interest was sparked by the space program. PEFCs were first deployed in the Gemini space program in the early 1960s using cells that were expensive and had short lifetimes due to the oxidative degradation of their sulfonated polystyrene–divinylbenzene copolymer membranes. These cells were considered too costly and short-lived for real-world applications. The advent of Nafion® by DuPont, a perfluorinated ionomer with much greater stability, in the late 1960s helped to generate interest in terrestrial applications. While methanol was considered as a fuel for fuel cells as early as 1962 [11], little progress was made until researchers at Jet Propulsion Laboratories and collaborators showed reasonable power density in the early to mid 1990s [12,13].

Research on PEFCs has skyrocketed in the past decade. The fall 2002 edition of Fuel Cells 2000s Fuel Cell Directory features nearly 1000 listings of fuel cell manufacturers, researchers and consultants, suppliers to the fuel cell industry, utilities, associations, and interested government agencies around the world. Current estimates are that auto manufacturers are currently investing $1 billion per year in the development of fuel cell technology for automotive applications [14]. Research efforts on DMFCs have seen similar increases, albeit at a far lower level of funding than hydrogen-based systems.

Figure 4.1 shows the number of hits for "DMFC" under Title/Subject/Abstract using SciSearch®, a search engine for scientific journal articles over the past 10 years. There has been a significant increase in publications in this area, doubling at an average rate of every 2 years, even showing signs of acceleration in 2004. Many people believe that DMFCs will represent the first widespread commercial products based on fuel cell technology. This assumption is supported by the fact that Smart Fuel Cell, a company based in Munich, Germany, already sells a 25-Watt DMFC system, the SFC A25 [15]. Additionally, MTI MicroFuel Cells, Inc., Toshiba, and NEC all have prototype DMFC units which were scheduled for commercialization by the end of 2004 according to company press releases [16–18].

Though fuel cells can theoretically be operated using any oxidizable species, hydrogen has been the primary fuel. Hydrogen makes an excellent fuel because it has a high theoretical power density, it is easily catalyzed, and it gives only water as a reaction product. The primary disadvantage of pure hydrogen is that it is a highly reactive, low density gas under normal conditions. Storage and distribution are therefore major problems. Solutions to hydrogen storage consist of cryogenic liquid storage, high pressure gas cylinders, metal hydrides, and chemical hydrides. These storage media add considerable weight, greatly reducing the effective energy density of the fuel, and can also require large volumes significantly increasing space requirements.

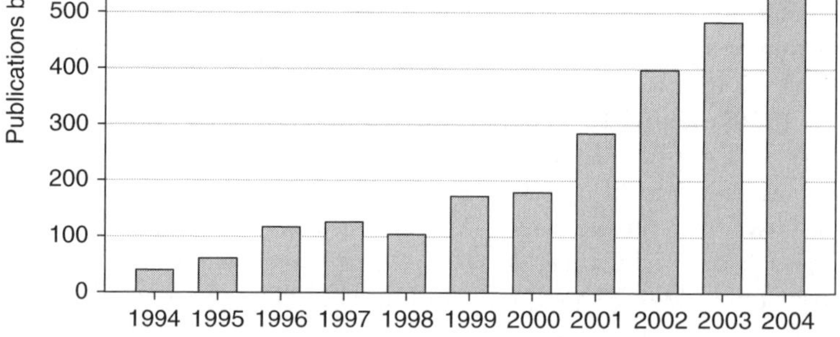

Figure 4.1 DMFC publications by year for 1994–2004.

There is great interest in running PEFCs using methanol instead of hydrogen. Methanol has many advantages over hydrogen. On a volume basis, methanol has a 50% greater energy density than liquid hydrogen. Methanol, an easily handled liquid under normal conditions, can be stored and distributed using existing equipment. Methanol can also be economically and efficiently produced from relatively abundant fossil fuels (coal and natural gas) as well as renewable resources (biomass).

While methanol is less electrochemically active than hydrogen, it still has high reactivity compared to other organic compounds. Because methanol is a simple molecule, the number of reactions occurring at the anode is small, decreasing the need for complex catalysts. Methanol contains only a single carbon atom, i.e., there is no carbon–carbon bond which decreases electrochemical activity. Methanol has a readily oxidizable (hydroxyl) group that increases electrochemical activity, but still has a reasonably high energy density compared to other oxygenated organic molecules. The combination of these factors make DMFCs attractive power sources, especially for low power systems where issues associated with fuel storage or the added weight and space of reformers cannot be tolerated.

4.3 Characterization of PEMs

In order to discuss performance of PEMs in DMFC, we must first present the properties required for a successful fuel cell electrolyte. The critical features for polymer electrolytes in fuel cells are high protonic conductivity; low permeability to reactants (for DMFCs methanol permeability is a critical concern); low electronic conductivity (almost never an issue); chemical stability; and good mechanical properties. Other properties such as water transport through diffusion and electro-osmosis, or water (or methanol) uptake are important when considering performance in an operating system. Also, the polymers need the ability to be made into high performance, durable MEAs. These properties and the methods most commonly used to obtain them are presented here as a general resource and as a background discussion for investigating alternative PEM performance. Finally, DMFC performance, the ultimate goal of any novel electrolyte, is presented as a method of PEM characterization with an emphasis on efficiency.

4.3.1 Proton Conductivity

Conductivity is central for any fuel cell electrolyte, and proton conductivity is the first characteristic considered when evaluating membranes for potential fuel cell use. Proton conductivity in polymer electrolytes is characterized primarily by alternating current (ac) impedance spectroscopy. (Although other techniques such as current interrupt, hydrogen pump or high frequency resistance are also used, particularly when conductivity is measured in fuel cells. Here, we focus here on measuring conductivity in free standing films which occurs almost exclusively by ac impedance.) Alternating Current impedance spectroscopy, also referred to as electrochemical impedance spectroscopy, involves applying an alternating current to a sample through a range of frequencies and separating the impedance features as a function of frequency [19–21]. This approach has been used under a number of different experimental conditions and configurations in order to determine the conductivity of polymer electrolytes. Experiments have been performed on free standing films and on MEAs, there have been both two and four point probe measurements performed in either the transverse (through-plane) or longitudinal (along-plane) directions and test conditions, such as water vapor equilibrated versus liquid or acid equilibrated have all been varied [e.g., 22–26].

While proton conductivity in polymer electrolytes is known to be highly dependent on humidification conditions [27–29], the case is somewhat simpler in DMFCs. Unlike hydrogen-based PEFCs, DMFCs typically operate under liquid saturated or near saturated conditions. The aqueous methanol solution at the anode keeps the membrane in a well hydrated state and the cathode also tends to have significantly more water due to diffusion from the methanol feed solution (at low current density) and elevated electro-osmotic drag (at high current density). In fact, DMFCs often have flooding issues at the cathode [30,31]. These factors make liquid equilibrated rather than vapor equilibrated membranes a more relevant test condition for DMFC properties including conductivity.

For all fuel cells, the resistive losses associated with proton conduction are directly proportional to current density. Ohmic resistance of membranes is directly proportional to the thickness of the membrane. Hydrogen fuel cell membranes tend to be thin, often limited by mechanical robustness of the membrane although gas crossover rates are also a concern. Methanol membranes have tended to be thicker, limited almost exclusively by high methanol crossover rates. DMFC membranes therefore typically have higher resistances associated with them. The relatively high resistance of the membranes and decreased anode performance has limited DMFCs to lower current densities where efficiency is maximized and ohmic losses are minimized.

While membrane conductivity can also be estimated from fuel cell measurements, specific factors must be taken into account when interpreting results. First, methanol within the membrane can affect membrane conductivity [32–34], however at most practical methanol feed concentrations this is a relatively minor effect. Second, measurements performed on MEAs in fuel cells have additional resistances associated with them due to catalyst layers, gas diffusion layers, flow fields, current collectors, and interfaces between these components. Membrane conductivities calculated from such measurements often lead to an underestimation of true conductivity. Specific experiments, such as membrane thickness studies performed on MEAs, have demonstrated that these resistances can be isolated from the bulk membrane resistance [35].

4.3.2 Methanol Permeability

Methanol permeability is a key concern for DMFC electrolytes, because methanol crossover is directly related to permeability and cell performance is directly related to crossover. While the exact relationship between permeability and crossover is complex, certain factors such as membrane thickness and methanol feed concentration affect crossover in predictable ways, i.e., the higher the methanol concentration or the thinner the membrane the higher the crossover rates. Methanol crossover can be measured directly in fuel cells, but often methanol permeability is measured on free standing films.

Related to permeability, diffusion coefficients or solubility parameters are often reported as well. At this point it is useful to define these terms and relate them to the methanol crossover rates (fluxes) of operating DMFCs. For this illustrative discussion, we will use terminology consistent with that reported by Cussler for separation processes [36].

For crossover in DMFCs, the species of concern is methanol. The simple form of Fick's law for diffusion across a membrane is

$$j = -D\frac{dc_1}{dz} \tag{4.1}$$

where j is equal to the flux; D is the diffusion coefficient, c_1 is the concentration of methanol, and z is the position within the membrane. This equation can be integrated to yield

$$j = \frac{DH}{l}\Delta c \tag{4.2}$$

where H is the partition coefficient (or the ratio of concentration in the membrane to that in solution), l is the thickness of the membrane and Δc is the concentration difference between the solutions in contact with the membrane. Methanol permeability (DH) is defined as the product of the diffusion coefficient and partition coefficient and takes into account both solubility and diffusion. This simple version of Fick's law is valid when the only driving force for transport is diffusion and convection can be ignored.

Operating fuel cells have a combination of other forces that can influence transport. Electromotive forces act on charged species, current passage creates convective transport through electro-osmotic drag, and hydraulic pressure can also result in species transport. Methanol is an uncharged molecule so that electromotive forces do not act on it directly. Additionally, DMFCs are typically run under ambient conditions with little or no pressure difference across the cell so that, in general, hydraulic permeation can be ignored. However, the effects of convective transport through electro-osmotic drag are important for methanol crossover and a modified version of Fick's law can be used that incorporates convection. This modified equation is

$$j = -D\frac{dc_1}{dz} + c_1 v^0 \tag{4.3}$$

where v^0 is a convective velocity introduced by the passage of current and the effects of electro-osmotic drag. This equation takes into account both diffusive and convective contributions and can be integrated to yield

$$j = \frac{Hv^0 \Delta c}{1 - \exp\left(\frac{-v^0 l}{D}\right)} \tag{4.4}$$

An expansion of the exponential term yields an equation similar to Eq. (4.2),

$$j = \frac{DH}{l}\Delta c\left(\frac{1}{1 - \frac{1}{2!}\left(\frac{-v^0 l}{D}\right) + \frac{1}{3!}\left(\frac{-v^0 l}{D}\right)^2 - \frac{1}{4!}\left(\frac{-v^0 l}{D}\right)^3 + \cdots}\right) \tag{4.5}$$

Equation (4.5) simplifies to Eq. (4.2) for the case where convective velocity (v^0) equals zero, and the relative importance of diffusion to convection is represented by the ratio of $v^0 l/D$. Interestingly, membrane thickness shows up in this ratio because diffusion fluxes depend on membrane thickness, while electro-osmotic drag is independent of membrane thickness. Additionally, the partition coefficient does not show up in this term because it impacts both diffusion and convection. Similar approaches have been shown to be valid for DMFCs [37] and related systems [38] when v^0 is defined as

$$v^0 = \frac{E_D i \hat{V}_{H_2O}}{F\theta_{H_2O}} \tag{4.6}$$

where E_D is the electro-osmotic drag coefficient, i is the current density, $\hat{V}_{H_2O}$ is the molar volume of water, F is Faraday's constant and θ_{H_2O} is the volume fraction of water within the membrane.

Equations (4.3)–(4.6) show the relative importance of permeability and diffusion coefficient to convective velocity due to electro-osmotic drag. At low current densities the effects of permeability dominate crossover. At high current densities, the effects of electro-osmosis become more important. For illustrative purposes we take the case of Nafion 117 (the most commonly used DMFC membrane) and apply Eqs. (4.2), (4.4), and (4.6) (using data presented later in Table 4.2, the molar volume of water ($18\,cm^3$/mole of water) and a diffusion coefficient taken from Ref. [41]) to determine the current density where the flux of methanol doubles based on convection. This current density can be used as a loose gauge where convection and diffusion play roughly an equal role on transport. For the case of Nafion 117 and the equations discussed above this occurs at $131\,mA/cm^2$ (above this current density, convective rather than diffusive transport plays a larger role in crossover). However, when the membrane is changed to a thinner membrane such as Nafion 112, the current density where this transition occurs is significantly higher ($524\,mA/cm^2$ for this analysis). Beyond membrane thickness, factors such as methanol diffusion coefficient and the electro-osmotic drag coefficient also impact the relative driving forces for methanol crossover when considering alternative membranes. Most alternative membranes have lower electro-osmotic drag coefficients and lower diffusion coefficients (and therefore are typically thin), the resulting importance of convection versus diffusion depends on the magnitude of these effects.

The equations presented are meant as an illustrative example of how different physical factors impact methanol crossover and they do not account for methanol transport resistance through the backing layers or methanol consumption within the anode layer for which more complicated models are needed [37,38]. For example, this analysis predicts increasing fluxes with current density, when under most operating conditions crossover rates actually decrease. This is because consumption and mass transport resistance are more important than increasing transport within the membrane due to convection. This approach also neglects the role of direct association of methanol with the proton or methanol electro-osmotic drag, a quantity recently reported in the literature [32,33]. This assumption is justified, because this approach has proven adequate to accurately model methanol crossover in operating DMFCs [37], and because at practical operating conditions ($<2\,M$) water (which is a better proton solvating species than methanol) is found in significantly larger quantities than methanol.

The most common approach to measure methanol permeability has been to use a diffusion cell, where two reservoirs are separated by a test membrane [39]. At the start of the experiment, one of the reservoirs contains pure water while the other reservoir contains methanol of known concentration. During the experiment, methanol permeates through the membrane from the methanol rich reservoir to the methanol poor reservoir. By tracking the change in concentration of one of the compartments and geometric data, methanol permeability for the membrane can be calculated. Because permeability is often a function of concentration, 1 M methanol is typically used in the methanol rich reservoir because it is believed to be a reasonable concentration for DMFC operation (allowing practical power densities while limiting crossover).

NMR has also been used to determine methanol diffusion in free standing films [40,41]. Care needs to be taken in interpreting or applying the data, because these measurements have been taken with vapor equilibrated and high methanol concentrations that are not relevant for typical DMFC operating conditions. Additionally, the diffusion coefficients reported do not take into account partition coefficients or membrane asymmetry.

Methanol permeability has also been calculated from MEAs using three separate techniques: limiting crossover current measurements [30,37], CO_2 measurements of cathode exhaust [42–44], and methanol mass balance measurements [45,46]. While, CO_2 measurements of cathode exhaust and methanol mass balance measurements can be used for determining methanol crossover at any DMFC current density. The effects of electro-osmotic drag and convective velocity make calculating methanol permeability easiest at open circuit conditions, and so open circuit measurements are usually used to determine an estimate of methanol permeability. Open circuit voltage (OCV) of a DMFC can also be used as a rough gauge of crossover; however issues with electrode performance lead to large experimental uncertainty [47].

A fairly straightforward method for determining methanol permeability is the limiting crossover current method, first suggested by Ren *et al.* [30,37]. In this experiment methanol solution is fed to the "anode" of the fuel cell at some known concentration and flow rate, and a humidified inert gas, typical nitrogen, is fed to the "cathode". A high potential is applied across the cell to oxidize the methanol crossing over the membrane to protons and CO_2, interestingly in this configuration the DMFC "cathode" actually behaves like an anode. The current produced by this reaction is easily tracked and the potential is reduced stepwise until a low, specified potential is reached. A typical curve is shown in Figure 4.2. The plateau region at high potential corresponds to the complete oxidation of all methanol that is transported through the MEA, termed the limiting methanol crossover current, and can be used to estimate methanol permeability of the membrane with additional information about the anode (see Ref. [37] for more detail).

CO_2 measurements on the cathode exhaust can also provide an estimate of methanol permeability when performed under open circuit conditions. In this experiment a sensor that can determine CO_2 concentration monitors the cathode exhaust. If bottled air or oxygen is used, the only source of CO_2 in the cathode exhaust is methanol crossover. For ambient air such as that from a compressor, the baseline CO_2 concentration needs to be accounted for and adds experimental uncertainty. The humidification of the cathode needs to be controlled as permeability may vary with water uptake of the membrane. From the resulting CO_2 production and various other physical factors permeability can be estimated. Figure 4.3 shows an example of methanol crossover current as a function of current density. This experiment ignores mass transfer resistances due to backing and catalyst layers, downstream flow field effects, and assumes complete oxidation to CO_2 of the crossover methanol and no

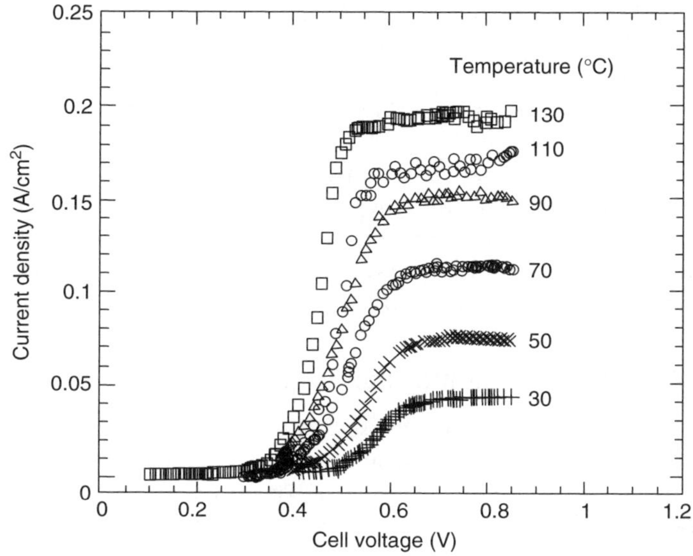

Figure 4.2 Voltammetric curves at various cell temperatures for the oxidation of methanol permeating through a Nafion membrane 117 exposed to a 1 M methanol feed (reproduced with permission from *J. Electrochem. Soc.* [37]).

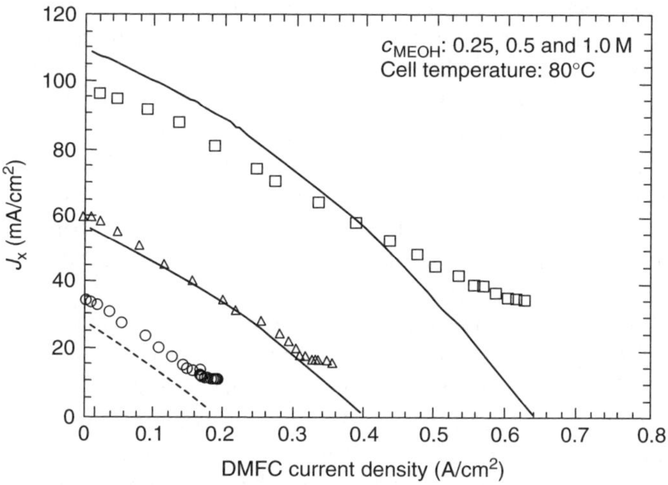

Figure 4.3 Rate of methanol crossover (J_x) measured for an operating DMFC as a function of methanol feed concentration (reproduced from Ref. [42]).

CO_2 crossover through the membrane. While CO_2 crossover can be a problem for estimating methanol crossover in DMFCs, it should not be an issue at open circuit conditions because methanol is not being oxidized on the anode [48,49].

The methanol mass balance is quite similar to the CO_2 crossover measurement, in that it is typically used to determine methanol crossover under DMFC operation, but it can also be used to estimate permeability under open circuit conditions. It is the most tedious and time consuming of the MEA methods to determine methanol permeability and is usually only attempted when CO_2 crossover is an issue. The experiment involves a recirculating methanol feed stream, where methanol consumption due to reaction is tracked with time. From differences between the initial and final concentration of methanol in the methanol feed stream and other physical factors methanol permeability can be estimated.

4.3.3 Water Related Membrane Properties

Water uptake of the membrane, the self-diffusion coefficient of water within the membrane and the electro-osmotic drag coefficient of water carried through the membrane are all properties of polymer electrolytes that have been studied in detail. These water related properties have been recognized as additional factors that either directly or indirectly impact fuel cell performance.

Among water related properties, water uptake is the simplest to obtain and most commonly reported. As transport processes in polymer electrolytes take place through water containing domains, it is not surprising that water uptake has been directly linked to properties such as proton conductivity and methanol permeability [50]. In fact, specific trends for transport properties have been demonstrated within specific polymer families as a function of water content [32,51,52]. While not as directly apparent, water uptake has also been shown to affect morphological/mechanical stability [53]. Relating water uptake to polymer properties across different polymer families would be useful; however, quantitative relationships across different copolymer families are difficult (if not impossible) due to the importance of specific water interactions in different chemical environments.

Water uptake of polymer electrolytes can be determined by measuring the weight increase of a dry sample of a PEM after immersion in liquid water [50,54]. Additional data can be obtained by equilibration of a sample with liquid water or with water vapor of a known relative humidity [55,56]. More precise measurements can be obtained using a Cahn microbalance in a quartz vessel [57] or a tapered element oscillating microbalance [58]. These techniques allow accurate measurement of equilibrium water sorption–desorption isotherms for PEMs at actual cell operating temperature, although they are of limited value for DMFC applications where systems typically remain well hydrated. Uptake of methanol or methanol/water mixtures has also been reported [59,60]. While these data are useful, because typical DMFCs operate at low methanol concentrations (<2 M) the impacts of absorbed methanol are often minor.

Other properties such as the water diffusion coefficient and electro-osmotic drag coefficient play a critical role in water management in operating cells. Water management in fuel cells is important because it affects membrane properties (conductivity, permeability) and reactant transport (i.e., cell flooding). Additionally, electro-osmotic drag affects methanol crossover rates in DMFCs due to the effects of convective transport (e.g., Eqs. (4.3)–(4.6) in the previous section).

Water self-diffusion coefficients have been estimated using various techniques, such as radiotracer measurement [61–64], water sorption/desorption in vapor phase [65], quasi-elastic neutron scattering (QENS) [66], and (most commonly) pulsed field gradient spin-echo NMR techniques (PFR-NMR) [55,67–70]. In radiotracer studies, membranes were immersed in tritium tagged water and the amount of radiotracer was monitored over time. The rates of isotropic-exchange were then analyzed to determine the self-diffusion coefficient of water in the membrane. Diffusion coefficients have also been determined from steady state permeability combined with solubility from equilibrium vapor sorption/desorption measurements and corrections for gas phase boundary layer resistance. In QENS, diffusion coefficients were estimated from the elastic incoherent structure factor as a function of scattering vector assuming either confined spherical or unbounded jump diffusion. Finally, PFG-NMR measurements have been used to determine diffusion coefficients. In this method, an intradiffusion coefficient for polymer electrolytes bearing the detected nucleus (typically water) was determined from the diffusional dephasing of a gradient-encoded magnetization. Every *et al.* further refined this technique to measure methanol diffusion coefficients [41].

Electro-osmotic drag, the number of solvent molecules transported across a membrane per ion, is also an important membrane property. For PEFCs, protons are the ion of interest and water is the solvent. For DMFCs, methanol is a potential solvent of concern, but due to the low methanol contents within operating electrolytes (required for low crossover rates and reasonably efficient operation); the discussion presented here will largely ignore this phenomenon. However, it should be noted values for methanol electro-osmotic drag have been reported [32,33].

At a local level, electro-osmotic drag impacts the level of hydration of the polymer electrolyte and the electrodes. DMFCs under typical operating conditions do not experience significant dehydration at the cathode. However, if water builds up at the cathode, it can lead to a flooded state that negatively impacts cell performance by impeding mass transport. At a system level, electro-osmosis also affects performance. System issues, such as feed gas flow rates and pressures, have to be addressed to ensure that individual cells remain properly hydrated but not flooded. This often requires that reactant streams be humidified and water condensed for reuse. Both of these operations can be energy intensive, thereby reducing the overall system efficiency. Thermal management of the stack is also influenced by electro-osmotic drag as evaporative cooling within the cells can be a major source of heat removal. Also, electro-osmotic drag can impact the power density of a fuel cell system as it influences the need for peripherals, such as condensers, or the amount of make up water needed to be carried with the system. For low power direct methanol fuel cell systems, electro-osmotic drag has been shown to be a key limiting factor for system power density [71].

A number of different techniques have been used to determine electro-osmotic drag coefficients. In fact, one of the author's of this chapter recently published a review of electro-osmosis in fuel cell electrolytes [72]. For a more complete discussion refer to this review. Here we focus on two techniques: an electro-osmotic drag cell and fuel cell derived electro-osmotic drag coefficients because they have been the most commonly used to measure electro-osmotic drag coefficients under fully hydrated or DMFC operating conditions. Traditionally, electro-osmotic drag was measured using a two compartment cell where palladium or Pt-based electrodes are used to pass a current [54,73–75]. Under this set-up, the amount of water dragged across the membrane was determined by measuring the volume change of the cell compartments as a function of current passed. Later, Ren *et al.* developed a method for measuring electro-osmotic drag in a DMFC configuration [43,44]. In this method, water flux across the membrane was measured in a DMFC cell which was operated with dilute methanol solution (e.g., 0.5 M) at the anode and dry O_2 at the cathode. The water flux across membrane was then plotted as a function of current density (Figure 4.4). Electro-osmotic drag was then calculated from the slope of water flux versus current density plot at high current densities where the effects of diffusion were suppressed.

Beyond water uptake, self-diffusion coefficient and electro-osmotic drag coefficient of water; water in polymer electrolytes, because it is so critical to transport properties, has been studied by other methods. Specifically, the state of water has been linked to the properties and morphology of different copolymers. The state of water within these systems has been investigated using techniques such as NMR, differential scanning calorimetry (DSC), and Raman and dielectric spectroscopy. Through these experiments researchers have determine the state of water within these systems and related the state of water to specific membrane properties [76–78]. Morphology of the polymers has also been identified as a factor that impacts the state of water within a polymer electrolyte and thereby membrane properties [6,33,51,79–84]. Different experimental approaches have investigated polymer morphology included neutron and X-ray scattering, atomic force microscopy (AFM), scanning electron microscopy (SEM), and scanning tunneling electron microscopy (TEM) [85–90].

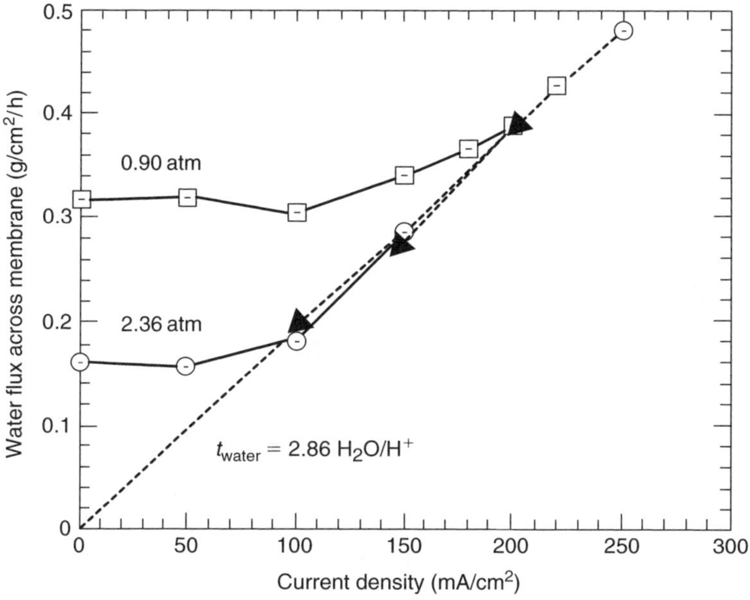

Figure 4.4 Water flux across a Nafion in a DMFC operated at 60°C for determination of electro-osmotic drag coefficient (t_{water}) (reproduced with permission from *J. Electrochem. Soc.* [44]).

4.3.4 Other Fuel Cell Relevant Membrane Properties

Other physical properties such as mechanical strength and chemical or thermal stability are critical for fuel cell performance, but are often overlooked by the research community. These properties have been investigated using a number of techniques including: dynamic mechanical analysis (DMA), thermal gravimetric analysis (TGA), DSC, fuel cell life testing and Fenton's reagent. While all of these techniques give important data regarding structure–property relationships within these materials, they will not be addressed in significant detail in this chapter, as we have focused this work on transport properties.

4.3.5 DMFC Performance

Improved DMFC performance is the ultimate goal for any DMFC researcher. Here we discuss how DMFC performance is typically presented. Later in this chapter we will review and compare reported DMFC performances. DMFC performance, like hydrogen fuel cell performance, is presented primarily in the form of polarization curves (voltage versus current density). This makes sense because it relates to cell power and stack/cell design concerns.

An example of a polarization curve taken at Los Alamos National Laboratory is presented in Figure 4.5 showing the dependence of fuel cell

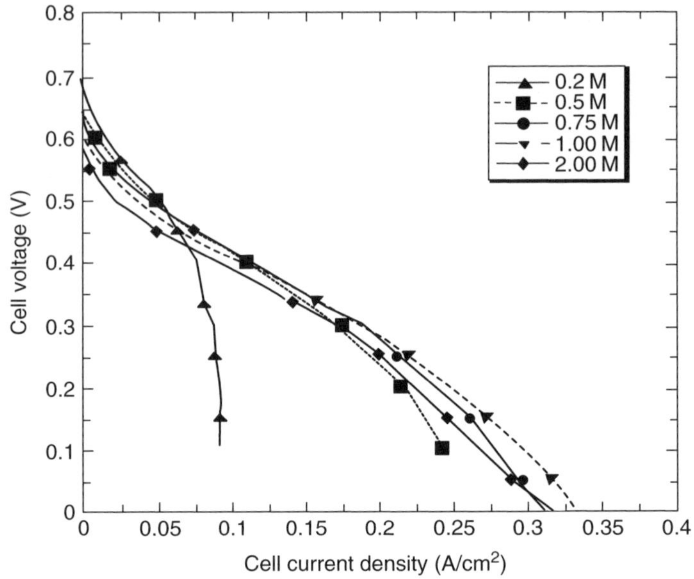

Figure 4.5 Polarization curves of a DMFC using Nafion 117 membrane at 80°C. The cell anode was fed with methanol solution of different concentrations and the cell cathode with 85°C humidified air at 2.1 atm at 0.6 l/min (reproduced with permission from *J. Electrochem. Soc.* [30]).

performance on methanol feed concentration. Fuel cell polarization is affected by overpotential at the electrodes (anode and cathode), ohmic losses, and mass transfer effects. DMFCs unlike hydrogen cells suffer from high anode overpotential and a mixed potential at the cathode due to the effects of methanol crossover. The slow kinetics of the methanol anode, in particular, are responsible for substantially lower power densities and the high catalyst loadings used in DMFCs. DMFCs typically utilize thick membranes to reduce methanol crossover resulting in increased ohmic resistances. The increased resistance of these systems is partially offset by the lower operating current densities compared to hydrogen cells. Mass transfer losses in DMFCs are possible due either to anode or cathode effects depending on operating conditions (methanol concentration, methanol or air stoichiometry, and temperature).

Although polarization curves are the most popular tool for evaluating DMFC performance, methanol crossover in these systems is important and cannot be fully interpreted by polarization curves. DMFCs suffer from significant fuel loss due to crossover even under optimized conditions; therefore, polarization curves alone are insufficient to fully characterize cell performance. Methanol crossover rates as a function of current density are commonly reported in conjunction with fuel cell polarization curves (e.g., Figure 4.3 shown earlier).

In order to combine the effects of fuel cell polarization and methanol crossover, overall efficiency has also been reported. Overall efficiency of a DMFC consists of a combined voltage and fuel efficiency. Voltage efficiency of the cell, ε_V, is determined as a function of cell current density from polarization curves, according to

$$\varepsilon_V = \frac{V_{cell}}{V_{Th}} \tag{4.7}$$

where V_{cell} is the cell voltage at a given current density, and V_{Th} is the theoretical maximum voltage of the cell ($V_{Th} = 1.21$ V at 25°C) based on free energy change of the overall reaction. Fuel efficiency takes into account the methanol crossover of a cell. The methanol that crosses over is oxidized directly at the cathode and results in an efficiency loss because the crossover methanol does not produce useful power. The fraction of methanol that reacts at the anode and produces useful power is usually termed fuel utilization. Fuel efficiency of a DMFC system is the amount of methanol that reacts at the anode, that used to produce useful power, versus the total methanol consumed by the system usually estimated to be the methanol reacting at the anode and that oxidized at the cathode. This ignores the possible methanol lost from CO_2 exhaust from the anode and unreacted methanol at the cathode. This assumption is justified for most efficient DMFC systems because these loss mechanisms are small compared to methanol reacting at the anode and cathode. The fuel efficiency, ε_{Rxn}, is expressed as

$$\varepsilon_{Rxn} = \frac{i_{cell}}{i_{cell} + i_{xover}} \tag{4.8}$$

where i_{cell} is the current density of the cell, and i_{xover} is the experimentally determined crossover current density. The fuel efficiency of these cells are simply the fraction of methanol that is reacted at the anode compared to the total amount of methanol consumed due to both reaction at the anode and methanol crossover. An example of fuel efficiency plotted versus cell current density is given in Figure 4.6. Combining voltage efficiency and fuel efficiency, an overall efficiency of the cell, $\varepsilon_{Overall}$, can be obtained:

$$\varepsilon_{Overall} = \varepsilon_v * \varepsilon_{Rxn} \tag{4.9}$$

The overall efficiency of a cell takes into account both cell polarization and methanol crossover, and will be presented later in our discussion of reported DMFC performance and in our comparison of optimized systems.

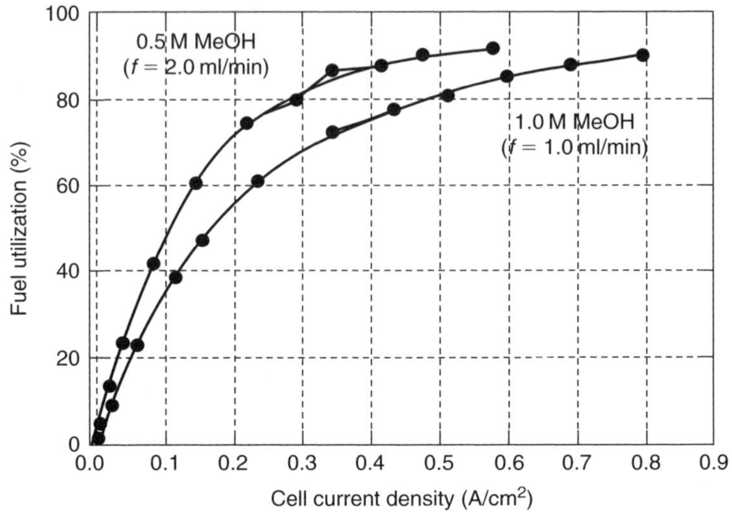

Figure 4.6 Fuel utilization (reaction efficiency) as a function of cell current density at 100°C (reproduced from Ref. [91]).

In designing practical DMFC systems, issues such as energy storage required, power demand, device size, and the ability to respond to dynamic system requirements will all be key issues in system optimization. It is vital to understand how fuel efficiency and power density affect these design issues. Cost, reliability, and response to environmental conditions are also important, but are beyond the scope of this discussion.

4.4 Polymer Electrolytes for DMFC

Having established a basis for evaluating DMFC electrolytes and fuel cell relevant properties, we shift our focus to the families of polymer electrolytes that have been investigated for DMFCs. There has been great effort in developing new materials for either cheaper or better performing fuel cell membranes. The discussion presented here is split between Nafion (or other perfluorinated ionomers), the standard for PEFCs, and alternative (non-perfluorinated) approaches. The alternative approaches presented include: sulfonated polyarylenes, which to date represent the most significant and promising research effort, PIs, PPZs, PSs, PVAs, and styrene block copolymers, as well as other membrane-based approaches. While some membrane/polymer properties are presented in this section, the discussion of general trends in fuel cell relevant properties and DMFC performance is presented in the following section.

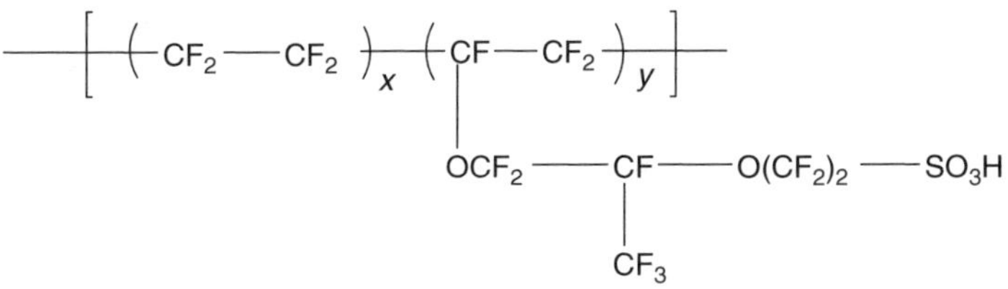

Figure 4.7 Chemical structure of Nafion®.

4.4.1 Perfluorinated Sulfonic Acid Ionomers

Nafion®, a perfluorinated sulfonic acid ionomer, has been the standard DMFC (as well as hydrogen fuel cell) membrane. Other perfluorinated ionomers such as Flemion, Aciplex, Gore-Select and Dow have been highly studied in hydrogen fuel cell systems, however rather little appears in the literature on the DMFC performance or methanol permeability of these materials [92–97]. The chemical structure of Nafion consists of a perfluorinated backbone with pendent vinyl ether side chains terminating with SO_3H, see Figure 4.7. Alternative perfluorinated ionomers share similar chemistry, usually with minor modifications to the ionomeric side chains. Equivalent weight of these ionomers has been varied by changing the ratio of backbone (tetrafluoroethylene) to side chain (perfluorovinylether) monomer in the polymerization step. 1100 equivalent weight Nafion is by far the most commonly used, and unless otherwise specified the use of the term Nafion here will refer to 1100 equivalent weight Nafion.

The effect of different equivalent weights and membrane thickness on methanol permeability has been investigated, showing improvement under the conditions studied for higher equivalent weight and increased membrane thickness [30,98]. While some benefits were found or might be expected using other systems, it might be anticipated that these benefits will be fairly modest due to chemical and structural similarities.

Nafion, like other perfluorinated ionomers, is quite resistant to chemical attack and has a highly phase separated morphology that imparts excellent proton conductivity with moderate water uptake (i.e., 30%) [99]. Nafion has been thoroughly studied in terms of conductivity, however a wide range of values have appeared in the literature. Values depend on test conditions, most importantly temperature and hydration, but also on pretreatment of the membrane [26,100–102]. Nafion is often cleaned before testing by boiling in peroxide to remove organics, acid to ensure full conversion to proton form, and water to remove acid introduced in the prior step. While many areas of DMFC characterization need standardization, proton conductivity of Nafion in liquid water at 25°C is listed as 0.083 S/cm on Du Pont's product information sheet [103].

The most significant drawback of perfluorinated sulfonic acid ionomers in DMFC applications has been their relatively high methanol permeability. While these membranes have shown high methanol permeability, DMFCs based on perfluorinated ionomers have been demonstrated with ~90% fuel utilization under optimized conditions [42,93], a value that makes DMFCs, even those based on Nafion, a potentially viable technology. Methanol permeability of Nafion has often been reported. Typical values for Nafion are in the range of $1.2–3.0 \times 10^{-6}\,cm^2/s$ at room temperature, although the permeability values reported have a large spread depending on test conditions [53,104–106].

Because Nafion has high methanol permeability, a thick membrane (Nafion 117, 7 mil thickness) has been typically employed for DMFCs to reduce the impact of methanol crossover. While thinner membranes (e.g., Nafion 112, 2 mil thickness) are generally applied to hydrogen/air applications to minimize ohmic losses, these membranes exhibit prohibitive methanol crossover rates under most operating conditions. Recently, a thicker (Nafion 1110, 10 mil thickness) was commercialized allowing a further decrease of methanol crossover rates but many researchers still use Nafion 117 as the standard DMFC membrane.

Electro-osmotic drag is also a concern as Nafion exhibits much higher drag coefficients than other membranes. This has been shown to be important for low power density systems for portable applications [44,71,107] and can also be important for cathode flooding or thermal balances in operating systems.

4.4.2 Alternative Polymer Electrolytes

Primarily due to the high methanol permeability of perfluorinated copolymers, significant research has been conducted on developing lower methanol permeable electrolytes. Figure 4.8 displays (as a pie chart) the type and number of alternative DMFC membranes reported in the literature from SciSearch® over the past 10 years. The pie chart is separated into various polymer types (polyarylenes, PIs, PPZs, radiation grafted polystyrenes, polystyrene block copolymers and PVAs) and various membrane approaches (inorganic–organic composites, polymer–polymer blends and layered structures). The chemical structures of the polymer electrolytes are shown in Figure 4.9 for reference.

Figure 4.8 highlights the significant amount of research performed on polyarylenes such as poly(arylene ether sulfone) (PES) [86,108–114], poly(ether ether ketone) (PEEK) [115–118], and other polyphenylene copolymers [119–121]. These polymers are traditional engineering polymers and are known for their good thermal/mechanical properties, oxidative stability, and processibility [122]. In their sulfonated form, they have shown good mechanical properties, proton conductivity and relatively low methanol permeability

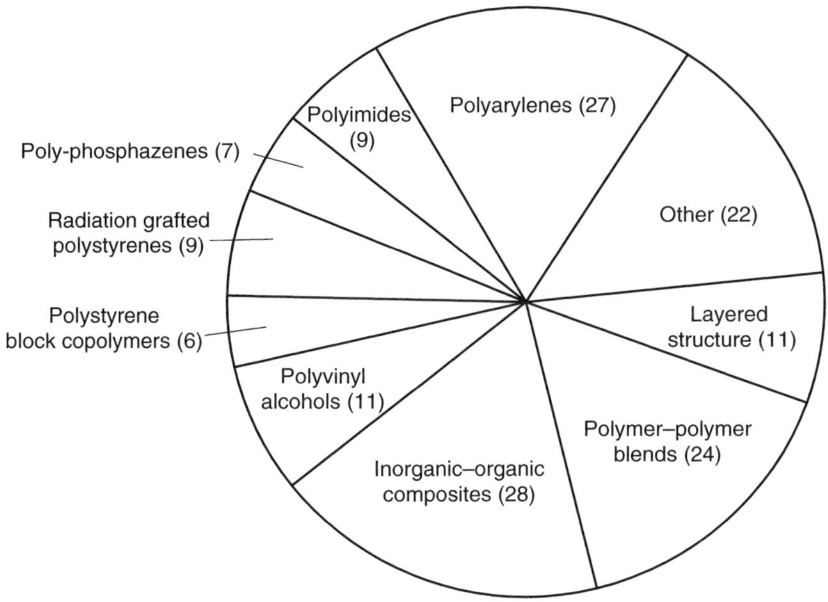

Figure 4.8 Type and number (in parenthesis) of DMFC alternative membranes papers appearing in open literature from SciSearch® for years 1994–2004.

compared to Nafion. Additionally, they have demonstrated fuel cell lifetimes up to 3000 h [123], making them compelling alternatives for Nafion particularly in the area of DMFCs. Water uptake of these polymers has been relatively high compared to perfluorinated ionomers of similar ion exchange capacity (IEC) or proton conductivity. The water uptake of these systems has been linked to membrane–electrode interfacial effects [124]. This has led to efforts to lower water uptake in these systems. Current approaches include direct monomer sulfonation [113,114,125], block copolymer synthesis [126, 127], introduction of fluorine/polar group moiety [128,129], and physical/chemical blending with other functional groups [130–133].

Sulfonated PIs are similar in chemical structure to the polyarylenes just presented, however they have the added presence of imide bonds within the polymer backbone. Sulfonated PIs have been found attractive for DMFC applications due to their even lower methanol permeability and water uptake compared to polyarylenes [134–140]. For these reasons and their slightly lower proton conductivities, sulfonated PIs having higher sulfonation levels have been developed for DMFCs. A major shortcoming of these materials has been hydrolytic instability. Over time under operating conditions, the chemical degradation of the PI backbone has caused a decrease in conductivity and a loss of mechanical properties, resulting in lifetimes too short for most practical applications [141]. Researchers have attempted to replace five-membered rings with the more stable six-membered ring of the naphtahalenic PI

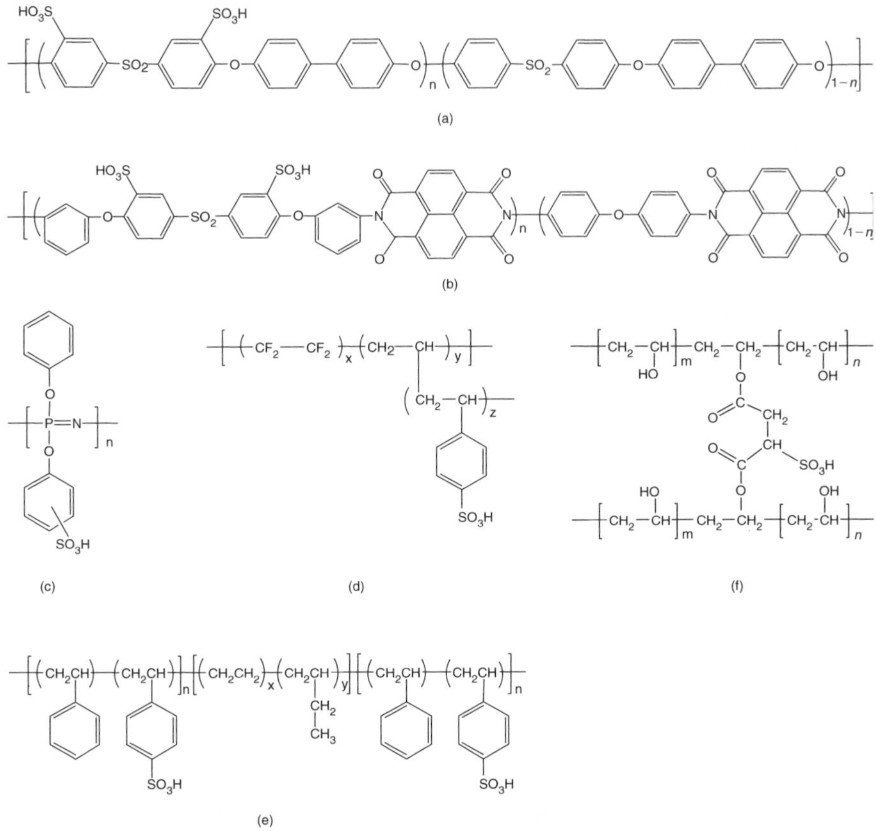

Figure 4.9 Chemical structure of alternative membranes (a). sulfonated polyarylene ether sulfone, (b). sulfonated six-membered ring PIs, (c). Sulfonated poly (bisphenoxy) phosphazene, (d). ETFE-*g*-PSSA, (e). Sulfonated SEBS block copolymer, (f). Sulfosuccinic acid crosslinked PVA.

[137,138] and have increased the electron density of the imide nitrogen atom [142] in an attempt to improve the hydrolytic stability of the imide bond. To this point, lifetime improvements reported have not been compelling for most applications.

Sulfonated PPZs have also been reported as potential DMFC polymer electrolytes [143–145]. These polymers have been reported to have good chemical and thermal stability. In addition, versatility of chemical synthesis has allowed a variety of chemical structures to be investigated. Membranes with an IEC < 1.5 mmol/g have exhibited reasonable mechanical properties with surprisingly good oxidation stability in hot hydrogen peroxide/ferrous ion solution [143]. Further dimensional and mechanical stability was explored by cross linking [143,146] and physical blending with other hydrophobic polymers such as polyacrylonitrile and polybenzimidazole [147,148]. Initial data on methanol permeability of sulfonated PPZ were obtained from the desorption kinetics and diffusion of methanol and water by NMR. These data indicated that these membranes showed significantly lower methanol permeability (e.g., 400 times lower than Nafion for sulfonated PPZ (IEC = 1.4 mmol/g)) [143]. However these measurements were problematic since these methods measured only self-diffusion coefficients by the corresponding thermodynamic factor. Later permeability results using a diffusion cell showed that methanol permeability of ultraviolet cross linked, sulfonated PPZ (IEC = 1.12 mmol/g) was approximately 2.4 times lower than that of Nafion 117 [145,146] significantly higher than values reported for similar materials earlier and casting doubt on the reported values. To this point, reported fuel cell performances for these materials have not been compelling and do not appear as promising as other alternative polymers.

Radiation grafted polymers have been proposed for DMFC polymer electrolytes as well [149–153]. Typically, polystyrene has been grafted to a relatively inert porous framework such as polyethylene–tetrafluoroethylene, PVDF or low density polyethylene by γ-radiation or electron beam irradiation. This process has been followed by sulfonation with chlorosulfonic acid to incorporate acid functionality into the polystyrene region. This procedure has resulted in reasonable proton conductivity and decreased methanol permeability compared to Nafion. Divinylbenzene has been used to create crosslinks between grafts and thus reduce water swelling of the membranes. Polymer degradation due to radiation treatment has created durability problems in specific instances [154]. While the primary concern with similar polymer systems has focused on losing IEC due to backbone degradation of styrene under fuel cell conditions [155,156], literature reports from these materials have shown reasonable lifetimes, up to 2000 h with little performance degradation in DMFC testing [153,157]. These data suggest these classes of materials still merit further exploration.

Several researchers have suggested that sulfonated styrene block copolymers could be used for DMFC membranes. Several block copolymers, such as sulfonated poly(4-vinylpyridinium-styrene-4-vinylpyridinium) [158], sulfonated poly(styrene-(ethylene-co-butylene)-styrene) [159–161], and sulfonated poly(styrene-isobutylene-styrene) [162–164] have been synthesized and characterized. Sulfonated polystyrene block copolymers have the advantage of controlled morphology by tailoring the block length and composition of the unsulfonated starting polymer. Through various chemical selections, a wide range of conductivity and methanol permeability can be obtained. Crosslinking [165] and physical blending techniques [166,167] have also been applied to this system in an attempt to further reduce methanol permeability. Inferior oxidative stability is a major concern for these electrolytes and has thus far restrained the operating temperature to less than 60°C. Very limited DMFC fuel cell performance has been reported using this type of polymer and what has been reported has been inferior to Nafion.

PVA is the final specific polymer family presented. PVA-based membranes have been investigated for DMFCs due to their good permselectivity of water over methanol. To impart proton conductivity and/or reduce water/methanol permeability, several techniques have been tried such as crosslinking with sulfur succinic acid [168,169], blending with conductive inorganic filler [170–173], and blending with highly conductive polymers [174,175]. However, chemical stability of these polymers is poor and no DMFC performance has been reported. Overall, they do not show promise as DMFC electrolytes.

Beyond pure polymer systems, several inorganic–organic composite membranes, where particles of suitable inorganic fillers are dispersed in the polymer matrix, have also been developed for DMFCs. These inorganic–organic composites have been based on layered metal phosphate, metal oxide, silica, or heteropolyacids in a conductive polymer matrix. The resulting membranes are made by casting a dispersion of filler particles in an ionomer solution or by growth of the filler particles within a preformed membrane [176]. Generally, the use of the inorganic materials has been aimed at increasing proton conductivity and/or decreasing methanol permeability. For highly conductive membranes such as Nafion, exfoliated inorganic particles have been used for lowering methanol permeability [177–182], while highly conductive inorganic fillers have been used for non-conductive membranes such as PVA and polyethylene glycol [171–173,183–186]. These composites represent a wide range of compositions from low contents of well-dispersed additives to high contents of interpenetrating phases. Improved DMFC efficiency has been reported with zirconium phosphate incorporated sulfonated polyarylenes [187,188], zirconium phosphate incorporated Nafion [189], and montmorillonite modified Nafion [190], although power outputs were sometimes low.

Polymer–polymer blends are another type of composite membrane that have been investigated for DMFCs. While many examples showed that physical blending of Nafion with highly selective polymers such as PVDF and PVA might have a positive effect on lowering methanol permeability, undesirably decreased proton conductivity often offset the apparent benefits from physical blending [191–193]. Blend membranes having acid–base interactions or covalent crosslinking appear more promising because of significantly reduced methanol permeability and water uptake with only modest decreases in conductivity [194–199]. The research group of Kerres reported membrane properties and improved DMFC performances at temperatures up to 130°C for a number of polymer–polymer blend membranes [195]. Brittleness of these materials, particularly in the dry state, remains a problem and work in this area continues.

A somewhat different multicomponent approach for DMFC membranes is layered composites. Layered composites consist of multiple layers of different materials designed to have high conductivity, low permeability, and good compatibility between layers and with the electrodes. This approach was originally designed for highly methanol permeable Nafion membranes. A number of methanol barrier layers have been reported with Nafion for improved DMFC performance including: palladium [200], polypyrrole [201], polybenzimidazole [202] and PVA [174]. This approach has also become popular for alternative membranes in efforts to improve compatability between alternative membranes and electrodes. Three-layer structures, in which a highly selective membrane is sandwiched between two Nafion membranes, have been reported for polyarylenes and PPZ [135, 203–205]. Additionally, hydrophobic maleic anhydride layers have also been used as the selective layer using a plasma polymerization technique [206]. The layered composites have been reported to improve cell performance when compared with pure polymer electrolytes and/or Nafion under DMFC conditions. Lifetime performances have not been reported.

Beyond the approaches and membranes listed previously, several other approaches to improved DMFC membrane performance have been attempted. For example, acid-doped polybenzimidazole membranes were studied for high temperature (200°C) DMFC operation [207–209]. Highly conductive polymers and acid-doped nanosized particles were used as fillers in inert porous substrates such as PVDF [210], crosslinked polyethylene [211], and SiO_2 glass [212]. In general, these other approaches showed limited potential or were restricted to very specific operating windows.

Property differences of the polymer electrolytes presented here are due to both chemical and structural effects. There have been several papers that have investigated the relationship between chemistry/structure and properties for these types of polymer electrolytes [51,111,213–215]. While a detailed discussion of these relationships is beyond the scope of this chapter, these factors are critical in determining DMFC performance.

4.5 Evaluation Criteria for DMFC Electrolytes

In the previous section, we briefly reviewed the various types of PEMs developed for DMFC applications. We now focus our attention on the evaluation and comparison of these membranes. This discussion covers both free-standing membranes and MEAs used in DMFC testing. These two separate comparisons are necessary due to the difficulty in preparing high performance MEAs from alternative polymer electrolytes and the dependence of DMFC performance on operating conditions.

Improved performance is the ultimate goal of any alternative DMFC electrolyte. Although there are a number of critical membrane properties for DMFCs, determining tradeoffs between individual properties is complex and is highly dependent on operating conditions. We begin our discussion with selectivity, a qualitative basis for screening DMFC electrolytes. We then discuss membrane–electrode interfacial issues and their impact on DMFC performance. We follow this with a presentation of literature reports of DMFC performance, highlighting alternative membrane systems. We finish with a discussion of relevant DMFC performance comparisons made under optimized conditions.

Selectivity, which is defined as the proton conductivity divided by the methanol permeability, has been suggested as a qualitative tool for screening the potential of one polymer versus another [39]. This criterion is reasonable because as an electrolyte for DMFC applications, a low ohmic resistance and a low methanol crossover are desired. Additionally, both of these quantities are independent of membrane thickness which affects resistance and crossover losses in opposite ways, i.e., thinner membranes have lower resistive losses but higher crossover losses. While selectivity does not take into account the effects of electro-osmotic drag or methanol consumption within the anode on crossover, it does combine two of the most critical DMFC performance properties, conductivity and permeability.

Numerous studies have investigated the methanol permeability and proton conductivity of alternative electrolytes and compared them to Nafion in an attempt to show improved promise for DMFCs. We have compiled the results of a number of these studies in Figure 4.10 as relative proton conductivity versus relative methanol permeability. In this study the conductivity and

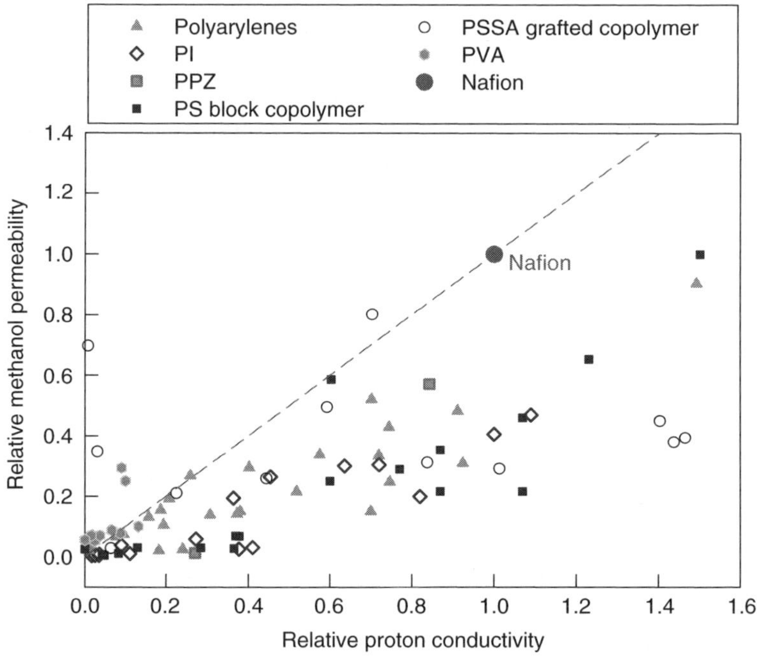

Figure 4.10 Relative methanol permeability versus relative proton conductivity for alternative PEMs [Refs. 80,113,115,117,125,136,137,148,150,157,161,162,165,182, 216–232].

permeability of alternative membranes have been normalized to the values reported for Nafion. Nafion has been defined as a relative conductivity and permeability of 1 (this approach will be carried throughout this paper and Nafion will always be given a relative value of 1). The dashed line in Figure 4.10 is a line of constant selectivity drawn through Nafion. Data that fall above the dashed line have poorer selectivity than Nafion and data below the dashed line have better selectivity than Nafion. Although the data scatter is significant, most alternative PEMs exhibit a clear trend toward higher selectivity than Nafion. This improved selectivity is primarily due to lower methanol permeability. While many of the alternative membranes have lower conductivity than Nafion, the relative decrease of permeability is larger than that of conductivity.

In order to help visualize selectivity in these systems, the data in Figure 4.10 has been plotted as relative selectivity (conductivity/permeability) versus relative proton conductivity in Figure 4.11. From this graph it is apparent that alternative membranes reported so far have selectivity much higher than Nafion up to a factor of 8 at low relative proton conductivity and routinely a factor of 2 to 3 times more selective for a wide range of polymers. This improvement in selectivity for alternative polymers has been rationalized in terms of microstructure and the chemical compositional differences between Nafion and alternative polymers [6,51,233].

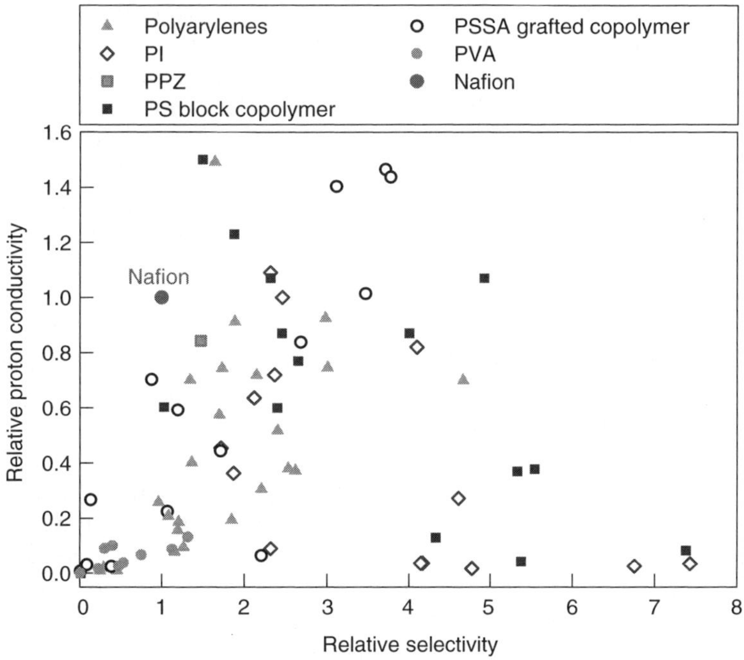

Figure 4.11 Relative proton conductivity versus relative selectivity for alternative PEMs [Refs. 80,113,115,117,125,136,137,148,150,157,161,162,165,182,216–232].

The emergence of highly selective PEMs is very encouraging because PEMs with greater selectivity have potential for improved DMFC performance. However, just because selectivity of a PEM is higher than Nafion does not mean improved DMFC performance can be achieved. One limitation of selectivity as a gauge of DMFC performance is due to the fact that a minimum conductivity is still required for DMFC operation, regardless of how low methanol permeability is. This minimum conductivity is related to the fact that the membrane itself has a minimum attainable thickness due to issues of mechanical robustness and reactant crossover. While single component membranes are preferred due to simplicity, it is possible to greatly reduce the thickness of highly selective membrane layers by applying them to support layers with higher, less selective transport properties. However, compatibility issues between these layers need to be addressed before this approach is viable, and processing concerns will still ultimately limit the minimum thickness of the selective layers, requiring some minimum thickness for application. These concerns would limit the useful membranes in Figure 4.11 to a relative conductivity greater than perhaps 0.4 for most applications.

Selectivity also neglects mechanical properties, gas permeability, electronic conductivity, thermal and chemical robustness, and the ability to fabricate high performance MEAs. Chemical stability was discussed in some detail in the prior section as a function of polymer architecture, and most of the effort in alternative membranes has focused on membranes that have shown good chemical stability. Gas permeability and electronic conductivity are not problems for typical DMFC membranes. Mechanical stability and the ability to fabricate high performance MEAs are two critical issues that are often overlooked in evaluating alternative DMFC electrolytes.

Several research groups have reported that modulus of PEMs significantly decreased upon hydration, and membranes with higher water contents (within a polymer family) exhibited less tensile strength and toughness [148,204,234]. Performance losses due to the deterioration of mechanical properties evidenced by pin-hole formation, membrane tearing, and membrane thinning have been reported during hydrogen fuel cell testing and are likely to occur in DMFC systems as well [130,147,150,174,203,235]. These mechanical property losses may be accelerated by chemical or thermal processes, but still have received only limited attention.

Another factor largely overlooked in the development of alternative polymer electrolytes is the development of highly performing MEAs. A few studies have reported the importance of the membrane–electrode interface on the performance of alternative MEAs with Nafion-bonded electrodes [5,150,236,237]. Kerres *et al.* pointed out that MEAs using poly(ether ether ketone) based copolymers exhibited relatively poor cell performance compared to Nafion MEAs due to the non-optimized interface between catalyst and membranes [5]. Scott *et al.* also observed poor bonding of the catalyst layer due to highly water swollen radiation grafted membranes, resulting in poor cell performance at high current density [150]. Recently, Holdcroft *et al.*

indicated that their highly conductive polystyrene grafted copolymers had significant contact resistance with Nafion-bonded electrodes caused by delamination of the catalyst layers [236].

Nafion-based electrodes have been optimized for performance in DMFC systems for over 15 years, have high reactant permeability (a trait desirable in the electrodes, but undesirable for the membrane) and, have good stability without adversely interacting with the catalyst. Alternative electrodes are an area of ongoing study [236,238–241], but performance equivalent (or relatively close) to that from Nafion-based electrodes has yet to be reported. These difficulties in preparing high performance MEAs due to membrane–electrode compatibility is one of the primary reasons that improved membrane properties has not always resulted in improved DMFC performance.

The origin of membrane–electrode compatibility has been a subject of significant effort within our research group over the past few years. We have attributed membrane–electrode interfacial resistance to differential swelling between the membrane and electrodes leading to electrode delamination [35]. Other factors such as mismatch of the electro-osmotic drag coefficients of the membrane and electrodes or poor adhesion between dissimilar polymers were also investigated but found to be of significantly lesser importance. Through modification of PEM chemistry improved membrane–electrode compatibility [128] and improved lifetime performance [237] in DMFCs has been demonstrated.

Water uptake of PEMs is important for both membrane–electrode compatibility and mechanical properties of the membrane. Figure 4.12 shows the relative water uptake of various PEMs as a function of relative conductivity. Because density data on the alternative polymers was not available in many cases, the relative water uptakes in Figure 4.12 have been reported on a mass basis. Nafion with a higher density than non-perfluorinated polymers would be shifted slightly higher compared to the other membranes when considered on a volume basis. Still, it is noted that Nafion competes favorably on a conductivity to water uptake basis, having the highest conductivity for membranes with similar water uptakes. This behavior in part describes why Nafion has been so successful as a hydrogen PEM where conductivity is of central importance and reactant permeability is a lesser concern. In some cases alternative PEMs having the same water uptake as Nafion have up to 5 times lower conductivity and membranes with similar conductivity can have up to 3 times the water uptake. Although there is no clear guideline for maximum allowable water uptake, higher water uptakes can lead to difficulties preparing robust and high performing MEAs or decreased durability during cycling between different levels of hydration.

Concerns with preparing highly performing MEAs have led to several strategies for controlling PEM properties. One approach has been introducing hydrophobic elements into the polymer backbone in an effort to increase phase separation. Polymer backbone fluorination with hexafluoro bisphenol A for polyarylenes and trifluorostyrene for styrene sulfonic acid

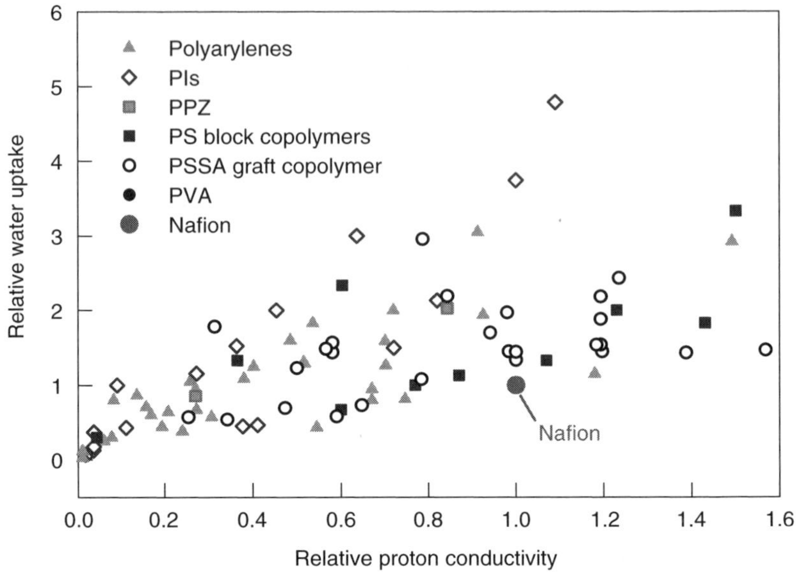

Figure 4.12 Relative water uptake versus relative proton conductivity for alternative PEMs [Refs. 80,113,115,117,125,136,137,148,150,157,161,162,165,182,216–232].

have both been demonstrated [111,242,243]. Another approach used specific interactions such as van der Waals, ionic interactions (acid–base) or covalent crosslinking [130,229]. Additionally, incorporating highly conductive additives such as heteropolyacid, Nafion, and metal oxide particles has been investigated for increased conductivity and decreased methanol permeability [184,205,244,245]. Finally, a number of controlled morphology approaches such as block or graft copolymer architectures and various thermal/solvent pretreatments have been reported [4,40,246–248]. Each of these approaches have showed various levels of promise and reflect general trends in the research direction of alternative PEMs.

While the search for improved membranes continues, a compilation of results reported to date has not been attempted. Table 4.1 displays select examples of DMFC performance using alternative PEMs from references that have reported as better than or comparable performance when compared to Nafion. While no suitable references using polystyrene block copolymers or PVA-based copolymers could be found, Table 4.1 does includes data from polyarylenes, PIs, radiation grafted polystyrene, polymer blends, layered composites, and inorganic–organic composites. Table 4.1 contains a significant amount of data for the membranes investigated. Presented are IEC; relative conductivity, selectivity and water uptake; membrane thickness; operating methanol concentration; cell temperature; OCV; and cell potential at $100\,mA/cm^2$ (an arbitrary measure of performance at a reasonable operating point for high efficiency).

Trends in membrane properties from Table 4.1 include alternative membranes with relative minimum conductivity greater than 0.4 (at least 40% the conductivity of Nafion), improved selectivity (up to 13.6 times greater than Nafion), and relative water uptake generally below 2 (although a single exception at 3 exists). These reported properties reflect our earlier discussion regarding membrane properties and needs of DMFC electrolytes. Alternative membrane thicknesses typically employed are thinner than Nafion due to their increased selectivity, but often lower proton conductivity. Typical IEC values of alternative PEMs are significantly higher, 1.2–2.3 meq/g, than Nafion, 0.9 meq/g. However, it should be noted that IEC and relative water uptake values are based on mass, as density of the polymers presented was often not available. Because fluoropolymers have higher density than hydrocarbon polymers, the gaps between Nafion and the alternative polymers in IEC and relative water uptake would decrease on a volume basis. In fact, it is a volume basis that is the more appropriate comparison basis because the transport properties presented occur over distances that are related to volume, but independent of mass. Still, the presentation in its current form is useful, showing general trends between alternative polymers and meaningful differences when compared with Nafion.

OCV and cell performance for these membranes are also shown in Table 4.1. While all the membranes presented showed at least equal performance to Nafion for the conditions presented; the DMFC cell performance reported was not quantitatively proportional to the relative selectivity of PEMs. For example, highly selective ETFE-g-PSSA (relative selectivity = 4.5) showed almost identical performance to Nafion, while the less selective sulfonated PES (selectivity = 1.6) exhibited a much improved performance. Also apparent is the wide scatter in cell performance (0.31–0.63 V at 100 mA/cm^2) and OCV reported (0.55–0.88 V). Unlike membrane properties, showing general trends and making comparisons is more difficult with DMFC performance. This is due to issues involving MEA fabrication and cell operating conditions and their effects on the observed DMFC performance, factors that we discuss in detail in the following paragraphs.

While we have already discussed membrane–electrode compatibility issues as a concern for DMFC performance, other electrode issues such as performance, durability, and reproducibility are also important. The problem arises in part because MEA fabrication techniques are often vague when reported or proprietary and not reported. MEA quality and performance, in general, is highly variable between various research groups making direct lab to lab comparison of DMFC performance difficult. By reporting data in Table 4.1 as we have, a comparison between Nafion and alternative membranes, we have removed some of the lab to lab comparison features. However, MEA preparation on the research level is typically done "by-hand" leaving serious questions about reproducibility from sample to sample even within a given research group. Industrial MEA producers certainly do use automated production of

Table 4.1 DMFC performance comparison of various copolymers [Ref. 128,130,138,148,150,157,174,203,204,234,235,249]

PEM	IEC (meq/g)	Relative Conductivity	Relative Selectivity	Water uptake	Thickness (μm)	MEOH concentration (M)	Cell temperature (°C)	OCV (V)	Cell voltage at 100 mA/cm² (V)	Reference
Sulfonated PEEK	1.4	0.4	13.6	1.3	82	2	80	0.63	0.42	[204]
Nafion 115	0.9				152			0.60	0.38	
Sulfonated PES	1.7	0.9	1.9	1.8	127	1	60	0.73	0.52	[235]
Nafion 117	0.9				180			0.70	0.45	
Sulfonated PES	1.3	0.8	1.6	1.3	137	0.5	80	0.88	0.58	[128]
Nafion 115	0.9				127			0.76	0.53	
Sulfonated PI	2.3	~1	~2	1.6	50	1	80	0.60	0.42	[249]
Nafion 112	0.9				50			0.55	0.40	
Sulfonated PI	1.8	0.6	2.1	3.0	112	0.5	80	0.81	0.57	[138]
Nafion 117	0.9				180			0.80	0.54	
PVDF-g-PSSA	2.0	1.5	3.4	1.5	50	2	80	0.68	0.46	[157]
Nafion 117	0.9				170			0.60	0.33	
ETFE-g-PSSA	1.4	1.9	4.5	1.7	50	2	80	0.59	0.31	[150]
Nafion 117	0.9				170			0.59	0.31	
SPEK/PBI/bPSU blend	0.7	–	–	1.0	60	1	110	0.82	0.63	[130]
Nafion 117	1.0				180			0.80	0.60	

SPPZ/PBI blend	1.2	0.6	1.6	1.4	77	1	60	>0.8	0.46	[148]
Nafion 117	0.9				178			0.63	0.46	
Nafion/PEEK/ Nafion tri-layer composites	–	0.4	1.0	0.8	15/70/15	1	60	–	0.49	[203]
Nafion 115	0.9				125			–	0.46	
Nafion/(PVDF–Nafion blend)/ Nafion tri-layer composites	–	<1.1	1.6	–	–	2	90	0.69	0.49	[174]
Three layers Nafion 112	0.9				150			0.65	0.46	
Nafion/PVDF/ Nafion tri-layer composites	–	0.4	1.4	<1.0	20/10/20	1	60	0.80	0.42	[234]
Nafion 117	0.9				175			0.80	0.40	
Nafion/ Montmorillonite composites		0.4	1.3	–	120	2	40	0.80	0.41	[190]
Nafion 115	0.9				120			0.80	0.40	

MEAs and quality control to ensure better reproducibility from sample to sample; however, this is of little use to the researchers investigating novel polymers and MEAs for improved performance [250,251].

For reference, we include a brief discussion about MEA fabrication techniques. Three methods of DMFC MEA fabrication have been reported: conventional paste [252,253], decal transfer [254,255], and direct membrane transfer [59]. All methods produce well-dispersed catalyst inks containing Nafion in a dispersion and catalyst particles. For the conventional paste method, the catalyst ink is transferred to gas diffusion layers by-hand painting or spraying. This step is then followed by pressing the PEM between two catalyst coated gas diffusion layers. For the decal transfer method, catalyst ink is coated onto decals (usually Teflon). The catalyst coated decals are then transferred by hot pressing on to the PEM. For the direct membrane transfer method, the catalyst ink is directly applied to the PEM by techniques such as spraying, hand painting, or screen printing. While performance comparisons between these MEA fabrication methods have been shown to be important, even changes to processing conditions using the same MEA fabrication method such as hot pressing temperature, time, and catalyst ink composition or processing can have a large effect on resulting DMFC performance [256–261]. These MEA fabrication factors play a large role in the observed DMFC performance.

Clouding the interpretation of DMFC performance results even further; cell operating conditions also strongly influence DMFC performance. The membranes presented in Table 4.1, show a large variability in membrane thickness (50–180 μm), methanol feed concentration (0.5–2 M), and cell temperature (60–110°C). A number of studies investigating the effects of operating conditions on DMFC performance have been reported. These include the effects of electrolyte thickness [157,262], methanol feed concentration [37], cell operating temperature [260], oxidant feed conditions [263], and cell humidification [264]. Additionally, design factors such as flow field design [265,266], different gas diffusion layers [250,267], and cell compression [251] have also been found to be important. Due to the dependence of DMFC performance on these variables and the previously mentioned issues, making reasonable and relevant comparisons between membranes based on DMFC performance is difficult at best.

A common approach, and the one presented in Table 4.1, is to compare performance of difference membranes by applying identical operating conditions during testing. This approach has limitations because operating conditions may not be equivalent for different PEMs. For example, high methanol feed concentrations or thin membranes may significantly lower the performance of a polymer like Nafion that has high methanol permeability at low current densities. While thick membranes with high selectivity but low conductivity may show significantly lower performance at high current densities. For hydrogen fuel cells, significant effort is being placed into establishing standardized testing protocols. This is possible because the role of hydrogen crossover in these systems can largely be ignored when comparing fuel cell performance. For DMFCs, standardized test conditions would not be sufficient due to tradeoffs between resistive and crossover losses. What is required for DMFCs is a comparison of optimized test conditions for each individual membrane, so that relevant comparisons can be made between different classes of polymer electrolytes. While not the primary focus of this chapter, DMFC performance optimization needs to be considered in the evaluation of alternative electrolytes. In the following paragraphs, we discuss performance optimization and give an optimized performance comparison between Nafion and an alternative polymer electrolyte (BPSH-30) as an illustrative example.

DMFC performance optimization of Nafion has been performed by several researchers [268–270]. An example from Meyers *et al.* simulated the performance of DMFCs as a function of membrane thickness and methanol feed concentration [269]. Figure 4.13 shows the effect of Nafion membrane thickness on overall performance of the DMFC (Nafion 112, 115 and 117 are approximately 50, 125 and 175 μm in thickness, respectively). Figure 4.13 illustrates the tradeoffs in methanol crossover and ohmic losses in DMFCs. At low

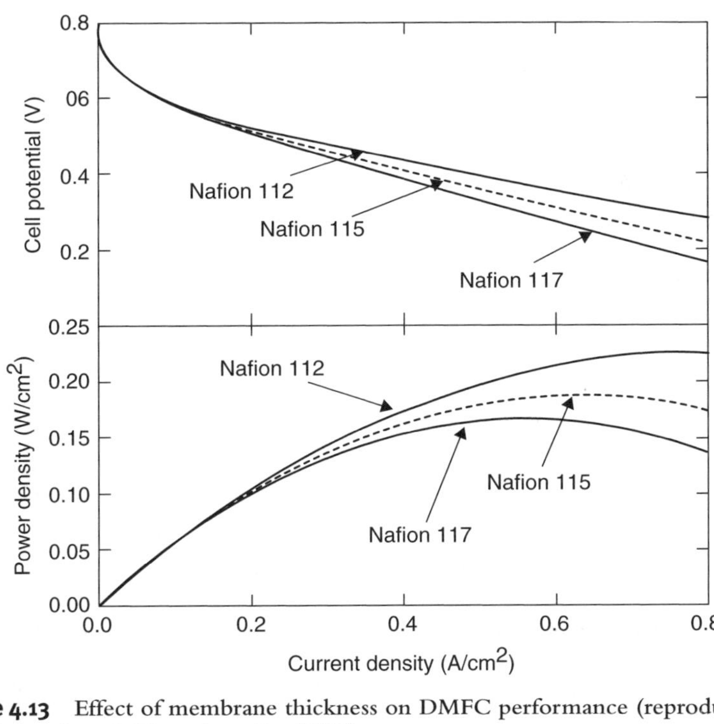

Figure 4.13 Effect of membrane thickness on DMFC performance (reproduced with permission from *J. Electrochem. Soc.* [269]).

current densities, where performance depends more on methanol crossover than conductance, thicker membranes have higher voltages, due to lower methanol crossover rates. At high current densities, ohmic losses become more important and thinner membranes yield better performance. These results illustrate the importance of membrane thickness even within the same polymer family. However, these performance take into account only voltage efficiency and not the reaction efficiency of the cell due to crossover effects.

Figure 4.14 combines reaction and voltage efficiency for a Nafion 117 membrane and shows overall efficiency and optimized methanol feed concentration as a function of power density. The results show that at low power requirements, a low feed concentration is best, and as power requirements increase, optimum methanol feed concentration increases. Xu *et al.* reported

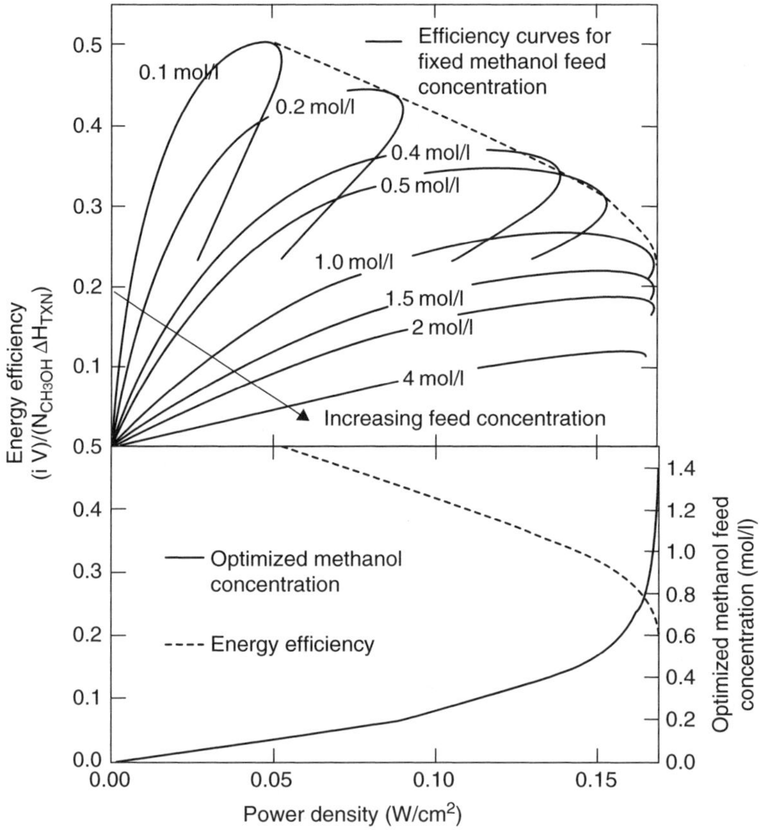

Figure 4.14 Selection of optimum methanol feed concentration. The upper graph shows energy efficiency as a function of power density for a series of fixed feed concentrations; the lower graph shows feed concentration and efficiencies which correspond to the envelope of peak energy efficiencies for a prescribed power density (reproduced from Ref. [269]).

similar results using a sensitivity analysis between cell voltage and methanol feed concentration [270]. These results make sense because DMFC systems require adequate methanol to meet power demands and avoid mass transport losses, while additional methanol raises crossover rates and lowers overall efficiency. These optimization analyses highlight the importance of required power and the ability to respond to transient demands versus fixed point operation. These are system concerns that depend on the given application, and are additional factors that need to be considered in PEM optimization.

The implementation of alternative PEMs in DMFC applications will require optimized conditions. In order to provide a fair comparison basis for Nafion and alternative polymers in DMFCs, we present an optimized comparison between Nafion and an alternative polymer electrolyte, disulfonated poly(arylene ether sulfone) with degree of disulfonation of 30%, (denoted as BPSH-30) [271]. Table 4.2 compares the electrochemical properties of BPSH-30 with Nafion. BPSH-30 has lower conductivity but higher selectivity than Nafion, similar to many of the alternative polymers presented in Table 4.1. The water uptake of BPSH-30 was only slight higher than that of Nafion when reported in vol. %. This led to a slightly higher interfacial resistance [35], but relatively good interfacial membrane–electrode compatibility.

Figure 4.15 compares the maximum efficiency of Nafion and BPSH-30 as a function of membrane thickness and methanol feed concentration at an operating temperature of 80°C. The curves in Figure 4.15 represent optimized performance as function of changing methanol feed concentration for membranes of a given thickness. The most striking feature of Figure 4.15 is the difference between Nafion and BPSH-30. For Nafion, thicker membranes give a better performance at low power density, while thinner membranes give higher performance at high power density. This performance is the result of

Table 4.2 Membrane property comparison of BPSH-30 and Nafion [Refs. 35,80,110, 184,214,272]

Property	BPSH-30	Nafion
IEC (meq/g)	1.34	0.92
Proton conductivity (mS/cm^2)	31	110
Methanol permeability ($\times 10^8 cm^2/s$)	36	167
Relative selectivity	1.3	1.0
Water uptake (vol.%)	40	38
Electro-osmotic drag coefficient	1.0	3.3
Interfacial resistance ($m\Omega\, cm^2$)	16	7

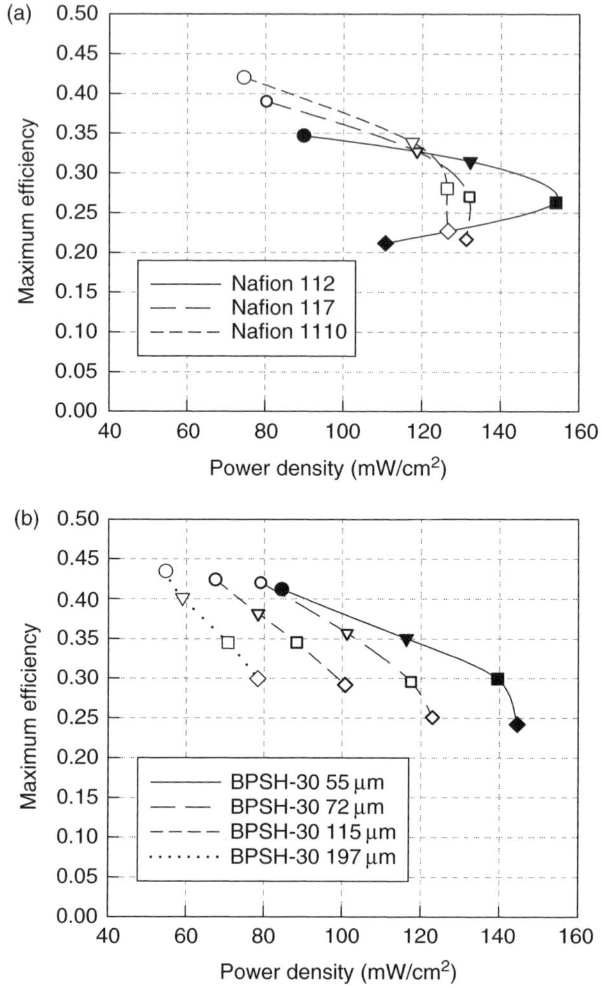

Figure 4.15 Maximum efficiency of (a) Nafion and (b) BPSH-30 as a function of membrane thickness and methanol feed concentration; Methanol feed concentration: circle: 0.3, triangle: 0.5, square:1, diamond: 2 M (cell temperature = 80°C) (reproduced from Ref. [271]).

tradeoffs between methanol crossover and ohmic losses. Interestingly at ~120 mW/cm^2, all three Nafion membrane thicknesses showed roughly equal performance. BPSH-30, on the other hand, exhibited higher power density at equivalent efficiency with decreasing membrane thickness. This trend resulted due to the characteristics of low conductivity, but high selectivity of BPSH-30. In other words, ohmic losses were more important than crossover for this system under optimized conditions.

Comparing the optimized efficiency of the thinnest BPSH-30 membrane (55 μm) to the thickest Nafion membrane (Nafion 1110, 250 μm), one finds that only modest gains ~1% maximum efficiency are apparent between power densities of 80 and 120 mW/cm^2. The 55 μm thick BPSH-30 membrane, due to lower ohmic losses, is capable of achieving somewhat higher power density; however the overall performance improvement is rather modest. While BPSH-30 does suffer from larger interfacial resistance than Nafion and has only modestly better selectivity; even with highly selective membranes and perfect interfacial compatibility expected performance gains are likely on the order of 5% maximum efficiency. While these gains are meaningful, they are unlikely critical from a performance standpoint for most fixed point of operation systems.

These fixed point of operation systems are not ideal for many applications, and higher selectivity membranes may provide significantly larger improvements for dynamic loads or when methanol feed concentrations are kept purposely high (such as passive systems). Additionally, these alternative membranes often have significantly decreased electro-osmotic drag coefficients. For example, the electro-osmotic drag coefficient of BPSH-30 is 1.0 while that of Nafion is 3.3. In circumstances where water balance is a key issue, significant advantages from using alternative PEMs may be realized.

4.6 Future Research Direction

The primary goals of this chapter were to provide an overview of polymer electrolytes investigated for DMFCs, a framework for evaluating electrolytes, and currently reported performance of those electrolytes. In our efforts to achieve these goals we were forced to provide a significant amount of more general DMFC background (and, in fact, we hope this chapter serves as a reasonable reference for those interested in DMFCs, but uninterested in polymer electrolytes). In much the same way as this chapter looked beyond polymer electrolytes, researchers who are interested in developing improved DMFC electrolytes need to also recognize electrolyte needs for other fuel cells (such as hydrogen and higher temperature/low humidity cells, namely the importance of improved conductivity and enhanced mechanical and chemical stability).

Additionally, our discussion of DMFCs focused on active systems (systems including pumps, condensers/active water recovery, elevated temperatures, and methanol concentration control). This was primarily because current electrolytes have been characterized under conditions relevant to active systems, but also because systems based on this technology have shown better energy density than competing technology (primarily lithium batteries). Due to issues with miniaturization of components and balance of plant efficiency, these active systems are limited to applications requiring a few to several watts or greater. For lower power, portable power applications (cell phones and Personal Digital Assistants PDAs); passive DMFC systems are being heavily pursued. The requirements of electrolytes in these applications are significantly different than the active system requirements. Issues such as water balance and crossover become much more critical and have to be addressed by engineering or design aspects, and will likely require advanced polymer electrolytes in order to enable the technology. Unfortunately, little work has been published on specific property requirements, due in large part to the industrial nature of this research.

For the further advancement of fuel cell electrolytes, a deeper understanding of the atomic environment and its impact on transport and mechanical properties is still required. Earlier (in Section 4.4) we commented to some degree on the viability of specific chemical approaches that have been presented; however, it is useful to discuss a few specific approaches/technologies we believe have special merit (for those interested, the authors' opinions are presented in greater detail elsewhere [51]). In particular, confined/constrained architectures, such as block copolymers, incorporation into porous supports, and specific polymer–polymer interactions (crosslinking, acid–base, or other weaker association) have shown interesting properties based on the ability to effect the local water environment. Computational modeling of these systems at the atomistic level is necessary to guide the development of future generation polymer electrolytes, and we believe this is an area that will grow in importance. Finally other self-assembling (possibly biologically inspired) structures may be of interest, as the structures that can currently be produced lack the ability to control structure down to the atomic level.

For active DMFC systems, a more selective membrane would be an improvement, but is not requisite for commercialization. The commercialization needs center on durability, reliability, fuel distribution, government regulation, and cost and have little to do with the membrane or membrane advances. As commercial markets arise incremental membrane advances will aid in lowering costs and improving performance, but for active DMFCs current generation membrane technology is acceptable from a cost and performance standpoint. For passive DMFC systems and hydrogen cells (including high temperature/low relative humidity), membrane advances are still needed in areas of water/methanol/proton transport, durability, and conductivity. These are the issues that should stay at the forefront of polymer electrolyte research for the foreseeable future.

REFERENCES

1. O. Savadogo, J. *New Mater. Electrochem. Syst.*, 1 (1998) 47.
2. J. Roziere and D. J. Jones, *Annul. Rev. Mater. Res.*, 33 (2003) 503.
3. M. A. Hickner, H. Ghassemi, Y. S. Kim, B. R. Einsla and J. E. McGrath, *Chem. Rev.*, 104 (2004) 4587.
4. Y. Yang and S. Holdcroft, *Fuel Cells*, 5 (2005) 171.
5. J. Kerres, *J. Membr. Sci.*, 185 (2001) 3.
6. K. D. Kreuer, *J. Membr. Sci.*, 185 (2001) 29.
7. W. L. Harrison, M. A. Hickner, Y. S. Kim and J. E. McGrath, *Fuel Cells*, 5 (2005) 201.
8. D. J. Jones and J. Roziere, *J. Membr. Sci.*, 185 (2001) 41.

9. R. Souzy and B. Ameduri, *Prog. Polym. Sci.*, 30 (2005) 644.
10. A. J. Appleby and R. L. Foulkes, *Fuel Cell Handbook*, Van Nostrand, New York, 1989.
11. D. B. Boies and A. Dravnieks, *J. Electrochem. Soc.*, 109 (1962) C198.
12. S. Surampudi, S. R. Narayanan, E. Vamos, H. Frank and G. Halpert, *Proceedings of the 8th Annual Battery Conference on Applications and Advances*, pp. 97–106, 1993, Long Beach, CA.
13. S. Surfampudi, S. R. Narayanan, E. Vamos, H. Frank, G. Halpert, A. Laconti, J. Losek, G. K. S. Prafkash and G.A. Olah, *J. Power Sources* 47 (1994) 377.
14. http://fuelcelltoday.com/FuelCellToday/IndustryInformation/IndustryInformation External/NewsDisplayArticle/0,1602,3199,00.html
15. http://www.smartfuelcell.de/en/presse/c030828.html
16. http://www.mtimicrofuelcells.com/news/article.cfm?a_id=13927
17. http://www.toshiba.co.jp/about/press/2003_03/pr0501.htm
18. http://www.nec.co.jp/press/en/0309/1701.html
19. J. E. Bäuerle, *J. Phys. Chem. Solids*, 30 (1969) 2657.
20. J. R. Macdonald, *J. Chem. Phys.*, 61 (1974) 3977.
21. T. E. Springer, T. A. Zawodzinski, M. S. Wilson, S. Gottesfeld and *J. Electrochem. Soc.*, 143 (1996) 587.
22. J. J. Fontanella, M. G. Mclin, M. C. Wintersgill, J. P. Calame and S. G. Greenbaum, *Solid State Ionics* 66 (1993) 1.
23. J. Halim, F. N. Buchi, O. Hass, M. Stamm and G. G. Scherer, *Electrochim. Acta*, 39 (1994) 1303.
24. C. L. Gardner and A. V. Anantaraman, *J. Electroanal. Chem.*, 395 (1995) 67.
25. Y. Sone, P. Ekdunge and D. Simonsson, *J. Electrochem. Soc.*, 143 (1996) 1256.
26. R. F. Silva, A. De Francesco and A. Pozio, *J. Power Sources*, 134 (2004) 18.
27. A. V. Anantaraman and C. L. Gardner, *J. Electroanal. Chem.*, 414 (1996) 115.
28. K. D. Kreuer, *Solid State Ionics*, 97 (1997) 1.
29. T. J. P. Freire and E. R. Gonzalez, *J. Electroanal. Chem.*, 503 (2001) 57.
30. X. Ren, T. E. Springer and S. Gottesfeld, *J. Electrochem. Soc.*, 147 (2000) 92.
31. G. Q. Lu and C. Y. Wang, *J. Power Sources* 134 (2004) 33.
32. K. D. Kreuer, in *Handbook of Fuel Cells – Fundamentals, Technology and Applications*, W. Vielstich, A. Lamm and H. A. Gasteiger (Eds.), Willey, Chichester, UK, 2003, p. 420.
33. K. D. Kreuer, S. J. Paddison, E. Spohr and M. Schuster, *Chem. Rev.*, 104 (2004) 4637.
34. Y. S. Kim and B. S. Pivovar, *206th Meeting of Electrochemical Society*, Abstract No. 1472, 2004.
35. Y. S. Kim and B. S. Pivovar, *J. Electrochem. Soc.*, submitted Oct 31, 2006.
36. E. L. Cussler, *Diffusion: Mass Transfer in Fluid Systems*, 2nd edn, Cambridge University Press, Cambridge, UK, 1997.
37. X. Ren, T. E. Springer, T. A. Zawodzinski and S. Gottesfeld, *J. Electrochem. Soc.*, 147 (2000) 466.
38. K. Mayer and D. Woermann, *J. Membr. Sci.*, 127 (1997) 35.
39. B. S. Pivovar, Y. X. Wang and E. L. Cussler, *J. Membr. Sci.*, 154 (1999) 155.
40. S. Hietala, S. L. Maunu and F. Sundholm, *J. Polym. Sci., Part B: Polym. Phys.*, 24 (2000) 3277.
41. H. A. Every, M. A. Hickner, J. E. McGrath and T. A. Zawodzinski, *J. Membr. Sci.*, 250 (2005) 183.
42. X. Ren, P. Zelenay, S. Thomas, J. Davey and S. Gottesfeld, *J. Power Sources,* 86 (2000) 111.
43. X. Ren, W. Henderson and S. Gottesfeld, *J. Electrochem. Soc.*, 144 (1999) L267.
44. X. Ren and S. Gottesfeld, *J. Electrochem. Soc.*, 148 (2001) A87.
45. J. T. Wang, S. Wasmus and R. F. Savinell, *J. Electrochem. Soc.*, 143 (1996) 1233.
46. V. M. Barragan, C. Ruiz-Bauza, J. P. G. Villaluenga and B. Seoane, *J. Power Sources*, 130 (2004) 22.
47. V. M. Barragan and A. Heinzel, *J. Power Sources*, 104 (2002) 66.

48. R. Z. Jiang and D. R. Chu, *Electrochem. Solid-State Lett.*, 5 (2002) A156.
49. R. Z. Jiang and D. R. Chu, *J. Electrochem. Soc.*, 151 (2004) A69.
50. T. A. Zawodzinski, T. E. Springer, J. Davey, R. Jestel, C. Lopez, J. Valerio and S. Gottesfeld, *J. Electrochem. Soc.*, 140 (1981) 7.
51. M. A. Hickner and B. S. Pivovar, *Fuel Cells*, 5 (2005) 213.
52. T. Okada, G. Xie, O. Gorseth, S. Kjelstrup, N. Nakamura and T. Arimura, *Electrochim. Acta*, 43 (1998) 3741
53. Y. S. Kim, L. M. Dong, M. A. Hickner, B. S. Pivovar and J. E. McGrath, *Polymer*, 44 (2003) 5729.
54. T. A. Zawodzinski Jr., C. Derouin, S. Radzinski, R. J. Sherman, V. T. Smith, T. E. Springer and S. Gottesfeld, *J. Electrochem. Soc.*, 140 (1993) 1041.
55. T. A. Zawodzinski Jr., M. Neeman, L. O. Sillerud and S. Gottesfeld, *J. Phys. Chem.*, 95 (1991) 6040.
56. J. T. Hinatsu, M. Mizuhata and H. Takenaka, *J. Electrochem. Soc.*, 141 (1994) 1493.
57. D. R. Morris and X. Sun, *J. Appl. Polym. Sci.*, 50 (1993) 1445.
58. N. H. Jalani, P. Choi and R. Datta, *J. Membr. Sci.*, 254 (2005) 31.
59. X. Ren, M. S. Wilson and S. Gottesfeld, *J. Electrochem. Soc.*, 143 (1996) L12.
60. E. Skou, P. Kauranen and J. Hentschel, *Solid State Ionics*, 333 (1997) 97.
61. H. Yeager and A. Steck, *J. Electrochem. Soc.*, 128 (1981) 1880.
62. M. W. Verbrugge and R. G. Hill, *J. Electrochem. Soc.*, 137 (1990) 893.
63. G. Suresh, Y. M. Scindia, A. K. Pandey and A. Goswami, *J. Membr. Sci.*, 250 (2005) 39.
64. H. R. Zelsmann, M. Pineri, M. Thomas and M. Excoubes, *J. Appl. Polym. Sci.*, 41 (1990) 1673.
65. D. Rivin, C. E. Kendrick, P. W. Gibson and N. S. Schneider, *Polymer*, 42 (2001) 623.
66. A. M. Pivovar and B. S. Pivovar, *J. Phys. Chem.*, B 109 (2005) 785.
67. K. D. Kreuer, T. Dippel, W. Meyer and J. Maier, *Mater. Res. Soc. Symp. Proc.*, 293 (1993) 273.
68. M. Legras, Y. Hirata, Q. T. Nguyen, D. Langevin and M. Metayer, *Desalination*, 147 (2002) 351.
69. V. Freger, E. Korin, J. Wisniak, E. Korngold, M. Ise and K. D. Kreuer, *J. Membr. Sci.*, 160 (1999) 213.
70. R. Carter, R. G. Evilia and P. N. Pintauro, *J. Phys. Chem. B*, 105 (2001) 2351.
71. B. Pivovar, M. Hickner, T. Zawodzinski, X. Ren, S. Gottesfeld, J. Neutzler, in *Direct Methanol Fuel Cells*, ECS Proceedings PV 2001–4, Electrochemical Society, Pennington, NJ.
72. B. S. Pivovar, *Polymer*, 47 (2006) 4194.
73. T. A. Zawodzinski, J. Davey, J. Valerio and S. Gottesfeld, *Electrochim. Acta*, 40 (1995) 297.
74. D. Weng, J. S. Wainright, U. Landau and R. F. Savinell, *J. Electrochem. Soc.*, 143 (1996) 1260.
75. B. S. Pivovar, W. H. Smyrl and E. L. Cussler, *J. Electrochem. Soc.*, 152 (2005) A53.
76. Y. S. Kim, L. Dong, M. Hickner, T. E. Glass and J. E. McGrath, *Macromolecules*, 36 (2003) 6281.
77. D. I. Ostrovskii, L. M. Torell, M. Paronen, S. Hietala and F. Sundholm, *Solid State Ionics*, 97 (1997) 315.
78. M. Elomaa, S. Hietala, M. Paronen, N. Walsby, K. Jokela, R. Serimaa, M. Torkkeli, T. Lehtinen, G. Sundholm and F. Sundholm, *J. Mater. Chem.*, 10 (2000) 2678.
79. J. F. Ding, C. Chuy and S. Holdcroft, *Macromolecules*, 35 (2002) 1348.
80. Y. S. Kim, M. A. Hickner, L. Dong, B. S. Pivovar and J. E. McGrath *J. Membr. Sci.*, 243 (2004) 317.
81. R. J. Bellows, M. Y. Lin, M. Arif, A. K. Thompson and D. Jacobson, *J. Electrochem. Soc.*, 146 (1999) 1099.
82. J. Elliot, S. Hanna, A. M. S. Elliot and G. E. Cooley, *Phys. Chem. Chem. Phys.*, 1 (1999) 4855.
83. L. Rubata, A. L. Rollet, G. Gebel and O. Diat, *Macromolecules*, 35 (2002) 4050.
84. G. Gebel and O. Diat, *Fuel Cells*, 5 (2005) 261.
85. R. S. McLean, M. Doyle and B. B. Sauer, *Macromolecules*, 33 (2000) 6451.
86. H. G. Haubold, T. Vad, H. Jungbluth and P. Hiller, *Electrochim. Acta*, 46 (2001) 1559.

87. C. S. Tasao, H. L. Chang, U. S. Jeng, J. M. Lin and T. L. Lin, *Polymer*, 46 (2005) 8430.
88. J. Chou, E. W. McFarland and H. Metiu, *J. Phys. Chem. B*, 109 (2005) 3252.
89. B. Smitha, S. Sridhar and A. A. Khan, *J. Membr. Sci.*, 225 (2003) 63.
90. P. J. James, J. A. Elliott, T. J. McMaster, J. M. Newton, A. M. S. Elliott, S. Hanna and M. J. Miles, *J. Mater. Sci.*, 35 (2000) 5111.
91. S. C. Thomas, X. M. Ren, S. Gottesfeld and P. Zelenay, *Electorchim. Acta*, 47 (2002) 3741.
92. X. Z. Du, J. R. Yu, B. L. Yi, M. Han and K. W. Bi, *Phys. Chem. Chem. Phys.*, 3 (2001) 3175.
93. M. Wakizoe, O. A. Velev and S. Srinivasan, *Electrochim. Acta*, 40 (1995) 335.
94. N. Yoshida, T. Ishisaki, A. Watakabe and M. Yoshitake, *Electrochim. Acta*, 43 (1998) 3749.
95. M. Yoshitake, M. Tamura, N. Yoshida and T. Ishisaki, *Denki Kagaku*, 64 (1996) 727.
96. A. Steck, in O. Savadogo, P. R. Roberge and T. N. Veziroglu (Eds.), *Proceedings of the First International Symposium on New Materials for Fuel Cell Systems*, Montreal, Canada, July 9–13, 1995, p. 74.
97. J. A. Kolde, B. Bahar, M. S. Wilson, T. A. Zawodzinski and S. Gottesfeld, in S. Gottesfeld, G. Halpert and A. Landgreve (Eds.), *Proceedings of the first International Symposium on Proton Conducting Membrane Fuel Cells*, The Electrochemical Society Proceedings, 95-23, 1995, p. 193.
98. S. Narayanan, A. Kindler, B. Jeffries-Nakamura, W. Chu, H. Frank, M. Smart, S. Surampudi, G. Halpert in S. Gottesfeld, G. Halpert and A. Landgrebe (Eds.), *Proton Conducting Membrane Fuel Cells*, Electrochemical Society Proceedings, 95-23, 1995, p. 261.
99. J. J. Sumner, S. E. Creager, J. J. Ma and D. D. DesMarteau, *J. Electrochem. Soc.*, 145 (1998) 107.
100. T. A. Zawodzinski, T. E. Springer, J. Davey, R. Jestel, C. Lopez, J. Valerio and S. Gottesfeld, *J. Electrochem. Soc.*, 140 (1993) 1981.
101. K. A. Mauritz and R. B. Moore, RB. *Chem. Rev.*, 104 (2004) 4535.
102. K. Lee, A. Ishihara, S. Mitsushima, N. Kamiya and K. Ota, *J. Electrochem. Soc.*, 151 (2004) A639.
103. http://www.dupont.com/fuelcells/pdf/nae101.pdf
104. V. Tricoli, N. Carretta and M. Bartolozzi, *J. Electrochem. Soc.*, 147 (2000) 1286.
105. R. A. Banaszak, S. A. Arbaugh, E. D. Steffee and R. Pyati, *J. Electrochem. Soc.*, 151 (2004) A1020.
106. A. B. LaConti, A. R. Fragala and J. R. Boyack in J. D. E.McIntyre, S. Srinivasan and F. G. Will, (Eds.), *Proceedings of the Symposium on electrode Materials and processes for Energy Conversion and Storage,* The Electrochemical Society Soft bound Proceedings Series, PV 77-6. The Electrochemical Society, Princeton, NJ, 1977, p. 354.
107. M. Kikerling, Y. I. Kharkats, A. A. Kornyshev and Y. M. Volfkovich, *J. Electrochem. Soc.*, 145 (1998) 2684.
108. R. Nolte, K. Ledjeff, M. Bauer and R. Mulhaupt, *J. Membr. Sci.*, 83 (1993) 211.
109. J. Kerres, W. Cui and S. Reichle, *J. Polym. Sci. Part A: Polym. Sci.*, 34 (1996) 2421.
110. F. Wang, M. Hickner, Y. S. Kim, T. A. Zawodzinski and J. E. McGrath, *J. Membr. Sci.*, 197 (2002) 231.
111. W. Harrison, F. Wang, J. B. Mecham, V. Bhanu, M. Hill, Y. S. Kim and J. E. McGrath, *J. Polym. Sci., Part A: Polym. Chem.*, 41 (2003) 2264.
112. D. Poppe, H. Frey, K. D. Kreuer, A. Heinzel and R. Mulhaupt, *Macromolecules*, 35 (2002) 7936.
113. L. Li and Y. X. Wang, *J. Membr. Sci.*, 246 (2005) 167.
114. L. E. Karlsson and P. Jannasch, *J. Membr. Sci.*, 230 (2004) 61.
115. L. Li, J. Zhang and Y. X. Wang, *J. Membr. Sci.*, 226 (2003) 159.
116. K. Kreuer, T. Dippel and J. Maier, *J. Proc. Electrochem. Soc.*, (1995) 241.
117. P. X. Xing, G. P. Robertson, M. D. Guiver, S. D. Mikahailenko, K. P. Wang and S. Kaliaguine, *J. Membr. Sci.*, 229 (2004) 95.

118. M. L. Hill, Y. S. Kim, B. R. Einsla, and J. E. McGrath, *J. Membr. Sci.*, 283 (2006) 102.
119. C. L. Ninivin, A. Balland-Longeau, D. Demattei, C. Coutanceau, C. Lamy and J. M. Leger, *J. Appl. Electrochem.*, 34 (2004) 1159.
120. K. Ramya, B. Vishnupriya and K. S. Dhathathreyan, *J. New Mater. Electrochem. Sys.*, 4 (2001) 115.
121. G. Van, G. P. Robertson, M. D. Guiver, S. D. Mikhailenko, X. Li and S. Kaliaguine, *Macromolecules*, 38 (2005) 3237.
122. S. Wang and J. E. McGrath, *Synthesis of Poly(arylene ether)s, Synthetic Methods in Step-Growth Polymers*, Wiley, New York, 2003, 327.
123. Y. S. Kim, B. S. Pivovar, *208th Meeting of the Electrochemical Society*, Abstract No. 1215, 2005.
124. Y. S. Kim, W. L. Harrison, J. E. McGrath, B. S. Pivovar, *205th Meeting of the Electrochemical Society*, Abstract No. 334, 2004.
125. M. Gil, X. L. Ji, X. F. Li, H. Na, J. E. Hampsey and Y. F. Lu, *J. Membr. Sci.*, 234 (2004) 75.
126. Y. S. Yang, Z. Q. Shi and S. Holdcroft, *Macromolecules*, 37 (2004) 1678.
127. H. Ghassemi, G. Ndip and J. E. McGrath, *Polymer*, 45 (2004) 5855.
128. Y. S. Kim, M. J. Sumner, W. L. Harrison, J. S. Riffle, J. E. McGrath, and B. S. Pivovar, *J. Electrochem. Soc.*, 151 (2004) A2150.
129. K. B. Wiles, C. M. de Diego and J. E. McGrath, *Polym. Prepr.* (Am. Chem. Soc., Div. Polym. Chem.), 228 (2004) U657.
130. L. Jorissen, V. Gogel, J. Kerres and J. Garche, *J. Power Sources*, 105 (2002) 267.
131. C. Manea and M. Mulder, *J. Membr. Sci.*, 206 (2002) 443.
132. S. Swier, V. Ramani, J. M. Fenton, H. R. Kunz, M. T. Shaw and R. A. Weiss, *J. Membr. Sci.*, 256 (2005) 122.
133. J. Schineisser, S. Holdcroft, J. F. Yu, T. Ngo and G. McLean, *Chem. Mater.*, 17 (2005) 387.
134. C. Genies, R. Mercier, B. Sillion, N. Cornet, G. Gebel and M. Pineri, *Polymer*, 42 (2001) 359.
135. Y. Yin, J. H. Fang, Y. F. Cui, K. Tanaka, H. Kita and K. Okamoto, *Polymer*, 4509 (2003) 44.
136. Y. Woo, S. Y. Oh, Y. S. Kang and B. Jung *B. J. Membr. Sci.*, 220 (2003) 31.
137. B. R. Einsla, Y. T. Hong, Y. S. Kim, F. Wang, N. Gunduz and N. J. E. McGrath, *J. Polym. Sci., Part A: Polym. Chem.*, 42 (2004) 862.
138. B. R. Einsla, Y. S. Kim, M. A. Hickner, Y. T. Hong, M. L. Hill, B. S. Pivovar and J. E. McGrath, *J. Membr. Sci.*, 255 (2005) 141.
139. T. Watari, J. H. Fang, K. Tanaka, H. Kita, K. Okamoto and T. Hirano, *J. Membr. Sci.*, 230 (2004) 111.
140. J. M. Song, N. Asano, K. Miyatake, H. Uchida and M. Watanabe, *Chem. Lett.*, 34 (2005) 996.
141. C. Genies, R. Mercier, B. Sillion, R. Petiaud, N. Cornet, G. Gebel and M. Pineri, *Polymer*, 42 (2001) 5097.
142. N. Asano, K. Miyatake and M. Watanabe, *Chem. Mater.*, 16 (2004) 2841.
143. Q. H. Guo, P. N. Pintauro, H. Tang and S. O'Connor, *J. Membr. Sci.*, 154 (1999) 175.
144. H. R. Allcock, M. A. Hofmann, C. M. Ambler and R. V. Morford, *Macromolecules*, 35 (2002) 3484.
145. M. V. Fedkin, X. Y. Zhou, M. A. Hofmann, E. Chalkova, J. A. Weston, H. R. Allcock and S. N. Lvov, *Mater. Lett.*, 52 (2002) 192.
146. H. R. Allcock, M. A. Hofmann, C. M. Ambler, S. N. Lvov, X. Y. Zhou, E. Chalkova and J. Weston, *J. Membr. Sci.*, 201 (2002) 47.
147. R. Carter, R. Wycisk, H. Yoo and P. N. Pintauro, *Electrochem. Solid-State Lett.*, 5 (2002) A195.
148. R. Wycisk, J. K. Lee and P. N. Pintauro, *J. Electrochem. Soc.*, 152 (2005) A892.
149. F. N. Buchi, B. Gupta, O. Haas and G. G. Scherer, *J. Electrochem. Soc.*, 142 (1995) 3044.
150. K. Scott, W. M. Taama and P. Argyropoulos, *J. Membr. Sci.*, 171 (2000) 119.

151. C. Chuy, B. I. Basura, E. Simon, S. Holdcroft, J. Horsfall and K. V. Lovell, *J. Electrochem. Soc.*, 147 (2000) 4453.
152. A. S. Arico, V. Baglio, P. Creti, A. Di Blasi, V. Antonucci, J. Brunea, A. Chapotot, A. Bozzi and J. Schoemans, *J. Power Sources*, 123 (2003) 107.
153. V. Saarinen, T. Kallio, M. Paronen, P. Tikkanen, E. Rauhala and K. Kontturi, *Electrochim. Acta*, 50 (2005) 3453.
154. E. A. Hegazy, I. Ishigaki, A. Rabie, A. M. Dessouki and J. Okamoto, *J. Appl. Polym. Sci.*, 28 (1983) 493.
155. R. A. Assink, C. Arnold and R. P. Hollandsworth, *J. Membr. Sci.*, 56 (1991) 143.
156. W. Becker and G. Schmidt-Naake, *Chem. Eng. Technol.*, 25 (2002) 373.
157. M. Shen, S. Roy, J. W. Kuhlmann, K. Scott, K. Lovell and J. A. Horsfall, *J. Membr. Sci.*, 251 (2005) 121.
158. J. P. Gouin, C. E. Williams and A. Eisenberg, *Macromolecules*, 22 (1989) 4573.
159. R. A. Weiss, A. Sen, C. L. Willis and L. A. Pottick, *Polymer*, 32 (1991) 1867.
160. G. Wnek, *Polym. Prepr.* (Am. Chem. Soc., Div. Polym. Chem.), 222 (2001) U308.
161. J. Kim, B. Kim and B. Jung, *J. Membr. Sci.*, 207 (2002) 129.
162. Y. A. Elabd, E. Napadensky, J. M. Sloan, D. M. Crawford and C. W. Walker, *J. Membr. Sci.*, 217 (2003) 227.
163. Y. A. Elabd and E. Napadensky, *Polymer* 45 (2004) 3037.
164. R. F. Storey, B. J. Chisholm and Y. Lee, *Polym. Eng. Sci.*, 37 (1997) 73.
165. J. Won, H. H. Park, Y. J. Kim, S. W. Choi, H. Y. Ha, I. H. Oh, H. S. Kim, Y. S. Kang and K. J. Ihn, *Macromolecules*, 36 (2003) 3228.
166. A. Nacher, P. Escribano, C. Del Rio, A. Rodriguez and J. L. Acosta, *J. Polym. Sci. Part A: Polym. Chem.* 41 (2003) 2809.
167. H. Bashir, J. L. Acosta and A. Linares, *J. Membr. Sci.*, 253 (2005) 33.
168. J. W. Rhim, C. K. Yeom and S. W. Kim, *J. Appl. Polym. Sci.*, 68 (1998) 1717.
169. J. W. Rhim, H. B. Park, C. S. Lee, J. H. Jun, D. S. Kim and Y. M. Lee, *J. Membr. Sci.*, 238 (2004) 143.
170. B. Ruffmann, H. Silva, B. Schulte and S. P. Nunes, *Solid State Ionics*, 162 (2003) 269.
171. H. Sawa and Y. Shimada, *Electrochem.*, 72 (2004) 111.
172. C. W. Lin, R. Thangamuthu and C. J. Yang, *J. Membr. Sci.*, 253 (2005) 23.
173. L. Li, L. Xy and Y. X. Wang, *Mater. Lett.*, 57 (2003) 1406.
174. Z. G. Shao, X. Wang and I. M. Hsing, *J. Membr. Sci.*, 210 (2002) 147.
175. J. L. Qiao, T. Hamaya and T. Okada, *Chem. Mater.*, 17 (2005) 2413.
176. G. Alberti and M. Casciola, *Annu. Rev. Mater. Res.*, 33 (2003) 129.
177. V. Baglio, A. S. Arico, A. DI Blasi, V. Antonucci, P. L. Antonucci, S. Licoccia, E. Tarversa, and F. S. Fiory, *Electrochim. Acta*, 50 (2005) 1241.
178. W. L. Xu, T. H. Lu, C. P. Liu and W. Xing, *Electrochim. Acta*, 50 (2005) 3280.
179. Y. S. Park and Y. Yamazaki, *Solid State Ionics*, 176 (2005) 1079.
180. D. J. Kim, M. A. Scibioh, S. Kwak, I. H. Oh and H. Y. Ha, *Electrochem. Comm.*, 6 (2004) 1069.
181. R. F. Silva, S. Passerini and A. Pozio, *Electrochim.*, Acta, 50 (2005) 2639.
182. M. L. Ponce, L. Prado, B. Ruffmann, K. Richau, R. Mohr and S. P. Nunes, *J. Membr. Sci.*, 217 (2003) 5.
183. S. M. J. Zaidi, S. D. Mikahailenko, G. P. Robertson, M. D. Guiver and S. Kaliaguine, *J. Membr. Sci.*, 173 (2000) 17.
184. Y. S. Kim, F. Wang, M. Hickner, T. A. Zawodzinski and J. E. McGrath, *J. Membr. Sci.*, 212 (2003) 263.
185. H. Y. Chang and C. W. Lin, *J. Membr. Sci.*, 218 (2003) 295.
186. S. P. Nunes, B. Ruffmann, E. Rikowski, S. Vetter and K. Richau, *J. Membr. Sci.*, 203 (2002) 215.
187. V. S. Silva, S. Weisshaar, R. Reissner, B. Ruffmann, S. Vetter, A. Mendes, L. M. Madeira and S. Nunes, *J. Power Sources*, 145 (2005) 485.

188. L. H. Melinda, Y. S. Kim, B. R. Einsla and J. E. McGrath, *J. Membr. Sci.*, January 2006. (submited)
189. F. Bauer and A. Willert-Posrada, *J. Power Sources*, 145 (2005) 101.
190. C. H. Rhee, J. K. Kim, H. Chang and J. S. Lee, *Chem. Mater.*, 17 (2005) 1691.
191. T. Shimizu, T. Naruhashi, T. Momma and T. Osaka, *Electrochem.*, 70 (2002) 991.
192. K. Y. Cho, J. Y. Eom, H. Y. Jung, N. S. Choi, Y. M. Lee, J. K. Park, J. H. Choi, K. W. Park and Y.E. Sung, *Electrochim. Acta*, 50 (2004) 583.
193. J. Liu, H. T. Wang, S. A. Cheng and K. Y. Chan, *J. Membr. Sci.*, 246 (2005) 95.
194. J. A. Kerres, *Fuel Cells*, 2 (2005) 230.
195. J. Kerres, A. Ullrich, M. Hein, V. Gogel, K. A. Friedrich and L. Jorissen, *Fuel Cells*, 4 (2004) 105.
196. J. Kerres, A. Ullrich, T. Haring, M. Baldauf, U. Gebhardt and W. Preidel, *J. New Mater. Electrochem. Sys.*, 3 (2000) 229.
197. C. W. Walker, *J. Electrochem. Soc.*, 151 (2004) A1797.
198. C. Manea and M. Mulder, *Desalination*, 147 (2002) 179.
199. B. Smitha, S. Sridhar and A. A. Khan, *Macromolecules*, 37 (2004) 2233.
200. C. Pu, W. H. Huang, K. L. Ley and E. S. Smotkin, *J. Electrochem. Soc.*, 142 (1995) L119.
201. N. Y. Jia, M. C. Lefebvre, J. Halfyard, Z. G. Qi and P. G. Pickup, *Electrochem. Solid-State Lett.*, 3 (2000) 529.
202. L. J. Hobson, H. Ozu, M. Yamaguchi and S. Hayase, *J. Electrochem. Soc.*, 148 (2001) A1185.
203. S. Z. Ren, C. N. Li, X. S. Z. M. Wu, S. L. Wang, G. Q. Sun, Q. Xin and X. F. Yang, *J. Membr. Sci.*, 247 (2005) 59.
204. B. Yang and A. Manthiram, *Electrochem. Solid-State Lett.*, 6 (2003) A229.
205. J. Sauk, J. Byun and H. Kim, *J. Power Sources*, 143 (2005) 136.
206. J. Won, S. W. Choi, Y. S. Kang, H. Y. Ha, I. H. Oh, H. S. Kim, K. T. Kim and W. H. Jo, *J. Membr. Sci.* 214 (2003) 145.
207. J. T. Wang, J. S. Wainright, R. F. Savinell and M. Litt. *J. Appl. Electrochem.*, 26 (1996) 751.
208. M. Kawahara, J. Morita, M. Rikukawa, K. Sanui and N. Ogata, *Electrochim. Acta*, 45 (2000) 1395.
209. H. T. Pu and Q. Z. Liu, *Polym. Inter.*, 53 (2004) 1512.
210. E. Peled, T. Duvdevani, A. Aharon and A. Melman, *Electrochem. Solid-State Lett.*, 3 (2000) 525.
211. T. Yamaguchi, H. Kuroki and F. Miyata, *Electrochem. Comm.*, 7 (2005) 730.
212. T. Ioroi, K. Kuaraoka, K. Yasuda, T. Yazawa and Y. Miyazaki, *Electrochem. Solid-State Lett.*, 7 (2004) A394.
213. T. Kallio, K. Kisko, K. Kontturi, R. Serimaa, F. Sundholm and G. Sundholm, *Fuel Cells*, 4 (2004) 328.
214. Y. S. Kim, W. Harrison, M. Sankir, J. E. McGrath, B. S. Pivovar, *Polymer*, 47 (2006) 4026.
215. J. F. Ding, C. Chuy and S. Holdcroft, *Adv. Funct. Mater.*, 12 (2002) 389.
216. N. Carretta, V. Tricoli and F. Picchioni, *J. Membr. Sci.*, 166 (2000) 189.
217. C. A. Edmondson, J. J. Fontanella, S. H. Chung, S. G. Greenbaum and G. E. Wnek, *Electrochim. Acta*, 46 (2001) 1623.
218. T. Kallio, M. Lundstrom, G. Sundholm, N. Walsby and F. Sundholm, *J. Appl. Electrochem.*, 32 (2002) 11.
219. M. M. Nasef, H. Saidi, H. M. Nor and O. M. Foo, *J. Appl. Polym. Sci.*, 76 (2000) 1.
220. S. Holmberg, P. Holmlund, R. Nicolas, C.E. Wilen, T. Kallio, G. Sundholm and F. Sundholm, *Macromolecules*, 37 (2004) 9909.
221. D. S. Kim, H. B. Park, J. W. Rhim and Y. M. Lee, *Solid State Ionics*, 176 (2005) 117.
222. B. Smitha, S. Sridhar and A. A. Khan, *J. Appl. Polym. Sci.*, 95 (2005) 1154.
223. H. B. Park, H. S. Shin, Y. M. Lee and J. W. Rhim, *J. Membr. Sci.*, 247 (2005) 103.
224. K. Okamoto, Y. Yin, O. Yamada, M. N. Islam, T. Honda, T. Mishima, Y. Suto, K. Tanaka and H. Kita, *J. Membr. Sci.*, 258 (2005) 115.

225. S. Ren, C. Li, X. Zhao, Z. Wu, S. Wang, G. Sun, Q. Xin and X. Yang, *J. Membr. Sci.*, 247 (2005) 59.
226. B. Bae and D. Kim, *J. Membr. Sci.*, 220 (2003) 75.
227. M. J. Canovas, J. L. Acosta and A. Linares, *Macromol. Chem. Phys.*, 206 (2005) 680.
228. K. B. Wiles, F. Wang, J. E. McGrath, *J. Polym. Sci., Part A: Polym. Chem.*, 43 (2005) 2964.
229. M. J. Sumner, W. L. Harrsion, R. M. Weyers, Y. S. Kim, J. E. McGrath, J. S. Riffle, A. Brink and M. H. Brink, *J. Membr. Sci.*, 239 (2004) 199.
230. D. S. Kim, H. B. Park, J. W. Rhim and Y. M. Lee, *J. Membr. Sci.*, 240 (2004) 37.
231. J. H. Chang, J. H. Paik, G. G. Park, C. S. Kim and O. O. Park, *J. Power Sources*, 124 (2003) 18.
232. B. Libby, W. H. Smyrl and E. L. Cussler, *AIChE J.*, 49 (2003) 991.
233. B. Pivovar, Ph.D Dissertation, University of Minnesota, Twin Cities, MN, 2000.
234. Y. Si, J. C. Lin, H. R. Kunz and J. M. Fenton, *J. Electrochem. Soc.*, 151 (2004) A463.
235. M. Hickner, Y. S. Kim, J. E. McGrath, P. Zelenay and B. Pivovar, *Electrochem. Soc. Ser.*, 31 (2005) 530.
236. E. B. Easton, T. D. Astill and S. Holdcroft, *J. Electrochem. Soc.*, 152 (2005) A752.
237. Y. S. Kim, B. S. Pivovar, *Advances in Materials for Proton Exchange Membrane Fuel Cell Systems 2005*, Abstract No. 35, February. 20–23, Pacific Grove, CA, 2005.
238. S. K. Zecevic, J. S. Wainright, M. H. Litt, S. L. Gojkovic and R. F. Savinell, *J. Electrochem. Soc.*, 144 (1997) 2973.
239. L. Zhang, C. Ma and S. Mukerjee, *J. Electroanal. Chem.*, 568 (2004) 273.
240. C. S. Ma, L. Zhang, S. Mukerjee, D. Ofer and B. D. Nair, *J. Membr. Sci.*, 219 (2003) 123.
241. T. Navessin, M. Eikerling, Q. P. Wang, D. T. Song, Z. S. Liu, J. Horsfall, K. V. Lovell and S. Holdcroft, *J. Electrochem. Soc.*, 152 (2005) A796.
242. D. S. Kim, G. P. Gilles, M. D. Guiver, Y. M. Lee, *J. Membr. Sci.* 281 (2006) 111.
243. V. I. Basura, C. Chuy, P. D. Beattie, S. Holdcroft, *J. Electroanal. Chem.*, 501 (2001) 77.
244. V. S. Silva, B. Ruffmann, H. Silva, Y. A. Gallego, A. Mendes, L. M. Madeira and S. P. Nunes, *J. Power Sources*, 140 (2005) 34.
245. M. Y. Jang and Y. Yamazaki, *Solid State Ionics*, 167 (2004) 107.
246. W. C. Choi, J. D. Kim, S.I. Woo, *J. Power Sources*, 96 (2001) 411.
247. M. Tsige, T. R. Mattsson and G. S. Grest, *Macromolecules*, 37 (2004) 9132.
248. J. Kim, B. Kim, B. Jung, Y. S. Kang, H. Y. Ha, I. H. Oh and K. J. Ihn, *Macromol. Rapid. Comm.*, 23 (2002) 753.
249. K. Miyatake, H. Zhou, T. Matsuo, H. Uchida and M. Watanabe, *Macromolecules*, 37 (2004) 4961.
250. A. Oedegaard, C. Hebling, A. Schmitz, S. Moller-Holst and R. Tunold, *J. Power Sources*, 127 (2004) 187.
251. C. Lim and C. Y. Wang, *J. Power Sources*, 113 (2003) 145.
252. M. Uchida, Y. Aoyanma, N. Eda and A. Ohta, *J. Electrochem. Soc.*, 142 (1995) 2.
253. M. Uchida, Y. Fukuoka, Y. Sugawara, H. Ohara and A. Ohta, *J. Electrochem. Soc.*, 145 (1998) 3708.
254. S. Gottesfeld and T. Zawodzinski, in C. Tobias, H. Gerischer, D. Kolb and R. Alkire (Eds.), Advances in Electrochemistry and Electrochemical Engineering, Vol. 5, Wiley/VCH, New York, 1997.
255. Z. B. Wei, S. L. Wang, B. L. Yi, J. G. Liu, L. K. Chen, W. J. Zhou, W. Z. Li and Q. Xin, *J. Power Sources*, 106 (2002) 364.
256. S. Q. Song, Z. X. Liang, W. J. Zhou, G. Q. Sun, Q. Xin, V. Stergiopoulos and P. Tsiakaras, *J. Power Sources*, 145 (2005) 495.
257. C. Song and P. G. Pickup, *J. Appl. Electrochem.*, 34 (2004) 1065.
258. Y. C. Liu, X. P. Qiu, Y. Q. Huang, W. T. Zhu and G. S. Wu, *J. Appl. Electrochem.*, 32 (2002) 1279.

259. H. N. Dinh, X. M. Ren, F. H. Garzon, P. Zelenay and S. Gottesfeld, *J. Electroanal. Chem.*, 491 (2000) 222.
260. A. S. Ariço, P. L. Antouncci, E. Modica, V. Baglio, H. Kim and V. Antonucci, *Electrochim. Acta*, 47 (2002) 3723.
261. J. Nordlund, A. Roessler and G. Lindgergh, *J. Appl. Electrochem.*, 32 (2002) 259.
262. X. Ren, T. E. Springer, S. Gottesfeld, *2nd International Symposium on Proton Conducting Membrane Fuel Cells II*, Penington, NJ, p. 341.
263. S. C. Thomas, X. Ren and S. Gottesfeld, *J. Electrochem. Soc.*, 146 (1999) 4354.
264. J. B. Ge and H. T. Liu, *J. Power Sources*, 142 (2005) 56.
265. A. S. Arico, P. Creti, P. L. Antouncci, J. Cho, H. Kim and V. Antonucci, *Electrochim. Acta*, 43 (1998) 3719.
266. A. S. Arico, P. Creti, V. Baglio, E. Modica and V. Antonucci, *J. Power Sources*, 91 (2000) 202.
267. J. C. Amphlett, B. A. Peppley, E. Halliop and A. Sadig, *J. Power Sources*, 96 (2001) 204.
268. M. Neergat and A. K. Shukla, *J. Power Sources*, 104 (2002) 289.
269. J. P. Meyers and J. Newman, *J. Electrochem. Soc.*, 149 (2002) A729.
270. C. Xu, P. M. Follmann, L. T. Biegler, M. S. Jhon, *Computers Chem. Eng.*, 29 (2005) 1849.
271. B. S. Pivovar Y.S. Kim, *Advances in Materials for Proton Exchange Membrane Fuel Cell Systems* 2005, Abstract. No. 10, February 20–23, Pacific Grove, CA, 2005; manuscript in preparation.
272. Y. S. Kim, F. Wang, M. Hickner, S. McCartney, Y. T. Hong, W. Harrison, T. A. Zawodzinski and J. E. McGrath *J. Polym. Sci. Part B: Polym. Phy.*, 41 (2003) 2816.

CHAPTER FIVE

Materials for State-of-the-Art PEM Fuel Cells, and Their Suitability for Operation Above 100°C

Frank A. de Bruijn[*†], Robert C. Makkus[*], Ronald K. A. M. Mallant[*] *and* Gaby J. M. Janssen[*]

Contents

[*]Energy Research Centre of the Netherlands, Petten, The Netherlands
[†]University of Technology Eindhoven, Eindhoven, The Netherlands

Advances in Fuel Cells
0080453945

Abstract

This review gives an overview of proton exchange membrane (PEM) fuel cells on the materials level. Materials for proton conducting membranes, electrodes and bipolar plates are described from the perspective of their suitability for long-term operation in automotive and stationary applications.

The state-of-the-art PEM fuel cells consisting of perfluorosulfonic acid membranes, platinum-based electrodes and graphite(-composite)-based bipolar plates are in a well advanced state of development. Although the main constituents are in use for many years, impressive improvements in performance and lifetime have been realized, leading to cost reduction, size reduction, improved robustness and mass manufacturing. For the sake of system simplification, increasing the temperature of operation of the proton exchange membrane fuel cells (PEMFC) would be beneficial. New materials are needed to enable operation at higher tem- peratures, and not less important, lower relative humidities. The suitability of present state-of-the-art materials for operation above 100°C is discussed, as well as a critical review of the suitability of emerging concepts. No alternative material that combines the virtues of the present materials with the ability of operation above 100°C has been developed so far. Therefore, the first commercial introduction of PEM fuel cells is likely to be based on the components that are at present in use, making use of perfluorosulfonic acid membranes.

5.1 General Introduction

Proton Exchange Membrane Fuel Cells (PEMFC), also called Polymer Electrolyte Fuel Cells (PEFC), are in development for their application in transport, stationary power generation and portable power. On different levels, efforts are being made to make the PEMFC suitable for market introduction. This review focuses on the development on the materials level.

State-of-the-art cells, operating in the temperature range 50–80°C, can achieve power densities in the range of 0.35–0.7 W/cm^2 in hydrogen–air systems at pressure levels between ambient pressure and 2.5 bar [1]. This has brought the power density at stack level above the 1 kW/l target that has been set for automotive applications. As long as the fuel and air are free of contaminants, and other cell conditions are ideal, power density can be sustained for thousands of hours.

Still, the challenges of further materials development are manifold. An important challenge is to lower the cost for materials use while maintaining the power density level and the durability of currently used materials. As fuel cell technology is meant to replace the internal combustion engine, the cost level of the fuel cell system must be in the same range as that of the internal combustion engine [2]. In the absence of tax relief measures for zero emission vehicles, PEMFC systems should cost less than 50 US$/kWe [2] in order to become commercially viable. For use in stationary applications, the allowable cost level of the fuel cell system is in the order of 400–750 US$/kWe [2] but depends strongly on the exact application. For systems to be operated on natural gas, the local electricity and gas prices as well as the compensation for electricity delivered to the grid determine the profitability of a combined heat and power system.

Each application imposes different constraints. Application in transport means that dynamic operation under extreme conditions must be anticipated. A lifetime of 5000 operation hours is generally used as target. Application in stationary power generation, most likely small combined heat and power units, means a required lifetime of 10–15 years, equivalent to 40,000 h of operation. Whereas extreme temperature excursions are not to be expected in indoor applications, it is in many countries common use to install heat and power equipment outdoors. Therefore the difference in requirements between stationary and automotive applications might be smaller than one is expecting on first sight. Fuel cell robustness and lifetime under realistic conditions and the ability to use realistic fuels are as important as cost reduction. Both for stationary as well as transport applications, a wider range of fuels than pure hydrogen would increase the willingness of users to adopt fuel cell technology. When operated on reformed fuels, cell and stack power densities become considerably lower, and long-term operation at a constant performance level becomes more problematic.

The current view is that increasing the operating temperature of PEM fuel cells is a key factor in further simplification of the fuel cell system. The heat management becomes easier and at the same time the sensitivity of the fuel cell for impurities in the anode and cathode feed is reduced. An increase of the operating temperature implies that the functioning of the fuel cell should depend less on the presence of liquid water in the cell. The operation at drier gas conditions would add to simplification of the fuel cell system but also requires that the role of water in the proton conduction should be diminished.

There is no consensus on the temperature range in which this new generation PEM fuel cells should work. At 120°C the gain in heat management and increased tolerance to impurities, such as carbon monoxide, is already appreciable. An upper limit of 120°C means that for automotive applications conventional cooling devices can still be used. However, for PEMFC to remain attractive for automotive applications the fuel cell should be able to operate below 100°C as well. For stationary applications a temperature window 120–180°C seems still advantageous, although the PEMFC will lose much of its distinctive features in comparison to SOFCs if operation below 100°C becomes impossible.

With respect to the PEM fuel cell materials, a substantial increase of the operating temperature window of operation starts with the development of new electrolytic membranes, which will function at less humid conditions.The other components, i.e. the electrodes, and bipolar plates, sealing materials and system components, must be suitable for this extended temperature window as well. The general approach of this review is to describe the available materials and components for application in PEM fuel cells, their suitability for operation at temperatures above 100°C, and the sensitivity with respect to the presence of water. First, the rationale for operation at higher temperatures and less humid conditions of PEM fuel cells is illustrated from the perspective of fuel cell system complexity.

5.2 The Impact of Materials Properties on Fuel Cell System Design

The PEM fuel cell system design is such, that on the one hand process fluids are supplied to the stack in the quantity and quality according to the envelope of operating conditions set by the stack developer; on the other hand, application requirements with respect to power, size, weight, operability, efficiency, emissions, safety, cost, etc. should be met. The system therefore is an important interface between real life requirements and the conditions at which fuel cell components should be operated in order to obtain a reasonable lifetime. Making the match between these two sides of the interface is not an easy task. As will be shown here, the system design is to a large extend dominated by the properties of a proton-conducting polymer material that was not developed to be applied in a fuel cell. In order to better understand this, a closer look at fuel cell systems is required.

5.2.1 Low Temperature PEMFC System Design

Current PEMFC systems are low-temperature systems. Two main classes of such systems can be discerned:

- *Pressurized systems*: Operating pressure typically is 3 bar, sometimes even higher. The operating temperature of the PEMFC stack is limited to approximately 80°C.
- *Non-pressurized systems*: The operating pressure typically is 1.1–1.3 bar, and the temperature is limited to 65–70°C.

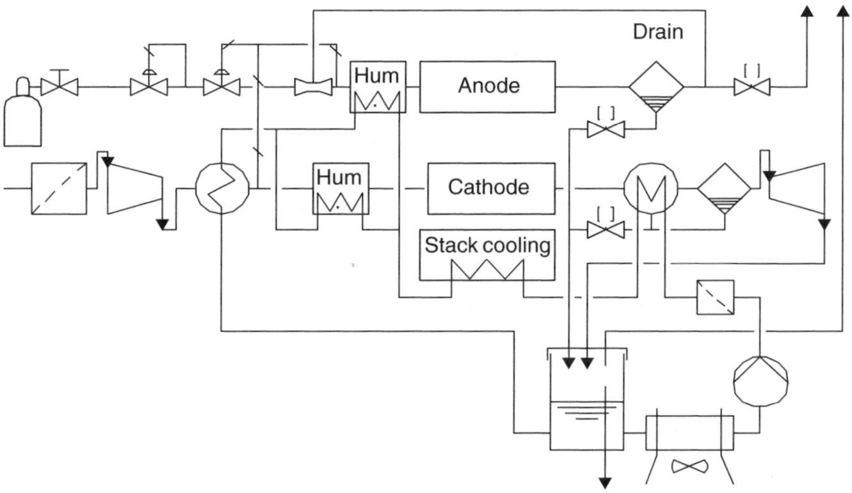

Figure 5.1 Schematic drawing of a H_2-fuelled PEMFC system. See text for detailed description. Not all components are shown, while some components in this drawing may not be needed, depending on the stack en system design. Additional components typically would comprise check valves, safety valves and sensors.

Low-temperature PEMFC systems are designed to keep the membrane in the PEMFC stack wet. Figure 5.1 shows a schematic drawing of a pressured H_2-fuelled PEMFC system. The system should be considered as being an example, many variations on the given design are possible. Three fluid circuits can be discerned: the H_2-circuit, the oxidant (air) circuit and the coolant circuit.

5.2.1.1 H_2-Circuit

The upper part of Figure 5.1 shows the H_2-circuit. In this case, hydrogen is stored in cylinders. Two pressure reducers lower the high pressure to a value that is acceptable to the stack. After passing an ejector and a humidifier, an excess amount of H_2 enters the anode compartment of the stack. Providing an excess amount of H_2, in combination with an adequate stack design, assures that a homogeneous flow distribution can be attained, even under part load conditions when H_2-consumption is low. The effect of the excess H_2 is to mobilize water droplets that would otherwise accumulate in the flow channels and thus cause local starvation of the cell. In addition to providing an excess amount of H_2, pressure pulses can be applied to mobilize water droplets [3,4]. Several options to create such pulses are given, an improvement of performance has been claimed for current densities over 400 mA/cm^2.

The H_2 that is not consumed in the stack is preferably not vented to the surroundings but is recirculated to the anode entrance, by means of a blower or an ejector.

Also included in the H_2-loop is a water separator or demister ("drain") that removes liquid water, and a purge valve that is intermittently used to purge the H_2-loop. Without such purge events, inert gases will build up. These gases may already be present in trace amounts in the fuel, or may enter the H_2-circuit from the cathode by diffusion across the membrane. Instead of recirculating excess hydrogen, the anode can also be operated in the so-called "dead-end" mode. In this case, a purge valve is used to occasionally vent the anode compartment to expel water and inert gases. Frequent purging of hydrogen contributes to the complexity of the system, as well as to losses in fuel efficiency. Other sophisticated methods, which comprise the activation of the purge valve by monitoring the decline in cell or stack voltage [5] suffer from a too high-voltage efficiency loss.

The humidifier in the H_2-loop in Figure 5.1 has been included for completeness. In this case the humidifier is based on a water exchanging membrane (e.g. Nafion) that facilitates the transfer of water and heat from the water-based coolant loop to the H_2-circuit. Such a humidifier is used in the Ballard Phase 4 bus engine [6]. GM/Opel reports that such a humidifier can be avoided by a clever combination of stack and system design features [7]. Systems design and built at the Energy Research Centre of the Netherlands (ECN) also do not include this humidifier. Hydration of the membrane has proven to be sufficient when H_2 is circulated and counter flow of anode and cathode gases is employed.

From the above it can be concluded that the presence of liquid water plays an important role in H_2-circuit of a fuel cell system. The design and control of the circuit is to a large extent dominated by the need to maintain high relative humidity (RH) while avoiding water build-up. Or, vice versa, the system could be simplified substantially if operating conditions in which the condensation of water would not occur. In the present configuration, however, both the electrolyte as well as the electrodes must be operated at a high RH and at a temperature not exceeding 80°C.

5.2.1.2 Oxidant Circuit

The middle part of Figure 5.1 shows the oxidant circuit. In this case, air is assumed to be the oxidant of choice. Typically, the components found in the oxidant circuit would be: air filter, compressor/blower, cooler, humidifier, fuel cell cathode, condenser, water separator/demister and expander. In the given arrangement the cathode exit gases are passed through the water buffer vessel of the water management/coolant system, thereby collecting the liquid water produced in the condenser. The expander will usually be combined with the compressor. In case of operation close to atmospheric pressure the air cooler and the expander will not be included. An additional air filter may be required if particulate matter and/or oil is potentially released by the compressor/blower.

Even more than in the H_2-circuit, water management issues dominate the layout and control of the air circuit. The penalty for choosing the wrong combination of pressure, temperature, air stoichiometry and heat exchanger quality is a net water loss from the system. This would necessitate the intake of water, potentially in substantial amounts, during refuelling of cars. A very high RH of the air is required in order to maintain good proton conductivity of the polymer electrolyte. Several authors have described the consequences of this demand on the operating temperature and pressure. For a fuel cell system comprising a methanol steam reformer, palladium membrane and PEMFC stack [8], it was concluded by Menzer and Höhlein that such a system can easily become a net water consumer. The authors describe the influence on water neutrality of external parameters, such as ambient temperature and RH, and of typical system parameters such as air stoichiometry, operating pressure, and quality of cooler and condenser. It is concluded that for typical PEMFC stack air stoichiometry numbers of 2–2.5, pressurized operation is required. This is especially the case if the system is supposed to be operated in the wide range of ambient conditions that should be faced if PEMFC vehicles are to have the same versatility as modern internal combustion engine powered vehicles.

A similar analysis was made by Ahmed [9] for automotive fuel cell systems comprising autothermal reforming (ATR) of hydrocarbon fuels. Instead of a membrane for hydrogen purification, water gas shift and preferential oxidation (PrOx) of carbon monoxide are used. In general, the study by Ahmed *et al.* confirms the conclusions made by Menzer and Höhlein: at non-pressurized operation, the system very easily becomes a net water consumer. The situation is worse for fuels having a low H/C ratio than for hydrocarbons such as ethylene glycol or methanol, which have a relatively high H/C ratio. In all cases, pressurised systems would be required to deal with relatively high ambient temperatures (>35°C), or relatively low ambient pressures (such as at Los Alamos, where the ambient pressure averages 0.76 atm. [9]).

For H_2-fuelled systems, the situation is somewhat different because no water is needed for reforming and water gas shift reactors. Yet, the conclusions in a qualitative way are more or less identical. An analysis of water and heat management issues for H_2-based systems [10,11] leads to the conclusion that if full humidification of the cathode air is required, atmospheric operation limits the operation temperature to about 65–70°C. At higher temperatures, the rapidly increasing water vapour pressure severely reduces the O_2-partial vapour pressure at the cathode, and therefore cell performance. In addition, maintaining water neutrality becomes a problem at this temperature level. If water neutrality *and* higher operating temperatures are required, an extra cooling effort is required to produce liquid water in the cathode outlet condenser. This condenser has to be cooled by ambient air, at a temperature difference that can be described as being relatively modest or even impractically low when seen in relation to the amount of heat that has to be rejected. Water neutrality can be achieved by pressurized operation [10,11] or low operating temperature. However, low operating temperatures are not attractive in cars because of heat rejection problems. Again, a completely new type of proton conducting material enabling to operate a fuel cell at >100°C without the need for humidification of the reactants, would lead to a much less water consuming system with higher flexibility in operating conditions.

The relation between water management and heat management also discussed in a paper describing the performance of direct hydrogen fuel cell systems for hybrid vehicles [2]. Again, one of the conclusions was that the fuel cell system, in comparison with an internal combustion engine, requires radiators with a much larger frontal area, larger blower fan, larger fin area and more ram air. According to the authors, these cooling requirements are likely to significantly enhance the aerodynamic drag of the fuel cell vehicle.

The conclusion therefore is that the properties of presently available proton-conducting polymers for the PEMFC place a high burden on system complexity, and therefore cost, as well as on overall vehicle efficiency (the latter due to the increased drag and the frequent use of a high-power cooling fan).

5.2.2 System Simplification As a Result of Dry Gas Operation

From all studies referred to here, it becomes apparent that the need for close to 100% RH has a significant effect on the design and the operation envelope of PEMFC systems. It is interesting to see what simplifications are possible if a membrane that can work under dry conditions would be available. Figure 5.2 shows the main components of a system *if no provisions for humidification and water retrieval are required* and *if the operating temperature would prevent that liquid water is formed.* The difference with the system in Figure 5.1 is clear. All components required for humidification and water retrieval can be omitted. If liquid water build-up no longer is an issue in the anode circuit,

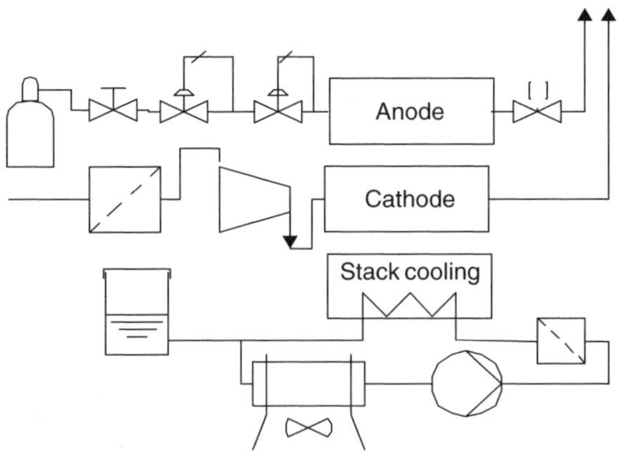

Figure 5.2 Atmospheric system without ancillaries for water management. Additional components such as safety valves and sensors may be included.

H_2-recirculation is no longer needed as well. A vent for releasing trace gases still is required though. Effective venting can be done very effectively with a minimal loss of H_2, when using concepts already applied in alkaline fuel cell systems [12].

One may argue whether the system should be operated at elevated pressure, for which a compressor/expander is needed, or that an operating pressure can be chosen that is sufficiently close to atmospheric pressure to allow for a simple type of air mover, such as a blower. The main reasons for operating a fully humidified system at elevated pressure are alleviating water management and increasing O_2-partial pressure in the water vapour diluted air. Of course, some dilution by water vapour will take place as water is produced in the electrochemical reaction. Figure 5.3 shows how the water vapour pressure in the air at the exit of the cathode depends on temperature and air stoichiometry. The graph is made up for a stack without humidifier, operating at atmospheric pressure (so, 1 bar), and a dew point temperature of the inlet air of 10°C. For typical air stoichiometry values of $\lambda = 2$, the water vapour pressure would be 200 mbar, *at the exit region of the cathode only.*

It has been suggested by Gasteiger and Mathias [13] that the operating pressure of the ideal system would be 1.5 bar.

At 1.5 bar operating pressure, the additional water vapour pressure at the cathode exit is almost 300 mbar (as opposed to 200 mbar for atmospheric systems), as can be concluded from Figure 5.4. So, the air partial pressure

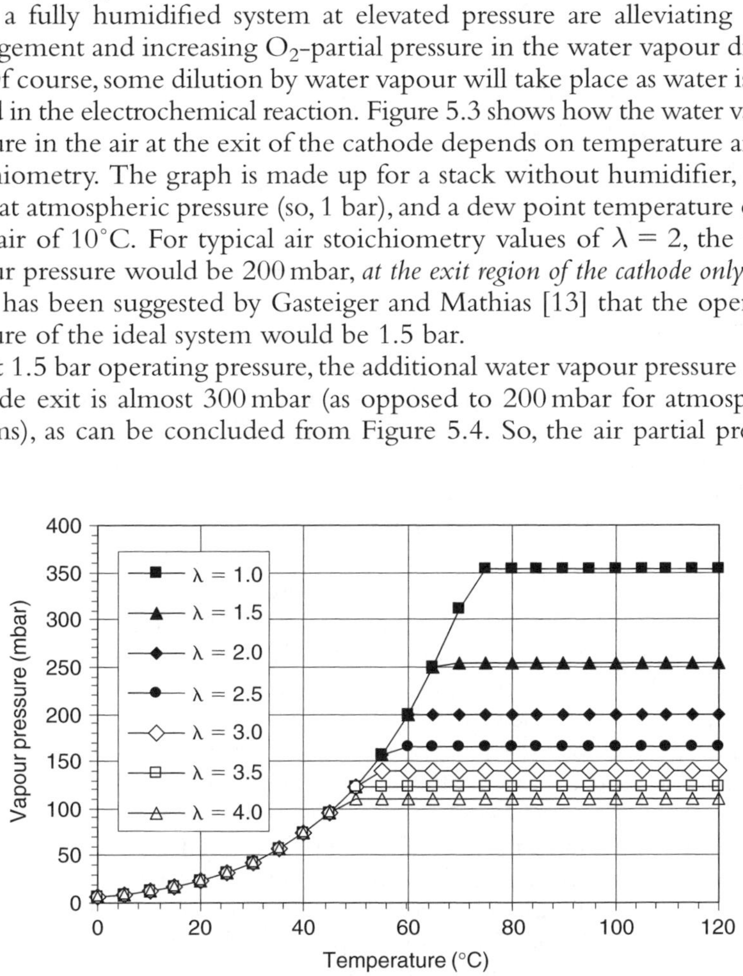

Figure 5.3 Partial water vapour pressure at various air stoichiometry values, as function of system temperature for a system pressure of 1 bar. The dew point of the inlet air is assumed to be 10°C. The curves are straight lines as long as the water vapour pressure is lower than the maximum water vapour pressure at the given temperature. Condensation takes place when the lines are no longer straight and horizontal, i.e. in the lower temperature area.

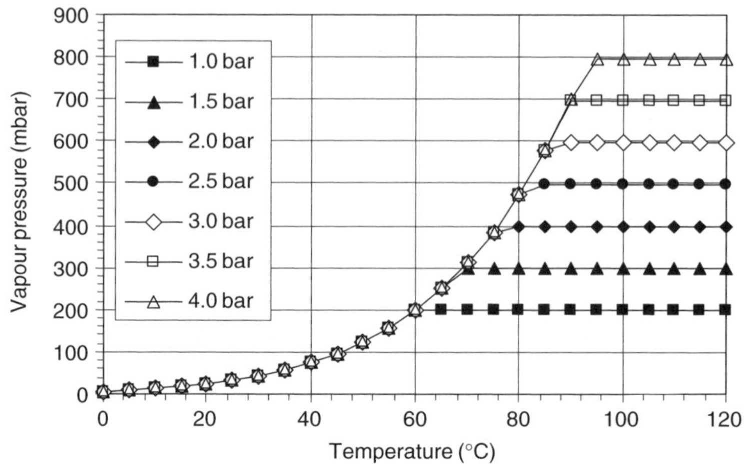

Figure 5.4 Partial water vapour pressure at the cathode exit, for various operating pressures and an air stoichiometry of 2. The dew point of the inlet air is assumed to be 10°C.

increases from 0.8 bar to 1.2 bar (again, at cathode exit area only). However, the minimal operating temperature that should be maintained to avoid liquid water formation increases from 60°C to almost 75°C (all at $\lambda = 2$).

The rationale of the system pressure of 1.5 bar is that when assuming an initial water vapour pressure of 0.5 bar, an effective air pressure (at the cathode inlet) of 1 bar is maintained. In this analysis by Gasteiger and Mathias, the ideal membrane would be able to operate at 120°C and 25% RH at 1.5 bar. The choice for 25% RH stems from the associated water vapour pressure which is then 0.5 bar, which corresponds to 100% RH at 80°C. By choosing this water vapour pressure, one still inherits a large portion of the water management problems of an 80°C, 100% RH system. The condenser cannot be omitted, in fact should be somewhat larger, and a humidifier still is necessary too. Liquid water will form in the stack if, e.g. during prolonged part load operation, the stack temperature drops below 95°C. Thus, both the stack and the system have to be designed such as to deal with two-phase flows.

The design depicted in Figure 5.2 can only be realized if there is no need for humidification of gases, and no risk of liquid water formation. However, condensation is difficult to avoid completely, especially at low operating temperatures (see Figure 5.3). For a typical stoichiometry value of $\lambda = 2$, the water vapour pressure in the region close to the cathode exit is approximately 200 mbar, and liquid water will form below 60°C, e.g. during cold start. This condensation will not take place throughout the whole stack, since the water vapour pressure will build up gradually along the air flow path. Possibly condensation during start-up can to a certain extent be avoided by high stoichiometry values. Still, liquid water formation during start-up is

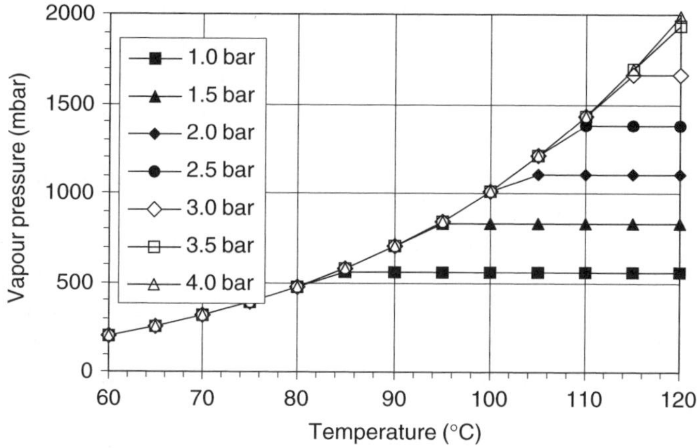

Figure 5.5 Partial water vapour pressure at the cathode exit, for an air stoichiometry of 2. The dew point of the inlet air is assumed to be 80°C. The curve for $P = 1.5$ bar corresponds to the Gasteiger and Mathias case. Below 95°C liquid water will form in the stack.

hard to avoid completely and should therefore be addressed in stack and system design (see Figure 5.5).

Apart from the simplification in system design, there are other advantages if water is no longer an issue in PEMFC operation. Control of the system is easier, since the PEMFC stack would no longer run the risk of drying out locally. And since, after the start-up phase, build up of liquid water is no longer an issue, this can no longer be the cause for an uneven distribution of gases. This will reduce the need for individual cell voltage monitoring. Also, the system no longer holds large amounts of pure water, thus freeze protection is no longer needed.

Thus, it is clear that the system would benefit substantially if a proton-conducting material would become available that performs well under ambient water vapour pressures, but that would still be resistant to the high RH values that occur during start-up. Nevertheless, the argument for new polymer development that is mostly used is that of the need to operate at higher temperatures.

5.2.3 Increasing the PEMFC Temperature

It has been demonstrated that the availability of a proton-conducting polymer capable of operation at low water vapour pressure would simplify the system considerably. The use of "standard" perfluorinated sulfonic acid (PFSA) membranes however makes dry operation impossible, RH values close to 100% have to be maintained. This in turn makes operating temperatures higher than 80°C virtually impossible, although there is a strong desire to be able to operate at temperatures at or slightly above 100°C [7,13]. Operating at temperatures higher than 120°C does not seem to increase attractiveness. For the automotive industry, the size and type of hardware needed for heat rejection would be comparable to that of internal combustion engine cars [2,7,13].

For stationary applications, the possibilities to use and particularly to effectively store the waste heat decrease rapidly when temperature falls below 90–100°C. Being at a level of 65°C, the waste heat of an atmospherically operating PEMFC stack is too low for buffering and efficient use in many of the existing water-based domestic heating systems. A system operating >100°C would enable optimal use of water-based heat buffers, allows for compact water-fed radiators and makes combination with absorption cooling systems for refrigeration and air conditioning possible.

So, at first glance, both stationary and automotive applications seem to benefit from an operating temperature closer to, or some tens of degrees above 100°C. Operating at even higher temperatures does not seem to be required. The situation becomes somewhat more complex if the use of hydrocarbons comes into play. The low-temperature PEMFC when combined with a fuel processor requires a device that reduces the concentration of CO to levels of several tens of ppm. In most cases, PrOx is used as a means to accomplish this. At 160°C and above, the Pt-based anode catalyst is able to perform even in the presence of about 1% of CO. At this level of CO the PrOx reactor in fuel-processing systems becomes obsolete. This increases system efficiency and operability, and reduces cost and maintenance. For this reason, the operating temperature should be selected in the 160–180°C region to fully exploit the benefits of CO tolerance. The advantages of a high-temperature membrane in terms of *heat rejection* are however less than for H_2-fuelled systems. The fuel processor needs water for both the reforming and water gas shift reactors. In order to maintain a neutral or slightly positive water balance, a condenser has to be included in the system. For a gasoline system with ATR fuel processor, it has been shown that the condenser should be operating at temperatures ranging between 47°C and 58°C, depending on system design [14]. The heat load for this condenser is considerable: approximately 30 kW for a 50 kW fuel cell system. Guaranteeing that a suitable heat sink is available at all times could again be a problem, both for automotive systems and for stationary systems.

5.2.4 Cooling of High-Temperature PEMFC Systems

So far, the issue of cooling PEMFC systems has been described in term of heat rejection and waste heat use. There is another issue though: the selection of the cooling fluid. An evaluation of cooling systems for the low-temperature PEMFC in transport application is given by Oseen-Senda *et al.* [15]. In additional to traditional automotive coolant systems, two requirements seem to be of paramount importance: the fluid should not introduce a short-circuit current between the individual cells and the fluid should not pose a threat to the membrane–electrode assemblies (MEAs) in the fuel cell stack in case of leakage. For this reason, ultra pure water has been the fluid of choice in many systems. With appropriate selection of wetted materials in the cooling loop, and the inclusion of ion exchange filters, the water purity can be maintained at a level where conductivity is sufficiently low ($\leqslant 50\,\mu S/cm$). Freezing of such systems now becomes an issue, which can be solved by the addition of anti-freezing compounds such as alkanetriols and alkylene glycols [15,16].

Little information on cooling of high-temperature PEMFC stacks is available. For temperatures slightly above 100°C, the problem seems to be relatively small. Conventional water/ethylene glycol cooling systems are also designed to accommodate coolant temperatures up to 120°C. This is done by pressurization of the fluid. Pressurization in case of the PEMFC may be unattractive though. The many seals present in a PEMFC stack in combination with pressurization of the fluid impose a high risk of intrusion of the cooling fluid into the cell, where it may harm the catalyst. The use of alkanetriols as anti-freeze agents is therefore proposed; alkanetriols are reported to be compatible with the catalysts of the PEMFC [15,17,18].

The use of certain types of vegetable oils up to temperatures of 140°C has been described in WO2004/054014 [19]. For higher temperatures, no solution seems to be readily available. Dielectric fluids with high boiling points will probably be unattractive because of the potential risk for the environment when spilled, and for the fuel cell catalyst when leaked internally. Possibly, cooling techniques employed in the phosphoric acid fuel cell (PAFC) can be used. These are air cooling or cooling with a two-phase water-based system.

5.2.5 Temperature and Humidity, Conclusions

In conclusion, it can be stated that, *from a systems point of view*, a big leap forward would be made when a membrane becomes available that allows for:

- Functioning at virtually zero water vapour pressure.
- Good proton conductivity in the 100–120°C temperature range, while still maintaining reasonable conductivity at sub-zero temperatures.

The capability to *also* operate at higher than 120°C will have additional benefits for reformate-fuelled systems, but will not enhance the attractiveness for H_2-fuelled automotive systems, unless the higher waste heat temperature is required for other processes in the system. Such processes could be to provide the heat for releasing H_2 from metalhydrides or to drive absorption-based air conditioners. The attractiveness of the new membrane would be reduced if:

- The operating temperature window is limited to the 160–180°C range (i.e. if operation at lower temperatures is not possible).

- The membrane still needs to be in contact with air having a water vapour pressure higher than that of the ambient air.
- The membrane is not compatible with high RH values that occur during start-up.

5.3 Membranes

5.3.1 Introduction

Proton-conducting polymer membranes offer huge advantages in comparison to all other fuel cell electrolytes. In comparison to liquid electrolytes (potassium hydroxide, phosphoric acid, molten carbonate), it offers the benefits of a solid electrolyte that it does not creep into other fuel cell compartments, it does not evaporate and that it is easy to handle in manufacturing processes. In comparison to other solid electrolytes (solid oxide), it offers higher mechanical strength, mechanical flexibility, the possibility to manufacture in large quantities and easy interfacing with the electrode material as it can be dispersed in liquid media.

The most important functions of the membrane are:

1. electrolyte,
2. gas separator,
3. electronic separator.

5.3.1.1 Electrolytic Function

The most important characteristic of the proton-conducting membrane is its proton conductance. The conductance is determined by the bulk conductivity of the membrane, which target value is in the range of 0.05–0.1 S/cm, and its thickness. In nearly all proton-conducting polymers, the conductivity strongly depends on the water content of the membrane. The acidic proton has to be dissociated, and when it is dissociated it has to be transported. Water plays a vital role, both as the Brønsted base as well as for the mobility of the proton.

In the majority of the proton-conducting polymers, the acidic functionality is introduced by means of sulfonation. The density of acid sites is determined by the number of SO_3H sites per repeating unit. In the proton-conducting membranes, the acidity is often given as the equivalent weight (grams polymer/mole sulfonic acid), which is (ion exchange capacity)$^{-1}$. In membranes which are in development for operation at higher temperatures, proton conductivity is introduced also by phosphonation or doping the membrane with a low volatility-acid donor, such as phosphoric acid.

For a membrane to be suitable for use in the PEMFC, a maximum real resistance of the membrane of $25\,m\Omega\,cm^2$ at steady-state conditions and $100\,m\Omega\,cm^2$ at 20°C is set for automotive applications [20]. For $25\,\mu m$

membranes, this translates into a minimum conductivity of 0.1 S/cm for steady-state operation and 0.025 S/cm at 20°C.

5.3.1.2 Gas Separating Function

In the PEMFC, hydrogen is used as the fuel and air as the oxidant. The membrane serves as the gas separator. When the separation is not ideal, hydrogen diffuses to the cathode compartment and oxygen to the anode compartment. On both sides, the mixture can react on the platinum electrode to water, which is accompanied by the production of heat. This can lead to small pinholes. The side effect of the presence of mixture of hydrogen and oxygen on both sides is the lowering of the cell voltage, which can amount to more than 100 mV. The maximum allowable permeability of the membrane for hydrogen, at full load (2 A/cm^2) is 8×10^4 cm^3 cm/m^2/s/kPa when accepting a 1% fuel efficiency loss due to crossover and 3×10^4 cm^3 cm/m^2/s/kPa corresponding to 10% fuel efficiency loss at low current load (0.05 A/cm^2) [20]. For oxygen, the maximum allowable permeabilities are 2.5×10^3 cm^3 cm/m^2/s/kPa at full load and 1×10^3 cm^3 cm/m^2/s/kPa at low-current load.

5.3.1.3 Electronic Separating Function

All components but the proton-conducting membrane are constructed such that their electronic conductance is maximal. In order to prevent the cell to be short circuited, the proton-conducting membrane has to be electronically insulating. The electronic conductivity is generally not an issue, no specific targets are found for this property. The electronic resistivity of Nafion, the current state-of-the-art membrane, is in the order of $2 \times 10^5\,\Omega$ cm [21].

5.3.1.4 Mechanical Strength

As the membrane separates the anode and cathode compartments, its mechanical properties are determining factors for the fuel cell lifetime. As part of the membrane electrode assembly, the water content of the membrane can vary widely, from a relative dry condition during still stand to water saturated during operation. All polymers exhibit a certain degree of swelling during water uptake. Excessive swelling can damage the membrane and the membrane–electrode interface leading to an increase in cell resistance.

During MEA manufacturing, one of the vital issues for optimal performance is the optimization of the interface between the membrane and the electrodes. In many cases, organic solvents used in electrode inks are applied on the membrane, which swells even more upon exposure to organic solvents than upon exposure to water. Hot pressing is another procedure mostly used for creating the right membrane–electrode interface. As automated manufacturing is used for all large-scale manufacturing, membranes are subjected to all sorts of physical forces. Tensile strength and modulus, puncture resistance and tear strength are relevant properties of the membranes which need attention in their selection for fuel cell use.

5.3.1.5 Water Management

High membrane conductance can be achieved by using thin membranes. Gas separation, mechanical strength and to a much lesser extent electronic separation however become poorer when using thinner membranes.

In addition to a higher conductance, thin membranes are also preferred for reasons of better water management within the fuel cell. This is illustrated in Figure 5.6.

Protons which are transported from the anode to the cathode are accompanied by water molecules, due to the electro-osmotic drag. For perfluorosulfonic acid membranes, typically 1–2 water molecules are transported per proton [22]. At high current densities, this can lead to local drying of the membrane at the anode side, leading to a local increase in membrane resistance at the anode. As stack manufacturers tend to operation at low degree of reactant humidification from a standpoint of system simplicity, this issue becomes even more important.

Diffusion of water, which is produced at the cathode by the reduction of oxygen in the presence of protons, to the anode is essential for the stable behavior of the PEMFC. As the diffusion of water is proportional to the water gradient in the membrane, and thus inversely proportional to the membrane thickness, it is easier to obtain stable fuel cell performance using thin membranes than if using thick membranes.

Impedance measurements have indicated that stable performance of 50 μm membranes under varying conditions is much simpler in comparison to 175 μm membranes of the same composition [23]. The thick membranes show dehydration behaviour when humidification is less then optimal and with increasing current densities, especially at cell temperatures lower than 60°C. The latter effect is related to the lower diffusion coefficient of water at decreasing membrane temperature. Thin membranes do not show a strong relationship between membrane resistance at increasing current densities and cell temperature [23].

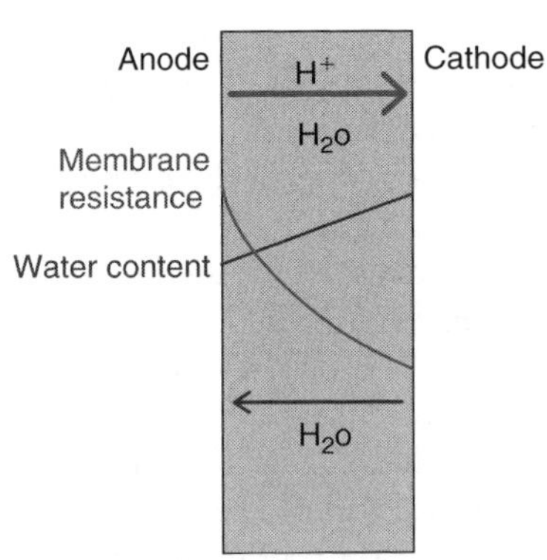

Figure 5.6 Water management in PEMFC.

Besides keeping the membrane hydrated at both sides, back-diffusion of water also regulates the water content of the cathode. A poor removal of water from the cathode through the membrane to the anode leads to flooding, which means that oxygen diffusion to the electrode/membrane interface is hindered by the presence of liquid water. Thick membranes were shown to suffer from this flooding phenomenon, when high current density are imposed at well humidified conditions and relatively low operating temperatures [23].

Summarized, robust operation of the fuel cell is much easier when using membranes with a thickness of approximately 50 μm than when using thick membranes.

5.3.1.6 Membrane Architecture

Electrolytic membranes used for PEM fuel cells consist of a polymer backbone with ionic goups attached to it. The polymer backbone is the determing factor for the mechanical properties of the membrane. The collection of the ionic clusters determine (?) the acidity (?) of the membrane. The phase separation between the polymeric phase and the ionic phase is an important factor for the membrane conductivity.

The degree of phase separation is determined by the nature of the polymer chain and the ionic group (Figure 5.7).

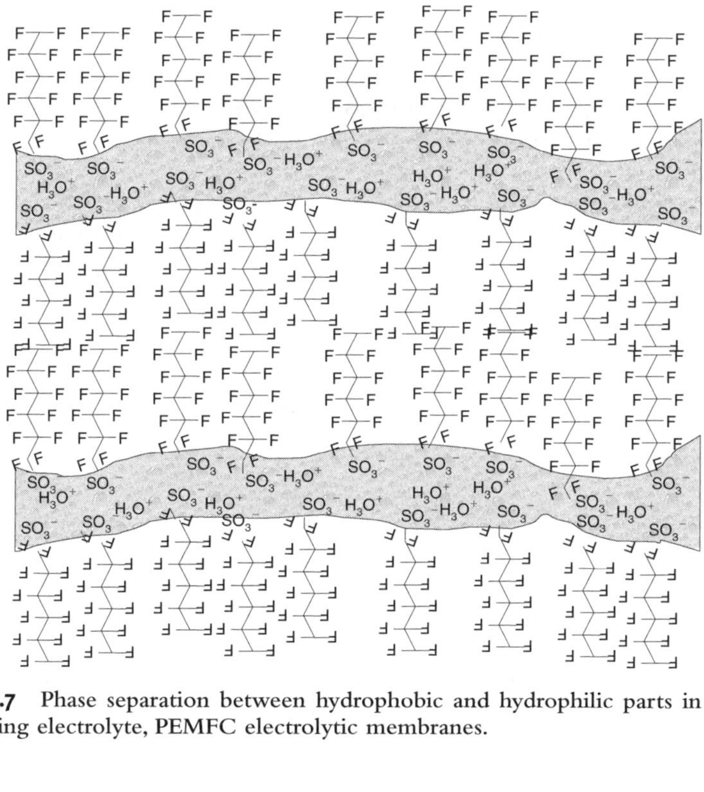

Figure 5.7 Phase separation between hydrophobic and hydrophilic parts in proton-conducting electrolyte, PEMFC electrolytic membranes.

5.3.2 Perfluorosulfonic Acid/Tetrafluoroethylene Copolymer Membranes

Currently the most used proton-conducting membrane is based on a perfluorosulfonic acid/tetrafluoroethylene copolymer. It is produced by DuPont under the trade name Nafion [24], by Asahi Kasei Chemicals under the trade name Aciplex [25], and by Asahi Glass under the name of Flemion [25]. In the past, also Dow [26], Chlorine Engineers [27], ICI [28,29] and Hanwa Chemical Corp [30,31] were reported to develop PFSA membranes.

The molecular structure of perfluorosulfonic acid/tetrafluoroethylene/copolymer membranes is shown in Figure 5.8.

The perfluorosulfonic acid/tetrafluoroethylene/copolymer is chemically very resistant, especially to oxidizing conditions, due to the substitution of the CH_2 by CF_2. According to Linkous [32], fluoridation of the hydrocarbons in the vicinity of the sulfonic acid leads to an increase in acidity. Generally, the fluorosulfonic acids like Nafion are considered to be super acids, i.e. the acid dissociation in water is 100%. The equivalent weight of the Nafion-type polymers is generally between 800 and 1100 g/mol.

5.3.2.1 Nafion

Nafion was originally developed for use as separator membrane for chlor-alkali electrolyser cells [22]. Later on, its use for fuel cells, electrodialysis cells, ion-selective barrier layers for electrochemical sensors has followed. As the commercial use of Nafion in the chlor-alkali industry was leading, its properties were not specifically tailored for fuel cell use: relatively thick membranes with high equivalent weight and poor water uptake characteristics from unsaturated gases.

The application of the Nafion-type polymers in fuel cells is widespread. A huge amount of literature exists on the use and the properties of Nafion membranes. A recent review on the morphological properties of Nafion and its characterization was published by Mauritz and Moore [33]. Another

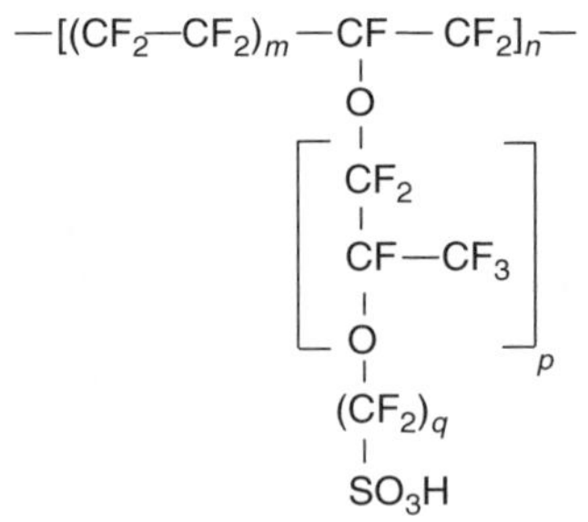

Figure 5.8 Molecular structure of perfluorosulfonic acid tetrafluoroethylene/copolymer membranes. $m = 6–10$; $n = 1000$; Nafion: $p = 1$, $q = 2$; Aciplex: $p = 0-1$, $q = 2-5$; Flemion: $p = 1$, $q = 2$.

useful review on the properties of Nafion and methods to characterize it is given by Doyle and Rajendran [22].

Until a few years ago, this application was limited to Nafion 117. The coding in this type of polymers is determined by the first two digits, which give the equivalent weight divided by 100, and the last digit giving its thickness in ml (1 mil = 25.4 μm) in the dry state. In the mid-1990s, the variety of Nafion-type membranes was extended to thinner membranes and to membranes with a lower equivalent weight. Most used membranes were Nafion 117, Nafion 115 and Nafion 105. Nafion 112 did exist but was not used very often for MEA and stack manufacturing. A shift can be noticed to the thinner membranes. Nafion 117 and 115 are becoming "out of fashion", while Nafion 112 and Nafion 1135 are becoming a sort of standard. Nafion 1135 has a thickness of 3.5 mil, 88 μm.

Membranes with an equivalent weight of 1000 are apparently too weak for further processing, as they are not offered anymore by DuPont [24]. Ion Power does offer lower equivalent weight membranes [34], in the thickness range from 25 to 125 μm [34].

The water content of a perfluorosulfonic acid polymer with low equivalent weight is higher than that of a comparable perfluorosulfonic acid with a high equivalent weight [35].

DuPont now even has developed a 25 μm Nafion N111 membrane, which shows improved performance in comparison to 50 μm Nafion N112, leading to a 10% higher power output at humidified conditions [36]. The gain is even higher when the cathode feed is less than fully saturated. In a 2500 h lifetime test, the 25 μm membrane shows stable behaviour over the complete test time, and a lower resistance than the 50 μm membrane [36].

The ideal Nafion membrane would thus be a very thin membrane with a low equivalent weight. The mechanical strength of a membrane however decreases with decreasing thickness and with decreasing equivalent weight. In both cases a substrate can be used to enhance the mechanical strength.

The crossover of hydrogen and oxygen increases with decreasing thickness, leading to a lower cell voltage at a given current density, especially at low-current densities. For wet Nafion membranes with an equivalent weight of 1100, the permeability for hydrogen amounts to $4.5 \times 10^5\,cm^3\,cm/m^2/s/kPa$ and for oxygen it amounts to $2.5 \times 10^5\,cm^3 cm/m^2/s/kPa$ [37]; For Nafion NRE-212, the hydrogen permeability amounts to less than $0.01\,ml/min\,cm^2$ [24] at an RH of 100%, 23°C and a pressure difference of 50 psi (3.44 bar). Comparing the open circuit voltage (OCV) of a Nafion 117 membrane to that of a Nafion 112 membrane, decreasing the membrane thickness with 125 μm leads to a drop in OCV of approximately 50 mV [28].

Increasing the lifetime of Nafion is obtained by decreasing the number of polymer end-groups containing hydrogen [36]. These end-groups are subject to the attack by $OH^{\bullet}$ and $OOH^{\bullet}$ radicals, which are decomposition products from H_2O_2 [37]. H_2O_2 can be formed either by the 2-electron reduction of oxygen at the cathode, or by the chemical reaction between hydrogen and oxygen as resulting from gas crossover through the membrane [37]. Membrane degradation following this path can be predicted by measuring the release of fluorine in the peroxide-Fe^{2+} stability test (Fenton's test), in which the membrane is subjected to 16–20 h boiling in an aqueous 30% H_2O_2 solution containing 20 ppm Fe^{2+} [36]. The iron salt leads to the formation of peroxy radicals, thus representing an accelerated test of the fuel cell environment.

A 10- to 25-fold decrease in fluorine release rate can thus be obtained using methodologies aiming at the decrease of H-containing polymer end-groups.

Another modification introduced by DuPont is the manufacturing of membranes by solution casting instead of extrusion. Although the method of solution casting has been applied since long time by researchers, it was not used by DuPont for making their commercial membranes. Differences in water retention and transport properties are introduced by these two different methods, which stem from differences in hydrophobic/hydrophilic phase separation. Annealing at high temperatures tend to diminish these differences.

The proton conductivity of Nafion highly depends on its water content. Figure 5.9 is illustrative for the properties of Nafion, which shows the water content and the proton conductivity in transverse direction as a function of the RH of the environment. Data are from Ref. [38].

In this study, 100% RH is experimentally established by immersion in liquid water. Several studies have addressed a certain discrepancy between the water uptake by Nafion from a fully saturated gas and from immersion in liquid water, although thermodynamically the water activity is the same [39,40]. Water adsorption from the gas phase typically leads to a maximum water

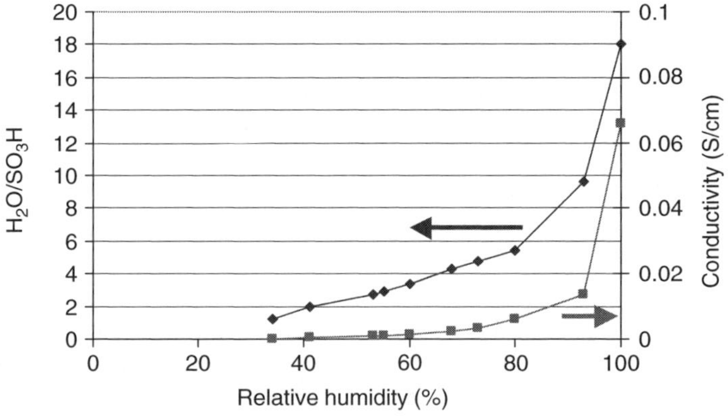

Figure 5.9 Proton conductivity and number of water molecules per sulfonic acid site versus RH for Nafion.

uptake of 14 water molecules per SO_3H group[39,40], while immersion in liquid water leads to an uptake of 22 water molecules per SO_3H [25,40].

Concluding from the data collected by Slade *et al.* [41], conductivity in membranes immersed in liquid water at 20–30°C is typically 0.1 S/cm, while membrane conductivity in membranes in equilibrium with water vapour is in the range of 0.06–0.08 S/cm. This is consistent with the data from Zawodzinski *et al.* [40], which show that 14 H_2O/SO_3H leads to a conductivity of 0.06 S/cm and 22 H_2O/SO_3H leads to a conductivity of 0.1 S/cm, both at 30°C.

This should be taken into account when translating membrane conductivity as measured *ex situ* on a membrane in equilibrium with partly humidified gases. In a fuel cell in operation, water formed at the cathode can directly be transferred into the membrane, probably leading to more easy absorption of liquid water in the membrane pores than from the gas phase. An *ex situ* measured conductivity can thus underestimate the membrane conductivity in the fuel cell under operation under apparently similar conditions.

The different water content in the membrane as imposed by liquid water absorption as compared to vapour phase adsorption is also expressed by different electro-osmotic water drag coefficients in the absence of water activity gradients, that of a membrane containing 11 H_2O/SO_3H amounting to 0.9, while a membrane containing 22 H_2O/SO_3H has an electro-osmotic water drag coefficient amounting to 2.5 or higher [40]. Another interesting observation of this study was the fact that recast Nafion membranes exhibit a higher electro-osmotic water drag coefficient than the extruded Nafion membrane, explained by a lower degree of crystallinity. At the same time it implies that recast membranes, as now commercialized by DuPont as NRE-211 and NRE-212, are expected to have better water absorption properties than the extruded membranes.

There are indications that the conductivity decreases with decreasing membrane thickness [41]. The effects are ascribed to the production process of the Nafion membranes, where the hydrophilic channels might be distorted during the extrusion process [41].

Fuel cells using standard Nafion membranes have proven to have a lifetime exceeding 11,000 h [42], without membrane failure, when operated under relatively ideal conditions. It illustrates the chemical stability of perfluorinated membranes and its resistance against radical attack. Chemically modified Nafion membranes will survive even longer [36], based on accelerated lifetime tests a factor of 2.

At slightly undersaturated conditions however, performance degradation was measured already after 1900 h, accompanied by an increase in hydrogen permeability of the membrane of a factor of 20 [43].

As the crossover of oxygen to the anode, and its subsequent 2-electron reduction by hydrogen two hydrogen peroxide is one of the membrane degradation causes, thinner membranes would be likely to suffer more from this degradation than thick membranes. On the other hand, water management is better in thin membranes, so that the above mentioned degradation at undersaturated conditions would become less severe. A good comparitive study on the lifetime of membranes with various thicknesses with the same electrodes, does not exist to our knowledge.

5.3.2.2 Aciplex, Flemion, 3M

Many similarities exist between Nafion, Aciplex, Flemion and 3M. They are treated separately in this review primarily because the majority of fuel cell and membrane studies are performed on Nafion. As stated in Section 5.3.2.1, Nafion is at present only offered commercially in the EW1100 series. Both Aciplex and Flemion are offered with lower equivalent weights, EW950 for Aciplex [44,45] and EW909 for Flemion [46]. This enabled Saito *et al.* [25] to study properties such as ion conductivity, swelling behaviour, water permeability and diffusion as a function of equivalent weight (EW909–EW1100) and membrane thickness (120–285 μm).

It appears that ion conductivity in this group of membranes is determined by the water mobility more than by ion concentration. As the lower equivalent weight membranes take up more water, the increase in the concentration of sulfonic acid groups in the dry state is compensated by the increase in water uptake, leading to almost identical ion concentrations in the wet membranes. At an equivalent weight of 1050 for Aciplex S-1112 and an EW of 1099 for Flemion LSH-180, the proton conductivity amounts to 0.13 S/cm at 25°C. Under the same conditions, proton conductivity increases to 0.18 S/cm at EW = 909 for Flemion SH-120. In the case of Flemion, the number of water molecules per SO_3^- group increases from 18 to 23 in this range of equivalent weights. More importantly, the lowering of the equivalent weight also leads to higher water uptake and higher conductivity at lower RH: at an RH of 30%, an EW770 Asahi Glass membrane has proton conductivity of 0.03 S/cm at 80°C, while a EW909 membrane has a proton conductivity of 0.017 S/cm under these conditions [47]. The mechanical strength clearly becomes poorer when the equivalent weight of the membrane is lowered [44].

The variation in equivalent weight is in fact nothing more than a change in backbone chain length per monomer, so that the ratio between the hydrophilic phase and the hydrophobic phase increases with decreasing equivalent weight, leading to higher water mobility on the one hand, and on the other hand to poorer mechanical properties, which are determined by the hydrophobic phase.

3M has developed its own perfluorosulfonic acid polymer, the structure of which is probably according to Figure 5.1 with $m = 0, p = 0$ and $q = 4$ [48]. The 3M membranes are 30 μm thick, have an equivalent weight of 980 and is claimed by 3M to have better thermo-mechanical properties than Nafion [49]. Whereas Nafion loses its strength above 80°C, 3M starts to soften above 110°C [49]. The proton conductivity is similar to that of Nafion, i.e. 0.17 S/cm at 70°C and RH = 100%. Its conductivity has the same dependency on RH as other perfluorosulfonic acid membranes. 3M sells complete MEAs, not the membranes on itself.

The 3M MEA has been operated during 17,000 h with a voltage degradation of 2 μV/hr [50].

5.3.2.3 Reinforced Perfluorosulfonic Acid Membranes

Reinforced membranes are applied to enable the use of thin membranes while preserving mechanical strength. Thin membranes have as advantage a low membrane resistance and a good water management, as earlier discussed. When using un-reinforced membranes, the mechanical strength becomes too poor, both for MEA manufacturing as well as for fuel cell operation. A limitation which holds both for un-reinforced membranes as well as for reinforced membranes is the crossover of hydrogen and oxygen. Besides leading to a lower cell voltage, which results in a lower fuel cell efficiency, the direct reaction between hydrogen and oxygen leads to local heat production which can cause damaging of the cell.

This leads to reinforced membranes of typical thickness of 25–35 μm, with a mechanical strength which is comparable to un-reinforced membranes of 175 μm. A key advantage of reinforced membranes is that they swell in only one direction. This prevents mechanical stress at the membrane electrode interface when exposed to organic solvents during MEA production, water uptake during fuel cell operation and during freeze–thaw cycles, likely to occur during daily operation in vehicles. While un-reinforced membranes rapidly deteriorate during freeze–thaw cycles, leading to complete loss of performance after seven freeze–thaw cycles, reinforced membranes show almost no performance loss under the same operation [51].

The most advanced reinforced membrane is at this moment the one developed by W. L. Gore. Gore uses a porous polytetrafluoroethylene (PTFE) matrix, which is filled with a perfluorosulfonic acid solution resulting in a membrane which consists of around 80% proton-conducting electrolyte. The membranes were used to be called Gore-Select, but at present especially MEAs are commercialized by Gore under the trade name Gore Primea. The thickness of the Gore membranes is 25–35 μm.

The lifetime of reinforced Gore MEAs, determined by the gas permeability in an accelerated fuel cell test, has been shown to be 2–4 times that of un-reinforced Nafion membranes of comparable thickness (25 μm) and 1.5–3 times that of Nafion membranes of 90 μm thickness [52]. Crack propagation in un-reinforced membrane is seen as one of its potential weaknesses which is circumvented by the introduction of a porous matrix as reinforcement. The increase in hydrogen permeability in reinforced membranes is a gradual process, whereas in the case of un-reinforced membranes, the hydrogen permeability increases very steeply from a certain point [52]. Under more- or less-standard conditions, the Gore-reinforced MEAs have a demonstrated lifetime of over 8000 h, based on its hydrogen crossover rate [52]. The voltage degradation rate amounts to 1.5–7 μV/h at current densities ranging from 500 to 800 mA/cm^2.

Besides the improved durability, the Gore MEA offers superior performance, at 0.7 V, its power output is more than 50% higher than a comparable MEA based on Nafion 1035 [52]. At present, a 25 μm Nafion 101 membrane is available and this would approach the fuel cell performance of Gore on the short term. The durability of such a thin un-reinforced membrane, both during testing as well during cell and stack manufacturing is questionable. At unsaturated conditions and 95°C, i.e. anode at RH = 75% and a dry cathode, the latest generation Gore MEAs show a power density of 0.4 W/cm^2, compared to 0.7 W/cm^2 at 80°C and fully humidified conditions [53].

Asahi Glass has developed reinforced membranes of 50 μm thickness by implementation of 2.7 wt% PTFE fibrils, dispersed uniformly in the ion exchange resin [46,54]. This reinforcement has a positive influence on the tear strength, amounting to a factor 5 in the machine direction and more than a factor 20 in transverse direction [46]. Also the creep behaviour upon water uptake is diminished by a factor of 3. There is no positive influence of this type of reinforcement on the tensile strength [54]. Constant fuel cell performance during 3000 h is reported [54].

Another commercially available reinforced membrane, sold by DSM Solutech, uses ultra-high molecular weight polyethylene as porous matrix, filled with Nafion perfluorosulfonic acid polymer [55]. The membrane thickness is 25 μm. Despite the fact that a non-fluorinated matrix is used as reinforcement, constant performance of over 3000 h has been demonstrated at ECN [56].

Reinforced membranes manufactured by impregnating a porous matrix with PFSA resins, need careful post-manufacturing quality controls, as incomplete impregnation of the porous matrix leads to beginning-of-life gas leakage through the membrane.

5.3.2.4 Operation of Perfluorosulfonic Acid Membranes at High-Temperature and Low-Humidity Conditions

Besides the limited thermal stability of perfluorosulfonic acid polymers itself, as a result of too low glass transition temperature, the limitations of the most widely used membranes with respect to high-temperature operation are primarily set by the dependence on water for proton mobility. If the proton conductivity is accomplished by acid dissociation coupled to proton transfer to a proton accepting and carrying medium, than the volatility of that medium determines the window of operation of the system. In the case of water, this is 100°C at atmospheric pressure and 120°C at 3 bars.

Nafion membranes can be modified by doping them with electrolytes, e.g. phosphoric acid [57] or heteropolyacids [58], with the purpose of improving their conductivity at temperatures above 80°C. The incorporated acid which has a low vapour pressure, acts as a brønsted base, such that the dissociation of the sulfonic acid site and the transportation of the proton is possible without the presence of water.

In the case of Nafion impregnated with 12-phosphotungstic acid dissolved in either acetic acid or tetra-*n*-butylammonium chloride [58], the fuel cell can be operated at 120°C, at a power density of 0.47 W/cm at 0.5 V. Long-term stability is unknown for this system, as follow-up work of Datta on this subject has not been discovered by us.

A second type of modification occurs via the incorporation of a surfactant [59]. Examples of suitable surfactants are: ammonium perfluoro-octane sulfonate, potassium perfluoro-octane sulfonate, potassium perfluoroalkyl-cyclohexyl sulfonate, sodium butane sulfonate and sodium octane sulfonate. The amount of dopant is up to 60 mol%. The effect of surfactant incorporation is an enhanced transport of water through the membrane, up to a 2-fold increase, depending on the type of surfactant. It is believed that the surfactant induces an alteration of the microscopic structure of the membrane, resulting in an improved polymer morphology with improved transport characteristics.

The third type of modification consists of the incorporation of platinum and silica or titania in a solution cast Nafion membrane [60]. The function of the platinum is to generate water from the crossover of molecular oxygen and hydrogen in the membrane itself, instead of at the anode and cathode. The water thus produced is held in the membrane by the hygroscopic silica or titania. The membrane resistance is lowered by the internally humidification, combined with a higher voltage due to the decreased gas crossover. Especially when working with non-humidified gases, the effect is remarkable: when no external humidification is applied, at 80°C and more or less open circuit conditions, the membrane resistance drops by a factor of 1000 by inclusion of both platinum and silica [60]. The silica alone roughly accounts for a decrease by a factor of 10.

The inclusion of silica into Nafion and Aciplex membranes has been demonstrated to enable their stable operation at 130°C, *albeit* at elevated pressure and humidification of hydrogen and oxygen at 130°C [61]. While the performance of a fuel cell using unmodified Nafion 115 immediately drops to zero current density within an hour, silica-modified membranes show promising behaviour during 50 h of operation. The thinner the membrane, the higher the current density [61], the highest current density shown by the fuel cell using a Aciplex 1004/Silica composite membrane. In the experimental work of Ref. [61], humidification is done by bubbling the reactant gases through humidifiers. From the rapid performance loss of the Nafion 115, it can safely be concluded that the gases are not water saturated. What the actual water content has been in the experiment is unknown, so that the establishment of the usefulness of the silica/perfluorosulfonic acid composite membranes needs further attention.

The inclusion of 10% zirconium in Nafion does lead to a higher water uptake and a higher conductivity [62]. At an RH of 40% at 90°C, the conductivity increases from 0.012 S/cm for Nafion 112 to 0.017 S/cm for Nafion containing 10% ZrO_2. This is still far too low for continuous operation at these conditions. Besides, fuel cell performance was worsened by inclusion of ZrO_2, probably by a sub-optimal MEA.

Song *et al.* [63] have tested Nafion 112 at 120°C, atmospheric pressure and an RH of 35%, by humidifying hydrogen and air at 90°C. The power density they measured was 0.24 W/cm^2 at 0.6 V, and a membrane resistance of 0.274 Ω cm^2. The latter corresponds to a conductivity of 0.018 S/cm. Although the demonstration of stable behaviour of Nafion at these dry conditions is remarkable on itself, enabling short-term excursions to higher operating temperatures without permanent damage of the fuel cell, it is too low for commercial application.

Asahi Glass has recently communicated that it has developed an alternative fluorinated composite membrane which can be used at 120°C and an RH of 50% [64]. While a conventional MEA shows a complete loss of performance within tens of hours, the loss of performance of this new MEA amounts to 60 μV/hr over 2500 h test time. The composition of this composite fluorinated membrane is not disclosed by Asahi Glass. In a recent patent submitted by Asahi Glass, it claims a class of perfluorinated polymers which are easier to manufacture and have much better mechanical properties above 100°C. This would however not explain the excellent performance at these dry conditions at 120°C [65].

Asahi Kasei is also developing alternative membranes for operation above 100°C and undersaturated conditions. Their so-called Thermo-stable Membrane-1 has an equivalent weight of 710, a thickness of 25 or 50 μm [45]. Despite its low equivalent weight, its mechanical strength is superior to Aciplex membranes with an equivalent weight of 950. It has a conductivity of 0.1 S/cm at 110°C and RH = 70%. At 110°C and RH = 50%, its conductivity has fallen to 0.03 S/cm [45].

5.3.3 Alternative Fluorinated Membranes

5.3.3.1 Radiation Grafted Membranes

The research on membranes alternative to Nafion-type membranes has started as soon as the development of the PEMFC had been demonstrated to become a viable option for application in fuel cell vehicles. The drivers for searching for alternatives has been various, but cost has always been an important one. Also the dependency of the proton conductivity of the Nafion membrane on the humidity of the reactant gases has been an important reason for developing alternatives, not necessarily aimed at operation at higher temperatures.

Perfluorinated and partially fluorinated membranes can be tailor-made synthesized by means of radiation–grafting. The fact that fluorinated commodity polymer films such as polytetrafluoroethyle (PTFE), poly (tetrafluoroethylene-co-hexafluoropropylene) (FEP), polyvinylidene fluoride (PVDF), poly(ethylene-alt-tetrafluoroethylene) (ETFE), which are manufactured in large quantities, can be modified into proton conducting membranes opens the door to cheaper membranes than Nafion and its analogues.

The method of radiation–grafting for synthesis of proton-conducting membranes has been extensively explored by Scherer *et al.* of Paul Scherrer Institut (PSI) in Switzerland.

The base polymer films are irradiated by an electron beam, creating reactive sites on the base polymer. Immersion of this radical-containing polymer in a solution containing the side-chain molecule of choice, styrene in the case of the PSI [66–68], leads to a modified membrane. Grafting levels are around 20%. Cross-linking by the addition of divinylbenzene to a degree of 12% during the grafting leads to improved mechanical properties enabling long-term operation. Proton conduction is introduced by sulfonating the grafted membrane with chlorosulfonic acid, preferably at room temperature to maintain the mechanical strength of the polymer. The conductivity of the resulting membranes lies between 0.03 S/cm at 40°C and 0.04 S/cm at 60°C, as measured *in situ* during fuel cell operation [69].

While in first instance both ETFE as well as FEP were used as base films [66–68], the FEP-based polymers seem to be the most advanced. The mechanical stability of the membrane is largely determined by the stability of the backbone, which is, in the case of fully fluorinated carbon atoms, comparable to the stability of Nafion. From the work of Büchi *et al.* [10] follows, however, that membrane degradation occurs by the loss of the grafted monomers from the backbone (Figure 5.10).

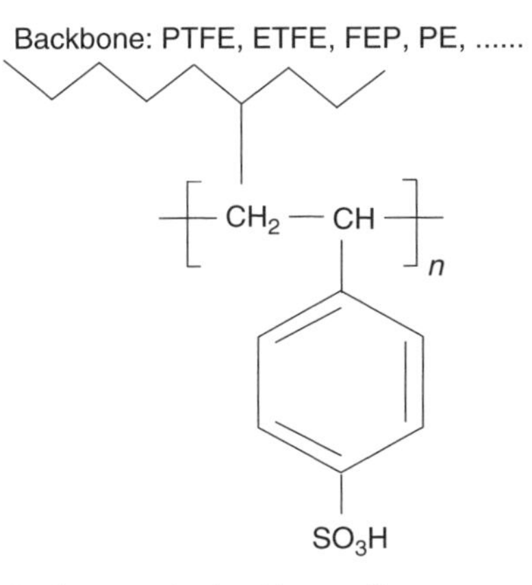

Figure 5.10 Structure of polymers obtained by grafting styrene on a polymer backbone.

Continuous effort at the PSI in Switzerland has led to grafted membranes with a performance that is comparable to that of Nafion membranes of equal thickness and a proven durability of up to 10,000 h [66–68]. Most long-term fuel cell tests has been done using FEP-based membranes [66–68]. As FEP is perfluorinated and ETFE is partially fluorinated, the resistance of FEP towards radical attack should be higher.

Membranes with a very low equivalent weights can be obtained by using higher grafting levels. Chuy *et al.* [70] synthesized ETFE-styrene sulfonic acid membranes with a grafting level of 32–46% leading to equivalent weights ranging from 470 to 300. With these equivalent weights, proton conductivities are considerably higher than Nafion: 0.2–0.3 S/cm [70]. According to Gubler *et al.* [69], membranes with such a high grafting level have a poor long-term stability.

Translating well-conducting grafted membranes into well-performing MEAs seems to be an issue on itself. The surface of the grafted membranes appears to be quite hydrophobic, leading to poor contact with Nafion impregnated gas diffusion electrodes [68,69]. Using the dispersed radiation grafted polymer for the electrode, in order to improve interface characteristics have not been pursued [5].

5.3.3.2 Blended Membranes

Atofina Chemicals has developed a blended membrane consisting of PVDF with acrylic and vinylic resins containing sulfonic acid groups [71,72]. The membranes have a thickness of 25 μm and a conductivity in liquid water of 0.06–0.08 S/cm at 25°C and 0.1–0.13 S/cm at 70°C [71]. Due to its reduced thickness, the areal resistance is the same as that of 50 μm perfluorosulfonic acid membranes.

Under fuel cell conditions, the performance is especially at high-current densities lower than when using Nafion-based MEAs. There is however much improvement possible, and the endurance, already more than 2000 h with a decay rate of 45 μV/h is encouraging [71].

5.3.3.3 Poly(α,β,β-trifluorostyrene)

Poly(α,β,β-trifluorostyrene) has received appreciable attention at Ballard. The polymer membrane was seen as a viable alternative for Nafion-type membranes, with sufficient chemical stability due to its fluorinated backbone and enabling fuel cell performance comparable to or even better than using Nafion membranes [73]. The membranes are also known as BAM-3G (Ballard Advanced Materials Corporation – 3rd Generation) (Figure 5.11).

Extensive fuel cell testing with poly(α,β,β-trifluorostyrene) has proved its suitability at low-temperature fuel cell conditions, i.e. operation at 70–85°C and water-saturated conditions. Long-term operation during more than 14,000 h with a degradation rate of 7 μV/h on single-cell level and 2200 h operation in a 75-cell stack has been established [73]. In Fenton's test,

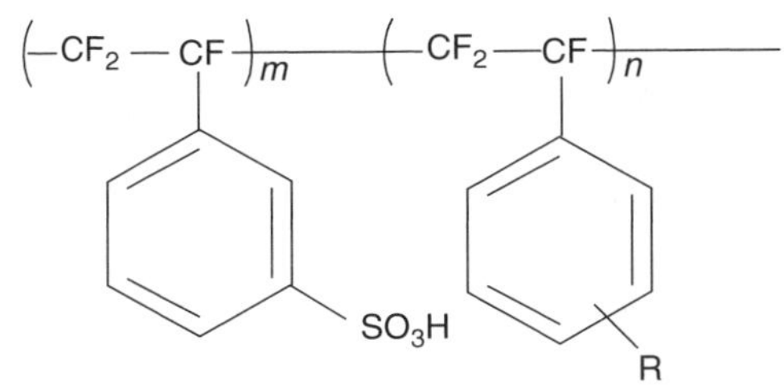

Figure 5.11 Structure of poly(α,β,β-trifluorostyrene).

polytrifluorostyrene sulfonic acid, which might not be exactly the same polymer as Ballard used, does have a slightly higher degradation rate than Nafion [37].

Despite the successful fuel cell demonstrations, Ballard has to our knowledge not pursued its poly(α,β,β-trifluorostyrene) membrane development for incorporation in Ballard's fuel cell stacks.

5.3.3.4 Bis[(perfluoroalkyl)sulfonyl]imide

A perfluorinated polymer bearing a different strong acidic group is the bis [(perfluoroalkyl)sulfonyl]imide. Instead of the sulfonic acid group, it bears a $SO_2^-NH\text{-}SO_2$ group. The group of Desmarteau at Clemson University has synthesized and characterized the polymer, and determined its proton conductivity as a function of humidity in comparison with Nafion [74]. The motivation for this work is the higher gas-phase acidity and a better thermal stability of the sulfonylimide group in comparison to the sulfonic acid group. The bis[(perfluoroalkyl)sulfonyl]imide behaves similar to Nafion membranes at both 80°C and 110°C. Over the whole humidity range, its proton conductivity is slightly lower than that of Nafion; the lower the RH, the bigger the difference [74]. Also in fuel cell tests, the MEAs using the sulfonyl imide behave similar to the MEAs using Nafion. Whether the sulfonyl imides provide a cost advantage is questioned by the authors themselves.

5.3.4 Non-Fluorinated Membranes

5.3.4.1 Sulfonated PEEK

Of the non-fluorinated polymers, sulfonated polyarylene ether ether ketone (S-PEEK) has probably received the most attention. The base polymer PEEK is so-called high-performance polymer, used for applications where chemical stability in combination with mechanical strength is required, e.g. as HPLC tubing. The PEEK polymer can readily be sulfonated by concentrated sulphuric acid [75,76]. The resulting polymer can be cast into membranes.

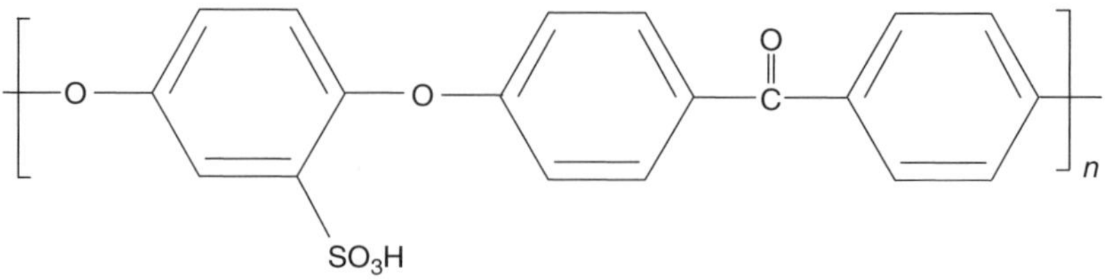

Figure 5.12 Structure of S-PEEK.

However, as it is believed that post-sulfonation of PEEK both leads to weakening of its mechanical and thermal properties, as well to a lack of control over degree and distribution of sulfonation, polymerization of sulfonic acid containing monomers is done as well [77,78].

Due to the absence of fluorine in the backbone, the acidity is weaker, its oxidative stability is less (it does not resist the Fenton's test) and phase separation is less pronounced when compared to PFSA polymers [63] (Figure 5.12).

The degree of sulfonation can be set, either by the time of the sulfonation process or by the polymerization reaction of presulfonized monomers. The degree of sulfonation leads to an ion-exchange capacity and thus to an equivalent weight. To obtain a reasonable conductivity, equivalent weights of 650–1000 are made. In the fully hydrated state, the proton conductivity of S-PEEK with EW650 approaches that of Nafion: 0.04–0.07 S/cm at 25–27°C [75,77,79]. It is however as dependant on the water content as in the case of Nafion and its analogues, and by that on the RH of its surroundings [79]. In fact, proton conductivity even falls sharper with decreasing water content. This effect is ascribed to the smaller pore diameter in S-PEEK, and the more eddy form of the pores. At lower water content, the electrostatic forces of the sulfonic anions hinder the mobility of the dissociated protons [79]. Swelling of S-PEEK is a concern [76]. With a too high sulfonation grade, the water uptake of S-PEEK becomes extreme, even far below 100°C [79]. This limits the sulfonic acid content and thus the proton conductivity. Reinforcements have been applied to control the swelling of S-PEEK [75]. Polymerization of sulfonated monomers leads to a lower degree of swelling than post-sulfonated polymers of comparable equivalent weight, as well as to higher conductivities [77].

As in the case of perfluorosulfonic acid membranes, the pre-conditioning of the membrane can have a profound effect on its properties. In the case of S-PEEK, the proton conductivity shows a high temperature dependence when no pre-treatment is done [75,76]. However, when membranes have been pre-swollen in boiling water, proton conductivity has a much weaker dependence on temperature [75].

Although the claim is that S-PEEK can be used at temperatures higher than 100°C, it still needs full saturated water conditions [75,80], so that this benefit only holds at elevated pressure, a condition which has a negative impact

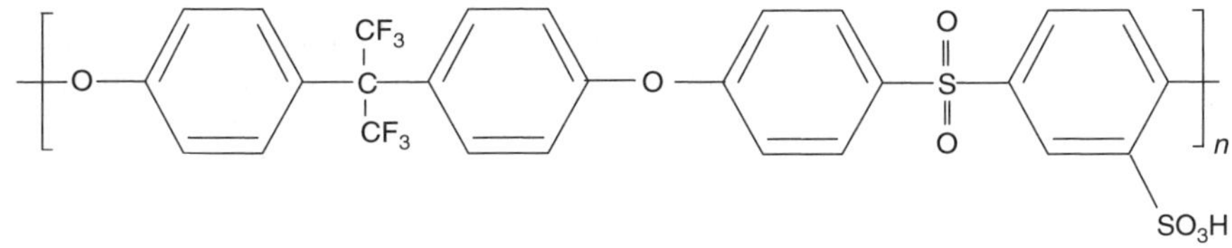

Figure 5.13 Structure of S-PES.

on fuel cell system efficiency and leads to a lot of water to be fed externally to the fuel cell system.

Fuel cell tests with S-PEEK are relatively scarce. Under the same conditions, a fuel cell based on 70 μm S-PEEK membrane gave comparable fuel cell performance as Nafion 115 (dry thickness 125 μm) [75].

Operation at 120°C at fully humidified conditions showed that a fuel cell comprising a sulfonated PEEK membrane is more stable than a fuel cell comprising a Nafion 117 membrane [80]. Although both membranes failed due to physical damage at the edge of the electrode, Nafion failed after 50 h of operation while the PEEK membrane lasted for almost 400 h [80].

Sulfonated PEEK membranes are commercialized by Fumatech GmbH in Germany [81]. Low equivalent weight membranes (EW675–850) with a conductivity ranging from 0.03 S/cm at 40°C to 0.06 S/cm at 80°C are offered.

The main benefit of S-PEEK is its relative low cost. With respect to performance, it appears that S-PEEK hardly offers an advantage in comparison to perfluorosulfonic acid membranes. As proton conductivity is determined by the sulfonic acid function, it is heavily dependent on water content of the membrane and the gas-phase RH. Lifetime is not sufficiently proved, and fuel cell tests relatively scarce.

5.3.4.2 Sulphonated Polyether Sulfone

Sulfonated poly(arylene)ethersulfone (S-PES), has been proved not to be stable enough for usage in fuel cells. Lifetime of S-PES membranes at fuel cell conditions was limited to 500 h, for which obvious reason Ballard decided not to continue the research with this membrane [73] (Figure 5.13).

5.3.5 Membrane Specifically Developed for Dry Operation Above 100°C

Schuster *et al.* [82] have performed a more basic study on the suitability of different acidic functionalities, being sulfonic acid, phosphonic acid and imidazole attached to a heptane chain. From this comparison it becomes clear that sulfonic acid offers a superior conductivity under wet conditions. The phosphonic acid containing molecule has a maximum conductivity of 0.01 S/cm in the temperature range of 100–200°C under wet conditions

($P(H_2O)$ = 1 bar). Under dry conditions, the phosphonic acid groups enable proton conductivity within a hydrogen bonded network formed by the phosphonic acid functions. But this network conductivity has a conductivity of 0.01 S/cm at 200°C, at 100°C it is only 1 mS/cm [82]. Alternative systems with shorter alkyl groups show significant higher conductivities, but again only in the temperature range far above 100°C [82]. The potential of imidazole remains unclear from this study. Although imidazole is in theory amphoteric, the density of charged species in the undoped systems is very low.

The research on polymers for operation above 100°C is still in an exploratory stage. Studies are on the synthesis of alternative structures, their characterization and measurement of proton conductivities as a function of temperature and humidity. These conductivities tend to be orders of magnitude lower than what is needed for practical fuel cell use. An overview of all these efforts can be obtained by reading reviews by Schuster and Meyer [83] and Savogado [84]. Hogarth *et al.* reviewed inorganic proton-conducting systems [85].

The present situation is that only polybenzimidazole (PBI) based systems offer an alternative to previously described low-temperature membranes, with respect to maturity and proven performance. All other systems are in the embryonic stage and need to prove at least whether membranes can be made and suitable MEAs can be developed.

5.3.5.1 Phosphoric Acid Doped PBI

The most mature membrane available for high-temperature operation is phosphoric acid doped PBI (H_3PO_4–PBI). PBI is an industrial polymer which has a very high thermal stability and has originally been developed for flame-retarding purposes. It was produced by Celanese, which sold its PBI business very recently to PBI Performance Products Inc.

At Case Western University, the development of proton-conducting PBI has taken place first by doping with sulphuric acid, later by doping with phosphoric acid [86,87]. Phosphoric acid serves two functions, that of proton donor as well as that of the proton conducting medium. Also at the technical university of Denmark, a continuous effort on H_3PO_4–PBI takes place [88].

To get membranes that are suitable for the PEMFC, high molecular weight PBI is needed, a minimal molecular weight of 50,000 is proposed by Wainright *et al*. [87]. The PBI synthesis by Celanese produces an average molecular weight of 27,000, the high molecular weight can be obtained by extracting the high molecular weight fraction from the product mixture [87]. This limited yield of around 70% [86] of the high molecular weight fraction has to be taken into account for cost calculations.

Two different routes exist for obtaining H_3PO_4–PBI. The first consists of membrane formation followed by acid doping, the second consists of membrane formation and acid doping simultaneously. Although the first method leads to initially stronger membranes, continued heating leads to brittleness. The latter leads to better mechanical properties at prolonged heating and to higher conductivities [87].

The doping level is an important factor for the proton conductivity of the membrane. Typically 5 mol H_3PO_4 per repeat unit is used. The first two moles lead to protonation of the benzimidazole groups, the other three moles are needed as the proton-conducting medium. Whereas the first two moles of H_3PO_4 are tightly bound, the other three moles of H_3PO_4 have a weaker interaction (Figure 5.14).

The proton conductivity follows a Grotthus mechanism involving the imide sites as well as the phosphoric acid groups [89], where the phosphoric acid molecules play a vital role in the mobility of the proton. It is clear that the distance between the adjacent nitrogen atoms is too large to bridge without the intermediate phosphoric acid molecules [89].

For the proton conductivity in this system, the presence of water is not needed, which makes it conducting in the temperature range well above 100°C. In practice, however, the RH does have a positive influence on the proton conductivity of H_3PO_4–PBI [87]: going from an RH of 5–30%, the proton conductivity increases by a factor of 2–4 in the temperature range between 80°C (from 0.005 to 0.02 S/cm) and 160°C (from 0.03 to 0.065 S/cm). The water content is believed to influence the conductivity through its control on the association behaviour of phosphoric acid into polyphosphorate clusters [87], H_3PO_4 being the dominant species in the presence of water.

On the other hand, one would also expect a contribution from proton conductivity in the hydrated phosphoric acid at higher RH, in which case the polybenzimidazole itself is not participating in the proton transfer, but instead water is acting as the Brønsted base and proton carrier. Indeed, also

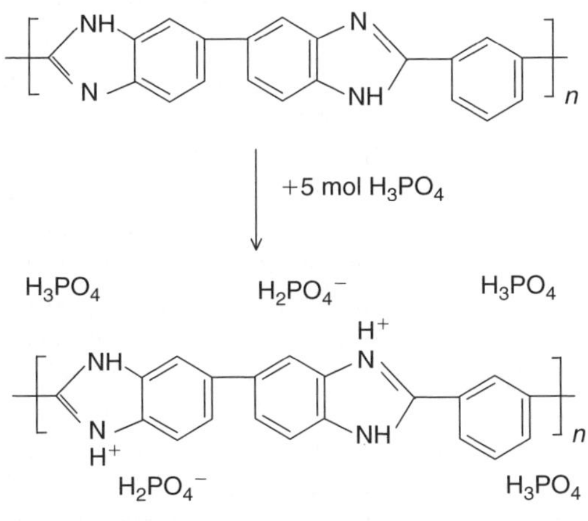

Figure 5.14 Polybenzimidazole and its phosphoric acid doped analogue (derived from [87]).

the conductivity of 100% phosphoric acid increases by the addition of a few percentage of water [88].

By comparing the conductivity in free phosphoric acid, PBI-doped phosphoric acid and an inert fibre impregnated with phosphoric acid [88], the overall picture is that at low doping levels, i.e. 2 mol H_3PO_4 per repeat unit, the N-site of the PBI structure contributes significantly to the proton conductivity. At high doping levels, however, it is primarily the free H_3PO_4 which contributes to the overall conductivity. This does not imply that whatever inert matrix can replace PBI, as the polymer-H_3PO_4 interaction is important in the containment of the phosphoric acid, preventing its leakage to the outside [88].

Fuel cell performance The performance of an MEA from PEMEAS, containing H_3PO_4–PBI is even at 160°C quite low. At 0.7 V, power density is 0.07 W/cm^2; at 0.6 V, it is 0.25 W/cm^2 [90]. This is at atmospheric pressure and without any humidification, so no alternative MEA will meet this performance under these conditions. With reformate containing 0.2% CO, the power density is 0.18 W/cm^2 at 0.6 V. Note that the performance is thus lower than that of the PAFC, which produces 0.14 W/cm^2 at 0.7 V, and 190°C [91].

Cold start capabilities Based on the data available [86,87,89,92], the conductivity of H_3PO_4–PBI at 20°C is negligible. The minimum conductivity of 0.025 S/cm for start-up is achieved by H_3PO_4–PBI at a temperature of 90°C at an RH of 30% and at 150°C at an RH of 5% [87]. This would imply, irrespective of the fuel cell application, that the fuel cell needs to be externally heated up for start-up.

Membrane stability PBI itself is thermally stable up to temperatures of 600°C. The stability under simulated fuel cell conditions was investigated by Samms *et al.* [93]. The conclusion of their investigation was that below 600°C, only weight loss due to water vaporization and phosphoric acid dehydration occurred. The experimental method is, however, doubtful. Simulated fuel cell conditions in this study meant exposure of platinum loaded H_3PO_4 to either hydrogen containing gas or air. Whereas this might impose a certain voltage to the membrane/platinum interface, it is not likely to lead to substantial peroxide formation, for which the oxygen reduction reaction is needed. Oxygen reduction with concurrent peroxide formation either takes place during the electrochemical reduction at the cathode or by the reaction between hydrogen and oxygen, diffused through the membrane, at the anode. The much simpler Fenton's test , as described in Section 2.2.1, is a more reliable prediction of long-term membrane stability under oxidative conditions. Li *et al.* state that H_3PO_4–PBI did not survive this Fenton's test in their laboratory [88], and mention the contradiction with reasonable observed lifetimes of over 5000 h. Different factors might play a role in this discrepancy. The absence of liquid water, slower oxygen reduction kinetics, a lower rate of formation of peroxide relative to the overall reduction rate, lower crossover of oxygen to the anode, might all play a role.

The Celtec-P 1000 MEA from PEMEAS, using H_3PO_4–PBI as membrane, has a demonstrated lifetime of more than 6000 h, at 160°C, atmospheric pressure and no humidification [90]. The voltage degradation amounted to 5 μV/hr. Li *et al.* [88] also demonstrated lifetime for more than 5000 h when operated under continuous operation at 150°C. Even start/stop cycles were not detrimental for the performance of a H_3PO_4–PBI containing fuel cell [88]. Some degradation was observed, whether special measures were taken to prevent the formation of liquid water is unclear.

Acid leaching The leaching of phosphoric acid is a major concern for H_3PO_4–PBI. Both evaporation of phosphoric acid at the high-temperature operation can lead to acid loss, as well as extraction by liquid water formed by condensation during shut-down. Especially the unbound phosphoric acid, present in considerable amounts at the doping levels generally used [88], is likely to be extractable from the polymer matrix.

Availability Although there is a much academic research on H_3PO_4–PBI membranes, they are not commercially available separately. At present, MEAs using the H_3PO_4–PBI are exclusively commercialized by the German Company PEMEAs, which does not sell the membranes.

ABPBI A polymer which strongly resembles PBI is ABPBI, poly(2,5-benzimidazole). ABPBI is simpler to synthesize than PBI [86,94]. Analogous to PBI, proton conductivity is obtained by doping with phosphoric acid, which it, however, absorbs more strongly than PBI. Sulfonation of ABPBI, followed by phosphoric acid doping leads to even higher proton conductivities in the entire temperature range from 80°C to 190°C [94]. Preliminary fuel cell tests do not show better fuel cell performance, which is ascribed to the absence of tailor-made MEAs for this new membrane type.

5.3.6 Comparison of Membrane Properties

5.3.6.1 Proton Conductivity

The membrane property which is decisive for the power density of the fuel cell, is the proton conductance. Direct comparison of fluorinated membranes is the easiest, as they all can be applied under comparable conditions. Table 5.1 gives the proton conductivity for the various membranes, and their respective thickness and equivalent weight.

Figure 5.15 summarizes the proton conductivity of the different classes of membranes described in this chapter under the conditions which are ideal for the respective membranes, necessarily not the same for each class.

Table 5.1 Properties of (per)fluorinated sulfonic acid membranes; thickness in dry state or at RH = 50%; proton conductivity at 23–25°C unless otherwise stated

Membrane	Thickness (μm)	EW (g/mol)	Conductivity H^+-form (S/cm) at RH = 100%	Conductivity H^+-form (S/cm) in liquid water
Nafion 117	175	1100	0.062 [74]	0.083 [24]
Nafion 115	125	1100		0.083 [24]
Nafion 112	50	1100		0.083 [24]
Nafion 1135	89	1100		0.083 [24]
Nafion NRE-211 [24]	25	1100		
Nafion NRE-212 [24]	51	1100		
Flemion SH-120	120	909		0.18 [25]
Flemion MSH-175	175	1000		0.16 [25]
Flemion LSH-180	180	1099		0.13 [25]
Aciplex S-1001X	25	950	0.21 [45]	
Aciplex S-1002	50	950	0.21 [45]	
Aciplex S-1004	100	950		
Aciplex S-1008	220	950		0.15 [25]
Aciplex S-1104	100	1050		
bis[(perfluoroalkyl) sulfonyl]imide	60	1075	0.052 [74]	
Aciplex S-1112	285	1050		0.13 [25]
PSI-FEP	32	763		0.045 [68]
3M	30	980	0.17 (70–80°C) [49]	
PVDF/acryl-vinyl resins M31 (25 μm)	25			0.078 [71]

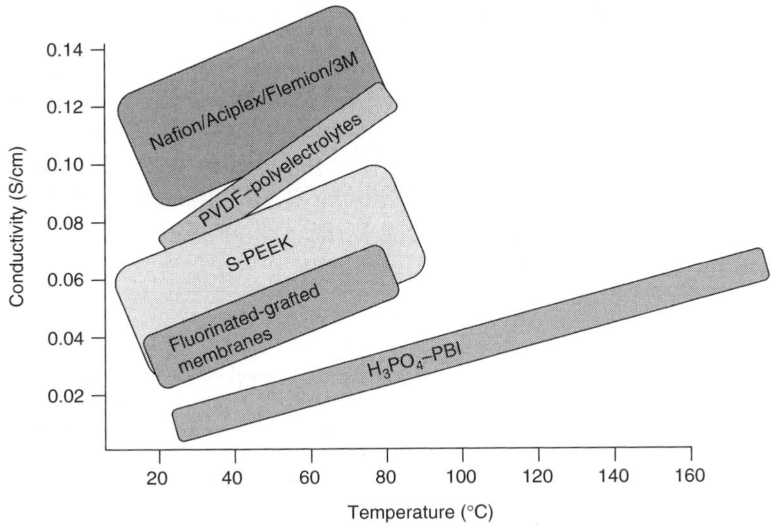

Figure 5.15 Qualitative comparison of membrane conductivity versus temperature for various membrane classes, under operating conditions.

At low temperatures, the proton conductivity of perfluorosulfonic acid/tetrafluoroethylene copolymer membranes is the highest. Their specific architecture leads to very effective proton transport, as long as the hydrophilic regions contain enough water. Alternative sulfonic acid containing membranes generally have somewhat lower conductivities under comparable conditions. The only available system not depending on sulfonic acid for its proton conductance is H_3PO_4–PBI. Its conductivity is low at room temperature, and steadily increases with temperature. But even at 180°C, its conductivity still is below that of the perfluorosulfonic acid membranes at 80°C.

5.3.6.2 Mechanical Properties

The mechanical properties of PEMFC membranes are important for both surviving MEA manufacturing as well as during fuel cell operation. Compared to data on proton conductivity and fuel cell performance, data on mechanical properties are relatively scarce. Table 5.2 gives the mechanical properties of some membranes which are not reinforced.

The mechanical properties of the membranes are strongly depending on the water content. For Nafion, going from 50% RH at 23°C to the fully wetted state at 100°C, tensile strength [24] and tear resistance [24] are almost halved, while the elastic modulus [43] decreases an order of magnitude. As described in Section 3.2.3, reinforced membranes offer the big advantage that their mechanical properties are not determined by the proton conducting polymer, but by the reinforcement. Table 5.3 shows the mechanical properties of 35 μm Gore membranes in comparison to the 175 μm Nafion membrane.

Table 5.2 Mechanical and hydrolytic properties of un-reinforced membranes of 50 μm thickness (unless otherwise stated)

Membrane	Tensile strength (MPa)	Tensile modulus (MPa)	Water uptake on dry basis (%)	Linear expansion during water uptake (%)
Nafion NRE-212 (50 μm) [24]	32 MD[a] 32 TD[a]	266 MD[a] 251 TD[a]	50[b]	15[c]
Nafion N112 (50 μm) [24]	43 MD[a] 32 MD[a]	249[d]	38[b]	15[c]
H_3PO_4–PBI [86]		1880[e] [86]		
S-PEEK EW660	49[d,f] [78]	1010[d,f] [78]	54[g] [77]	
PVDF/acryl-vinyl resins M31 (25 μm) [71]	45[a,d]		68[h]	27[h]
3M (30 μm) [49]	30[a,d] [49]	301[a,d] [49]	40[g] [49]	

[a] 23–25°C, RH=50%; [b] in water at 100°C; [c] From 23°C, RH = 50% to water at 100°C; [d] direction not specified; [e] doping level H_3PO_4 = 100 mol% versus 500 mol% for PEMFC-suitable membrane; [f] water saturated membranes; [g] in water at 25°C; [h] temperature not reported, but probably similar to [b] for water uptake and [c] for linear expansion, based on values of Nafion in same report.

Table 5.3 Mechanical of reinforced membranes under dry and wet conditions [26]

Membrane	Tensile strength dry (MPa)	Tensile strength wet (MPa)	Shrinkage (%)
Nafion N117 (175 μm)	30 MD 25 TD	14 MD 10 TD	11 MD 12 TD
Gore-Select (35 μm)	34 MD 24 TD	33 MD 18 TD	3 MD 3 TD

MD = machine direction; TD = transverse direction.

5.3.6.3 Membrane Cost

The cost estimates throughout this review are taken directly from the references without adjusting for price inflation. Given the uncertainties stemming from extrapolation to large quantities and different manufacturing methods, the adjustment towards a standard price index might have limited added value.

The price of the membrane should amount to 5 U$/kW [2] for automotive applications, according to the DoE program goals, which the DoE translates to US$40/m^2.

A thorough analysis of membrane cost has been made by Gebert *et al.*, comparing the cost of Nafion, BAM3G, sulfonated PEEK and H_3PO_4–PBI [95]. Materials costs and production costs were taken into account. BAM3G, sulfonated trifluorostyrene, does not have a cost advantage in comparison to Nafion.

Non-fluorinated alternatives as sulfonated PEEK and H_3PO_4–PBI could be cheaper, as for both membranes cheaper materials can be used. In the case of sulfonated PEEK, the production process is simpler and total processing time comparable to that of Nafion. The production process for H_3PO_4–PBI is estimated to cost twice the time of that of Nafion, but this is more than compensated by the cheaper materials costs.

For Nafion, the membrane cost (taking both materials cost and manufacturing costs) would be in the range of US$55–105 per kg membrane, translating to US$5.5–10.5 per m^2 for Nafion 112 [95]. This is not prohibitively high for the introduction of fuel cell technology. DuPont itself claims that the cost of Nafion will not be decisive for the market penetration of fuel cells, and could drop to $50/$m^2$ with sufficient production volumes [22]. In this cost statement, DuPont uses 0.2 m^2/kW fuel cell power and 0.19 kg Nafion/m^2 membrane [22], i.e. Nafion 1135.

The solution casting process, enabling the cost effective manufacturing of thin membranes, 25 µm and 50 µm thick, opens the way to even further cost reductions. At an annual production volume of 1000 m^2, the cost range for Nafion NR-112 is 40–70 $/$m^2$, for NR-111 it is 25–45 $/$m^2$ [96].

The cost of ETFE-based radiation-grafted membranes is estimated by the researchers themselves to be in the range of US$6/$m^2$, based on US$5/m^2 for the ETFE-base film and an additional $1/$m^2$ for the styrene and irradiation procedure [66]. As the analysis is not comparable to the analysis by Gebert *et al.* the cost figures should be compared with care. Nonetheless, it is a confirmation that as long as fluorinated materials are used, a strong cost reduction might be hard to achieve.

5.3.6.4 Commercial Availability

For membranes to be used for further cell and stack manufacturing, it is a prerequisite that a commercial company takes up the manufacturing and sales of these membranes. If an alternative membranes stays in the research and development (R&D) environment, MEA- and stack manufacturers will be very hesitant in using these membranes for their product development. Table 5.4 gives the present situation with respect to companies and the type of membranes they sell.

The number of suppliers which manufacture and sell PFSA-based membranes is large, leading to product improvement through competition and a healthy pricing for clients. Alternative membranes are either sold by only one supplier, or are not offered commercially at all.

Table 5.4 Commercial membrane suppliers

Company	Product range	Trade name
DuPont	Perfluorosulfonic acid membranes	Nafion
Asahi Chemical	Perfluorosulfonic acid membranes	Aciplex
Asahi Glass	Perfluorosulfonic acid membranes	Flemion
3M	Perfluorosulfonic acid membranes	3M MEA
Fumatech	Perfluorosulfonic acid membranes Sulfonated PEEK	F-series P-series
PEMEAs	Phosphoric acid doped polyBenzimidazole (only as part of MEAs)	Celtec
Gore Fuel Cell Technologies	Reinforced perfluorosulfonic acid membranes (only as part of MEAs)	GoreSelect
DSM Solutech	Reinforced perfluorosulfonic acid membranes	Solupor

5.3.7 Conclusions on Membranes

The present state-of-the-art PEM fuel cell is based on perfluorosulfonic acid membranes, manufactured and sold in various thicknesses and equivalent weights by DuPont, Asahi Kasei, Asahi Glass, 3 M and Gore. Stack and system experience is based on this membrane type. It offers excellent cold-start capabilities thanks to its high proton conductivity at 20°C and even lower. Its chemical stability is high, long-term fuel cell performance is likely to be determined by other factors than the stability of the membrane. When going to very thin membranes, reinforcements help to maintain a high mechanical strength, which contributes to membrane durability during normal operation, freeze–thaw cycles and MEA manufacturing.

With respect to cost, high manufacturing volumes will lead to perfluorosulfonc acid membrane cost of $\$50/m^2$ or less. With cell power densities exceeding $0.5\,W/cm^2$, this corresponds to \$10/kW.

The operation of perfluorosulfonic acid membranes is limited by the dependence of proton conductivity on the presence of water. Water is needed for acid dissociation and for proton mobility. This determines the window of operation of perfluorosulfonic acid membranes to those conditions which guarantee certain water content of the membrane. In the temperature regime between 20°C and 80°C, stable operation is possible by either humidifying anode and or cathode feed, or by applying dedicated MEA characteristics which keep the water generated by the fuel cell reaction inside the membrane. Above 80°C, more far-reaching measures are necessary, such as increasing the cell pressure and oversaturating the anode and cathode feed. From a system point of view, these measures have serious impact on efficiency and workability.

For operation in the 20–80°C range, no alternative membranes are present which offer stack manufacturers a better performance than the perfluorosulfonic acid membranes. Higher proton conductivities in this temperature range are scarce, chemical stability is questionable and experimental track records are not comparable to that of the perfluorosulfonic acid membranes.

For operation in the 80–120°C range, modifications of the perfluorosulfonic acid membranes are at present the most advanced. Membranes with a low equivalent weight, around 700, with improved mechanical properties, are in development at Asahi Glass and Asahi Kasei. These membranes enable fuel cell operation at 110–120°C and an RH of 50%, with a voltage degradation which is still too high, but offers perspective for further optimization.

For operation in the 120–180°C range, the only membrane available is H_3PO_4–PBI. In this temperature window, the proton conductivity of the membrane is still relatively low; below 100°C, it is so low that external heating is needed for start-up. It leads to fuel cell power densities which are even at these high operating temperatures too low for mobile applications. For stationary applications, its proton conductivity might be high enough, given the fact that it enables long-term operation on reformed fuels at a steady point of operation, and frequent cooling down might be avoided. For long-term endurance, issues which need to be clarified for this membrane are the level of acid leakage, and its distinctive features when compared to PAFCs, whose commercialization is hindered by tough cost forecasts.

Notwithstanding these conclusions, a considerable amount of new concept membranes is under development for operation above 100°C. The stage of development is generally embryonic. No commercial alternative which is comparable to either the perfluorosulfonic acid membranes or even the H_3PO_4–PBI membranes can be expected in the coming 5 years. Systems which need water for acid dissociation and proton mobility are likely to fail above 120°C. Systems which use other non-volatile Brønsted bases should be able to cope with the water generated at the cathode and stay contained within the membrane matrix. The ideal membrane might possess both acidic as well as basic sites with such a small distance in between that effective proton transfer is possible without an extra proton-carrying phase.

5.4 ELECTRODES

5.4.1 Function and Requirements

In the electrode of the PEMFC the electrochemical reactions take place, i.e. at the cathode the oxygen reduction (ORR) and at the anode the

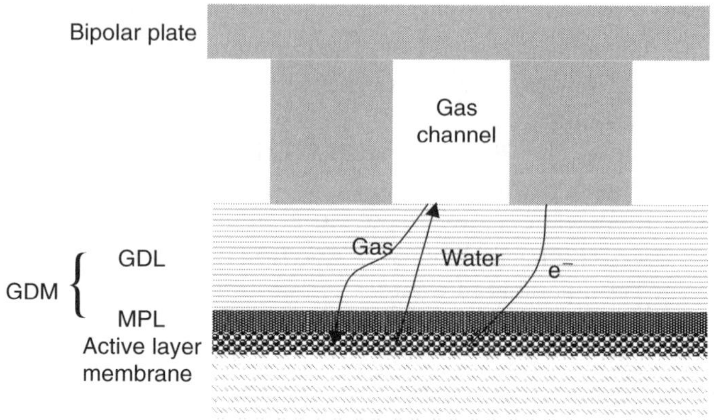

Figure 5.16 Schematic drawing of the structure of a PEMFC electrode.

hydrogen oxidation (HOR). These reactions are catalysed by Pt or Pt-alloys. In addition to providing sufficient catalytic sites the electrodes must be designed such that optimal use can be made of the available sites. Explicitly this means that there must be good:

- access for the reactant oxygen or hydrogen molecules,
- proton transport between catalytic sites and the membrane,
- removal of water formed in the oxygen reduction reaction to the gas channel,
- electronic connectivity of the catalyst sites with the bipolar plate,
- removal of heat produced during the reaction to the bipolar plate.

In order to achieve this the electrode is usually divided into at least two sub-layers: the so-called active layer next to the membrane, and the gas diffusion medium (GDM) next to the bipolar plate, as shown in Figure 5.16. In the active layer the electrochemical reaction takes place. The catalyst particles are dispersed on an electronically conducting carbon support that is connected to the GDM. Between and on the carbon support there is a proton-conducting phase, usually recast Nafion, which is in contact with the Pt or Pt-alloy. There is a considerable void space to allow for gas and water transport. The advantage of using supported catalyst material is that it enables a high surface area of the catalyst. The penalty for this is that the active layer has a considerable thickness in the order of 5–15 μm.

In the GDM no electrochemical reactions take place, and it therefore does not contain Pt(-alloy) or any proton-conducting material. The GDM has a thickness of 200–400 μm and consists of macroporous, electronically conducting material that is partly hydrophobic. This layer contributes to providing a uniform distribution of gases, electric potential and temperature under the gas channels and the ribs of the bipolar plate. In addition it provides mechanical stability to the MEA by averaging differences in pressure

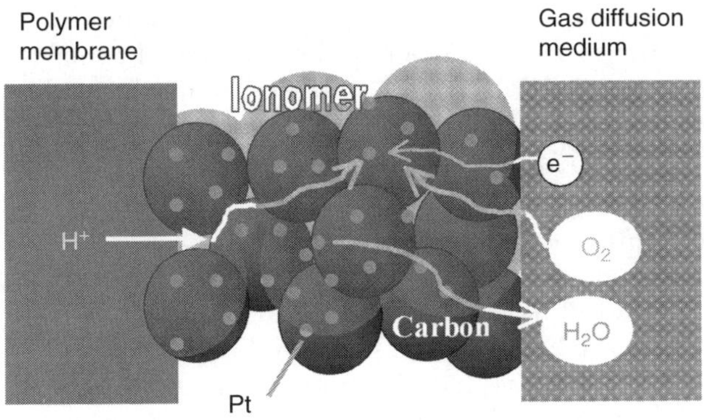

Figure 5.17 Schematic drawing of the structure of the cathode active layer.

encountered by the presence of ribs and gas channels, ensuring permanent good contact with the ribs and compensating dimensional changes in the membrane.

Also in the water management of the MEA the GDM has an important function. At steady-state operation, water needs to be removed continuously from the cathode active layer, either through the GDM at the cathode side or through the membrane and GDM at the anode side. As under PEMFC-operating conditions water is often produced in the liquid state this means that the GDM must allow the transport of liquid water in addition to water vapour without becoming flooded. On the other hand the GDM plays a role in keeping the membrane and active layers humidified when gas conditions are sub-saturated. In order to optimize the water management requirements a hydrophobic, microporous[1] (pore sizes 100–500 nm) carbon layer (MPL) with thickness 30–50 μm is often applied on the side facing the active layer (see Figure 5.17). The GDM then is a bilayer with the macroporous part denoted by gas diffusion layer (GDL).

The performance of the MEA is to a large extent determined by the performance of the electrodes. In H_2–air PEMFCs the cathode is critical as the kinetics of the ORR are much slower than of the HOR. At low-current density the ORR activation losses are already in the order of 300 mV. The transport of oxygen through humid air is much slower than hydrogen transport and water accumulates first at the cathode before it can be transported to other parts of the cell. At high-current density the transport of oxygen determines the limiting current density of the fuel cell [97,98]. When the PEMFC is not operated with hydrogen but with reformate gas substantial losses at the anode may arise. Even with modern fuel reformers

[1] The term microporous is commonly used for this layer although according to IUPAC definitions microporous materials have pores sizes <2 nm, mesoporous materials 2–50 nm and materials with larger pores are macroporous [330].

the reformate gas can contain 10–50 ppm carbon monoxide which due to adsorption on the surface is a severe poisoning for a Pt catalyst in the temperature range up to 100°C [99]. Moreover, it has been shown that also CO_2, present in the order of 25 vol% in the reformate gas, is not inert but can react with adsorbed hydrogen atoms on the catalyst to adsorbed CO in a kind of reverse shift reaction [100–102]. Under those conditions PtRu alloy presently is the catalyst of choice at the anode.

Both for reasons of cost and of availability of Pt, the reduction of the required amount of Pt has always been the main focus of PEMFC electrode R&D. In the past, the use of carbon supported Pt [103] and the development of thin film electrodes [104] have been major breakthroughs in this respect. State-of-the-art cathodes have a Pt loading in the order of 0.4 mg/cm^2 [98,105]. At the anode less Pt can in be used, e.g. experiments by Gasteiger [106] showed that with a reduction of the Pt loading at the anode from 0.4 to 0.05 mg/cm^2 additional voltage losses were limited to 10 mV at 1 A/cm^2. Still, the use of reformate gas requires in the order of 0.2 mg/cm^2 PtRu [106]. In Table 5.5, a summary of performances recently reported by commercial manufacturers [107–111] and in the literature [98,105] is given. Due to the lack of details, e.g., notice that in most cases the type of membrane has not been specified, a detailed analysis of these data is not possible but it can be inferred that a power density in the order of 0.7 W/cm^2 at a cell voltage of 0.65 V is achievable at near-ambient pressures. Higher power densities can

Table 5.5 Performances of MEAs reported by commercial manufacturers and in the literature

	E-TEK [107]	JM [111]	TKK [110]	Gore [109]	3M [108]	Kocha [98]	Gasteiger [105]
Power (W/cm^2) at 0.65 V	0.52	0.58	0.65	0.48	0.55	0.78	0.71
Temperature (°C)	65	n.s.	80	70	70	65	80
P anode (bar)	1	n.s.	1	1	1	1	1.5
P cathode (bar)	1	n.s.	1	1	1	1	1.5
Pt loading cathode (mg/cm^2)	n.s	n.s	n.s	n.s.	n.s	0.4	0.4
Stoichiometric air	2.3	n.s.	3	2	2.5	2	2
Stoichiometric H_2	1.4	n.s.	1.4	1.2	1.5	1.25	2
Membrane	n.s.	n.s	n.s	n.s	n.s.	n.s.	N112SC

be achieved at lower cell voltage or at higher gas pressures, but this reduces the efficiency of the fuel cell system.

At present there are two main drivers behind further electrode research. The first is a further reduction of the Pt loading. Calculations by Gasteiger [105] indicated that the target for hydrogen/air fuel cells in automotive applications should be 0.8–0.9 W/cm^2 with a total Pt loading of the MEA of 0.15 mg Pt/cm^2 at cell voltages of 0.65 V or higher, i.e., 0.2 g Pt/kW. However, this means that the performance still has to be improved but also that it has to be achieved at a lower Pt loading. Also the requirements for higher operating temperatures in order to obtain better system efficiency will have to be considered in relation with the electrode performance. In the following sections these issues will be addressed for subsequently the cathode active layer, the anode active layer and GDM.

5.4.2 Cathode Active Layer

5.4.2.1 Carbon Supported Pt Catalysts

Morphology The state-of-the-art cathode catalysts are Pt supported on carbon black materials with high Pt weight percentages, order of 20–60 wt%. The main suppliers are E-TEK [107], Tanaka Kikinzoku International (TKK) [110] and Johnson Matthey [112]. Carbon blacks are prepared by thermal decomposition (acetylene blacks) or by partial oxidation of hydrocarbons (furnace blacks). Typical examples of furnace black are Vulcan, Ketjenblack and Black Pearls. The different methods result in carbon blacks with different rates of graphite and amorphous carbons and with specific surface area in the order 10–2000 m^2/g [113]. Such high specific surface areas are achieved because carbon black consists of very small primary particles (10–40 nm), which during the production of carbon black form aggregates with a diameter 20–100 nm. The porosity of carbon black consists of the pores between the primary particles and the pores on the surface of the primary particle.

Up to now, the most widely used support in fuel cells is Vulcan XC-72, a carbon black with a specific surface area of 250 m^2/g. Recently, both Johnson Matthey and TKK have brought Pt catalysts on the market that are dispersed on a carbon blacks with a much higher specific-surface area. The high surface area support of TKK has a specific-surface area of 800 m^2/g [110,114]. The new carbon black support from Johnson Matthey is called AC01 [112]. Brunauer–Enmett–Teller (BET) measurements in our laboratory showed that it also had a surface area of 800 m^2/g.

The size of the Pt particles determines the specific surface area (SA) of platinum through the relation:

$$SA = \frac{6}{\rho d} \tag{5.1}$$

Here d is the particle diameter and ρ the specific density of platinum. Uchida [116] has shown that Pt crystallites tend to have similar inter-particle separations on various types carbon blacks. This implies that smaller particle sizes can be obtained on supports with larger carbon surface area, at a given wt% of Pt on carbon. On the other hand, the particle size will increase with increasing wt% of Pt. This is illustrated by Figure 5.18, which shows the particle size and surface area of platinum on Vulcan, Ketjen Black (880 m^2g) and AC01. At present E-TEK catalysts of Pt on Vulcan have particle sizes between 1.8 nm at 5 wt% Pt and 5 nm at 80 wt% [107]. This corresponds to surface areas between 150 and 55 m^2/g Pt. On the new supports Pt particles with size less than 3 nm can be deposited at loadings as high as 60 wt%.

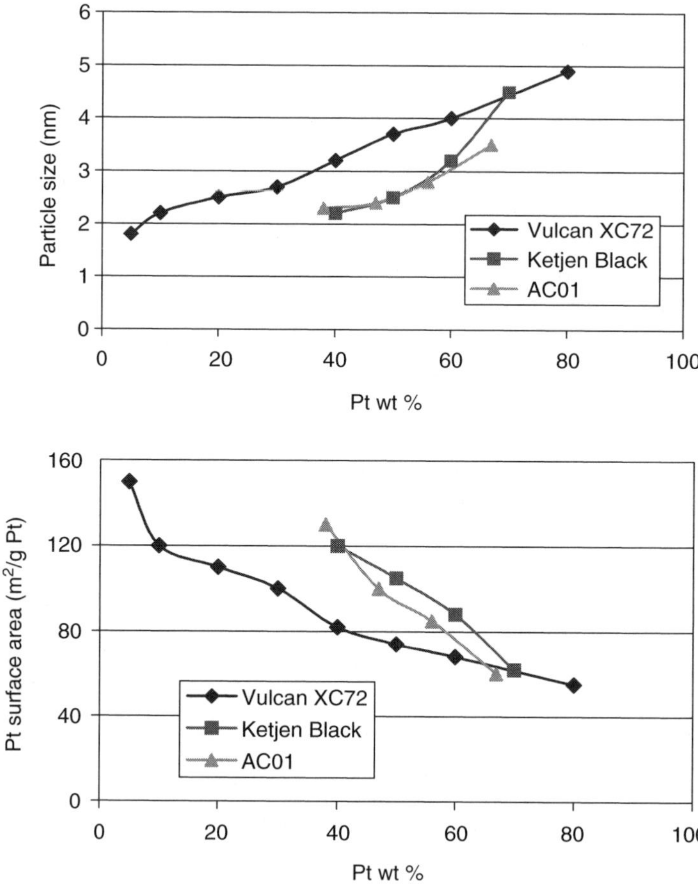

Figure 5.18 Particle size and specific area of Pt on Vulcan [107], on Ketjenblack [115] and AC01 [112].

Kinetic activity The kinetic activity of electrocatalysts can be characterized by the exchange current density for the ORR and the Tafel slope, which indicates how the overpotential increases with current density. For comparison between catalysts the exchange current density is not ideal since it involves extrapolation of the data over several decades of current density and depends strongly on the Tafel slope. More convenient measures are derived from the activity at 0.9 V, IR-free. IR-free means the cell performance corrected for the ohmic losses. In particular the mass activity i_m expressed in A/g Pt and the specific activity i_s in A/cm^2 Pt are used. The latter value is calculated with respect to the electrochemically active surface area of Pt (in cm^2 Pt). Hence this value reflects the activity of catalysts sites that can be used in the reaction. The two current densities are related through the *specific* electrochemical surface area (ECSA) expressed in cm^2/g according to:

$$i_m = i_s \times ECSA \tag{5.2}$$

Activity in PEMFCs In Table 5.6, these activities have been compared for a number of Pt/C catalysts in PEMFC systems. Also the Tafel slope has been given. The theoretical Tafel slope for the ORR on Pt at low-current density (i.e., on oxide-covered surfaces) is RT/F i.e., 67 mV/dec at 65°C and 70 mV/dec at 80°C, in good agreement with the data found here. At high current density (potential lower than *ca.* 0.8 V) the Pt surface is oxide-free and the ORR has a Tafel slope of $2RT/F$ [117,118]. The mass activity and the specific activity in Table 5.6 have, when needed, been corrected for the oxygen partial pressure, referring all values to 1 bar O_2. For this correction a linear dependence on the oxygen partial pressure of the activity at fixed potential was assumed, as was found by most researchers [119–121].

Table 5.6 shows that at 80°C and 1 bar oxygen the specific activity of Pt in Nafion-based PEM systems is in the order of 0.2 mA/cm^2. The question is, whether from a kinetic point of view this is enough to meet the requirements mentioned earlier. This question cannot be answered without taking the ohmic resistance of the cell into account. A power density of 0.91 W/cm^2 at 0.65 V corresponds to a total current density j in the MEA of 1.4 A/cm^2. The current produced at the cathode will be determined by the cathode potential V_{cat}. Neglecting transport losses in the cathode and completely neglecting anode losses, the cathode potential can be put equal to the IR-free cell potential. The required ECSA at a given Pt loading (L_c in g/cm^2) and given oxygen pressure (p_{ox}) can then be calculated from:

$$ECSA = \frac{j}{L_c \cdot i_s \cdot (p_{ox}/p_{ref})} \times 10^{\frac{V_{cat}-0.9}{b}} \tag{5.3}$$

Here b is the Tafel slope and p_{ref} is the oxygen pressure for which i_s was determined, i.e. 1 bar.

Table 5.6 ORR activity of Pt/C catalysts in Nafion-based PEMFCs

	T (°C)	b (mV/dec)	i_s at 0.9 V (mA/cm² Pt)	i_m at 0.9 V (A/mg Pt)	Remarks
20 wt% Pt/Vu	80		0.17	0.11	[105][a]
40 wt% Pt/Vu	80		0.20	0.06	
46 wt% Pt TKK	80	78	0.21	0.16	
47 wt% Pt TKK	80	70	0.18	0.11	
10 wt% Pt/Vu	80	57	0.24	0.01	[328][b]
20 wt% Pt/Vu	80	59	0.21	0.03	
40 wt% Pt/Vu	80	57	0.17	0.03	
20 wt% Pt/Vu	80	64	0.33	0.05	[120][b]
20 wt% Pt/Vu	70	65	0.20	0.03	
20 wt% Pt/Vu	70	62	0.48	0.07	[329][b,c]
20 wt% Pt/Vu	70	62	0.43	0.10	
20 wt% Pt/Vu	70	61	0.47	0.14	

[a] Thin film electrode.
[b] Nafion impregnated electrode, published results normalized to 1 bar oxygen.
[c] Nafion content varied: 0.6, 1.3 and 1.9 mg/cm² from top to bottom, respectively.

The results are shown in Figure 5.19. The results show that with the standard loading of 0.4 mg/cm^2 the target can be met with ECSA values around 100 m^2/g and ohmic resistances of the cell of 0.08 Ω cm^2, or lower. At lower loading the requirements for the maximum ohmic resistance are much more severe. Moreover, it must be noted that these data were obtained assuming there are no transport losses.

Particle size effect It might be suggested that an increase of the ECSA by increase of the surface area, i.e. by decreasing the Pt particle size, would result in more efficient active layers. It must, however, be assumed that there is a limit to increasing the mass activity of Pt by further reduction of the particle size, due to the so-called particle size effect. For the PAFC, the particle size effect is well documented (see e.g. the review by Kinoshita [122]). The specific activity decreases with decreasing particle size and correspondingly the mass activity has a maximum around 3 nm. This was ascribed to the changes in the distribution of crystal faces on the particle surface that occur when the size varies. Both the specific adsorption of anions and the activity for the ORR in a PAFC are structure dependent. In the case of PAFC adsorption of phosphate anions was assumed to be the

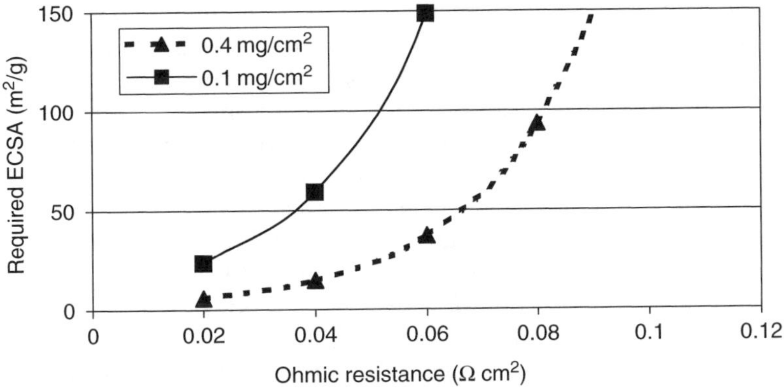

Figure 5.19 Required ECSA to obtain 0.91 W/cm^2 @ 0.65 V with a cathode containing 0.1 mg/cm^2 Pt or 0.4 mg/cm^2 Pt as a function of the ohmic resistance of the cell. The calculation was done neglecting transport losses in the active layer and assuming a Tafel slope of 70 mV/dec, i_s = 0.2 mA/cm^2 and p_{ox} = 0.2 bar.

main cause of the particle size effect. In non-adsorbing electrolytes such as Nafion and $HClO_4$ no or a much smaller effect was expected. Still, a similar effect seems to occur in PEMFC electrodes, with also a maximum in mass activity at 3 nm [123,124]. Results by Gasteiger [105] showed that with $HClO_4$ as an electrolyte the specific activity decreased by an order of magnitude when the particle size decreased from 20 to 2 nm. The mass activity was four times higher for 3 nm particles than for 20 nm particles. Studies by Markovic and by Mukerjee [125,126] showed that OH adsorbed on Pt suppresses the kinetics of the ORR, and that this adsorption, like the activity for the ORR, is structure sensitive. Moreover, with decreasing particle size the average Pt–Pt distance increases [127] and according to Hammer and Nørskov's theory [128,129] this causes a shift upwards in energy of the Pt d-levels, which leads to stronger interaction energies with adsorbents like OH. The specific activity values reported in Table 5.6 were measured using particle sizes in the order of 3–5 nm, hence the variation is only a factor of 2.

Mass activity of Pt An ECSA value of 100 m^2/g with the specific activity found in Nafion systems results in a mass activity of 0.2 A/mg. The values in Table 5.6 are usually well below this value implying that these ECSA values were not met. First of all because the average particle size was larger than 3 nm, resulting in lower values of surface area, and secondly because not all Pt surface area could be used in the reaction, i.e. limited utilization. The utilization is defined as the ratio between the ECSA and surface area. Table 5.6 shows the improved results for the mass activity obtained with thin film electrodes compared to impregnation of Pt/C structures with Nafion. In the latter only good Pt utilization could be obtained at high Nafion loading.

To increase the surface area of Pt, a catalyst can be used with a lower wt% Pt on carbon, or a carbon with a higher specific surface. The first method has the disadvantage that it results in thicker electrodes, which enhances proton and gas diffusion losses. A higher surface area carbon black may be more beneficial, as high weight percentages can be obtained with still small (near 3 nm) size particles. Work by Uchida has shown, however, that in electrodes made of Ketjenblack supported Pt the fraction of Pt surface area that is electrochemical active, i.e. the utilization, is relatively low [116,130]. This was ascribed to the large amount of Pt particles that is located in small pores inaccessible to the ionomer. In carbon-supports with low-specific surface area the average pore size is much larger and hence the utilization is much better. This means an optimum for utilization and Pt particle size had to be found, which was met by Vulcan. However, with the new carbon supports such as the AC01 from Johnson Matthey [112] and the carbon support developed by TKK [114], high ECSA values at high loadings are becoming possible. As pointed out earlier, utilization loss can also result from particles being encapsulated by Nafion, i.e. not connected to the electron-conducting system.

Electrolyte effects The data in Table 5.6 refer to Nafion-based systems. It is, however, well known that also the electrolyte affects the kinetics of the ORR [131]. The solubility of oxygen, the acidity and the adsorption of electrolyte on the catalytic surface depend on the nature of the electrolyte. Kinetic data on non-Nafion-based PEMFCs are scarce, but the ORR has been extensively studied in electrochemical cells with microelectrodes or rotating-disk electrodes (RDEs). Table 5.7 summarizes the data on the specific activity and the oxygen solubility. First of all it must be noted that not all authors report the specific activity at 0.9 V. In some cases these data were calculated from reported values for the Tafel slope, the exchange current density, the reversible potential and roughness factor. There are large variations in the values for the specific activity, which can be due to the inaccuracy of the procedure, especially if there was uncertainty about the exact value of the reversible potential. Moreover, the effective oxygen pressure was not always reported. Meaningful comparisons are therefore only possible for data reported within the same paper. Work by Beattie *et al.* [132] showed that the ORR kinetics in the oxide-free region at the Pt BAM 407 interface (i.e. BAM3G with an equivalent weight of 407 g/mol) is somewhat slower than on Nafion. The concentration of O_2 in BAM 407 was also lower than in Nafion. This is mainly due to the lower water content in BAM 407. In the hydrophobic parts of the polymers O_2 dissolves more easily [133]. A similar result was found by Zhang *et al.* [121,134] for the kinetics at Pt/S-PES interfaces. Here as well a lower concentration of oxygen was observed.

Table 5.7 ORR activity of Pt in various electrolytes, near room temperature and around 60–80°C

	near RT T(°C)	b (mV/dec)	i_s at 0.9 V (mA/cm² Pt)	$C \times 10^6$ (mol/cm³)	60–80°C T(°C)	b (mV/dec)	i_s at 0.9 V (mA/cm² Pt)	$C \times 10^6$ (mol/cm³)
Beattie *et al.* microelectrode 3 atm oxygen [132]								
Nafion 117	30	57.4	0.112	9.19	70	60.4	0.201	7.81
BAM 407	30	63.4	0.014	2.08	70	58.2	0.043	1.65
Zhang *et al.* microelectrode 1 atm oxygen [134]								
Nafion 117	40	70	0.021	7.74	70	71.4	0.036	4.74
S-PES-40	40	83.7	0.012	3.16	70	88.7	0.017	1.62
Zhang *et al.* microelectrode 3 atm oxygen [121]								
Nafion 117	30	69.9	1.136	11.24	80	67.1	2.149	7.08
S-PES-SP	30	71.5	0.171	11.58	80	77	0.548	5.09
S-PES-40	30	72.6	0.097	9.84	80	67.9	0.268	3.90
Gottesfeld RDE oxygen saturated liquids [135]								
0.5 M H_2SO_4	25		0.270	1.13				
85% H_3PO_4	25		0.009	0.130				
recast Nafion/H_2SO_4	25		0.340	4.2				
recast Nafion/H_3PO_4	25		0.017	3.8				

A clear relation with the oxygen concentration is not always observed. Gottesfeld [135] compared bare Pt electrodes and Pt electrodes covered with a recast Nafion film. The presence of the film results in larger oxygen concentrations but the increase in the specific activity was much lower. In the case of H_3PO_4 the concentration increased by a factor of 20 but the activity only by a factor of 2. Several explanations have been suggested for the weak dependency on the oxygen concentration near the electrode surface and the oxygen reduction rate, given the undisputed fact that the relation is first order in oxygen concentration. Gottesfeld suggests that only the perfluorinated region of the Nafion contains the higher concentration, and this phase is not in contact with the Pt. Furthermore, blockage by the polymer film or organic contaminants may play a role. Yeager *et al.* [131] have pointed out that the chemical potential of oxygen rather than the actual concentration may be determining for the rate of the ORR.

A blocking by inactive parts was given as the reason for decreasing activity of Pt at recast Nafion interfaces with decreasing humidity observed by Uribe [136]. The relation between the water activity or humidity on the kinetic activity at Pt/Nafion interfaces has recently been further investigated by Neyerlin *et al.* [137] and Xu *et al.* [138]. In both studies it was found that above 50–60% RH the rate of the ORR did not depend on the RH, but below these values a distinct decrease of the ORR rate was observed. In both papers this was first of all ascribed to a reduced proton activity in the Nafion. In Nafion membranes the H_2O/SO_3H ratio drops below 3 at such a low RH (Figure 5.9) and the acid dissociation is reduced. A similar effect probably occurs in the Nafion phase in the electrode. Xu *et al.* found no reduction of the electrochemically active surface area at low RH but there seemed to be a change in the reaction mechanism at RH $<50\%$ reflected by an increase in Tafel slope and a decrease of the reaction order in oxygen. This may be a result of the lower proton activity but also of changes at the Pt surface, e.g. reduced OH adsorption. Murthi *et al.* found evidence for a change in the mechanism as a result of decreased water activity in a study of the ORR in concentrated trifluoromethane sulfonic acid (TFMSA) [139]. TFMSA is often considered to be a Nafion look alike. In this system an enhancement of the kinetics was observed at lower water activity, ascribed to the suppression of OH adsorption. Decreased proton activity due to incomplete acid dissociation does probably not play a role here.

High-temperature effects Operation of PEMFCs at elevated temperatures ($>80^\circ C$) and/or less humid conditions will affect the kinetics of the ORR at the Pt catalyst sites in several ways. Due to the positive activation energy the exchange current density of the reaction will increase, but so will the Tafel slope. Moreover, due to the high water saturation pressure at temperatures $>90^\circ C$, the oxygen partial pressure will be low, even at low RH. Additional effects of low humidity on the kinetics may be reduced proton

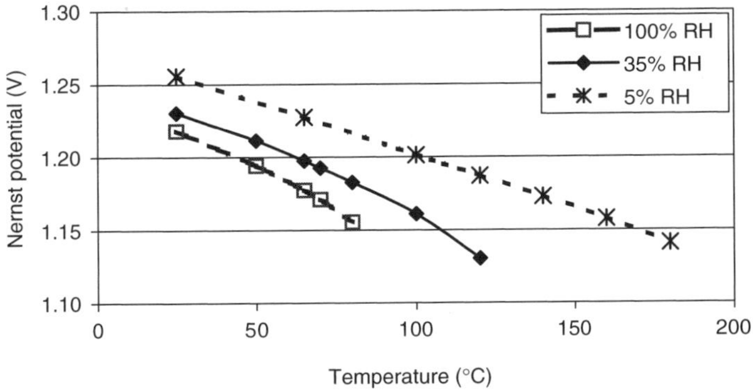

Figure 5.20 Nernst potential as a function of temperature for H_2/air, atmospheric conditions, with saturated gases and gases with 35% RH and 5% RH, respectively.

activity and suppressed OH adsorption. For some of the effects listed here quantitative data are available. This enables a first estimate of the IR-free polarization curves of a PEMFC at elevated temperatures and reduced humidity, taking also into account the change in the Nernst potential.

For a fixed value of the RH the Nernst potential decreases with temperature (Figure 5.20). Decreasing the humidity leads to higher values for the Nernst potential. Figure 5.20 shows that at atmospheric conditions similar values for the Nernst potential are obtained at 80°C 100% RH, 110°C 35% RH and 160°C 5% RH. For the activation energy of the exchange current density in the oxide-covered region of Pt/Nafion systems a value of 73.2 kJ/mol has been reported by Parthasarathy [140], in agreement with data by Mukerjee [120]. However, more recent measurements report lower values between 36 and 55 kJ/mol [121,132,134,138]. For BAM 407 Beattie *et al.* report a much lower value of 23.4 kJ/mol [132] and for S-PES values in the order of 45 kJ/mol were found [121,134]. The Tafel slope of the ORR on Pt/C at oxide-covered and well-humidified conditions is RT/F, i.e. there is a linear increase with temperature.

In Figure 5.21, a few IR-free polarization curves are shown, assuming zero overpotential for the anode and omitting transport effects in the cathode. The base case was 65°C and water-saturated conditions. For this case an i_o^0 value of 1 μA/cm^2 (geometric) was assumed. The IR-free curves were calculated from:

$$V_{IR\text{-}free} = E_{Nernst}(p_{H_2}, p_{ox}, p_{vapour}, T) - b \log \frac{j}{i_0^0 (p_{ox}/p_{ref})^{\gamma}} \quad (5.4)$$

with

$$p_{ox} = 0.21(p_{cat} - p_{vapour}) \quad (5.5)$$

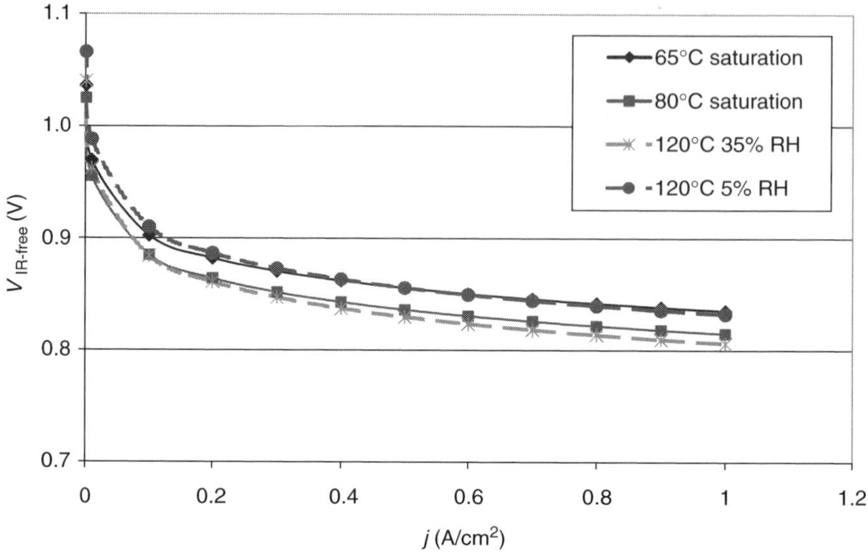

Figure 5.21 Calculated IR-free polarization curves at various temperatures and humidity.

and p_{ref} being 1 bar and p_{vapour} the water pressure at the relevant temperature and humidity. For γ the value 0.75 was used corresponding to a factor unity for the reaction order of the activity at fixed potential [98]. For the base case (65°C, 100% RH) this results in a curve for a well-performing MEA, operating in the oxide-covered region up to 1 A/cm^2. At higher temperatures the i_o^0 was calculated using the high value for the activation energy, i.e. 73.2 kJ/mol. Operation at 80°C results in a worse performance due to the high water vapour pressure at saturated condition. The results at 120°C fall in the same range, but suggest a better performance at reduced humidity. However, as shown by Neyerlin [137] and Xu [138] changes in the reaction mechanism and reduced proton activity further increase the kinetic losses at low RH cancelling the effect of the higher partial oxygen pressure. With state-of-the-art materials the cathode losses will still increase when the temperature is increased from 80°C to 120°C [138,141].

In PA doped-PBI systems the cathode losses are high due to the limited kinetic activity of Pt in phosphoric acid (Table 5.7). The largest inhibiting factor is formed by the adsorption of phosphate anions on the Pt surface. The presence of a polymer, PBI or Nafion, increases the local oxygen concentration but can also be effective in suppressing the adsorption of phosphate ions, as suggested by Li *et al.* [142]. The beneficial effect of the presence of PBI seems to be enhanced in concentrated H_3PO_4 compared to diluted

H_3PO_4. The activation energies of the exchange current density, measured on Pt catalysts in H_3PO_4, are between 40 and 80 kJ/mol [122]. Also in PAFCs, at high temperatures, was the Tafel slope reported to be RT/F [122]. In the case of PAFC typical mass activity values are in the order 0.03–0.05 A/mg Pt at operating temperatures 170–190°C [91,122]. Mass activity data for PBI/PA do not seem to be available, but cathode losses of PBI/PA systems at 150–180°C [143] are higher than for well-humidified Nafion systems 80°C, i.e., 400 mV at low-current density [88,143].

Stability of the Pt/C catalyst Loss of activity of the Pt/C catalyst in the PEMFC due to decrease of active surface area was already reported by Wilson *et al.* [144] in 1993. From the research on the PAFC it was well known that Pt/C catalysts lose their activity due to both Pt dissolution and Pt sintering. The Pourbaix diagrams [145] show that Pt is not stable between 1 and 1.2 V versus NHE and $pH < 1$. It was found that Pt/C already dissolves at lower potentials in H_3PO_4 and H_2SO_4 at temperatures in the order of 170–200°C [122,146]. In this temperature range the solubility of Pt increases by two orders of magnitude when the potential increases by 0.1 V [146]. Pt sintering can occur either by migration of the Pt particles on the carbon support and coalescence, or by dissolution and re-precipitation of the Pt atoms, also called Ostwald ripening. Both mechanisms lead to an increase of the average particle size with an asymptotic particle size distribution but the coalescence mechanism has a log-normal distribution (tail at large sizes) and the Ostwald ripening has a tail at the smaller particle sizes but with a maximal particle size cut-off [147]. Pt ions can be absorbed in the ionomer phase in the electrode and in the membrane where they replace H^+ ions [148]. In an operating fuel cell the electric field is against metal ion migration from the cathode into in the membrane, but Pt can be transported by diffusion [149,150]. Loose Pt particles can also be washed out of the cell.

Although the loss of Pt active area has been well known, there have been few studies to the exact mechanisms in the PEMFC. Only recently has this subject gained more interest as requirements for durability and for low Pt loadings have become more demanding [151]. Coalescence of particles seems to be enhanced by hydrogen cross-over, e.g. resulting from membrane degradation at low humidity conditions [152]. In a study by United Technologies Corporation (UTC) Fuel Cells [153] Pt/Vulcan cathodes (0.4 mg Pt/cm^2) were repeatedly cycled at 65°C between 0.87 and 1.2 V (H_2 on the anode, nitrogen on the cathode). After 3000 cycles the ECSA had decreased from 67 to 33 m^2/g with the sharpest decrease in the first 1200 cycles, to 40 m^2/g. This fast initial decay was explained assuming that the dissolution rate of small particles is faster [149]. Also a coalescence mechanism can be responsible: as Xie *et al.* [154] point out coalescence of smaller particles is easier and faster. Electron probe microanalysis EPMA measurements made before and after the cycling showed that initially the Pt particles are evenly distributed over the cathode layer. After 2400 cycles a band of Pt was observed in the Nafion 112 membrane about 25% of the membrane thickness separated from the membrane/catalyst surface. This band was thought to be Pt precipitated after dissolution from the cathode and subsequent reaction with dissolved H_2. Pt ions are not expected to move fast towards the anode because of the opposing electric field in the membrane but both Pt ions and Pt particles may be transported with the back diffusion of water [154].

So far other studies seem to corroborate that ECSA loss and Pt precipitation occurs upon potential cycling in PEMFC electrodes. Shao-Horn *et al.* [155] report an increase of the Pt particle size (established by XRD) from 2.3 to 10.5 nm after 10,000 cycles between 0.6 and 1 V versus NHE at 80°C, with 60% of the initial ECSA lost. It is still unknown how the decay rates measured with potential cycling correspond to real operating conditions and times of the PEMFC, but as was pointed out by Kocha *et al.* [156] cycling between OCV (1.0 V) and peak performance (0.65 V) is part of the normal operation of automotive stacks.

At elevated temperature the Pt dissolution will be faster. Estimates by General Motors [151], based on Bindra's results in H_3PO_4 at 196°C [146] and data from the Pourbaix diagram at 25°C [145], showed that between 80°C and 120°C the solubility increases by an order of magnitude. Combined with the observation (see above) that the dissolution of Pt increases rapidly with the cathode potential Pt dissolution may become a serious issue for automotive applications, as was recently pointed by General Motors. The limited stability of Pt at high potentials might put an upper limit to the temperature window in which PEMFC can be operated in automotive applications. Further investigations on the rate of Pt solubility in polymers such as Nafion have to be carried out, and on ways to reduce the Pt solubility. Research on the PAFC has shown that heat treatment (at 900°C) of the catalysts led to decreased surface area of the catalyst but also to a decreased decay rate of the ECSA and an increased cell performance [122]. In PAFC operation high voltages are avoided, which is feasible for stationary applications. For H_3PO_4/PBI systems similar dissolution rates of Pt as in PAFC must be expected. However, as long as these are used in stationary applications high potentials can also be avoided here. As mentioned in previously, the cathode activation overpotential in H_3PO_4 /PBI systems is high, which reduces the risk attaining high cathode potentials at normal operation.

Also the role of the carbon support has to be further investigated. Loosely attached Pt particles will sooner come off the support [154,155] into the ionomer phase or migrate more easily over the surface and coalescence. A final mechanism for ECSA loss is corrosion of the carbon support, also a well-known issue from PAFC development. Corrosion of carbon would result in Pt becoming disconnected from the electron conductive phase, and thus made inactive. The Pourbaix diagrams [145] show carbon is thermodynamically unstable at the cathode conditions. Still the carbon blacks exhibit a reasonable stability because the carbon oxidation reactions are slow. Kinoshita [113] showed that in acid environments ordered, highly graphitized carbon is more stable than disordered carbon. Similar results were recently reported for carbon supports in PEMFC MEAs [157]. Thermal degradation of carbon supports in dry air has been studied at temperatures 125–195°C, and was shown to increase with Pt content and temperature [158]. A similar dependence on temperature and Pt content was found for carbon electro-oxidation in phosphoric acid. Despite the instability, studies in PAFCs have shown that carbon corrosion contributes less than other factors to the degradation of the cathodes at 190°C [113]. Still, as PEMFCs are expected to operate at higher voltages than PAFCs, and are expected to undergo more potential cycles in automotive applications, carbon corrosion is still a reason for concern.

5.4.2.2 Alternative Catalysts

Pt alloys From the work on PAFC it is well known that several Pt binary and ternary alloys show up to 3-fold increase in mass and specific activity for the oxygen reduction reaction. The reasons given for enhanced activity are quite diverse and include Pt–Pt distance decrease, reduced sintering rate, increased surface roughening, reduced adsorption of anions and oxygen species. A good survey of alloys considered for PAFC can be found in [91]. For the PEMFC also several alloys have been identified with promising activities for the ORR in the temperature range 60–80°C [120,125]. Tests in electrochemical cells showed enhancements in mass activity by a factor of 1.5–4, for alloys such as PtCr, PtMn, PtCo, PtNi. In active alloys the Pt–Pt distance is usually reduced. This may be more favourable with respect to O_2 dissociation, and it also leads to a downward shift of the Pt d bands, which reduces the OH adsorption on Pt. As mentioned above, Markovic showed that OH adsorption reduces the activity for the ORR [125]. Also Mukerjee *et al.* and Antolini *et al.* [159,160] ascribed the better activity to suppression of oxide formation. It should also be noted that the Pt alloys of this type have larger particle sizes than obtained for Pt [161]. This also contributes to reduced OH adsorption and higher specific activity [160]. Other explanations given for the better reactivity are the rougher surface due to leaching of the base metal and a better stability to sintering [115,124].

However, most transition metals are unstable in the operating region of the PEMFC cathode (pH $\approx$ 0, E = 0.8–1.1 V versus NHE) [145]. Leaching of metal can be more detrimental in a PEMFC than in a PAFC since it does not only lead to loss of functionality. The metal ions can contaminate the proton-conducting phase (both in electrodes and membrane), where they replace H^+ ions. This reduces both the proton conductivity and the rate of the ORR [148]. In principle, well-ordered Pt–M alloys can be more stable under these conditions than the constituent metals, and in view of the potential for much higher mass activity much research has been done on Pt alloys as catalysts for the PEMFC cathode. In recent years also combinatorial methods have been developed to identify ternary and binary alloys for the ORR [162,163].

Johnson Matthey [115] investigated some promising catalysts for the use in fuel cells. PtFe and PtMn were not stable under the PEMFC conditions and leached base metal into the cell. PtCr therefore was the material selected for further study. In the PEMFC, however, it was found that the kinetic gain was lost in air at high current density. This was ascribed to the hydrophilic properties of the oxides formed on the base metal which led to flooding of the cathode at high current density. Although it was reported that inclusion of a third metal could reduce the flooding problem, both stability and hydrophilic properties it remain a matter for concern. Recently, Xie [154,164] reported on the morphology and stability of Pt_3Cr catalysts from E-TEK in the PEMFC. The PtCr particles were larger and less uniformly distributed over the active layer than Pt in a similarly made electrode. A large increase of particle size and loss of active surface area in the first 500 h of operations was observed. After 1000 h of operation an increase of Pt_3Cr particles was found near the catalyst/membrane interface. The composition of the Pt_3Cr particles was the same as in the fresh material. Although no Pt or Cr was detected in the membrane near the cathode, Cr was observed on the anode. Moreover, Cr and Pt were detected in the outlet water, with especially in the first 800 h an excess of Cr.

Also PtCo has been a promising candidate for some time. Both fuel cell experiments and electrochemical cell experiments have shown that the mass activity is 2.5-fold higher than for platinum [105,125,165]. Unlike with PtCr the performance gain seems to be maintained at higher current density [105,153]. Moreover, the particle size seems to be more stable than for Pt. Yu *et al.* [153] performed potential cycling experiments for PtCo electrodes similar to the ones described in Section "Stability of the Pt/C catalyst" for Pt. In a fresh electrode PtCo particles were larger than Pt particles (average 4.6 versus 2.5 nm), but the PtCo particle size increased less rapidly during the cycling. After 2400 cycles the overall performance loss for the PtCo MEA was only 10 mV, whereas it was 67 mV for the MEA containing a Pt electrode. However, in the first 1200 cycles additional performance loss was found which was ascribed to Co dissolution and corresponding loss of activity. Additional tests in $HClO_4$ showed dissolution of Co in the electrolyte, while dissolution of Pt could not be detected. X-ray diffraction (XRD) analysis before and after testing showed an increase of the lattice parameter, indicating loss of Co. There seemed to be no detrimental effect on the membrane conductivity. After 3000 cycles it was observed in EPMA measurements that an enrichment of PtCo particles at the catalyst/membrane interface had occurred. Unlike with the Pt electrode but, like Pt_3Cr, PtCo did not seem to migrate into the membrane.

Gasteiger recently reported qualitatively similar results [105]. It was shown that pre-leaching of Co from PtCo alloys removes non- or poorly alloyed Co from the catalysts, and that this reduces subsequent leaching of Co. A 2.5 mass activity increase was obtained with PtCo catalysts compared to Pt. In a 12 cell stack tests several catalyst were compared for 1000 h. The activity gain of PtCo compared to Pt was maintained at high current density under air and during the whole operation time. The degradation rate for MEAs containing Pt or pre-leached PtCo was similar (55 versus 60 μV/h). A detrimental effect on the membrane conductivity was not observed. Recent results by Murthy and Sisofo [157] showed even better stability for (unspecified) Pt alloys compared to Pt in potential cycling experiments at 80°C.

From these data it can be concluded that alloying can be effective in attaining the goal of increased power/g Pt. Careful preparation of the alloy and of the microstructure seems a prerequisite, but they can be more stable than Pt. Although the dissolution processes are faster at elevated temperatures, the experience from PAFC is that stable alloys can be developed [91]. However, it is not certain that the beneficial effect of alloys in PEMFC will be maintained at higher temperature. Results by Murthi [139] obtained in concentrated TFMSA solutions suggested that alloys do not outperform Pt at reduced water activity, as there is no longer a difference in OH adsorption. This may imply that reduced humidity has an even more negative effect on the ORR kinetics for alloys than for Pt. This aspect, like the stability, needs further clarification.

Pt-free catalysts In view of the above reported stability problems the perspectives for catalyst based on non-noble metals are not very bright. Most metals are excluded and so are most oxides. Moreover, additional durability issues arise in a PEMFC since most non-Pt catalysts seem to increase the amount of peroxide formed.

A few classes of materials have been identified which may survive in the PEMFC. Best known are the macrocycle catalysts where a transition metal ion is coordinated by nitrogen atoms forming part of an organic ligand of a highly aromatic nature [166]. In a recent evaluation of these materials Gasteiger [105] showed that although the activity per catalytic site may just be sufficient for automotive applications, the site density of these catalysts is probably too low. Large volumes of catalyst would be required leading to increased mass transport problems. Compared to Pt one order of magnitude activity per volume unit less might be acceptable, but not more.

Another class are the chalcogenides, as proposed by the group of Alonso-Vante. Chalcogenides containing Ru, Se and small amounts of Mo were shown to have an activity for the ORR [167,168]. However, in fuel cells 200 mV additional overpotential is expected, too high for feasible operation. On the other hand these catalysts are much more methanol tolerant than Pt and therefore are considered as possible replacement of Pt in the direct-methanol fuel cell (DMFC) [169].

Alternative supports Several authors have explored the possibility to modify the surface of the carbon support to enhance the performance in a fuel cell. Such modification should aim at reduction of transport losses and increasing the ECSA. The group of Kaufman/Pickup [170–172] have shown that the attachment on the carbon support surface of sulfonated silane, phosphonic or sulfonic acid groups on the carbon support may result in higher utilization and proton conductivity. In the case of sulfonated Pt/Vulcan an increase in performance of 60% was reported with 10 wt% Nafion in stead of 30 wt%. While this demonstrates the potential of surface modification of the carbon support, it must be noted that the stability of these groups under the harsh fuel cell conditions is still unknown. The ethane sulfonic acid groups used in Ref. [172] were reported to be stable for a week but in general aliphatic groups are not considered to be stable for a long time. Saab *et al.* [173] investigated uncatalysed carbon powders modified with phenylsulfonic acid groups used in an ink applied on a substrate. It was found that this enhances the proton conductivity but decreases the electronic conductivity, i.e. it seems carbon–carbon (C–C) contacts are lost. A slightly different approach was followed by Mizuhata [174], who investigated the grafting of polymer ionomers on the carbon support. The basic idea was that the ionomer solution would penetrate into the smaller pores of the carbon support. Polymerization takes place in the support. This approach resulted in higher active areas but since the electrodes were tested under oxygen only the effect on diffusion is not yet clear.

The use of carbon nanotubes or nanofibres instead of carbon black has recently been introduced in the field of PEMFC. Joo *et al.* [175] showed that with Pt dispersed on ordered single-wall carbon nanotubes (SWNT) a mass activity for the oxygen reduction reaction at 0.9 V in an electrochemical cell was obtained that was several times higher than measured with Pt on Vulcan under similar conditions. The enhanced activity was ascribed to the more uniform and smaller size of the Pt particles on nanotubes. The nanotubes used by Joo were obtained using an aluminosilica template, which resulted in an ordered mesoporous carbon. Other methods to obtain nanotubes include arc discharge, laser ablation and chemical vapour deposition. The resulting carbon powders can be single-wall, multi-wall or disordered-wall nanotubes. Several authors have reported good activity for the oxygen reduction in a fuel cell [176–180] as well as better utilization and mass transport [181]. These results look promising but more research and electrode optimization has to be done. In particular, these materials offer the possibility to tailor the carbon support to the requirements of the fuel cell.

5.4.2.3 Proton-Conducting Phase

Structural aspects In an optimized electrode the amounts of catalyst material and proton-conducting phase (usually perfluorosulfonate ionomer, PFSI) must be carefully matched, to optimize utilization, proton conductance, gas and water transport. The proton resistance of the electrode is proportional to the thickness of the electrode and depends on the fraction of Nafion present in the electrode. For this volume dependence several relations have been formulated [182–184]. A general formula for the proton resistance R_p is given by:

$$R_p = \frac{L}{\sigma_{bulk} f(\varepsilon_N)} \tag{5.6}$$

Here L is the thickness of the electrode, σ_{bulk} the proton conductivity of the bulk proton- conducting material and $f(\varepsilon_N)$ a function of its volume fraction ε_N. Boyer [182] and Perry [183] state that this function is linear in ε_N. However, recent measurements by Ma *et al.* [185] show that the proton resistivity of the proton-conducting phase can be larger than the membrane value by two orders of magnitude. Ma *et al.* ascribe this to the effect that the microstructure has on the proton conductivity. The aggregates of the recast PFSI have a microstructure very different from the corresponding membrane.

At high-current density, gas diffusion losses become large. The gas diffusion has to take place though the entire thickness of the electrode (order μm) as well as through the agglomerates. Where addition of PFSI may help to improve both the utilization and the proton conductivity in the electrode, addition of too much PFSI would block gas pores resulting in large diffusion losses. Rigorous expressions of the oxygen transport properties in the electrode are not available, but in all electrode models the oxygen transport is characterized by the oxygen permeability of the ionomer phase, i.e. the product of the oxygen concentration or solubility C and the diffusivity D. This quantity $(C \times D)_{eff}$ is dependent on the oxygen pressure through the dependence of the concentration on the pressure given by Henry's law. The subscript "*eff*" is appended on the permeability because the permeability is not a simple function of the bulk ionomer permeability and volume fraction ε_N. Also the gas phase fraction ε_V and the geometry play a role.

There is still considerable uncertainty about the structure of the catalyst layer. Based on porosimetry results on layers consisting of Flemion as PFSI and acetylene black as carbon medium, Uchida *et al.* [186] concluded that the carbon forms agglomerates, with small intra-agglomerate, or primary, pores with size 20–40 nm and larger, 40–1000 nm, inter-agglomerate, or secondary, pores. From their data Uchida *et al.* concluded that the PFSI phase (Flemion) fills the secondary pores only, leaving the primary pores open. This structure was believed to result from the large molecular weight of the PFSI which cannot enter small pores. The resulting structure has carbon agglomerates (100–200 nm) consisting of carbon grains (20–40 nm) and covered by Nafion. The void space is divided over the intra-agglomerate pores and the secondary pores, where these are not occupied by PFSI. Recent tunneling electron microscopy (TEM) images presented by workers at LANL and

ORNL show Nafion located between the carbon agglomerates [154]. The effective permeability applicable in modelling studies based on this geometry is usually much higher than through the ionomer fraction alone [184,187], indicating additional gas diffusion in voids.

On the other hand, work by the group at KTH (Stockholm) [188,189] suggested that Nafion, like a liquid electrolyte, fills the agglomerates. These results were also based on porosimetry, and on elemental mapping of samples containing Nafion and Vulcan XC-72 supporting Pt particles. In such a structure, the agglomerates would contain both the percolating carbon structure and the percolating Nafion structure. Gas diffusion takes place by species dissolved in Nafion within the agglomerates and in voids between the agglomerates. This work also showed that increasing the volume fraction of Nafion resulted in a decrease of the void volume fraction, with the Pt/C volume fraction remaining relatively constant. Only at high Nafion volume fractions a decrease of the Pt/C fraction was observed. This indicates that the carbon phase in the electrode determines the structure, with the Nafion filling voids. In models based on this geometry a much smaller effective permeability value is found than would be expected from diffusion in filled agglomerates alone, indicating the presence of ionomer phase outside the agglomerates [190].

Finally, the water management in the catalysts layer can seriously affect the oxygen transport. Water produced at the cathode will first of all be adsorbed in the hydrophilic part of the proton-conducting phase. Next, large hydrophobic pores in the carbon phase will be filled. Both effects will result in a reduction of gas voids and decrease of permeability. Only small hydrophobic pores in the carbon will remain water-free up to high water content. The water content and management of the catalyst layer are strongly influenced by the GDM, as will be discussed in Section 5.4.4, and by the membrane used [191].

Optimal Nafion content Usually the proton-conducting phase in thin film electrodes consists of recast Nafion 1100 [104]. For thin film electrodes containing 20 wt% Pt/Vulcan XC-72 and a total Pt loading of 0.4 mg/cm^2 the optimum is 30–33 wt% Nafion, or 0.9–1 mg Nafion/cm^2 as has been reported by several authors [192–195]. Considering the expressions for the proton conductivity and the parameters characterizing the gas transport, it seems likely that the volume fractions of each component determine the transport properties. Not many papers specify these data but from those cases were they have been [97,188], it can be concluded that the optimal Pt/C volume fraction is about 21–27%, the Nafion fraction is 15–20% and the porosity is 50–65%.

For other weight percentages, loadings and supports little information is available, but a guide would be that these volume percentages remain about the same. This means that the optimum amount of Nafion decreases with increasing wt% Pt/C. According to Antolini and Passalacqua the optimal wt% Nafion is independent of the total Pt loading [192,193]. This was recently questioned in papers by Sasikumar [196,197] but since their conclusions were based only upon measurements in oxygen, the mass transport effects were not sufficiently taken into account. Still, as the loading affects the thickness of the electrode and this has an effect on the proton resistance and the gas permeability of the electrode, a dependence of the optimum Nafion content on the loading cannot be ruled out. Also the surface area of the carbon and the wettability of the surface by ionomer can affect the optimal Nafion loading.

Li and Pickup [195] as well as Song *et al.* [198] have suggested that a grading of the Nafion content, in particular increasing the Nafion content near the membrane interface at the expense of the content near the backing may be beneficial. Experiments by Xie *et al.* [199] confirmed this, although only tests with oxygen were carried out. Earlier experiments by Li and Pickup did not show an improvement, possibly due to bad contact between the GDM and the Nafion-poor phase of the active layer.

Alternative ionomers The use of alternative ionomers in the active layer has not been well documented. Effects on the kinetics of the ORR have been discussed in Section "Kinetic activity", but a different ionomer will also affect the structure and transport properties of the active layer. To influence especially these properties, forms of Nafion with a lower equivalent weight have been considered. Gasteiger [105] uses Nafion 1000 and Nafion 900, and Boyer [182] Nafion 960. These types of Nafion have shorter side chains and a larger water uptake per volume unit and in principle a better conductivity [200]. However, flooding and consequently mass transport problems might also occur more easily [201].

Non-perfluorinated ionomers for use in electrodes have not often been investigated. Recently Zhang *et al.* published studies on the use of S-PES in the electrode layer [121,134,185]. These studies showed the important effect of the microstructure in the electrode layer, i.e. data obtained for the bulk membrane cannot one-to-one be transferred to the electrode layer. Still, it must expected that the performance at larger current density will be determined by the transport of gases in the proton-conducting phase and that the permeability of the bulk material can be an indication. Permeability data of different membranes and electrolytes have been listed in Table 5.8. It is found that Nafion 1100 has a very good permeability compared to other polymers and liquid electrolytes, due to high solubility in the hydrophobic part and the high diffusion in the hydrophilic part. Only BAM 407 has a similarly high permeability, but due to the much higher IEC oxygen in this material has lower solubility but a higher diffusion constant.

Table 5.8 Solubility, diffusion and permeability of oxygen in various media (at one bar oxygen pressure)

	T (°C)	$C \times 10^6$ (mol/cm³)	$D \times 10^6$ (cm²/s)	$C \times D \times 10^{12}$ (mol/(cm s))	Reference
H_2O	25	1.30	29.50	38.35	[133]
PTFE	25	26.80	0.14	3.75	[133]
Nafion 120	30	6.50	0.29	1.89	[133]
Nafion 117	30	3.75	3.06	11.46	[121][a]
S-PES-SP	30	3.86	1.95	7.53	[121][a]
S-PES-40	30	3.28	1.24	4.07	[121][a]
Nafion 117	30	3.06	5.96	18.26	[132][a]
BAM 407	30	0.69	26.41	18.31	[132][a]
0.5 M H_2SO4	25	1.13	18.00	20.34	[135]
85% H_3PO4	25	0.13	4.00	0.52	[135]
recast Nafion/H_2SO_4	25	4.20	1.50	6.30	[135]
recast Nafion/H_3PO_4	25	3.80	0.10	0.38	[135]
0.1 M H_2SO_4	25	1.18	19.00	22.42	[202]
0.1 M H_3PO_4	25	1.18	19.00	22.42	[202]
0.1 M $HClO_4$	25	1.18	19.00	22.42	[202]
PBI/0.1 M H_2SO_4	25	1.42	7.60	10.76	[202]
PBI/0.1 M H_3PO_4	25	1.53	6.27	9.62	[202]
PBI/0.1 M $HClO_4$	25	2.01	7.03	14.10	[202]

[a] concentration re-calculated to 1 bar oxygen.

It must be noted that the permeability through PBI films immersed in H_3PO_4 compares unfavourably with the one in Nafion and even to that in concentrated phosphoric acid. The PBI presents an additional barrier for diffusion through the concentrated acid, which is already slow due the strongly netted structure of anions in H_3PO_4 [202].

Effects of elevated temperatures With increasing temperature the oxygen concentration in Nafion decreases but this is compensated by the increase in the diffusion constant, resulting in an increasing permeability. For other polymers similar effects are obtained [121,132,134]. On the other hand a reduction of the humidity results in higher concentrations but lower diffusion constants and a decreased permeability. Flooding effects will be diminished at atmospheric conditions above 100°C. Due to the low humidity the proton resistance in the electrode will increase [141]. The net effect of changes in the transport in the electrode and kinetic activity as a result of elevated temperature and reduced humidity seems to be an increase of the cathode losses [141].

Stability of the proton-conducting phase In the studies by Xie [154,164] as well as by Yu *et al.* [153] it was found that loss of ECSA was not the only source of performance degradation. This suggests that the proton resistance and mass transfer limitations in the MEA had increased. As pointed out by Xie *et al.* [164] the proton-conducting phase, i.e. the recast Nafion ionomer may be susceptible to the same chemical degradation and dissolution processes as the membrane and even more so since the recast Nafion is mechanically less stable and located near the site where the ORR takes place and peroxide radicals can be formed. Dissolution of Pt into the ionomer phase or membrane can cause a reduction in proton conductivity both as a result of decreased proton concentration as of dehydration of the ionomer [148]. Hydrogen peroxide can react with contaminating ions such as Fe^{2+} to form peroxide radicals that chemically attack the membrane and ionomer phase in the active layer. In RDE studies it was found that at potentials >0.6 V the hydrogen peroxide formation during the ORR at Nafion covered Pt/C is limited to less than 0.2% [203]. In a PEMFC different values may be found due to the different electrolyte and catalyst loading [204], but considering the high-current density the absolute amount of peroxide formed can be substantial. Xie found fluoride and sulfate anions in the outlet water of the cathode [164], but a distinction between ionomer and membrane degradation could not be made.

5.4.3 Anode Active Layer

5.4.3.1 State-of-the-Art-Catalysts

Pt/C catalyst The kinetics of the HOR on Pt are faster than of the ORR by an order 6–7 [106,205]. For operation under hydrogen Pt/C catalysts are used, similar to the ones for the ORR [107,110,112]. Until recently Vulcan XC-72 was the standard support but the novel carbon supports with high SA (600–800 m^2/g) from Johnson Matthey and Tanaka TKK are beneficial here as well, resulting in higher Pt surface areas. At the anode thickness of the electrode is less critical because of the faster diffusion of hydrogen. Moreover, due to the fast kinetics the HOR takes place near the membrane interface, and proton transport losses are small. Therefore at the anode a lower wt% of Pt can be used to achieve higher surface areas. Also the Nafion content can be higher at the anode than at the cathode to optimize the utilization, since this will not be counterbalanced by mass transport effects. Moreover, lower Pt loadings should be applicable (order 0.05 mg/cm^2) resulting in a thinner electrode. However, it must be noted that with ultra-low loadings stability issues are becoming critical. This will be discussed in the next section.

As stated Pt catalysts are not suited for operation with reformate gas at temperatures of 80°C or lower. However, at temperatures >80°C the coverage of Pt by CO decreases rapidly with temperature [99,206,207]. Moreover, Dhar *et al.* found only a dilutional effect of CO_2 at temperatures >110°C [206,207]. This was corroborated by the group at TUD who tested PBI/PA-based fuel cells with Pt catalysts and found that with 3% CO at 200°C the CO-induced loss was less than 10 mV up to 0.8 A/cm^2 [208]. At 125°C 0.1 % (1000 ppm) CO in H_2 resulted in 10 mV loss at 0.3 A/cm^2. At 175°C 25% CO_2 had only a dilutional effect on the cell performance. These measurements were all done with a loading of 0.5 mg/cm^2 Pt at the anode. PEMEAS claim similarly high reformate tolerance at similar conditions but do not specify the Pt loading [143]. So, although CO tolerance of Pt in the order of more than 1000 ppm at temperatures >130°C is undisputed, the Pt loading necessary in at these conditions needs further investigation.

In the temperature range 90–120°C and at reduced humidity the situation is less clear. Only a few data are available for Nafion-based systems at these conditions. Gubler [209] showed data, obtained at saturated gas conditions, suggesting that the CO tolerance of E-TEK Pt and PtRu catalysts increased fast between 80°C and 100°C, with at 100°C E-TEK Pt performing about equally well as E-TEK PtRu. In a study by Jiang *et al.* [210] it was found that at 35–60% RH PtRu was superior to Pt in the temperature range 80–120°C.

PtRu/C catalyst At present PtRu/C is the only reformate tolerant catalyst offered by commercial manufactures. PtRu is much more tolerant towards CO [211,212] and CO_2 [101,102,213,214] than Pt. The CO tolerance of PtRu is based on two mechanisms. The first is that the adsorption energy of CO on Pt is reduced by a so-called ligand effect. Pt alloying with Ru results in a down-shift of the d-bands in the platinum that inhibits the adsorption [129,215,216]. The second mechanism is the bifunctional mechanism. Water is activated on Ru atoms in the form of adsorbed OH. This OH can then react with CO adsorbed on the platinum in an oxidation reaction. However, this requires a certain overpotential because at low potentials the OH adsorption on Ru competes with the CO-adsorption on Ru. For PtRu with the optimal atomic ratio (1:1) this overpotential is still in the order of 0.2 V versus NHE at fuel cell conditions [211]. For fuel cell operation this means that at low CO content the ligand mechanism will be responsible for the CO-tolerance. The bifunctional mechanism will only be active at high CO content where the anode overpotential is so high that efficient operation is no longer feasible.

A review of preparation and characterization of carbon supported PtRu catalysts been given by Antolini [217]. It was concluded that in most cases there is no complete alloying of PtRu. However, it has been found that complete alloying is not a pre requisite for the CO-tolerance of PtRu/C [218,219]. The presence of PtRu boundaries may be sufficient to enable the diffusion of CO from Pt to Ru where it is more strongly adsorbed [216]. Furthermore, oxidation of CO on Pt by OH on Ru can take place at these boundaries. Also the hydrogen oxidation rate can be enhanced on Pt sites in non-alloyed PtRu compared to sites on alloyed PtRu [219].

E-TEK, Johnson Matthey and Tanaka TKK all supply PtRu/Vulcan XC-72 [107,110,112]. Johnson Matthey and TKK also supply PtRu on their novel high SA supports. Comparison of reformate tolerance between different PtRu catalysts should be made with care. The total metal loading and the temperature are both critical. Moreover, the presence of CO_2 and/or N_2 in the fuel and the utilization of the fuel can affect the tolerance. The presence of large amounts of carbon dioxide and (inert) nitrogen results at high fuel utilization in a strongly decreasing partial hydrogen pressure along the gas channel, which will increase poisoning effects at the end of the gas channel [213,220].

Tanaka TKK claims the best reformate results at present for their 54 wt% PtRu (1:1) on high surface area carbon. With 0.5 mg/cm^2 PtRu and at 80°C the voltage loss at 0.5 A/cm^2 due to 25% CO_2 in H_2 is only 8 mV and with as much 100 and 300 ppm CO in this H_2/CO_2 mixture the loss increases to only 15 and 32 mV, respectively [110]. A reduction of the metal loading with 50% resulted in a doubling of the voltage drop. This is much better than the results from Johnson Matthey [111] who (at unspecified loading and temperature) report about 150 mV voltage loss at 0.5 A/cm^2 with 40 ppm CO in the same gas. The superior results of TKK were confirmed by Gubler *et al.* [209] in a comparison of PtRu from TKK and from E-TEK. Gubler *et al.* reported that X-ray photoelectron spectroscopy (XPS) measurements showed both more alloyed PtRu and more hydrous Ru oxides in the TKK electrodes. These oxides favour the bifunctional mechanism at low overpotential. The better alloying in the TKK material is probably due to the larger particle size in the TKK material. The better alloying reduces the CO-adsorption energy [114]. E-TEK and Johnsson Mathey report PtRu particle sizes in the order 2–3 nm at 20 wt%. Particle sizes from TKK were not specified.

Of course an increase in PtRu particle size is counterbalanced by a decrease in ECSA. Since a high electrochemically active surface area is also important for PtRu catalysts as due to CO adsorption a certain percentage will always be blocked, there is an optimum in the particle size and hence the wt% PtRu. Data from TKK show that on Vulcan the optimum wt% is 33%, on high SA carbon 54% [114]. Optimization of the electrode structure as suggested above to increase the utilization will certainly be beneficial in the case of reformate. Tada of Tanaka TKK [114] suggested that the ionomer:carbon ratio should be higher than found in optimized cathodes. Increasing the ionomer fraction in the anode will also increase the proton conductivity in the anode. When reformate is used, the hydrogen oxidation reaction will, like the ORR at the cathode, be distributed over the whole volume of the active layer, and not just take place near the membrane interface. So proton transport losses will be comparable to those at the cathode.

For stable and reliable operation with reformate gas at acceptable catalyst loadings, an air-bleed is usually applied [221]. A small amount of air (2–3 vol%) is added to the reformate gas in order to chemically oxidize the CO. Experiments by Gasteiger, also using catalyst material from TKK showed that with 0.2 mg $PtRu/cm^2$, an air-bleed of 2%, a feed of 40% H_2, 20% CO_2 (balance nitrogen) and 100 ppm CO no poisoning effect was found up to 0.8 A/cm^2. At a loading of 0.1 mg/cm^2 the losses rapidly increased with the current density, up to 200 mV. So it seems that for fuel cell applications the PtRu requirement at present is 0.2 mg/cm^2 minimum, using an air-bleed [106]. This is almost three times the Pt required for operation under pure hydrogen. It must be noted, however, that an air-bleed reduces the system efficiency since a major part of the oxygen reacts with hydrogen and not with CO. Moreover, there are concerns about the effects on the lifetime of an MEA (see below).

A substantial reduction of the Pt loading required for operation under reformate was reported by Sasaki *et al.* [222]. They tested a catalyst of carbon-supported 2–3 nm size Ru particles on which Pt islands had been deposited. This disposition is spontaneous and therefore relatively simple [223]. The Pt islands correspond to 1/8 of a monolayer on the Ru particle surface, with overall catalyst composition of $PtRu_{20}$. The CO bonding on Pt on Ru is reduced compared to on Pt and CO can spill over on to the Ru sites, on which it has a higher adsorption energy [224,225]. Subsequently CO oxidation can take place at low overpotential in the bifunctional mechanism. Sasaki *et al.* showed that there was no voltage loss using 0.2 mg/cm^2 $PtRu_{20}$ (i.e. 0.02 mg/cm^2 Pt) in H_2/50 ppm CO and 4% air compared to pure H_2. Moreover, the performance was stable over 1000 h. The strong segregation of Pt from the Ru particle is thermodynamically favourable and therefore inherently stable.

Studies, including electrochemical impedance spectroscopy (EIS) measurements, on the CO tolerance of PtRu/Vulcan XC-72 (supplied by E-TEK) at elevated temperature and reduced humidity were recently published by the group of Fenton at the University of Connecticut [210,226]. With increasing temperature the CO adsorption becomes suppressed and the activation of water on Ru shifts to a lower potential. Both these effects enhance the CO tolerance at elevated temperatures. However, the OH activation is again suppressed at low humidity, e.g. the tolerance at 120°C 35% RH was not much better than at 105°C 58% RH.

5.4.3.2 Stability of the Anode Active Layer

The stability issues at the anode are similar to those at the cathode, i.e. sintering and dissolution of the catalyst, carbon corrosion and degradation of the ionomer network. The potential of the anode is at normal conditions much lower than at the cathode and this will reduce the Pt dissolution rate. Still, significant increase of Pt particle size at the anode was observed both under low- [227] and high-humidity [164] conditions. This was ascribed to coalescence of smaller particles. At high humidity Xie *et al.* [164] also found Pt in the membrane, possibly carried by the electro-osmotic water flux. At elevated temperatures (>160°C) in H_3PO_4 it was found that Pt showed considerable sintering rates but below 0.8 V the dependence on the potential was small. Several strategies, including heat treatment of the Pt and treatments of the support have been devised to reduce sintering of Pt at these conditions [122].

In the case of PtRu the sintering effect will be reduced, as the presence of Ru seems to hinder particle growth [161]. Moreover, Ru should be fairly stable under anodic fuel cell conditions [145]. Still, leaching at certain conditions of Ru oxides that may be present in the alloy cannot be completely ruled out [228].

Due to the lower potential carbon corrosion is also less severe at the anode than at the cathode. An exception must be made for cases of fuel starvation, when the cell is operated at constant current density like in a fuel stack. Carbon can then, in a reaction catalysed by Pt and at overpotentials in the order of 1.4 V, be electrochemically oxidized by water to form protons [229].

The use of an air-bleed brings about additional stability problems. The chemical oxidation reaction is not at all selective for CO oxidation. Most of the O_2 reacts directly with H_2 producing heat. This in turn can enhance Pt sintering. For this reason it has been proposed to put additional air-bleed oxidation catalyst layers in the MEA either between the GDM and the catalyst layer [230] or on the GDL directly facing the gas channel [231,232]. The air-bleed reaction, i.e. oxidation of CO by oxygen at the anode catalytic sites was investigated by Jusys *et al.* [233,234] and Stamenkovic *et al.* [235] in electrochemical cells. It was found that the cleansing of the CO-covered catalyst surface area is always incomplete, due to the competition with the reaction of oxygen with hydrogen. The rate of this reaction is about four orders of magnitude faster. At high CO coverage there was a much-increased yield of peroxide, probably due to the impossibility of splitting the O–O bond on a poisoned surface. This reduction of oxygen to hydrogen peroxide can become the dominant reaction at poisoned surfaces. When the stable CO coverage was reached, however, the peroxide yield was comparable to that on a CO-free surface. Note that at these low potentials the H_2O_2 yield is much larger than at potentials >0.6 V, i.e. 5–6% in the electrochemical cell experiments [203,233,234]. Qualitatively the same results were obtained for Pt, PtRu and Pt_3Sn, with the peroxide yield, however, being lowest on

PtRu [235]. As explained in Section "Stability of the proton-conducting phase" hydrogen peroxide can enhance the degradation of the ionomer phase in the active layer and the membrane. It must be expected that at elevated temperatures this effect is reduced since CO poisoning becomes less severe.

As the authors of the above studies point out there is no difference between oxygen in the anode active layer resulting from an air-bleed or from permeation through the membrane. Hence, even when pure hydrogen is used as a fuel hydrogen peroxide can be formed, due to the low potential. Moreover, also adsorbed hydrogen can block the breaking of the O–O bond. For this reason, Stamenkovic *et al.* [235] recommend PtRu as a catalyst also for operation under hydrogen. Notice, that Laconti *et al.* [37] suggest that indeed membrane degradation starts at the anode side.

5.4.3.3 Alternative Catalysts

Extensive research has been carried out to in the past 10 years to find catalysts that are more CO tolerant. The focus has been at catalysts that would enhance the bifunctional mechanism at lower overpotentials. The advantage of bifunctional mechanism compared to the ligand mechanism is of course that CO is actually oxidized to CO_2, which is less poisonous. Pt_xMo_y [236–239] and Pt_3Sn [240,241] have been reported to be much better CO oxidation electrocatalysts than PtRu. The advantage of these materials seems to be that the CO does not adsorb on Mo sites, which at operational conditions are in the oxyhydroxide form [238,239,242], or on Sn sites [224,225] and therefore OH activation can take place at low overpotential. Still, for a number of reasons they have not yet replaced PtRu.

PtMo was proven to be a more CO-tolerant catalyst for hydrogen oxidation than PtRu [237,243,238,239]. However, PtMo was shown also to be very CO_2 intolerant [214,244]. It seems this can be overcome by using a bilayer system [244,245]. The bilayer would consist of a PtMo layer adjacent to the GDM and a PtRu layer next to the membrane. Electrochemical oxidation would take place in the PtMo layer and hydrogen oxidation in a CO_2-rich environment at the PtRu layer. Johnson Matthey claim good results with this concept up to 5000 ppm CO [111] but have not yet brought on the market. Moreover, the stability of PtMo is an issue. The Pourbaix diagram [145] shows that Mo can dissolve at the PEMFC operating conditions. The stability probably depends strongly on the preparation method, with a high degree of alloying favouring the stability [239,242,246].

The CO tolerance of Pt_3Sn as a hydrogen oxidation catalyst on the other hand seems to be less good than for PtRu [247,248]. The onset of CO oxidation may start at low overpotentials, but for sufficient removal higher overpotential is necessary. Also Pt_3Sn has been tested recently in a bilayer concept, with Pt_3Sn as a filter for electrochemical CO oxidation between the PtRu/C catalyst layer and the GDM [248]. Such a configuration led to somewhat better results than obtained with a single PtRu layer with a similar total metal loading. However, also for Pt_3Sn surface segregation of the oxophilic Sn and subsequent leaching may be an issue [240].

Research is ongoing again also with combinatorial methods to find binary or ternary alloys that do provide the right reformate tolerance and stability [162,163].

5.4.4 Gas Diffusion Media

5.4.4.1 Materials for GDM

GDM usually consist of a macroporous (pore sizes 10–30 μm) GDL consisting of carbon paper or cloth with thickness in the order of 200–400 μm, and optionally a hydrophobic microporous (pore sizes 100–500 nm) layer with thickness 30–50 μm. In general similar GDMs are used at the anode and cathode side although especially the requirements at the cathode side are more demanding as the cathode side is usually more humid, and oxygen (in air) has a lower partial pressure and a slower diffusion than hydrogen.

An excellent review of materials for GDLs and their manufacturing processes is given by Mathias *et al.* [249]. All materials are based on carbon fibres prepared from polyacrylnitrile (PAN). Four types can be distinguished:

1. Carbon fibre paper, i.e. fibres webbed by a resin.
2. Wet-laid filled papers, carbon paper without resin but filled with carbon and bound by PTFE.
3. Carbon cloth, i.e. woven fibres.
4. Dry-laid paper webbed by a hydroentangling process, i.e. a mechanically bonded, non-woven fabric.

All materials have undergone a high-temperature treatment in order to carbonize or even graphitize the carbon. All materials can subsequently be coated by PTFE, or further filled with carbon.

For the microporous layer (MPL) carbon black powder mixed with PTFE is used. Carbon blacks have been used for this purpose include Acetylene Black, Vulcan, Ketjenblack and Black Pearls [250–252]. Good results have been reported with Acetylene Black and Vulcan which have moderate pore sizes [251].

The main suppliers of GDM are Toray, E-TEK, Spectracorp and SGL Carbon. Table 5.9 gives an overview of commercially available materials and specifications supplied by the manufacturers. Most manufacturers offer a wider range of GDMs than shown here where we have limited the selection to the most representative ones. Toray produces carbon fibre paper with various thicknesses, without an MPL. Spectracorp also supplies carbon fibre papers. A GDM consisting of wet-proofed Toray or Spectracorp paper and an MPL is supplied by E-TEK (not listed). E-TEK also supplies carbon cloth covered with an MPL on one or both sides of the cloth. The data in Table 5.9 refer to the new type of cloth introduced by E-TEK in 2004 to

Table 5.9 Specifications of commercial GDM

Manufacturer			Toray	Spectracorp	E-TEK GDL	E-TEK GDL	Sigracet	Sigracet	Sigracet	Sigracet
Code			TGP-H-30 TGP-H-60 TGP-H-90 TGP-H-120	2050 HF	LT 1400-W	LT 2500-W	10 BA	10 BB	30 BA	31 BC
Fabric			Paper	Paper	Woven	Woven	Dry laid	Wet laid	Wet laid	Wet laid
Thickness		mm	0.11–0.37	0.26	0.4	0.43	0.38	0.42	0.31	0.34
Gaspermeability Gurley		cm^3/($s\ cm^2$)	789–141[a]	114	0.9	0.8	90	3	40	2
Electrical resistivity	In plane	mΩ cm	4.7–5.8	14						
	Through plane		80	90	500	550	<260	<360	<320	<440
Thermal conductivity	In plane	W/(m K)	21–23							
	Through plane		1.7							
Porosity		%	78–80				88	84	81	82
	Bubble point	μm			30	15				
	Mean pore size			38	7	3				
Microporous layer			No	No	Yes	Yes	No	Yes	No	Yes
	Thickness	μm			30	30				

[a] Depending on the thickness.

replace the older one that had been on the market for more than 10 years. Carbon cloth and carbon fibre paper are non-isotropic materials since the fibres are usually aligned in the plane parallel to the membrane. SGL manufactures a dry-laid paper filled with carbon (Sigracet 10 series), which is claimed to have a true 3-dimensionally (3D) isotropic structure. For a PEMFC, however, they recommend the Sigracet 30/31 series based on a wet-laid carbon paper, which has a 2D structure like the fibre papers and cloths. The Sigracet materials listed here are all wet-proofed with 5 wt% PTFE. W.L. Gore and Associates used to supply its own type of cloth-based GDM named Carbel CL, but this seems now to be only available in ready-made MEAs.

5.4.4.2 Physical Properties of the GDM

As already remarked by Mathias [249] the relation between GDMs and the performance of the PEMFC has received relatively little attention in the literature. This means few quantitative data is available although recent studies by Ihonen *et al.* [253] and Williams *et al.* [254] have already made a significant contribution to filling this gap. Here we will subsequently discuss the gas transport, transport of heat and electrons, mechanical stability and water management, mostly based on the data from these papers.

Transport of reactant gases The transport of gases takes place both through the plane and in the plane. A thicker GDM will increase the through-plane transport losses but increase the in-plane uniformity of the gas distribution. Gas transport is especially critical at the cathode where it determines the limiting current density, i.e. the current density where the oxygen partial pressure at the active layer becomes zero due to the finite transport rate. At the anode the partial hydrogen pressure is generally higher and the hydrogen diffusion is faster.

Williams [254] reported for GDMs consisting of an MPL and different GDLs (Toray paper, E-TEK-cloth, Sigracet) a good correlation between the through-plane air permeability and the limiting current density. At atmospheric conditions, 80°C and 75% RH the limiting current densities were between 1.7 and 2.2 A/cm^2. Notice that these data were obtained at conditions where flooding is not expected but that transport limitations in the active layer were also included. Table 5.9 shows that the presence of an MPL decreases the through-plane permeability by more than an order of magnitude. This seems to outweigh the benefits that this layer has for the water management, as will be explained below. The air permeability data in Table 5.9 for the systems without MPL correspond to values for the Darcy coefficient at atmospheric conditions in the order of $10^{-12} m^2$. Ihonen [253] calculated that the MPLs have a Darcy coefficient in the order of $10^{-14} m^2$.

To obtain uniformity of the gas composition under both the channel and the rib of the flow plate, a good in-plane permeability is required. Here the value for the macroporous GDL is relevant as the GDL is much thicker than the MPL. Measurements by Ihonen *et al.* [253] showed that for material like Sigracet 10 BA the in-plane permeability was comparable to the through-plane permeability as was to be expected for a 3D isotropic material without an MPL. But also for Toray paper it was estimated that both are similar [249]. Measurements by Ihonen showed that for the in-plane permeability of E-TEK cloth (the older version) was about 60 times lower than the value for Sigracet [253].

The correlation between the permeability and the limiting current density may be somewhat surprising as the oxygen transport is mostly diffusion driven and not pressure driven, except with interdigitated flow fields. In this respect porosity and tortuosity would be more relevant as they affect the effective diffusion constant directly. However, Williams did not obtain a correlation between the limiting current density and the porosity (which was similarly high for all materials). The permeability, however, reflects also the pore size. In the GDL hydrophilic pores can exist, the smaller of which can get flooded already at sub-saturated conditions. In addition, Knudsen diffusion will limit the diffusion of gases in pores with size smaller than 7 μm [254].

The above makes clear that it is not easy to establish a quantitative requirement for the permeability. Moreover, in a PEMFC the permeability values will decrease with clamping pressure, see also below. However, it is generally assumed that in the absence of flooding the permeability of these materials is sufficient for the transport of reactant gases [249,253,254].

Transport of electrons and heat The transport of electrons and of heat occurs through the carbon fibres and therefore these two phenomena are discussed here together. Like for the gas diffusion a thicker GDM makes the transport through-plane slower, but results in a more uniform in-plane distribution of temperature and potential. Due to the 2D structure of most materials the in-plane heat and electrical conductivity are substantially higher than the through-plane values, i.e. they differ in by an order of magnitude. For commonly used flow plates and GDL thickness this results in comparable in-plane and through-plane losses as shown by Mathias [249]. With Toray paper the total ohmic losses are small order of 2–4 mV [249]. However, as shown in Table 5.9 and in the study by Williams [254] materials not based on carbon fibre paper show higher resistivities and correspondingly larger losses.

It should be noted that contact resistances at the interface between GDL and bipolar plates can give rise to comparable losses. The MPL has a minor effect on the through-plane conductivity as it forms only a thin layer. The effect is usually to increase the resistance which is somewhat surprising as the MPL is much denser and should result in lower contact resistances [249,255]. The presence of the isolating PTFE is probably responsible for this effect. In the past it was also shown that coating or wet-proofing the material with

PTFE resulted in lower conductivity, probably also due to increased contact resistances [256].

Data on the heat conductivity are only supplied by Toray. As expected these data show a larger value for in-plane conductivity than for through-plane conductivity. Temperature measurements with working cells are not easy. Although there are no reasons to assume that the thermal conductivity seriously limits the PEMFC performance this area needs more research in order to obtain quantitatively reliable data [249,253]. In the paper by Ihonen it is suggested that thermal impedances affect the water management.

Mechanical stability All GDM are compressed when a pressure is applied, at 2.75 bar the paper strain is about 24% and the cloth strain 52% [249]. This has a direct effect on the permeability, which decreases with clamping pressure and faster for cloth than for paper. The electric and heat conductivity will increase and contact resistances will be less. Ihonen [253] found that for an open material like Sigracet 10 high clamping pressures resulted in better performance up to 30 bar whereas at for Sigracet 30, 10 bar was optimal. For cloth-based materials the performance dropped at 5 bar or lower. The clamping pressure also affects the flooding discussed below.

Also of importance is the flexural behaviour. Carbon cloth bends easily and can intrude into the gas channels, increasing the pressure drop along the channels. On the other hand it is more effective in compensating any unevenness in flow fields and membranes. Moreover, this material is easier to handle and therefore easier to use in manufacturing processes than stiffer papers and Sigracet materials [249].

Water management Maybe the most critical process in the GDM is the water management. As said in the introduction product water must be removed without clogging all gas pores and at the same time drying-out of the membrane and active layers must be prevented. The MEA seems to operate best when the net water flux through the membrane is close to zero [255,257]. The water management involves all components in the cell, as the liquid water phase, and the liquid water pressure are continuous in the cell [258,259]. It is therefore very difficult to single out one component of the cell as being the most important. However, the MPL seems to have large effect on the water management.

Already in 1995 Wilson pointed out that a microporous hydropohobic layer might prevent flooding in PEMFCs [260]. Passalacqua *et al.* showed that the insertion of a thin layer of carbon black mixed with PTFE between the GDL and the active layer substantially improved the performance at higher current density [261]. Further research has corroborated this result [262,263]. It has been found that the MPL reduces differences obtained with different carbon papers [264]. The pore-size distribution is more important than the porosity itself [250,265]. The amount of PTFE is critical where it affects this pore size and only a minimal amount is required for hydrophobicity [266]. The pore size distribution is determined by the PTFE content but most of all by the type of carbon black used. Passalacqua *et al.* [251] found that materials with a large volume of small pores performed best, e.g. acytelene black and Vulcan XC-72.

The functioning of the MPL at the cathode was recently mathematically described by Weber and Newman [255] and Pasaogullari *et al.* [267]. The main effect of the MPL is that it increases the back flow of water to the anode, and on the other hand affects the saturation level of the GDM near the active layer. Under dry conditions the level at the interface can be higher than in the GDL and gas channel, i.e. keeping water in the system, and under high humidity the saturation level is lower than in the GDL and gas channel. This mechanism functions best if there is a large difference in permeability of the two layers, but the requirement for gas transport puts a lower limit to the pore size and the permeability. It seems further understanding of the functioning of this layer will result in improvements of the structure. Although the MPL at the cathode will be the most important, usually one is applied at the anode as well where it will have a similar function of keeping the active layer drier. E-TEK recommends for very wet conditions a second MPL at the gas channel side.

As mentioned above the gas permeability of the GDM is sufficient at dry conditions but to have the same situation for wet or partially flooded conditions a good permeability must be maintained. SGL Carbon therefore recommend their more open materials for use at wet conditions. Wet-proofing the GDL beyond a first layer to make the surface sufficiently hydrophobic usually does not pay-off as the extra PTFE reduces the permeability [249, 263]. It was found that a combination of hydrophilic pores to carry the liquid water and smaller pores to transport gases is beneficial and can indeed be found in the commercial materials [263,268]. Also the effect of the clamping pressure on the permeability affects the flooding and has to be matched with the materials and conditions [253,269]. The interplay between thermal properties of the GDM and the MEA performance still needs to be clarified.

5.4.4.3 Stability and High-Temperature Aspects

Few data seem to be available on the stability of GDMs. However, a gradual loss of hydrophobicity has been reported [270]. The PTFE is high-temperature resistant and a mechanical removal of PTFE seems to be the main cause. Contaminants may also enhance this degradation [229]. Unlike the carbon black in the active layer the GDM is under normal conditions not in intimate contact with the acid electrolyte, nor is Pt available to enhance carbon corrosion reactions.

For high-temperature operation it seems that electrical conductivity and thermal conductivity will be worse, and need to be critically evaluated. At the cathode it must be expected that the oxygen partial pressures will be much lower at 120°C than at 80°C. However, flooding will not occur.

E-TEK already offers a material to be used at temperature >95°C. The specifications, however, for low-temperature and high-temperature material do not reveal which modifications have been made.

Williams investigated the performance of GDMs at several conditions, i.e. 80°C 75% RH, 100°C 70% RH and 120°C 35% RH, with corresponding inlet partial oxygen pressure of 0.013, 0.064 and 0.064 bar, respectively. The limiting current density decreased with about 300 mA/cm^2 due to reduction of partial oxygen pressure but even more at elevated temperature, which was ascribed to additional transport losses in the cathode active layer [254].

For the PBI/PA-based MEAs similar materials have been used so far as for the low-temperature systems. Notice that in the PAFC systems carbon fibre paper coated with PTFE is used. High PTFE loading is needed to keep the liquid PAFC out of the GDM in that case. For PBI/PA this creep of PA is much reduced.

5.4.5 Conclusions Regarding the Electrodes

- With present-day commercial Pt catalysts on high SA carbon supports the high automotive requirements regarding low Pt loadings can maybe just be met. For the cathode this requires minimization of transport losses. These should come from further structure improvement.
- The stability of Pt at elevated temperatures and high cathode potentials is a real concern and should be further investigated.
- For the oxygen reduction reaction humidified Nafion seems to be the most favourable electrolyte. However, operation at higher temperatures will result in lower performances due to the decreased humidity and decreased partial oxygen pressure.
- The use of Pt alloys, i.e. PtCo, can improve the performance at lower Pt loading. The beneficial effect at higher temperatures is not certain. In addition, stable, probably well-ordered alloys must be made for the required stability. But such alloys can be more stable than Pt.
- For the anode PtRu is the preferred catalyst material up to 120°C for operation with reformate gas. At more elevated temperatures Pt can be used.
- The GDM have to be optimized for operation at drier conditions to keep the active layers and membrane well humidified. This can be done by adjusting the microprous layer. Moreover, transport of heat, electrons and oxygen becomes more critical.

5.5 Bipolar Plates

5.5.1 Introduction

The bipolar plate, or separator plate in a PEM fuel cell stack connects two adjacent cells. It often consists of two flow plates, one for the cathode and one for the anode side, which do not need to be of the same material. A flow plate has a number of functions:

- Provide room for the flow pattern to obtain an even distribution and removal of reactants and products.
- Provide room for manifolding of reactants and cooling fluid.
- Conduct electrical current from cathode to anode.
- Separate electrode compartment from cooling compartment.
- Provide room for cooling flow field.
- Give a stack solidity.

At the cathode side the flow plate is located in a highly oxidizing environment, being polarized to a maximum of about +1 V (versus the normal hydrogen electrode) while being exposed to air or oxygen. At the anode side the flow plate is constantly exposed to hydrogen. In addition the flow plates are in constant contact with highly acidic solutions (pH less than 0 when in direct contact with the membrane) containing F^-, SO_4^{2-}, SO_3^-, HSO_4^-, CO_3^{2-}, and HCO_3^- for Nafion-based PEM fuel cells, and PO_4^{3-} for PBI-based PEM fuel cells. Hence, flow plate materials must be resistant to acids, oxidation and hydrogen embrittlement in the fuel cell environment.

Flow plates make up for the major part of the weight, in the order of 80–90% [271] and volume of a fuel cell. So they are for a major part determining the power density of a PEM fuel cell stack.

The functional aspects can be translated to the following functional requirements for a flow plate:

Requirement	Specification
• Gas tightness	Close to 100% in order to minimize the gas leakage.
• Deformability	None, unless functional.
• Thickness	As thin as possible. If integrated with both flow fields, then depending on flow-pattern dimensions, and cooling flow-field dimensions, otherwise, depending on material properties: strength, density.
• Conductivity	The resistance of the flow plate including the flow fields should be less than $20\,m\Omega\,cm^2$.
• Thermal conductivity	Should be high in order to limit the temperature difference between the cell side and cooling side to less than 1°C

Furthermore the material and the manufacturing method should be low cost.

With respect to the functional requirements, and the environmental conditions, materials suitable for use in PEMFC should have the following material properties [272]:

- Electrical conductivity $\geqslant 10\,S/cm^1$.
- Thermal conductivity $\geqslant 20\,W/m/K^1$.
- Permeability of $H_2 \leqslant 2 \times 10^{-6}\,cm^3/s^1/cm^2$.
- Resistance to corrosion in acid and, at the same time, exposure to O_2, heat and humidity; this translates to:
  - Low total contact resistance: target value less than $50\,m\Omega\,cm^2$. The contact resistance is a function of the compression force, which can vary from 20 to $140\,N/cm^2$. A high compression force yields a low contact resistance but on the other hand results in a deformation of the GDL.
  - Low release of MEA-contaminating ions: target value less than $8 \times 10^7\,mol/cm^2$ during 5000 h for a Nafion 117 membrane.

Presently several types of materials are being used in flow plates. The main materials are according to Middelman *et al.* [273] and the review articles by Mehta *et al.* [274], Brett *et al.* [275] and Hermann *et al.* [276]:

- Electrographite
- C–C composite
- Expanded graphite foil
- Graphite polymer composite
- Metal.

Each material type will be discussed; in addition an outlook will be given with respect to their potential for use in a high-temperature PEM fuel cell, operating in the temperature range of 120°C to approximately 180°C.

5.5.2 Graphite

Graphite has a high conductivity, is chemically stable in the PEM fuel cell environment, and can endure high temperatures making it suitable for the high-temperature PEMFC. However, it is brittle and lacks mechanical strength, and is also relatively permeable to hydrogen, which results in a minimal plate thickness of 5–6 mm. Machining is the only suitable production technique, which is costly and time consuming. The material is available from, among others, POCO Graphite (USA), and SGL Carbon (Germany).

In order to make the material gas-impermeable POCO impregnates the bulk material with a resin. This, however, reduces the maximum service temperature to 150°C [277].

A post-machining method to render a graphite flow plate gas-impermeable is through pyrolytic impregnation, sealing the pores in the top layer of the plate [278].

Although graphite flow plates are expensive, for historic reasons and the fact that they have a low contact resistance, a high conductivity and a high durability, they are the benchmark to which all other flow plate materials are to be compared to.

5.5.3 C–C Composite

A C–C composite is a porous structure made of carbon and carbon fibres. The manufacturing method to make plates from C–C composite material has been developed by Oak Ridge National Laboratory [279]. The process consists of two steps. In the first step, a slurry containing carbon fibres with dimensions $400 \times 10\,\mu m$ suspended in water containing a phenolic resin is vacuum moulded to form a pre-form, which is subsequently heated for a few minutes to 150°C. The phenolic resin serves as a binder providing strength and stability. In a second step, a flow field and other features can be pressed, stamped or machined into the pre-form. In this stage the component has an open porosity. In order to make the component gas-impermeable it is subjected to a chemical vapour infiltration (CVI) process. During this process the component is exposed to argon/methane at a temperature of 1500°C for about 4 h and carbon is deposited, sealing the surface. As a by-effect the phenolic resin present in the component is pyrolyzed. So a fully C–C composite flow plate is produced, which can also be used in high-temperature PEM fuel cells.

Due to its porous nature the density of the material is low, in the order of 1 g/cm^3. The bulk conductivity is about 1/3 of that of POCO graphite, 200–300 S/cm.

Porvair licensed the technology from ORNL and further improved the manufacturing process [280], yielding a density in the order of 1.2 g/cm^3 and an improved bulk conductivity of 500–700 S/cm. The minimum web thickness is less than 0.4 mm, which can lead to thin and lightweight flow plates. During the manufacturing process shrinkage of the pre-form occurs, which is a source of dimensional variability. Improvements in the process are being made by Porvair in order to minimize the tolerances [281].

Estimates by ORNL suggest a total cost of one plate of 500 cm^2 of \$2/plate for a high volume production, i.e. approximately 10–20 \$/kW.

5.5.4 Expanded Graphite Foil

Expanded graphite foil is a porous foil made of high surface area graphite particles. In a high-temperature process graphite is processed into expanded natural graphite flakes, which have a high surface area. Subsequently these flakes are locked together into a foil in a calendering process. Thickness and porosity depend on the calendering process parameters. The porous flexible foils are generally used as gasket in automotive and high-temperature applications. The porous foils can also serve as a component forming the flow field [282,283].

For use as PEMFC flow plates the foils need to be gas-impermeable which is achieved by impregnating the porous flexible foil with a resin. SGL Technologies provides blank impregnated plates. Graftech has developed a production process in which prior to calendering a resin is mixed with the exfoliated flakes. Subsequently via compression moulding the plate is structured and then cured [284]. Alternatively the foils can also be structured via embossing [285]. Flow plates can be as thin as 0.6–1.3 mm. Grafcell flow plates are being used by Ballard Power Systems in their Mark 902 stack series.

5.5.5 Graphite Composite

Graphite composite material is a mixture of different types of graphite and carbon powders, fibres and a polymeric resin. The carbon to polymer ratio is normally in the order of 80%/20%, depending on type of resin. The material properties depend, amongst others, on the type of polymer resin and on the relative amounts of graphite powder and polymer resin used. More graphite results in higher electric conductivity, while increasing the polymer resin percentage will result in higher mechanical strength. The polymeric resin is either a thermoplastic or a thermoset material.

5.5.5.1 Thermoplastic Materials

The advantages of a thermoplastic material are the low cost, high availability, high chemical resistance, good mechanical properties and gas-impermeability [286] and short cycle times in an injection moulding process [286,287]. Possible thermoplastic materials contain polypropylene (PP), polyethylene (PE), or polyvinylidinedifluoride (PVDF). SGL Technologies GmbH for instance uses polypropylene for its PPG86 material, and PVDF for its BMA5 material. Other producers of flow plates using a thermoplast as polymeric binder are Nedstack [273], and ZBT [286]. BASF [288] uses a mixture of two polymers, a polyamide and polyethersulfone as binder. A major disadvantage of thermoplastic-based graphite composites is their low maximum application temperature, less than 80°C for PP-based material and less than 120°C for PVDF-based composite [289]. This excludes these materials from application in a high-temperature PEMFC. Another weak point is the limited creep resistance.

DuPont [290] uses a liquid crystal polymer (LCP) as the polymeric binder. This material has an excellent high-temperature resistance, making it suitable for application in a high-temperature PEMFC, >120°C.

5.5.5.2 Thermosetting Materials

Possible thermoset resins include phenolic, epoxies, polyesters and vinyl esters. Advantages of thermoset polymers are the good chemical resistance, the creep resistance and low cost for phenolic and epoxy resins [287]. The materials have a high maximum application temperature. On the other hand cycle times can be rather high.

Los Alamos National Laboratory has developed a mouldable compound using a vinyl ester as binder [291]. The process is exclusively licensed to Bulk Moulding Compounds (BMC). A disadvantage of this process route is the release of volatiles during the manufacturing process. Schunk Kohlenstoftechnik GmbH uses an epoxy as binder [287], whereas SGL Technologies GmbH [292] has developed a phenolic compound, Sigracet BBP4, which can either be compression moulded or injection moulded.

5.5.5.3 Manufacturing Techniques

For the production of prototype flow plates commonly computer numerical control (CNC) milling is used to shape a blank plate. For two-sided structured plates the machining time can go up to 1 h. Tool wear due to the polymeric binder is an important issue when machining graphite composite material. This makes machining an expensive production method. Alternatively Morgan Fuel Cell has developed a technique called ElectroEtch, which uses a high-precision grit blasting technique. It reduces the production time drastically to a few minutes.

For production of high volumes either compression moulding or injection moulding can be used. Injection moulding is the best-suited process for thermoplastics, and LCP-based compounds. Granules of pre-mixed graphite polymer compound are fed into a heated chamber, where the compound melts. Subsequently the compound is injected into the not-heated mould. Through one or more openings the nest is filled with the compound. The compound solidifies during cooling down and the plate is formed. Very short cycle times can be achieved, as low as 30 s [286]. During cooling down a polymer-rich skin forms on the surface of the component. This can be removed by abrasive blasting [292] or by stepwise abrasion while measuring the resistance [293].

Compression moulding is best suited for thermoset-based compounds. A heated mould is filled the compound, closed and compressed for a certain time for curing the binder. After curing the plates are removed from the hot mould. When thermoplastic compounds are compression moulded the mould has to cool down, resulting in high cycle times. Nedstack [273] has developed a process in which a pre-form of its thermoplastic compound is heated and then placed in a cold mould. Cycle times of 10 s are claimed.

Graphite composite plates can be manufactured with a high accuracy, better than 0.05 mm [292]. A drawback, however, is the needed minimal plate thickness of 0.5–1 mm in order to provide sufficient strength and gas-impermeability. This leads to an effective flow plate thickness of about 2 mm, including the flow fields.

The endurance of the material is considered to be high. Most materials have a through-plane conductivity better than 10 S/cm, and a thermal conductivity close or better than 20 $W/m^1/K^1$. Price estimates for injection moulded [286] or compression moulded [273] plates are in the order of 10 €/kW for high-volume production.

5.5.6 Metal-Based Bipolar Plates

Metal-based flow plates can be made from thin sheet material as thin as 0.1 mm, yielding an effective flow plate thickness of 0.6 mm including the space needed for the gas and cooling channels. Metal flow plates are superior to other types with respect to mechanical strength and ductility. Due to the thin base material light and compact plates can be made, when compared to graphite, and the composites. Hence, metal separators can significantly reduce volume and weight of stacks and therefore result in stacks with a high-power density, 2.5 kW/l as reported by Intelligent Energy [294]. Moreover, the higher ductility of metal separators compared to carbon-based separators has additional advantages for the automotive end-use (potentially the biggest market for the PEMFC) with respect to shock absorption, vibration and stack-life.

However, the chemical stability is limited in the fuel cell environment, especially in contact with the acidic electrolytic membrane. Corrosion of the metal flow plate leads to a release of multi-valent cations, which can both lead to an increase in membrane resistance [51] and to poisoning of the electrode catalyst. Direct contact between the metal and the membrane can, of course, be avoided by using a coating or a gasket at the edges of the cell. The formation of a passivating layer on the surface, although protecting the bulk material from continued corrosion, results in an increasing contact resistance.

Materials that are used in a PEMFC include aluminium, titanium and stainless steel, with or without a protective coating.

Aluminium plates can be manufactured by several techniques such as machining, casting and etching. Aluminium has a stable oxide layer, which results in a high corrosion resistance, but also a very high contact resistance. Therefore, aluminium flow plate should be coated with a suitable material to improve contact resistance and maintain corrosion resistance. This coating must be 100% and flawless. A direct contact with the membrane should be avoided. DeNora (now Nuvera) [295] claims the use of aluminium.

Titanium has also an excellent corrosion resistance resulting from a stable oxide layer, but it is more expensive. An advantage of titanium is it can be diffusion bonded to make a flow field from thin foils. It is being used in a reversible PEMFC by ECN [296].

5.5.6.1 Corrosion and Passivation Properties of Stainless Steels

The corrosion behaviour of stainless steels is a function of the composition, environment and potential. A typical electrochemical polarization curve for an iron-based stainless steel in an electrolyte is shown in Figure 5.22.

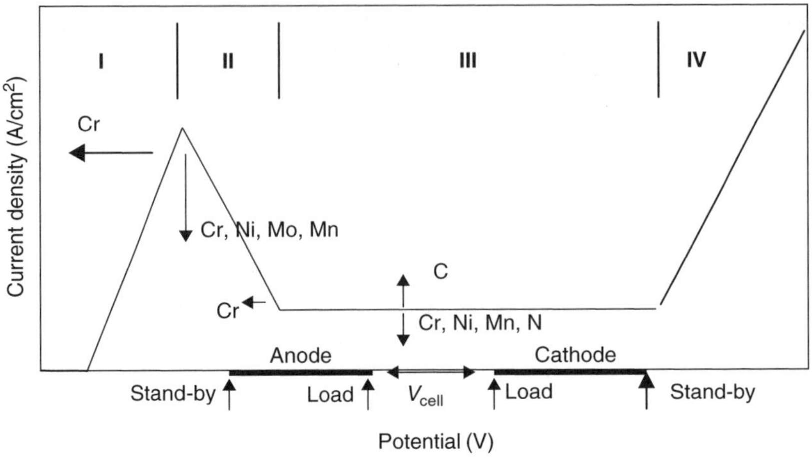

Figure 5.22 A schematic polarization curve of a metal in an electrolyte, based on Ref. [297].

Four distinct regions can be seen:

I Active region: In this potential range metal dissolves into the electrolyte, leaving a bare metal surface.
II Active/passive transition region: The current reduces and a built-up of a oxide layer starts.
III Passive region: A closed oxide layer is formed, the current is constant. The layer dissolves slowly via a chemical reaction or by transport through the oxide film. The current depends on the properties of the oxide layer, in particular the conductivity.
IV Transpassive region: The metal is oxidized to a higher valency (Cr(III) to Cr(VI)) and dissolves.

In a fuel cell the MEAs are in good electrical contact with the flow plates. Electrode, GDL and separator will have an equal, or almost equal potential.

Stainless steel at the cathode The potential of the cathode ranges from approximately 1000 mV (versus NHE) in stand-by mode (no current, fuel fed to anode, air to cathode) down to 600–700 mV in current producing conditions. For most stainless steels both potentials lie within the passive region, as is depicted in Figure 5.22. However, for some alloys depending on the electrolyte the stand-by potential might be in the transpassive region.

In general, the operational potential window of the cathode lies within the passive region. This means that a closed oxide layer is formed on top of the stainless steel. Initially the thickness of the layer increases. This increase is diminished by a slow dissolution of the oxide layer into the water present in the cathode compartment. Eventually equilibrium is established between the growth of the oxide layer and the dissolution rate, resulting in a constant thickness and a constant corrosion current. The dissolution rate is dictated by the local properties of the water. The oxide layer growth and the associated corrosion current depend on the physical and chemical properties of the oxide layer, such as the conductivity of the layer, the diffusion coefficients of metal and oxide ions. A low conductivity results in a low corrosion current, but this brings about a high contact resistance between flow plate and the adjacent gas diffusion component, which has a negative influence on the performance of the fuel cell. Therefore, the passive layer should have good conducting properties and low diffusion rates. These physical properties depend on the composition of the alloy, pre-treatment and environment.

Stainless steel at the anode The potential of the anode ranges from approximately 0 mV in stand-by mode (no current, fuel fed to anode, air to cathode) up to 100–200 mV in current producing conditions, as is depicted in Figure 5.22.

The active and passive regions of stainless steels coincide with this potential range, depending on electrolyte. The stand-by condition is the worst condition. During operation an oxide layer will be built-up which subsequently dissolves during stand-by. Due to the fact that no passive layer is built-up, the contact resistance will remain low.

Alloy elements can affect the polarization behaviour of an iron-based stainless steel. The typical effects of elements alloyed in the stainless steel are also shown in Figure 5.22.

5.5.6.2 Stainless Steel: Function of Various Elements

Chromium Chromium is essential in the formation of the passive film. Chromium shifts the active passive/transition potential to lower values as the oxidation potential is lower than that of iron. The formation of the passive layer will decrease the corrosion current.

The content can vary from 10% to 20% for austenitic stainless steels and up to 29% for ferritic stainless steels. Higher chromium content lowers the mechanical properties of the material, the manufacturability and weldability.

Nickel Nickel is an element that stabilizes the austenitic structure, which is beneficial for the mechanical properties of the material, and the manufacturability. In a reducing environment, such as at the anode, it can promote repassivation. Typical nickel content ranges from 10% to 30%.

Manganese Manganese has a similar effect as nickel. In addition it can react with sulphur present in stainless steels to form manganese sulphides. These sulphides are beneficial for increasing the pitting corrosion resistance.

Molybdenum Molybdenum stabilizes the passive layer in combination with chromium. It increases considerably the pitting corrosion resistance.

Carbon Carbon has a negative influence on the corrosion resistance.

Nitrogen Nitrogen enhances the pitting corrosion resistance. Furthermore it strengthens the steel.

Titanium and niobium These elements stabilize the steel by forming carbides, thereby preventing chromium carbides to form.

5.5.6.3 Suitability of Stainless Steels for Application in PEMFC

Stainless steels that are researched in the PEMFC environment are typically high in nickel and chromium content. ECN was one of the first to explore the possibility of using stainless steel as flow plate material [298] besides Los Alamos National Laboratory [299], followed by Siemens [300]. Others, including ECN, further continued these investigations on different types of stainless steel. Makkus *et al.* [301] of ECN have tested various stainless steel types directly in single-cell PEMFC tests. The following conclusions were drawn:

1. The anode environment is more aggressive for the separator than the cathode environment. At the cathode side, the passivating oxide layer formed on the surface protects the underlying steel. At the anode side, however, no protective layer is formed due to the hydrogen-containing atmosphere and the anode potential. This is in accordance with the basic theory. Li *et al.* [302] confirmed the observations by showing a lack of passivation with SS316 in a simulated anode environment.
2. Direct contact between steel and the membrane should be avoided, e.g. by smart gasket design. The acidic membrane is simply too aggressive for the steel.
3. The compaction pressure is of considerable influence on the contact resistance. Higher pressure means lower resistance.
4. The nature of the surface is of influence on the corrosion behaviour and the contact resistance.

Davies *et al.* [303] have shown that there is a general trend whereby the contact resistance of stainless steels decreases with increasing nickel and chromium content, being the lowest for the 904 L. Wang *et al.* [304] have measured the interfacial contact resistance between different types of stainless steel under air at different compaction force. They found the resistance to decrease with increasing chromium content, in the order 316 L, 317 L, 904 L, 349. The polarization behaviour of these stainless steel in a simulated PEMFC environment showed a similar trend, the alloy with the highest chromium content, 349, had the best behaviour in both anode and cathode environment. Using a slightly different technique Kim *et al.* [305] add that also the Mo content plays a role in the passivation behaviour.

Concluding it can be stated that bulk properties are important with respect to the behaviour of a stainless steel in the PEMFC environment, although this behaviour is influenced by the surface properties. A pre-treatment affects the surface properties as Makkus *et al.* have shown [301]. Despite the good results with alloy B (SS904L, the use of which is patented by Siemens [306]) in their investigation in a long-term test an increasing degradation observed after 2200 h of operation.

So it can be concluded that for a long endurance and low contact resistance a coating would be best.

Possible coatings are carbon-based, metal-based or resulting from a surface treatment. Carbon-based coatings include: graphite, conductive polymer, diamond-like carbon, organic self-assembled monopolymers. Metal-based coatings include: noble metals, alloys, metal nitrides and metal carbides.

Apart from the durability issue, a critical design criterion for coated separators is the difference between the coefficient of thermal expansion between the coating and the substrate. As the PEMFC stack is subjected to temperature cycles, the thermal cycling/fatigue behaviour of the coating/substrate assembly might be at least as important as the protective properties of the coating. Also the mechanical properties are important if the coating is applied prior to the forming process of the flow plate.

Scientists from the INRS, Canada, have experimented with directly pyrolyzing sprayed high-carbon polymer layers on SS316 [307]. These layers are sprayed from solutions of exotic aromatic/high carbon polymers that have in fact benzene as the main building block. Good films with good adhesion and ultra-low contact resistance are obtained but the coatings are not pinhole-free. Therefore a commercial carbon spray (TV-Koat: graphite particles and a binding polymer), which is conductive by itself, is applied as a second layer. Pyrolyzing TV-Koat alone produces no adhesion to the metal. On the other hand, pyrolyzed TV-Koat adheres very well on the pyrolyzed high carbon polymer coat but its surface is rather fragile. A third top layer of the high carbon polymer coat solves this problem and results in a coating without pinholes.

Fuel cell tests have demonstrated that the triple-layers coating works, but long-term stability (>1000 h) still has to be demonstrated. The total coating thickness is in the order of 100 μm which is rather thick compared substrate having the same thickness. Joseph *et al.* [308] have investigated conducting polymers as polyanilin and polypyrrole, which were deposited using cycling voltammetry. The coatings can show pinholes and more importantly the contact resistance is far too high. H. J. Davis of Avantcell Technologies Inc. has developed a double layer coating consisting of a noble metal intermediate layer and a conductive polymer top layer, comprising carbon or noble metal particles as conductive fillers [309].

ICI, later Ineos Chlor, has developed a coating FC5, which was tested on titanium by Adcock [310]. It is available under the trade name PEMCOAT. ICI patented an electrocatalytically active coating to be used on titanium and stainless steels [311]. Ineos Chlor [312] has furthermore developed a treatment for stainless which results in an increase of the chromium content in the surface layer, thereby reducing the interfacial resistance. In addition treated stainless steels can also be coated with the electrocatalytically active coating disclosed in the prior patent [311]. General Motors [313,314] has developed a conductive polymer coating consisting of a polymer filled conductive fibres, such as carbon, graphite, Au or others, on steel. Similar polarization curves as with graphite separators are obtained, indicating that at least the contact resistance is acceptably low with such a coating. But no data on durability and lifetime are presented. Wind *et al.* from Dornier [315] have tested several metallic coatings among which gold on 316L, that yield a performance similar to a cell with graphite flow plates. Dornier [316] has patented a low-ohmic chemically stable coating, which is a multi-phase coating, at least in the contacting outer surface with the MEA. Multi-phase means: an alloy of a metallic phase (consisting preferably of one of the elements Cr, Ti, Mo, W, V) and a nitride or carbide. This alloy has low contact resistance (due to the metal) and high corrosion resistance (due to the carbides/nitrides). The durability of the coating is further enhanced by creating a gradient in the composition of the coating (more metal at the MEA contact surface, more binder in the layers beneath). Extreme thin coatings in the order of magnitude of 1 μm are claimed without compromising on corrosion resistance (several months in 0.1 N H_2SO_4) and conductivity. Because passivation reactions are (almost) absent the contact resistance, under extreme corrosion conditions, remains low as well. No results in an actual PEMFC environment are shown. Delphi Technologies Inc. has claimed [317] the use of doped Y, La, Ti, Cr, Cu, Sn and Ni which form highly conductive oxide passivation layers. GM claims the use of a fluorine doped tin oxide [318]. The coating is deposited on the substrate via a physical vapour deposition (PVD) process. Very thin layers in the order of 5–100 nm can be applied. The coating yields a low contact resistance and has shown to have a low corrosion current. Scanning electron microscopy (SEM) micrographs revealed the coating to be very dense without observable defects. Delphi and GM only show *ex situ* test results.

Tawfik and Hung have [319] patented a coating made by high-velocity oxyfuel (HVOF) spraying of Ni-based, Cr-based and/or carbide-based alloy powders. The resulting coating has a thickness of 200–250 μm. Cycle times are expected to be very high. Cladding of a material, titanium [320] or niobium [321], seems to be a possible option in terms of endurance and cost.

A number of companies (GM, Honda) [322–325] have claimed the use of noble metal coatings on stainless via various application techniques. As with other materials the coating needs to be flawless. Otherwise the less noble base material is exposed to acidic water acting as an electrolyte and a local element can be formed resulting in the oxidation and dissolution of base material constituents.

Lee *et al.* [326] have shown that an electrochemical surface treatment of 316 L results in a 2-fold increase of the chromium content in the surface layer. Due to the treatment the corrosion behaviour is improved as well as the contact resistance. The treatment has the same effect as the method invented by Ineos Chlor [312]. Oak Ridge National Laboratory has developed a thermal nitridation technique to protect flow plates [327]. The process works well for a model alloy Ni-50Cr yielding a low contact resistance and corrosion current but nitrided SS349 showed unacceptable high corrosion current densities.

Many options for coating stainless steel exist. However, when the coating is applied prior shaping the flow plate a post-processing step is required in order to protect uncoated edges.

Metallic flow plates can be made via several production routes, depending on plate thickness and needed volumes. For thick plates and low volume production machining and etching are options. For large volume production and thin plates different types of forming methods can be used among which: stamping, embossing and hydroforming.

5.5.7 Suitability of Metal Plates for the PEMFC at 100°C and Higher

Metallic flow plates are being applied in high-temperature fuel cells. In the so-called low-temperature solid oxide fuel cell, operating at 700°C, special alloys are used for the interconnects. In the Molten Carbonate Fuel Cell 310S and 316L are applied at the cathode side, whereas nickel–clad material is used at anode side. In the PAFC, however, metallic flow plates are not applicable due to the presence of a liquid electrolyte (phosphoric acid).

In view of the above it is conceivable that metallic flow plates can be applied in the high-temperature PEMFC, with an operating temperature between 120°C and 180°C, depending on the specific properties of the PEM. In general the corrosion rate will increase if the circumstances are not changed, except for the temperature. But the absence of liquid water during operation might cause the corrosion rate to decrease. Furthermore it is likely that the contact resistance will be lower than at 70°C.

For cells and stacks employing PBI membranes filled with phosphoric acid such as PEMEAS a metallic flow plate is not suited due to presence of the liquid electrolyte which might leach out of the PBI matrix.

5.5.8 Comparison of Bipolar Plate Materials and Conclusions

In Table 5.10, the relevant properties for different bipolar plate materials, are summarized.

Table 5.10 Summary of relevant bipolar plate material properties

	Graphite	C–C composite	Expanded foil	Graphite composite (thermoplast)	Graphite composite (thermoset)	Metal
ρ (g/cm^3)	1.8–2	1.2	1.8	1.8–2	1.8–2	8
σ (S/cm)	500	500	10 (Sigraflex)	18	40	13,000
λ (W/(m K))	80	35	4 (Sigraflex)	14	20	15
E-modulus (GPa)	11		14	17	15	200
Applicable at $T > 100^{\circ}C$	Yes	Yes	Yes	No	Yes	Yes
Advantages	Low contact resistance	Low contact resistance	Low contact resistance Thin	Low contact resistance	Low contact resistance	Very thin High strength
Disadvantages	Brittle	High tolerances Expensive?				High contact resistance Low chemical stability

For a long endurance, such as in stationary applications, the graphite-based composite materials are best suited. For applications where size and weight are more important than endurance, metal plates and expanded foil are an attractive option. If strength is an additional criterion than only metal plates are a viable possibility. In terms of cost all materials, except for pure graphite, have an outlook on 10 €/kW, but these are all extrapolations with large uncertainties and assumptions.

5.6 General Conclusions

The state-of-the-art PEM fuel cells consisting of perfluorosulfonic acid membranes, platinum-based electrodes and graphite(-composite)-based bipolar plates are in a well advanced state of development. Although the main constituents are in use since many years, impressive improvements in performance and lifetime have been realized.

The perfluorosulfonic acid membranes, have been chemically modified, resulting in improved resistance to radical attack and to higher glass transition temperature enabling operation close to 100°C or higher. In addition membranes can be manufactured with reduced thickness for higher power densities. Reinforcements have lead to better mechanical properties, such that freeze–thaw cycles can be survived and lifetime is extended.

Improvements in catalysts and electrodes have diminished the noble metal loading of the fuel cell electrodes, especially when hydrogen is used as fuel. Although platinum is still the main constituent for both the anode and the cathode catalysts, alloying can lead to higher performance. At the anode, PtRu is the optimum choice for operation on reformate. At the cathode, PtCo can lead to lower noble metal loading and improved stability.

Bipolar plates made of graphite composites can be mass-manufactured by either injection moulding or compression moulding, leading to substantial cost reduction of the bipolar plates.

However, the state-of-the-art fuel cell, still does not seem to possess all the properties which are needed for successful introduction in stationary and transport applications. Simpler, more robust systems, capable of operation in a wider range of conditions, require materials which differ much from the above described. The material needing the most radical change is the proton-conducting electrolyte, present in both in the membrane as well in the electrodes. It should be able to operate from room or even sub-zero temperature up to 120°C or above, at low relative humidities. The analysis presented here shows that although progress is being made in that direction on the basis of modification of existing materials, a material that really meets the specifications has not emerged up till now. In the temperature range 100–200°C, operation is possible with H_3PO_4–PBI. However, with this electrolyte operation below 100°C is not possible, high RH no longer is required but instead has become a threat, and power densities are too low for automotive applications. Although many research groups are involved in the quest for an attractive alternative, and many novel electrolytes are in development, they are all in an embryonic state. A next generation of PEM fuel cells, stacks and systems based on completely new electrolytes enabling operation at low relative humidities and elevated temperatures, is not to be expected in the next 5–10 years.

With respect to the electrodes, apart from the electrolyte phase the constituent materials in state-of-the-art PEMFC are still applicable. The high-temperature operation might introduce new concerns with respect to the stability of the noble metal particles. Especially the operation at high cell voltages, attractive for efficiency purposes, could become impossible when long-term stability is required. Research on novel alloys might become necessary, when a new electrolyte has indeed been developed. The gas diffusion media will have to be re-optimized for less humid conditions.

A number of the presently used graphite composite-based bipolar plate materials are expected to be suitable for operation above 120°C. The suitability of metal concepts is unclear, as on the one hand the higher temperature leads to higher corrosion rates, but the absence of liquid water might make life much easier. In any case, H_3PO_4–PBI is most likely incompatible with metal-based bipolar plates.

The first commercial introduction of PEM fuel cells is likely to be based on the components that are at present in use, making use of perfluorosulfonic acid membranes. At least all the limitations imposed by this material are known. The road to a new generation PEM fuel cell, capable of operation at low RH above 100°C is long and uncertain.

REFERENCES

1. F. A. de Bruijn, *Green Chem*, 7 (2005) 132.
2. R. K. Ahluwalia and X. Wang, *J. Power Sources*, 139 (2005) 152.
3. A. Tsukada and P. Rodatz, in A.Wokaun and C. Daum (Eds.), *Paul Scherrer Institute Scientific Report 2001 Volume V: General Energy*, Paul Scherrer Inst. (PSI), Villigen (Switzerland), 2001.
4. F. N. Büchi, J. F. Affolter, S. Camenzind, N. Chmielewski, P. Dietrich, M. Hockel, M. Ruge, and M. Santis, *Chimia*, 58 (2004) 869.
5. L. Dumercy, M.-C. Péra, R. Glises, D. Hissel, S. Hamandi, F. Badin, and J.-M. Kauffmann, *Fuel Cells*, 4 (2004) 352.
6. http://www.eere.energy.gov/hydrogenandfuelcells/tech_validation/h2_manual.html (module 5) (2005).
7. D. A. Masten and A. D. Bosco, in W. Vielstich, H. A. Gasteiger, and A. Lamm (Eds.), *Handbook of Fuel Cells – Fundamentals, Technology and Applications*, Vol. 4, John Wiley and Sons, 2003, Ch. 53. Chichester, UK.
8. R. Menzer and B. Hoehlein, *J. Power Sources*, 71 (1998) 294.
9. S. Ahmed, J. Kopasz, R. Kumar, and M. Krumpelt, *J. Power Sources*, 112 (2002) 519.
10. R. K. A. M. Mallant, *J. Power Sources*, 118 (2003) 424.
11. R. K. A. M. Mallant, *Extended Abstracts Fuel Cell Seminar Palm Springs*, 2002.

12. K. Strasser, *J. Power Sources*, 29 (1990) 149.
13. H. A. Gasteiger and M. F. Mathias, Paper presented at the *Workshop for HT PEM fuel cells*, 11 December, Pennsylvania State University http://www.mri.psu.edu/conferences/htpemfc/ 2003.
14. R. K. Ahluwalia, E. D. Doss, and R. Kumar, *J. Power Sources*, 117 (2003) 45.
15. K. M. Oseen-Senda, J. J. Pauchet, M. Feidt, and O. Lottin, *Fuel Cell Science, Engineering and Technology*, ASME, 2003, Ch. FUELCELL2003-1753.
16. B. Wenderoth, S. Dambach, L. Meszaros, and U. Fidorra, Patent number WO 02/101848 (2002).
17. R. D. Breault, L. L. Van Dine, D. A. Condit, and M. M. Steinbugler, Patent number WO 0147051 (2002).
18. R. D. Breault, L. L. Van Dine, and A. P. Grasso, Patent number WO 0147052 (2001).
19. M. H. Abd Elhamid, Y. M. Mikhail, R. H. Blunk, D. J. Lisi, and S. Jodeh, Patent number WO 2004/054014 (2004).
20. C. Wieser, *Fuel Cells*, 4 (2004) 245.
21. T. E. Springer, M. S. Wilson, and S. Gottesfeld, *J. Electrochem. Soc.*, 140 (1993) 3513.
22. M. Doyle and G. Rajendran, in W. Vielstich, A. Lamm, and H. A. Gasteiger (Eds.), *Handbook of Fuel Cells – Fundamentals, Technology and Applications*, Vol. 3, John Wiley and Sons, 2003, Ch. 30. Chichester, UK.
23. T. J. P. Freire and E. R. Gonzalez, J. Electroanal. Chem., 503 (2001) 57.
24. DuPont Nafion PFSA membranes, DuPont product information, www.dupont.com/fuelcells (2005).
25. M. Saito, N. Arimura, K. Hayamizu, and T. Okada, *J. Phys. Chem. B.*, 108 (2004) 16064.
26. J. A. Kolde, B. Bahar, and M. S. Wilson, *Proceedings of the 1st International Symposium on Proton Conducting Membrane Fuel Cells*, 1995, p. 193.
27. S. Gottesfeld and T. A. Zawodzinski, in E. C. Alkire, H. Gerischer, D. M. Kolb, and C. W. Tobias (Eds.), *Advances in Electrochemical Science and Engineering*, Vol. 5, Wiley, VCH, 1997.
28. K. V. Lovell and C. H. Page, *Membrane Electrolyte Technology for Solid Polymer Fuel Cells*, ETSU F/02/00110/REP 1997 (1997).
29. D. J. Barnes and J. P. Gilday, Patent number US 5595676 (1997).
30. Y.-C. Chun, C.-S. Kim, D.-H. Peck, D. H. Jung, and D.-R. Shin, *Fuel Cell Seminar 1998*, 1998.
31. Y. G. Chun, D.-H. Peck, C.-S. Kim, and D.-R. Shin, *J. New Mater. Electrochem. Syst.*, 4 (2001) 31.
32. C. A. Linkous, Development of solid electrolytes for water electrolysis at higher temperatures, Proc. DoE Hydrogen Energ. Program Rev. 1 (1996).
33. K. A. Mauritz and R. B. Moore, *Chem. Rev.*, 104 (2004) 4535.
34. Ion-Power product information, www.ion-power.com (2005).
35. M. Wakizoe, O. A. Velev, and S. Srinivasan, *Electrochim. Acta*, 40 (1995) 335.
36. D. E. Curtin, R. D. Lousenberg, T. J. Henry, P. C. Tangeman, and M. E. Tisack, *J. Power Sources*, 131 (2004) 41.
37. A. B. LaConti, M. Hamdan, and R. C. McDonald, in W. Vielstich, H. A. Gasteiger, and A. Lamm (Eds.), *Handbook of Fuel Cells-Fundamentals, Technology and Applications*, Vol. 3, John Wiley and Sons, 2003, Ch. 49. Chichester, UK.
38. A. V. Anantaraman and C. L. Gardner, *J. of Electroanal. Chem.*, 414 (1996) 115.
39. T. A. Zawodzinski Jr., M. Neeman, L. O. Sillerud, and S. Gottesfeld, *J. Phys. Chem.*, 95 (1991) 6040.
40. T. A. Zawodzinski Jr., C. Derouin, S. Radzinski, R. J. Sherman, V. T. Smith, T. E. Springer, and S. Gottesfeld, *J. Electrochem. Soc.*, 140 (1993) 1041.
41. S. Slade, S. A. Campbell, T. R. Ralph, and F. C. Walsh, *J. Electrochem. Soc.*, 149 (2002) A1556.
42. J. St-Pierre and N. Jia, *J. New Mater. Electrochem. Syst.*, 5 (2002) 263.

43. J. Yu, T. Matsuura, Y. Yoshikawa, M. N. Islam, and M. Hori, *Electrochem. Solid State Lett.*, 8 (2005) A156.
44. M. Wakizoe and M. Watanabe, *Fuel Cell Seminar 2000*, 2000, p. 27.
45. M. Wakizoe, T. Kodani, and T. Ota, *Fuel Cell Seminar 2003*, 2003, p. 764.
46. T. Ishisaki, K. Umemura, Y. Kunisa, I. Terada, and M. Yoshitake, *Fuel Cell Seminar 2000*, 2000, p. 23.
47. M. Yoshitake, I. Terada, H. Shimoda, A. Watakabe, K. Yamada, K. Min, and Y. Kunisa, *Fuel Cell Seminar 2002*, 2002, p. 826.
48. M. A. Guerra, Patent number US 6624328 (2003).
49. L. Rivard, D. Pierpont, H. Freemeyer, A. Thaler, and S. Hamrock, *Fuel Cell Seminar 2003*, 2003, p. 73.
50. M. Hicks and D. Ylitalo, *Fuel Cell Seminar 2003*, 2003, p. 97.
51. M. Wakizoe, H. Murata, and H. Takei, *Fuel Cell Seminar 1998*, 1998.
52. W. Liu, K. Ruth, and G. Rusch, *J. New Mater. Electrochem. Syst.*, 4 (2001) 227.
53. S. Cleghorn, J. Kolde, R. Reid, and O. Teller, *Fuel Cell Seminar 2003*, 2003, p. 832.
54. S. Hommura, Y. Kunisa, I. Terada, and M. Yoshitake, *J. Fluorine Chem.*, 120 (2003) 151.
55. R. K. A. M. Mallant, F. A. de Bruijn, and G. H. M. Calis, Patent number WO98/20063 (1998).
56. F. A. de Bruijn, unpublished work.
57. R. F. Savinell and M. Litt, Patent number WO 96/13872 (1996).
58. S. Malhotra and R. Datta, *J. Electrochem. Soc.*, 144 (1997) L23.
59. J. Pellegrino, R. D. Noble, R. Rabago, and C. Koval, Patent number US 5417832 (1998).
60. M. Watanabe, H. Uchida, Y. Seki, M. Emori, and P. Stonehart, *J. Electrochem. Soc.*, 143 (1996) 3847.
61. K. T. Adjemian, S. Srinivasan, J. Benzinger, and A. B. Bocarsly, *J. Power Sources*, 109 (2002) 356.
62. T. M. Thampan, N. H. Jalani, P. Choi, and R. Datta, *J. Electrochem. Soc.*, 152 (2005) A316.
63. Y. Song, J. M. Fenton, H. Russel Kunz, L. J. Bonville, and M. V. Williams, *J. Electrochem. Soc.*, 152 (2005) A539.
64. E. Endoh, *Fuel Cell Seminar*, 2005, p. 180.
65. A. Watakabe, Patent number US2005/0037265 (2005).
66. H. P. Brack, F. N. Buechi, J. Huslage, and G. G. Scherer, *Proceedings of the 2nd Symposium on Proton Conducting Membrane Fuel Cells*, 1998, p. 52.
67. L. Gubler, N. Beck, F. Guersel, F. Hajbolouri, D. Kramer, A. Reiner, B. Steiger, G. G. Scherer, A. Wokaun, B. Rajesh, and K. Ravindranathan Thampi, Chimia, 58 (2004) 826.
68. L. Gubler, H. Kuhn, T. J. Schmidt, G. G. Scherer, H. P. Brack, and K. Simback, *Fuel Cells*, 4 (2004) 196.
69. L. Gubler, S. A. Guersel, and G. G. Scherer, *Fuel Cells*, DOI:10.1002/fuce200400078, in press.
70. C. Chuy, V. I. Basura, E. Simon, S. Holdcroft, J. Horsfall, and K. V. Lovell, *J. Electrochem. Soc.*, 147 (2000) 4453.
71. K. T. Adjemian, M. Foure, S. Gaboury, L. Hedhli, P. Piccione, S. Mullapudi, M. Petch, and J. Reinkingh, *Fuel Cell Seminar*, 2004, p. 77.
72. L. Hehdli and L. Billon, Patent number WO01/60872 (2001).
73. A. E. Steck and C. Stone, in O. Savadogo and P. R. Roberge (Eds.), *New materials for Fuel Cell and Modern Battery Systems: Proceedings of a Symposium*, Ecole Polytechnique, Montreal, PQ (Canada), 1997.
74. S. C. Savett, J. R. Atkins, C. R. Slides, J. L. Harris, B. H. Thomas, S. E. Creager, W. T. Pennington, and D. D. Desmarteau, *J. Electrochem. Soc.*, 149 (2002) A1527.

75. B. Bauer, D. J. Jones, J. Rozière, L. Tchicaya, G. Alberti, M. Casciola, L. Massinelli, A. Peraio, S. Besse, and E. Rammuni, *J. New Mater. Electrochem. Syst.*, 3 (2000) 93.
76. R. W. Kopitzke, C. A. Linkous, H. Randolph Anderson, and G. L. Nelson, *J. Electrochem. Soc.*, 147 (2000) 1677.
77. M. Gil, X. Ji, H. Na, J. E. Hampsey, and Y. Lu, *J. Membr. Sci.*, 234 (2004) 75.
78. X. Li, Z. Whang, H. Lu, C. Zhao, H. Na, and C. Zhao, *J. Membr. Sci.*, 254 (2005) 147.
79. K. D. Kreuer, in W. Vielstich, A. Lamm, and H. A. Gasteiger (Eds.), *Handbook of Fuel Cells – Fundamentals, Technology and Applications*, Vol. 3, John Wiley and Sons, 2003, Ch. 33. Chichester, UK.
80. B. Lakshmanan, W. Huang, D. Olmeijer, and J. W. Weidner, *Electrochem. Solid State Lett.*, 6 (2003) A282.
81. Fumatech membranes, www.fumatech.com (2005).
82. M. Schuster, T. Rager, K. D. Noda, K. D. Kreuer, and J. Maier, *Fuel Cells*, 5 (2005) 355.
83. M. Schuster and W. H. Meyer, Ann. Rev. Mater. Res., 33 (2003) 233.
84. O. Savadogo, *J. Power Sources*, 127 (2004) 135.
85. W. H. J. Hogarth, J. C. Diniz da Costa, and G. Q. Lu, *J Power Sources*, 142 (2005) 223.
86. R. F. Savinell, J. S. Wainright, and M. Litt, *Proceedings of the 2nd Symposium on Proton Conducting Membrane Fuel Cells*, 1998, p. 81.
87. J. S. Wainright, M. H. Litt, and R. F. Savinell, in W. Vielstich, A. Lamm, and H. A. Gasteiger (Eds.), *Handbook of Fuel Cells – Fundamentals, Technology and Applications*, Vol. 3, John Wiley and Sons, 2003, Ch. 34. Chichester, UK.
88. Q. Li, R. He, J. O. Jensen, and N. J. Bjerrum, *Fuel Cells*, 4 (2004) 147.
89. R. Bouchet and E. Siebert, *Solid State Ionics*, 118 (1999) 287.
90. J. Pawlik and C. Henschel, *Fuel Cell Seminar*, 2004, p. 89.
91. D. A. Landsman and F. J. Luczak, in H. A. Gasteiger and A. Lamm (Eds.), *Handbook of Fuel Cells – Fundamentals, Technology and Applications*, Vol. 4, John Wiley and Sons, 2003. Chichester, UK.
92. J. S. Wainright, R. F. Savinell, and M. H. Litt, in O. Savadogo and P. R. Roberge (Eds.), New *Mater. Fuel Cell Mod. Battery Syst. II*, 1997, p. 808.
93. S. R. Samms, S. Wasmus, and R. F. Savinell, *J. Electrochem. Soc.*, 143 (1996) 1498.
94. J. A. Asensio and P. Gomez-Romero, Fuel Cells, DOI:10.1002/fuce.200400081, in press.
95. M. Gebert, B. Hoehlein, and D. Stolten, *J. Fuel Cell Sci. Technol.*, 1 (2004) 56.
96. D. E. Curtin and M. A. Watkins, *Fuel Cell Seminar 2002*, 2002, p. 834.
97. H. A. Gasteiger, W. Gu, R. Makharia, M. F. Mathias, and B. Sompalli, in W. Vielstich, H. A. Gasteiger, and A. Lamm (Eds.), *Handbook of Fuel Cells – Fundamentals, Technology and Applications*, Vol. 3, John Wiley and Sons, 2003, Ch. 46. Chichester, UK.
98. S. S. Kocha, in W. Vielstich, H. A. Gasteiger, and A. Lamm (Eds.), *Handbook of Fuel Cells-Fundamentals, Technology and Applications*, Vol. 3, John Wiley and Sons, 2003, Ch. 43. Chichester, UK.
99. W. Vogel, J. Lundquist, P. Ross, and P. Stonehart, *Electrochim. Acta*, 20 (1975) 79.
100. T. R. Ralph and M. P. Hogarth, *Platinum Met. Rev.*, 46 (2002) 117.
101. F. A. de Bruijn, D. C. Papageorgopoulos, E. F. Sitters, and G. J. M. Janssen, *J. Power Sources*, 110 (2002) 117.
102. D. C. Papageorgopoulos and F. A. de Bruijn, *J. Electrochem. Soc.*, 149 (2002) A140.
103. I. D. Raistrick, in J. W. Van Zee, R. E. White, K. Kinoshita, and H. S. Burne (Eds.), *Proceedings of the Symposium on Diaphragms, Separators, and Ion Exchange Membranes*, The Electrochemical Society, 1986.
104. M. S. Wilson and S. Gottesfeld, *J. Appl. Electrochem.*, 22 (1992) 1.
105. H. A. Gasteiger, S. S. Kocha, B. Sompalli, and F. T. Wagner, *Appl. Catal. B–Environ.*, 56 (2005) 9.
106. H. A. Gasteiger, J. E. Panels, and S. G. Yan, *J. Power Sources*, 127 (2004) 162.
107. http://www.etek-inc.com/custom/index.php (2005).

108. http://www.3m.com/about3m/technologies/fuelcells/index.jhtml (2005).
109. http://www.gore.com/en_xx/products/electronic/fuelcells/series_56_mea_fuel_cells.html (2005).
110. http://www.tanaka.co.jp/products-e/products1/f_5.html (2005).
111. http://www.johnsonmattheyfuelcells.com/HiSPEC_Datasheet.pdf (2005).
112. http://www.johnsonmattheyfuelcells.com/HiSPEC_Customer_Chart.pdf (2005).
113. K. Kinoshita, *Carbon, Electrochemical and Physicochemical Properties*, John Wiley & Sons, Inc., New York, 1988.
114. T. Tada, in W. Vielstich, H. A. Gasteiger, and A. Lamm (Eds.), *Handbook of Fuel Cells – Fundamentals, Technology and Applications*, Vol. 3, John Wiley and Sons, 2003, Ch. 38. Chichester, UK.
115. T. R. Ralph and M. P. Hogarth, *Platinum Met. Rev.*, 46 (2002) 3. Chinchester, UK.
116. M. Uchida, Y. Fukoaka, Y. Sugawara, N. Eda, and A. Ohta, *J. Electrochem. Soc.*, 143 (1996) 2245.
117. D. B. Sepa, M. V. Vojnovic, and A. Damjanovic, *Electrochim. Acta*, 26 (1981) 781.
118. D. B. Sepa, M. V. Vojnovic, L. Vracar, and A. Damjanovic, *Electrochim. Acta*, 32 (1987) 129.
119. A. Parthasarathy, S. Srinivasan, and A. J. Appleby, *J. Electrochem. Soc.*, 139 (1992) 2856.
120. S. Mukerjee and S. Srinivasan, *J. Electroanal. Chem.*, 357 (1993) 201.
121. L. Zhang, C. S. Ma, and S. Mukerjee, *J. Electroanal. Chem.*, 568 (2004) 273.
122. K. Kinoshita, *Electrochemical Oxygen Technology*, John Wiley & Sons, Inc., New York, 1992.
123. O. Antoine, Y. Bultel, and R. Durand, *J. Electroanal. Chem.*, 499 (2001) 85.
124. D. Thompsett, in W. Vielstich, H. A. Gasteiger, and A. Lamm (Eds.), *Handbook of Fuel Cells – Fundamentals, Technology and Applications*, Vol. 3, John Wiley and Sons, 2003. Chichester, UK.
125. N. M. Markovic, T. J. Schmidt, P. N. Ross, and V. Stamenkovic, *Fuel Cells*, 1 (2001) 105.
126. S. Mukerjee and J. McBreen, *J. Electroanal. Chem.*, 448 (1998) 163.
127. Y. Takasu, N. Ohashi, X.-G. Zhang, Y. Murakami, H. Minagawa, S. Sato, and K. Yahikozawa, *Electrochim. Acta*, 41 (1996) 2595.
128. B. Hammer and J. K. Nørskov, *Surf. Sci.*, 343 (1995) 211.
129. B. Hammer and J. K. Nørskov, *Adv. Catal.*, 45 (2000) 71.
130. M. Uchida, Y. Fukoaka, Y. Sugawara, H. Ohara, and A. Ohta, *J. Electrochem. Soc.*, 145 (1998) 3708.
131. E. Yeager, M. Razaq, D. Gervasio, A. Razaq, and D. Tryk, in D. Scherson (Ed.), *Structural Effects in Electrocatalysis and Oxygen Electrochemistry, PV 92-11*, The Electrochemical Society Proceedings Series, Pennington, NJ, 1992.
132. P. D. Beattie, V. I. Basura, and S. Holdcroft, *J. Electroanal. Chem.*, 468 (1999) 180.
133. Z. Ogumi, Z. Takehara, and S. Yoshizawa, *J. Electrochem. Soc.*, 131 (1984) 769.
134. L. Zhang, C. S. Ma, and S. Mukerjee, *Electrochim. Acta*, 48 (2003) 1845.
135. S. Gottesfeld, I. D. Raistrick, and S. Srinivasan, *J. Electrochem. Soc.*, 134 (1987) 1455.
136. F. A. Uribe, T. E. Springer, and S. Gottesfeld, *J. Electrochem. Soc.*, 139 (1992) 765.
137. K. C. Neyerlin, H. A. Gasteiger, C. K. Mittelsteadt, J. Jorne, and W. Gu, *J. Electrochem. Soc.*, 152 (2005) A1073.
138. H. Xu, Y. Song, H. R. Kunz, and J. M. Fenton, *J. Electrochem. Soc.*, 152 (2005) A1828.
139. V. S. Murthi, R. C. Urian, and S. Mukerjee, *J. Phys. Chem.*, (2004).
140. A. Parthasarathy, S. Srinivasan, and A. J. Appleby, *J. Electrochem. Soc.*, 139 (1992) 2530.
141. Y. Song, J. M. Fenton, H. R. Kunz, L. J. Bonville, and M. V. Williams, *J. Electrochem. Soc.*, 152 (2005) A539.
142. Q. Li, H. A. Hjuler, and N. J. Bjerrum, *Electrochim. Acta*, 45 (2000) 4219.
143. http://www.pemeas.com/ (2005).
144. M. S. Wilson, F. H. Garzon, K. E. Sickafus, and S. Gottesfeld, *J. Electrochem. Soc.*, 140 (1993) 2872.
145. M. Pourbaix, *Atlas of Electrochemical Equilibria in Aqueous Solutions*, National Association of Corrosion Engineers, New York, 1974.
146. P. Bindra, S. J. Clouser, and E. Yeager, *J. Electrochem. Soc.*, 126 (1979) 1631.

147. P. Ascarelli, V. Contini, and R. Giorgi, J. Appl. Phys., 91 (2002) 4556.
148. T. Okada, in W. Vielstich, H. A. Gasteiger, and A. Lamm (Eds.), *Handbook of Fuel Cells – Fundamentals, Technology and Applications*, Vol. 3, John Wiley and Sons, 2003, Ch. 48. Chinchester, UK.
149. R. M. Darling and J. P. Meyers, *J. Electrochem. Soc.*, 152 (2005) A242.
150. A. Honji, T. Mori, K. Tamura, and M. Hishinuma, *J. Electrochem. Soc.*, 135 (1988) 355.
151. M. Mathias, H. Gasteiger, R. Makharia, S. Kocha, T. Fuller, and J. Pisco, *Prepr. Symp. – Am. Chem. Soc., Div. Fuel Chem.*, 49 (2004) 471.
152. J. Yu, M. N. Islam, T. Matsuura, M. Tamano, Y. Hayashi, and M. Hori, *Electrochem. Solid-State Lett.*, 8 (2005) A320.
153. P. Yu, M. Pemberton, and P. Plasse, *J. Power Sources*, 144 (2005) 11.
154. J. Xie, D. L. Wood, K. L. More, P. Atanassov, and R. L. Borup, *J. Electrochem. Soc.*, 152 (2005) A1011.
155. Y. Shao-Horn, P. J. Ferreira, and G. J. la O', *Abstract for 2nd International Conference on Polymer Batteries and Fuel Cells*, June 12–17, 2005, Las Vegas, Nevada, USA (2005).
156. S. S. Kocha, F. T. Wagner, and H. A. Gasteiger, *Abstract for 2nd International Conference on Polymer Batteries and Fuel Cells*, June 12–17, 2005, Las Vegas, Nevada, USA (2005).
157. M. Murthy and N. Sisofo, in U. Bossel (Eds.), *Proceedings of the 3rd European PEFC Forum*, Lucerne, Switzerland, 2005, p. B074.
158. D. A. Stevens and J. R. Dahn, *Carbon*, 43 (2005) 179.
159. S. Mukerjee, S. Srinivasan, M. P. Soriaga, and J. McBreen, *J. Electrochem. Soc.*, 142 (1995) 1409.
160. E. Antolini, J. R. C. Salgado, M. J. Giz, and E. R. Gonzalez, *Int. J. Hydrogen Energ.*, 30 (2005) 1213.
161. E. Antolini, Mater. *Chem. and Phys.*, 78 (2003) 563.
162. E. S. Smotkin and R. R. Diaz-Morales, *Ann. Rev. Mater. Res.*, 33 (2003) 557.
163. S. Guerin, B. E. Hayden, C. E. Lee, C. Mormiche, J. R. Owen, A. E. Russell, B. Theobald, and D. Thompsett, *J. Comb. Chem.*, 6 (2004) 149.
164. J. Xie, D. L. Wood, D. M. Wayne, T. A. Zawodzinski, P. Atanassov, and R. L. Borup, *J. Electrochem. Soc.*, 152 (2005) A104.
165. U. A. Paulus, A. Wokaun, G. G. Scherer, T. J. Schmidt, V. Stamenkovic, V. Radmilovic, N. M. Markovic, and P. N. Ross, *J. Phys. Chem. B*, 106 (2002) 4181.
166. J. H. Zagal, in W. Vielstich, H. A. Gasteiger, and A. Lamm (Eds.), *Handbook of Fuel Cells – Fundamentals, Technology and Applications, Vol. 2*, John Wiley and Sons, 2003, Ch. 37. Chichester, UK.
167. N. Alonso-Vante, H. Tributsch, and O. Solorza-Feria, *Electrochim. Acta*, 40 (1995) 567.
168. M. Bron, P. Bogdanoff, S. Fiechter, M. Hilgendorff, J. Radnik, I. Dorbandt, H. Schulenburg, and H. Tributsch, *J. Electroanal. Chem.*, 517 (2001) 85.
169. T. J. Schmidt, U. A. Paulus, H. A. Gasteiger, N. Alonso-Vante, and R. J. Behm, *J. Electrochem. Soc.*, 147 (2000) 2620.
170. B. E. Easton, Z. Qi, A. Kaufman, and P. G. Pickup, *Electrochem. Solid State Lett.*, 4 (2001) A59.
171. Z. Q. Xu, Z. G. Qi, and A. Kaufman, *Chem. Commun.*, (2003) 878.
172. Z. Q. Xu, Z. G. Qi, and A. Kaufman, *Electrochem. Solid State Lett.*, 6 (2003) A171.
173. A. P. Saab, F. H. Garzon, and T. A. Zawodzinski, *J. Electrochem. Soc.*, 150 (2003) A214.
174. M. Mizuhata, S. Nakao, and T. Yamaguchi, *J. Power Sources*, 138 (2004) 25.
175. S. H. Joo, S. J. Choi, I. Oh, J. Kwak, Z. Liu, O. Terasaki, and R. Ryoo, *Nature*, 412 (2001) 169.
176. Z. L. Liu, L. M. Gan, L. Hong, W. X. Chen, and J. Y. Lee, *J. Power Sources*, 139 (2005) 73.
177. F. L. Yuan and H. J. Ryu, *Nanotechnology*, 15 (2004) S596.
178. F. L. Yuan, H. K. Yu, and H. J. Ryu, *Electrochim. Acta*, 50 (2004) 685.
179. T. Matsumoto, T. Komatsu, H. Nakano, K. Arai, Y. Nagashima, E. Yoo, T. Yamazaki, M. Kijima, H. Shimizu, Y. Takasawa, and J. Nakamura, *Catal. Today*, 90 (2004) 277.

180. N. Rajalakshmi, H. Ryu, M. M. Shaijumon, and S. Ramaprabhu, *J. Power Sources*, 140 (2005) 250.
181. X. Wang, M. Waje, and Y. S. Yan, *Electrochem. Solid State Lett.*, 8 (2005) A42.
182. C. Boyer, S. Gamburzev, O. Velev, S. Srinivasan, and A. J. Appleby, *Electrochim. Acta*, 43 (1998) 3703.
183. M. L. Perry, J. Newman, and E. J. Cairns, *J. Electrochem. Soc.*, 145 (1998) 5.
184. M. Eikerling, A. S. Ioselevich, and A. A. Kornyshev, *Fuel Cells*, 4 (2004) 131.
185. C. S. Ma, L. Zhang, S. Mukerjee, D. Ofer, and B. D. Nair, *J. Membr. Sci.*, 219 (2003) 123.
186. M. Uchida, Y. Aoyama, N. Eda, and A. Ohta, *J. Electrochem. Soc.*, 142 (1995) 4143.
187. M. Eikerling and A. A. Kornyshev, *J. Electroanal. Chem.*, 453 (1998) 89.
188. P. Gode, F. Jaouen, G. Lindbergh, A. Lundblad, and G. Sundholm, *Electrochim. Acta*, 48 (2003) 4175.
189. A. Lundblad, *J. New Mater. Electrochem. Systems*, 7 (2004) 21.
190. F. Jaouen, G. Lindbergh, and G. Sundholm, *J. Electrochem. Soc.*, 149 (2002) A437.
191. T. Navessin, M. Eikerling, Q. P. Wang, D. T. Song, Z. S. Liu, J. Horsfall, K. V. Lovell, and S. Holdcroft, *J. Electrochem. Soc.*, 152 (2005) A796.
192. E. Antolini, L. Giorgi, A. Pozio, and E. Passalacqua, *J. Power Sources*, 77 (1999) 136.
193. E. Passalacqua, F. Lufrano, G. Squadrito, A. Patti, and L. Giorgi, *Electrochim. Acta*, 46 (2001) 799.
194. J. M. Song, S. Y. Cha, and W. M. Lee, *J Power Sources*, 94 (2001) 78.
195. G. Li and P. G. Pickup, *J. Electrochem. Soc.*, 150 (2003) C745.
196. G. Sasikumar, J. W. Ihm, and H. Ryu, *J. Power Sources*, 132 (2004) 11.
197. G. Sasikumar, J. W. Ihm, and H. Ryu, *Electrochim. Acta*, 50 (2004) 601.
198. D. Song, Q. Wang, Z. Liu, M. Eikerling, Z. Xie, T. Navessin, and S. Holdcroft, *Electrochim. Acta*, 50 (2005) 3347.
199. Z. Xie, T. Navessin, K. Shi, R. Chow, Q. Wang, D. Song, B. Andreaus, M. Eikerling, Z. Liu, and S. Holdcroft, *J. Electrochem. Soc.*, 152 (2005) A1171.
200. J. Huslage, F. N. Buechi, G. G. Scherer, R. v. Helmoldt, and M. Waidhas, in C. Daum and J. P. Leuenberger (Eds.), *Report: NR ETDE-CH-0001 Paul Scherrer Institute Scientific Report 1999.* Volume V: General Energy, 2000.
201. T. Navessin, S. Holdcroft, Q. Wang, D. Song, Z. Liu, M. Eikerling, J. Horsfall, and K. V. Lovell, *J. Electroanal. Chem.*, 567 (2004) 111.
202. S. K. Zecevic, J. S. Wainright, M. H. Litt, S. Lj. Gojkovic, and R. F. Savinell, *J. Electrochem. Soc.*, 144 (1997) 2973.
203. U. A. Paulus, T. J. Schmidt, H. A. Gasteiger, and R. J. Behm, *J. Electroanal. Chem.*, 495 (2001) 134.
204. M. Inaba, H. Yamada, J. Tokunaga, and A. Tasaka, *Electrochem. Solid State Lett.*, 7 (2004) A474.
205. N. M. Markovic, B. N. Grgur, and P. N. Ross, *J. Phys. Chem. B*, 101 (1997) 5413.
206. H. P. Dhar, L. G. Christner, and A. K. Kush, *J. Electrochem. Soc.*, 134 (1987) 3021.
207. H. P. Dhar, L. G. Christner, A. K. Kush, and H. C. Maru, *J. Electrochem. Soc.*, 133 (1986) 1574.
208. Q. Li, R. He, J.-A. Gao, J. O. Jensen, and N. J. Bjerrum, *J. Electrochem. Soc.*, 150 (2003) A1599.
209. L. Gubler, N. Beck, S. A. Gursel, F. Hajbolouri, D. Kramer, A. Reiner, B. Steiger, G. G. Scherer, A. Wokaun, B. Rajesh, and K. R. Thampi, *Chimia*, 58 (2004) 826.
210. R. Jiang, H. R. Kunz, and J. M. Fenton, *J. Electrochem. Soc.*, 152 (2005) A1329.
211. H. A. Gasteiger, N. M. Markovic, and P. N. Ross, *J. Phys. Chem.*, 99 (1995) 16757.
212. H. A. Gasteiger, N. M. Markovic, and P. N. Ross, *J. Phys. Chem.*, 99 (1995) 8290.
213. T. Gu, W. K. Lee, J. W. Van Zee, and M. Murthy, *J. Electrochem. Soc.*, 151 (2004) A2100.
214. S. Ball, A. Hodgkinson, G. Hoogers, S. Maniguet, D. Thompsett, and B. Wong, *Electrochem. Solid State Lett.*, 5 (2002) A31.

215. B. Hammer, Y. Morikawa, and J. K. Nørskov, *Phys. Rev. Lett.*, 76 (1996) 2141.
216. M. T. M. Koper, T. E. Shubina, and R. A. van Santen, *J. Phys. Chem. B*, 106 (2002) 686.
217. E. Antolini, *J. Appl. Electrochem.*, 34 (2004) 563.
218. E. Auer, A. Freund, T. Lehmann, K. A. Starz, R. Schwarz, and U. Stenke, Patent number US6007934 (1999).
219. D. C. Papageorgopoulos, M. P. de Heer, M. Keijzer, J. A. Z. Pieterse, and F. A. de Bruijn, *J. Electrochem. Soc.*, 151 (2004) A763.
220. G. J. M. Janssen, *J. Power Sources*, 136 (2004) 45.
221. S. Gottesfeld and J. Pafford, *J. Electrochem. Soc.*, 135 (1985) 2651.
222. K. Sasaki, J. X. Wang, M. Balasubramanian, J. McBreen, R. R. Adzic, and F. Uribe, *Electrochim. Acta*, 49 (2004) 3873.
223. S. R. Brankovic, J. X. Wang, and R. R. Adžic, *Electrochem. Solid State Lett.*, 4 (2001) A217.
224. T. E. Shubina and M. T. M. Koper, *Electrochim. Acta*, 47 (2002) 3621.
225. P. Liu, A. Logadottir, and J. K. Nørskov, *Electrochim. Acta*, 48 (2003) 3731.
226. Y. Si, R. Jiang, J. C. Lin, H. R. Kunz, and J. M. Fenton, *J. Electrochem. Soc.*, 151 (2004) A1820.
227. J. R. Yu, T. Matsuura, Y. Yoshikawa, M. N. Islam, and M. Hori, *Phys. Chem. Chem. Phys.*, 7 (2005) 373.
228. P. Piela, C. Eickes, E. Brosha, F. Garzon, and P. Zelenay, *J. Electrochem. Soc.*, 151 (2004) A2053.
229. S. D. Knights, K. M. Colbow, J. St-Pierre, and D. P. Wilkinson, *J. Power Sources*, 127 (2004) 127.
230. A. T. Haug, R. E. White, J. W. Weidner, and W. Huang, *J. Electrochem. Soc.*, 149 (2002) A862.
231. P. A. Adcock, S. V. Pacheco, K. M. Norman, and F. A. Uribe, *J. Electrochem. Soc.*, 152 (2005) A459.
232. F. A. Uribe, J. A. Valerio, F. H. Garzon, and T. A. Zawodzinski, *Electrochem. Solid State Lett.*, 7 (2004) A376.
233. Z. Jusys and R. J. Behm, *J. Phys. Chem. B*, 108 (2004) 7893.
234. Z. Jusys, J. Kaiser, and R. J. Behm, *J. Electroanal. Chem.*, 554–555 (2003) 427.
235. V. Stamenkovic, B. N. Grgur, P. N. Ross, and N. M. Markovic, *J. Electrochem. Soc.*, 152 (2005) A277.
236. B. N. Grgur, N. M. Markovic, and P. N. Ross, *J. Electrochem. Soc.*, 146 (1999) 1613.
237. B. N. Grgur, G. Zhuang, N. M. Markovic, and P. N. Ross, *J. Phys. Chem. B*, 101 (1997) 3910.
238. S. Mukerjee and R. C. Urian, *Electrochim. Acta*, 47 (2002) 3219.
239. S. Mukerjee, R. C. Urian, S. J. Lee, E. A. Ticianelli, and J. McBreen, *J. Electrochem. Soc.*, 151 (2004) A1094.
240. H. A. Gasteiger, N. M. Markovic, and J. Ross, *J. Phys. Chem.*, 99 (1995) 8945.
241. K. Wang, H. A. Gasteiger, N. M. Markovic, and J. Ross, *Electrochim. Acta*, 41 (1996) 2587.
242. B. N. Grgur, N. M. Markovic, and P. N. Ross Jr., in S. Gottesfeld, T. F. Fuller, and G. Halpert (Eds.), *Proton Conducting Membrane Fuel Cells (2nd International Symposium), PV 98-27*, The Electrochemical Society Proceedings Series, Pennington, NJ, 1999.
243. B. N. Grgur, G. Zhuang, N. M. Markovic, and P. N. Ross, *J. Phys. Chem. B*, 102 (1998) 2494.
244. G. J. M. Janssen, M. P. de Heer, and D. C. Papageorgopoulos, *Fuel Cells*, 4 (2004) 169.
245. S. Ball, S. Cooper, K. Z. Dooley, G. A. Hards, and G. Hoogers, *Reformate Tolerant Anode Development in Practical Solid Polymer Fuel Cells*, ETSU F/02/00160/REP (2001).
246. N. P. Lebedeva and G. J. M. Janssen, *Electrochim. Acta*, 51 (2005) 29.
247. B. N. Grgur, N. M. Markovic, and P. N. Ross, *Electrochim. Acta*, 43 (1998) 3631.

248. D. Lee, S. Hwang, and I. Lee, *J. Power Sources*, 145 (2005) 147.
249. M. F. Mathias, J. Roth, J. Fleming, and W. Lehnert, in W. Vielstich, H. A. Gasteiger, and A. Lamm (Eds.), *Handbook of Fuel Cells – Fundamentals, Technology and Applications*, Vol. 3, John Wiley and Sons, 2003, Ch. 42. Chichester, UK.
250. L. R. Jordan, A. K. Shukla, T. Behrsing, N. R. Avery, B. C. Muddle, and M. Forsyth, *J. Power Sources*, 86 (2000) 250.
251. E. Passalacqua, G. Squadrito, F. Lufrano, A. Patti, and L. Giorgi, *J. Appl. Electrochem.*, 31 (2001) 449.
252. J. Yu, M. N. Islam, T. Matsuura, M. Tamano, Y. Hayashi, and M. Hori, *Electrochem. Solid State Lett.*, 8 (2005) A320.
253. J. Ihonen, M. Mikkola, and G. Lindbergh, *J. Electrochem. Soc.*, 151 (2004) A1152.
254. M. V. Williams, E. Begg, L. Bonville, H. R. Kunz, and J. M. Fenton, *J. Electrochem. Soc.*, 151 (2004) A1173.
255. A. Z. Weber and J. Newman, *J. Electrochem. Soc.*, 152 (2005) A677.
256. E. Antolini, R. R. Passos, and E. A. Ticianelli, *J. Appl. Electrochem.*, 32 (2002) 383.
257. G. J. M. Janssen and M. L. J. Overvelde, *J. Power Sources*, 101 (2001) 117.
258. G. J. M. Janssen, *J. Electrochem. Soc.*, 148 (2001) A1313.
259. A. Z. Weber and J. Newman, *Chem. Rev.*, 104 (2004) 4679.
260. M. S. Wilson, J. A. Valerio, and S. Gottesfeld, *Electrochim. Acta*, 40 (1995) 355.
261. E. Passalacqua, F. Lufrano, G. Squadrito, A. Patti, and L. Giorgi, *Electrochim. Acta*, 43 (1998) 3665.
262. B. Mueller, T. Zawodzinski, J. Bauman, F. Uribe, S. Gottesfeld, M. De Marinis, and E. De Castro, in S. Gottesfeld, T. F. Fuller, and G. Halpert (Eds.), *Proton Conducting Membrane Fuel Cells (2nd International Symposium), PV 98-27*, The Electrochemical Society Proceedings Series, Pennington, NJ, 1999.
263. G. Lin and T. Van Nguyen, *J. Electrochem. Soc.*, 152 (2005) A1942.
264. Z. G. Qi and A. Kaufman, *J. Power Sources*, 109 (2002) 38.
265. C. S. Kong, D. Y. Kim, H. K. Lee, Y. G. Shul, and T. H. Lee, *J. Power Sources*, 108 (2002) 185.
266. L. Giorgi, E. Antolini, A. Pozio, and E. Passalacqua, *Electrochim. Acta*, 43 (1998) 3675.
267. U. Pasaogullari, C. Y. Wang, and K. S. Chen, *J. Electrochem. Soc.*, 152 (2005) A1574.
268. J. Benziger, J. Nehlsen, D. Blackwell, T. Brennan, and J. Itescu, *J. Membr. Sci.*, 261 (2005) 98.
269. W. Lee, C.-H. Ho, J. W. Van Zee, and M. Murthy, *J. Power Sources*, 84 (1999) 45.
270. B. Thoben and A. Siebke, *J. New Mater. Electrochem. Syst.*, 7 (2004) 13.
271. A. S. Woodman, E. B. Anderson, K. D. Jayne, and M. C. Kimble, *AESF SUR/FIN 1999 Proceedings*, 1999, p. 21.
272. G. O. Mepsted and J. M. Moore, in W. Vielstich, A. Lamm, and H. A. Gasteiger (Eds.), *Handbook of Fuel Cells – Fundamentals, Technology and Applications*, Vol. 3, John Wiley and Sons, 2003, Ch. 24. Chichester, UK.
273. E. Middelman, W. Kout, B. Vogelaar, J. Lenssen, and E. D. Waal, *J. Power Sources*, 118 (2003) 44.
274. V. Mehta and J. Smith Cooper, *J. Power Sources*, 114 (2003) 32.
275. D. Brett and N. Brandon, *Fuel Cell Rev.*, (2005) February/March.
276. A. Hermann, T. Chaudhuri, and P. Spagnol, *Int. J. Hydrogen Energ.*, 30 (2005) 1297.
277. www.poco.com/us/graphite/resinimpregnation.asp (2005).
278. K. Rossberg and V. Trapp, in W. Vielstich, A. Lamm, and H. A. Gasteiger (Eds.), *Handbook of Fuel Cells – Fundamentals, Technology and Applications*, Vol. 3, John Wiley and Sons, 2003, Ch. 26. Chichester, UK.
279. T. M. Besmann, J. W. Klett, J. J. Henry Jr., and E. Lara-Curzio, *J. Electrochem. Soc.*, 147 (2000) 4083.
280. www.porvairfuelcells.com/bipolar.htm (2005).

281. D. Haack, *DOE Hydrogen Program FY 2004 Progress Report* (2005).
282. J.-J. Hwang and H.-S. Hwang, J. *Power Sources*, 104 (2002) 24.
283. M. Ruge and M. Hoeckel, in U. Bossel (Eds.), *Proceedings of the 3rd European PEFC Forum*, Lucerne, Switzerland, 2005, p. B096.
284. www.graftechaet.com/grafcell (2005).
285. E.R. Gallagher, Patent number US 2003-0160357 (2003).
286. A. Heinzel, F. Mahlendorf, O. Niemzig, and C. Kreuz, *J. Power Sources*, 131 (2004) 35.
287. G. Rinn and S. Bornbaum, DKG, 82 (2005) E33.
288. I. Hennig, T. Kessler, C. Mehler, M. Senge, and M. Weber, Patent number EP1401039 A (2004).
289. http://www.sglcarbon.com/sgl_t/fuelcell/ (2005).
290. N.J. Whitty, *Fuel Cell Seminar*, San Antonio, 2004.
291. M. Wilson and D. Busick, Patent number US 6,248,467 (2000).
292. A. Mueller, P. Kauranen, A. v. Ganski, B. Hell, and N. Berg, *Fuel Cell Seminar*, San Antonio, 2004, 2004.
293. D. Chopra, J. Guolla, and Y. Cai, Patent number US2004191608 (2004).
294. D. P. Davies and P. L. Adcock, Lightweight, *High Power Density Fuel Cell Stack* (2002).
295. G. Faita and C. Manteggaza, Patent number US 5,565,072 (1995).
296. A. J. Grootjes and R. C. Makkus, *Fuel Cell Seminar*, San Antonio, 2004, 2005.
297. M. Ise, H. Schmidt, and M. Waidhas, *1st European PEFC Forum*, 2001, p. 285.
298. R. K. A. M. Mallant, F. Koene, C. W. G. Verhoeve, and A. Ruiter, *Fuel Cell Seminar*, San Diego, 1994, 1994, p. 503.
299. C. Zawodzinski, M. Wilson, and S. Gottesfeld, *Fuel Cell Seminar*, Orlando, 1996, 1996, p. 659.
300. R. Hornung and G. Kappelt, *J. Power Sources*, 72 (1998) 20.
301. R. C. Makkus, A. H. H. Janssen, F. A. de Bruijn, and R. K. A. M. Mallant, *J. Power Sources*, 86 (2000) 274.
302. M. C. Li, C. L. Zeng, S. Z. Luo, J. N. Shen, H. C. Lin, and C. N. Cao, *Electrochim. Acta*, 48 (2003) 1735.
303. D. P. Davies, P. L. Adcock, M. Turpin, and S. J. Rowen, *J. Power Sources*, 86 (2000) 237.
304. H. Wang, M. A. Sweikart, and J. A. Turner, *J. Power Sources*, 115 (2003) 243.
305. J. S. Kim, W. H. A. Peelen, K. Hemmes, and R. C. Makkus, *Corros. Sci.*, 44 (2002) 635.
306. R. Hornung, S. Birkle, and M. Waidhas, Patent number US 6,300,001 B1 (2001).
307. N. Cunningham, D. Guay, J. P. Dodelet, Y. Meng, A. R. Hlil, and A. S. Hay, *J. Electrochem. Soc.*, 149 (2002) A905.
308. S. Joseph, J. C. McClure, R. Chianelli, P. Pich, and P. J. Sebastian, *Int. J. Hydrogen Energ.*, 30 (2005) 1339.
309. D. H. John, Patent number US 2003027028 A1 20030206 (2003).
310. D. R. Hodgson, B. May, P. L. Adcock, and D. P. Davies, *J. Power Sources*, 96 (2001) 233.
311. B. May and D. R. Hodgson, Patent number WO 2000022689 (2000).
312. D. R. Hodgson and E. Farndon, Patent number GB 2 385 332 (2003).
313. M. H. Abd Elhamid and Y. M. Mikhail, *Proceedings of the AICH Spring Meeting*, March 11–14, 2002, 2002, p. 460.
314. M. H. Abd Elhamid, R. H. Blunk, D. J. Lisi, and Y. M. Mikhail, Patent number US 2003152821 A1 20030814 (2003).
315. J. Wind, R. Spaeh, W. Kaiser, and G. Boehm, *J. Power Sources*, 105 (2002) 256.
316. D. Repenning, R. Spaeh, W. Kaiser, and J. Wind, Patent number WO 2001078175 (2001).
317. N. R. Aukland, D. S. Eddy, J. V. Mantese, M. P. Thompson, and S. S. Wang, Patent number US2004058205 A (2004).
318. I. Busenbender, H. A. Gasteiger, Y. Mikhail, and G. Vyas, Patent number US2004081881 A1 (2004).
319. Y. Hung and H. Tawfik, Patent number US2004101738 A (2004).

320. K. Yamaga, H. Yamauchi, K. Takahashi, and M. Komachiya, *Fuel Cell Seminar*, San Antonio, 2004, 2004.
321. J. G. Kaiser, L. Prokop, and P. Lynch, *Fuel Cell Seminar*, San Antonio, 2004, 2004.
322. M. Seido, T. Tonogi, and M. Washima, Patent number US2004081879 A (2004).
323. Y. Kaneta, M. Shinkawa, Y. Suzuki, S. Par, and H. Yoshida, Patent number US6699593 (2004).
324. T. Kuwayama, T. Ohtani, M. Tsuji, and M. Utsunomiya, Patent number US2004197661 A (2004).
325. M. H. Abd Elhamid, Y. Cheng, Y. M. Mikhail, and G. Vyas, Patent number US 6866958 (2005).
326. S. J. Lee, C. H. Huang, J. J. Lai, and Y. P. Chen, *J. Power Sources*, 131 (2004) 162.
327. M. P. Brady, K. Weisbrod, C. Zawodzinski, I. Paulauskas, and R. A. Buchanan, *Electrochem. Solid State Lett.*, 5 (2002) A245.
328. E. A. Ticianelli, C. R. Derouin, and S. Srinivasan, *J. Electroanal. Chem.*, 251 (1988) 275.
329. S. J. Lee, S. Mukerjee, J. McBreen, Y. W. Rho, Y. T. Kho, and T. H. Lee, *Electrochim. Acta*, 43 (1998) 3693.
330. J. Rouquerol, J. D. Avnir, C. W. Fairbridge, D. H. Everett, J. M. Haynes, N. Pernicone, J. D. F. Ramsay, K. S. W. Sing, and K. K. Unger, *J. Pure Appl. Chem.*, 66 (1994) 1739.

CHAPTER SIX

Analytical Models of a Direct Methanol Fuel Cell

A. A. Kulikovsky

Contents

Institute for Materials and Processes in Energy Systems (IWV–3), Research Center "Jülich", D–52425 Jülich, Germany

Advances in Fuel Cells
0080453945

Abstract

The Chapter summarizes recent analytical and semi-analytical results obtained within the scope of one-dimensional (1D) and quasi two-dimensional (Q2D) isothermal models of a direct methanol fuel cell. 1D model takes into account the diffusive transport of methanol and oxygen to the catalyst sites, methanol permeation through the membrane (crossover), Tafel rate of feed molecules conversion in the electrochemical reactions and the direct catalytic methanol–oxygen combustion on the cathode side. Quasi-2D model in addition accounts for a variation of methanol and oxygen concentrations along the respective channel and the effect of gaseous bubbles in the anode channel. The procedure of models construction is described in detail and the physical processes, which stand behind model equations are discussed. The most striking result yields Q2D model, which shows that crossover leads to localization of current production close to the inlet of the oxygen channel. At constant oxygen stoichiometry the cell generates finite local current even at vanishingly small total current in the load. This effect explains lowering of cell open-circuit voltage (mixed-potential). Recent experimental studies of Ye, Zhao, Yang and Prabhuran (*Electrochem. Solid State Lett.* **8** (2005) A52), Muller, Dohle and Kulikovsky (*Electrochem. Solid State Lett.* **9** (2006) L7) and Kulikovsky et al (*Electrochem. Comm.* **8** (2006) 754) show that at open circuit the role of the load plays oxygen-depleted part of the cell, which turns into electrolysis regime. CO_2 bubbles in the anode channel dramatically reduce cell performance; however, under the shortage of oxygen the bubbles increase performance due to reduction of methanol crossover. The effect of non-Tafel kinetics of methanol oxidation on the anode catalyst layer performance is studied; the respective model reveals a novel regime of the active layer operation.

6.1 INTRODUCTION

6.1.1 General Remarks

Direct methanol fuel cell (DMFC) belongs to a family of low-temperature fuel cells with polymer electrolyte membrane. DMFC has a number of advantages over hydrogen cell, yet a "big brother" of that family.

Energy density of methanol is roughly three orders of magnitude higher, than that of atmospheric pressure hydrogen. Modern technologies enable mass production of hydrogen pressurized up to 700 bar; even at this pressure energy density of hydrogen is 3 times smaller, than that of a liquid methanol.

Methanol is easy to store and transport. "Charging" of DMFC reduces to a simple replacement of methanol cartridge or refilling a tank. These benefits stimulated growth of DMFC studies all over the world [1,2].

However, DMFC has a number of drawbacks, which still hinder its commercialization. First is sluggish kinetics of methanol oxidation, much slower than that of hydrogen. In contrast to a hydrogen cell, where the anode voltage loss is negligible, in a DMFC this loss constitutes a significant fraction of the overall voltage loss.

Second, DMFC performance suffers from methanol crossover through the polymer electrolyte membrane. Crossover dramatically lowers cell voltage, especially close to open-circuit conditions.

Third, methanol oxidation is accompanied by release of a large amount of gaseous CO_2, which disturbs the flow in the anode flow field and hinders methanol transport to the catalyst cites. Last but not least, high flux of liquid water through the membrane retards oxygen transport to the catalyst sites thereby increasing voltage loss on the cathode side.

In this chapter we will consider the models of DMFC in a 1–100 W range with the active surface of 10–1000 cm^2. The cells of that type can be assembled in a stack of a power on the order of 0.1–1 kW [3,4]. The processes specific to micro-DMFCs for portable applications (e.g., flow in the micro-channels) and membraneless cells [5] are not considered here. However, many of the results below are useful for understanding operation of all types of DMFCs.

6.1.2 Types of Voltage Losses in a DMFC

The sketch of a conventional DMFC is shown in Figure 6.1. Liquid methanol – water mixture and oxygen (or air) are supplied to the anode and the cathode channels, respectively. Feed molecules (methanol and oxygen) are then transported through the backing layers to the active layers, where the electrochemical reactions occur.

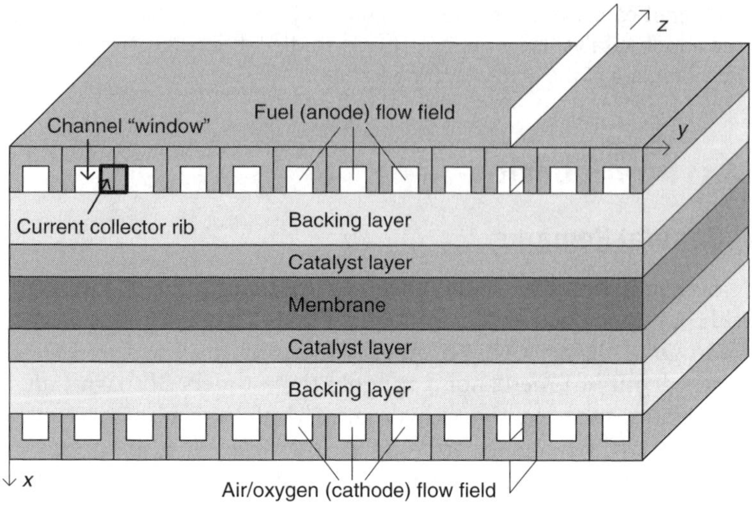

Figure 6.1 Sketch of the cell cross section and the system of coordinates.

If membrane is ideal (no methanol crossover), open-circuit voltage of DMFC V_{oc} should coincide with thermodynamically equilibrium value V_{oc}^T given by Nernst equation. At 25°C and typical methanol and oxygen concentrations V_{oc}^T is about 1.2 V [6]. However, methanol crossover through the membrane reduces V_{oc} down to 500–900 mV. Below we will show that the reason for this lowering is generation of finite local current in a cell even at vanishingly small total current in the external load (Section 6.3).

With the growth of current in the load kinetic and transport losses in a cell increase. The larger the current the larger the total voltage loss.[1] In other words, current from the cell costs some potential; thermodynamic value V_{oc}^T is the "capital" at our disposal.

The portrait of a fuel cell is polarization (voltage current) curve, which shows how much potential is left when the cell generates given current I. It is convenient to introduce mean current density $J = I/A_a$, where A_a is active cell surface. Polarization curve is then $V_{cell}(J)$, where V_{cell} is cell voltage.

Total voltage loss $V_{loss} = V_{oc}^T - V_{cell}$ is determined by all transport and kinetic processes in a cell. Identification of the most important processes and calculation of their contributions into V_{loss} is the main subject of this chapter.

The electrochemical reactions on the anode and the cathode sides of a cell are

$$CH_3OH + H_2O \rightarrow CO_2 + 6H^+ + 6e^- \tag{6.1}$$

$$\frac{3}{2}O_2 + 6H^+ + 6e^- \rightarrow 3H_2O \tag{6.2}$$

[1] The specific voltage loss due to crossover decreases with the current.

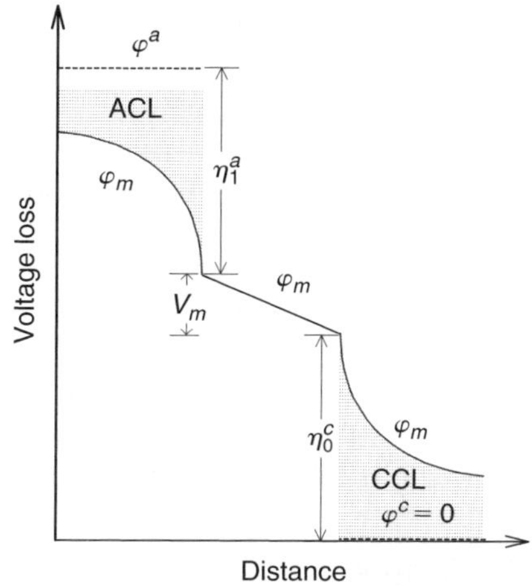

Figure 6.2 Sketch of voltage losses in a DMFC. ACL and CCL denote the anode and the cathode catalyst layers, respectively. Solid line: potential of the membrane phase, dashed lines: potential of the carbon phase. Shaded areas display local polarization voltage η. Polarization voltages of the anode and the cathode sides are η_1^a and η_0^c, respectively. V_m is voltage loss in membrane.

respectively. These reactions occur in a high field of a double layer at the catalyst particle/electrolyte interface. In the active (catalyst) layers tiny catalyst particles are mixed with polymer electrolyte. This mixture partially fills the voids in the matrix of carbon threads, which provide electronic contact between particles.

The driving forces for protons and electrons in this environment are modeled by continuous distributions of two potentials: the electrolyte potential φ_m and the carbon phase potential φ. φ_m induces proton current in the electrolyte phase and φ drives electrons in the carbon phase.

Polarization voltages (overpotentials) $\eta^a = \varphi^a - \varphi_m$ and $\eta^c = \varphi_m - \varphi^c$ determine the rate of the respective electrochemical reaction. Hereinafter the superscripts "a" and "c" mark the anode and the cathode side, respectively. Figure 6.2 sketches the distribution of potentials across the catalyst layers and membrane.

In the catalyst layers, electronic conductivity of the carbon phase is usually much larger than the conductivity of electrolyte (membrane) phase. For that reason the variation of φ across the active layer can be neglected (Figure 6.2).

In the anode catalyst layer (ACL) methanol oxidation reaction (6.1) generates charges; in the cathode catalyst layer (CCL) these charges disappear in the oxygen reduction reaction (6.2). Activation of electrochemical reactions requires significant portion of available voltage. Furthermore, transport of feed molecules (methanol and oxygen) to the catalyst sites also costs some potential. The sum of activation and transport losses on each side constitutes polarization voltage η_1^a and η_0^c of the anode and the cathode sides, respectively (Figure 6.2). Note that η_1^a and η_0^c is the difference (with the proper sign) of φ and φ_m at the catalyst layer/membrane interface (Figure 6.2).

Due to a large amount of liquid water in the anode feed the membrane in DMFC is usually well humidified; voltage loss in membrane V_m (Figure 6.2) then linearly depends on current.

Oxygen and methanol are supplied through the channels (flow field), which cover cell active area (Figure 6.1). The geometry of the flow field may vary from a single serpentine channel to a complicated multi-channel structures. One of the tasks of theory is calculation of the expenses (in terms of potential) for feed transport in the flow field.

6.1.3 Characteristic Scales

Conventional cell shown in Figure 6.1 has three characteristic space scales. The smallest scale is associated with the thickness of the catalyst layer ($\simeq 10^{-3}$ cm), typically the thinnest element in a cell sandwich. The medium scale is determined by alternating channel "windows" and current collector ribs (lands) in a flow field (Figure 6.1). This scale is usually on the order of 0.1 cm, the characteristic size of the "windows" and ribs. The large scale is determined by the length of the feed channel (10–100 cm). The small-scale effects are thus due to the transport properties of the porous layers; the medium- and large-scale effects are determined by geometry of the flow field.

Quasi-2D models described below deal with the transport of reactants and potential distribution in the x–z plane (Figure 6.1). These models enable to describe the interplay between small- and large-scale phenomena in a cell.

Medium-scale effects are out of the scope of these models. Below we will neglect the variation of reactant concentrations and potentials between "windows" and ribs (Figure 6.1). At present it seems difficult to take into account medium-scale effects in analytical models; these effects require "heavy artillery" of numerical approach.

6.1.4 Numerical Models

One-dimensional (1D) numerical models of DMFC ignore medium- and large-scale phenomena and consider transport of reactants and charges across the cell only [7–15]. The usual argument in favor of 1D approach is uniformity of feed concentration in the channels under large stoichiometry of the respective flow. This argument is correct for hydrogen cells, where hydrogen is supplied in a gas phase; however, it does not work for most DMFC designs and operating conditions. The reason is gaseous CO_2 bubbles on the anode side.

In Section 6.4 we will show that already at moderate current gaseous bubbles dramatically reduce local concentration of methanol in the channel, which is equivalent to reduction of stoichiometry of anodic flow. Due to this effect 1D approximation is justified only in the two cases: (i) at room temperatures, when solubility of CO_2 in water is large and (ii) at small current densities. In these two situations the rate of bubbles formation is negligible; in all other practically interesting cases 1D approach is not applicable.

Nevertheless, 1D models are very useful for understanding the fundamentals of DMFC functioning. Furthermore, these models provide the basis for construction of more realistic quasi-2D models, which take into account two-phase effects in the anode flow.

Evidently, a more accurate approach to cell modeling is 2D [16–21] or even 3D. Ironically, the drawbacks of 2D and 3D models stem from the power and versatility of numerical methods. Having a tool for solution of equations of virtually arbitrary complexity it is difficult to resist temptation to take into account all imaginable processes in a cell. The resulting models include up to 60 kinetic, transport and operational parameters [18]; many of them are poorly known. Understanding, interpretation and experimental validation of results is then difficult and computationally expensive task.

The approach described below aims at construction of simple, but physically realistic analytical models of DMFC. The basic idea is to make analytically as much work as possible. The models of that type contain minimal number of parameters and at least in the limiting cases are solvable. These models give new insights into DMFC function and provide useful hints for real cell design.

6.1.5 Model Splitting: 1D+1D (Quasi-2D) Approach

The large ratio of large and small scales makes it possible efficient splitting of a 2D model of DMFC. The idea of splitting is as follows.

The simplest and most often used flow field for transport of reactants is a single serpentine channel. To model the cell equipped on both sides with this flow field we will ignore the effects due to channels curvature and consider a cell with the straight channels of equivalent length (Figure 6.3). This transformation is valid provided that the transport of reactants between two adjacent turns of the serpentine through the backing layer is not large. This is justified when pressure gradient between adjacent turns is small. Evidently, cell with n parallel serpentines on both sides can be cut and transformed into n cells with the straight channels.

The length of the channel is typically two to three orders of magnitude larger, than the thickness l_{MEA} of membrane-electrode assembly (MEA). The prototype of DMFC, shown in Figure 6.3, is thus a two-scale object: the characteristic scale of variation of any parameter along the channel is much larger than l_{MEA}. Furthermore, hydraulic permeability of the porous layers in the MEA is typically much smaller than permeability of the channel. Therefore, to a good approximation we may neglect z-component of all fluxes *in the MEA*

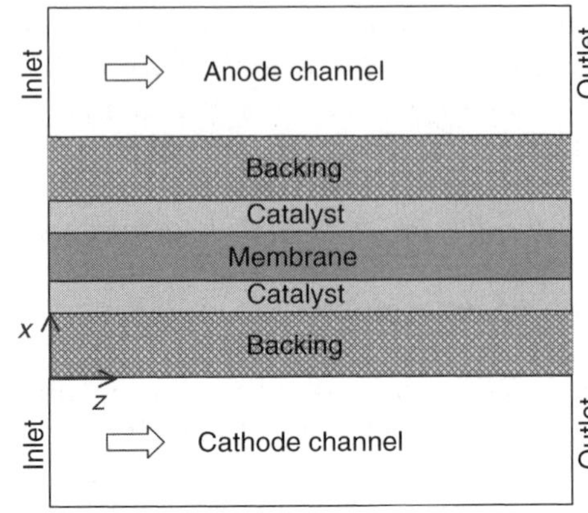

Figure 6.3 Cross section of the cell prototype and the system of coordinates.

and assume that transport in the MEA occurs only in the through-plane direction (Figure 6.3). These qualitative arguments have recently been supported by accurate analysis of leading terms in a full system of hydrodynamic equations for DMFC [20,21].

Fully 2D problem of DMFC in a plane, shown in Figure 6.3, can thus be split into 1D problem across the MEA (internal problem) and 1D, 2D or even 3D model of the flow in the feed channel (channel problem). The channel problem provides the boundary conditions for internal problem; the latter in turn gives local current density $j(z)$ required for solution of equations along the channel.

According to commonly accepted terminology this is 1D+1D (or quasi-2D) approach. This idea was used in numerical calculations of a hydrogen cell [22] and DMFCs [20,21]. Here we will utilize this approach to construct analytical models of DMFC. We begin with the internal (through-plane) problem.

6.2 1D Model of a DMFC

6.2.1 Performance of Generic Catalyst Layer

The basis for construction of internal model forms the problem of performance of "generic" catalyst layer. Consider for definiteness the CCL of a thickness l_t. Typical distribution of overpotential $\eta \equiv \eta^c$ across the active layer is sketched in Figure 6.4 (in this section we will omit the superscript "c").

Our goal is calculation of voltage loss η_0 at the CCL/membrane interface as a function of proton current in the membrane j_0 (Figure 6.4).

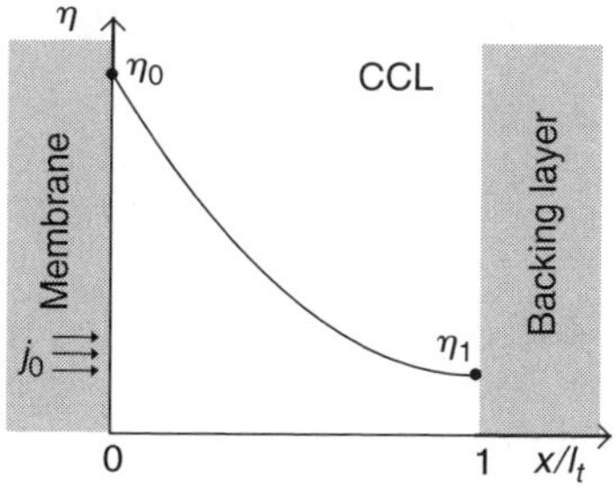

Figure 6.4 Sketch of the problem and the typical distribution of overpotential across the CCL.

Function $\eta_0(j_0)$ relates the boundary conditions in the problem of CCL performance. We will formulate this problem using the following assumptions:

1. Variation of feed molecules (oxygen) concentration across the active layer is small.
2. Proton conductivity of electrolyte phase in the catalyst layer is constant.
3. The rate of electrochemical reaction obeys Tafel law with the first-order dependence on feed molecules concentration.

The first assumption holds if the catalyst layer is thin enough and/or diffusivity of feed molecules through the layer is high. The second assumption is justified if electrolyte phase is evenly distributed over the CCL volume. Tafel law is a reasonable approximation for any analytical model; a more subtle effects due to non-Tafel kinetics of electrochemical reactions will be discussed in Section 6.6.

Under these assumptions CCL performance is described by two equations:

$$\frac{\partial j}{\partial x} = -Q(x) \tag{6.3}$$

$$j = -\sigma_t \frac{\partial \eta}{\partial x} \tag{6.4}$$

Here j is the proton current density, x is the coordinate across the CCL, σ_t is the proton conductivity of electrolyte phase in the CCL and local polarization voltage $\eta = \varphi_m - \varphi$ (Figure 6.2).

The first equation describes consumption of proton current at a rate Q. The second equation is Ohm's law, which establishes linear relationship between local gradient of overpotential and proton current density. Since φ is almost constant along x, we have $\partial\eta/\partial x \simeq \partial\varphi_m/\partial x$, which justifies appearance of η instead of φ_m in Eq. (6.4).

Following assumptions, for Q we will employ Tafel law:

$$Q = i_\star \left(\frac{c_t}{c_{ref}} \right) \exp\left(\frac{\eta}{b} \right) \tag{6.5}$$

Here $i_\star$ is the exchange current density (per unit volume), c_t and c_{ref} is available in the catalyst layer and the reference oxygen molar concentration, respectively, and b is Tafel slope.

Introducing dimensionless variables

$$\tilde{x} = \frac{x}{l_t} \quad \tilde{j} = \frac{j}{j_\star} \quad \tilde{\eta} = \frac{\eta}{b} \quad \tilde{Q} = \frac{l_t Q}{j_\star} \tag{6.6}$$

where

$$j_\star = \frac{2\sigma_t b}{l_t} \tag{6.7}$$

is characteristic current density, Eqs. (6.3)–(6.5) take a form

$$\frac{\partial \tilde{j}}{\partial \tilde{x}} = -\tilde{Q} \tag{6.8}$$

$$2\tilde{j} = -\frac{\partial \tilde{\eta}}{\partial \tilde{x}} \tag{6.9}$$

$$\tilde{Q} = \tilde{k}_t \exp(\tilde{\eta}) \tag{6.10}$$

where

$$\tilde{k}_t = \frac{k_t}{j_\star} = \frac{i_\star l_t}{j_\star} \left(\frac{c_t}{c_{ref}} \right) \tag{6.11}$$

$$k_t = i_\star l_t \left(\frac{c_t}{c_{ref}} \right) \tag{6.12}$$

Solution to the system (6.8)–(6.10) leads to [23]

$$\tilde{j}_0 = \sqrt{\tilde{Q}_0 - \tilde{j}_0^2} \tan\left(\sqrt{\tilde{Q}_0 - \tilde{j}_0^2} \right) \tag{6.13}$$

With $\tilde{Q}_0 \equiv \tilde{Q}(0) = \tilde{k}_t \exp(\tilde{\eta}_0)$ this is the general implicit relation for polarization curve of the catalyst layer $\tilde{\eta}_0(\tilde{j}_0)$.

In the limits of small and large current (6.13) can be simplified [23]:

$$\tilde{\eta}_0 = \ln \tilde{j}_0 - \ln \tilde{k}_t \qquad \tilde{j}_0 \ll 1 \tag{6.14}$$

$$\tilde{\eta}_0 = 2 \ln \tilde{j}_0 - \ln \tilde{k}_t \qquad \tilde{j}_0 \gg 1 \tag{6.15}$$

Physically, in the case of $\tilde{j}_0 \ll 1$ overpotential and reaction rate are almost constant across the catalyst layer. Total proton current density j_0 converted in the layer is then simply a product of reaction rate (6.5) by active layer thickness l_t; in dimensionless form this gives (6.14).

In the opposite limit of $\tilde{j}_0 \gg 1$ the first logarithm on the right side of (6.15) contains factor 2, which is equivalent to twice larger Tafel slope (see also Refs. [24,25]). Physically, $\tilde{j}_0 \gg 1$ is equivalent to small proton conductivity of the CCL. The reaction is then concentrated close to membrane, where the protons are "cheaper" [23]. In the rest part of the active layer the rate of electrochemical conversion is small. Equation (6.15) shows that this regime of active layer operation requires higher polarization voltage η_0.

Now consider the ACL. In the ACL protons are generated and $\eta = \varphi - \varphi_m$ (Figure 6.2). Thus, the problem for ACL is obtained if we change the sign of the right sides in Eqs. (6.3) and (6.4). Furthermore, since proton current in the ACL has opposite direction we have to change also the sign of x-coordinate in Eqs. (6.3) and (6.4). These transformations leave the system (6.3)–(6.5) intact and we again come to (6.13)–(6.15).

Equations (6.13)–(6.15) are thus valid for both ACL and CCL in a DMFC. Hereinafter the subscript "0" will be omitted, so that $j \equiv j_0$ is current density in the cell and $\tilde{\eta} \equiv \tilde{\eta}_0$ is polarization voltage of the catalyst layer. To distinguish the anode and the cathode sides we will equip the respective variables with the superscripts "a" and "c".

The concentration of feed molecules in the catalyst layer c_t appears in the Tafel law (6.5) and further in the parameter $\tilde{k}_t$ (6.11). The concentrations of oxygen c_t^c and methanol c_t^a in the respective catalyst layer depend on j. Furthermore, these concentrations are coupled due to methanol crossover. To calculate c_t^c and c_t^a we have to construct a model of oxygen and methanol transport across the MEA.

6.2.2 Feed Molecules Concentration in the Catalyst Layers

Figure 6.5 sketches the profiles of methanol and oxygen concentration across the cell. As before, c_t^a and c_t^c are assumed to be constant across the respective

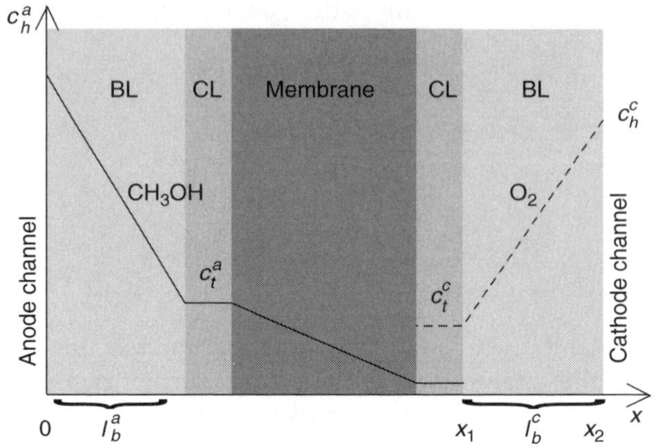

Figure 6.5 The sketch of methanol and oxygen concentration profiles across the cell. BL: backing layer and CL: catalyst layer.

active layer (Figure 6.5). However, c_t^a and c_t^c depend on current density and on the rate of methanol permeation through the membrane; consider the respective fluxes.

6.2.2.1 Methanol

Diffusion flux of methanol in the anode backing layer should cover the expenses for current production and crossover:

$$D_b^a \frac{\partial c^a}{\partial x} = -\frac{j}{6F} - N_{cross} \tag{6.16}$$

where D_b^a is the diffusion coefficient of methanol in the anode backing layer, c^a is methanol molar concentration and N_{cross} is the molar flux of methanol through the membrane (crossover flux).

To calculate N_{cross}we assume that (i) methanol is transported through the membrane due to diffusion and electroosmosis and (ii) the diffusion coefficient of methanol in membrane D_m is constant [26]. Under these assumptions

$$N_{cross} = -D_m \frac{\partial c^a}{\partial x} + n_d \frac{j}{F} \frac{c_t^a}{c_w^a} \tag{6.17}$$

where n_d is the drag coefficient (number of molecules of methanol–water mixture, transported by one proton), c_w^a is molar concentration of water. Here we also assumed that proton does not distinguish methanol and water molecules; the drag term in (6.17) is then proportional to methanol molar fraction in the ACL.

Neglecting methanol concentration in the CCL as compared to c_t^a, we can write

$$N_{cross} \simeq \frac{D_m c_t^a}{l_m} + n_d \frac{j}{F} \frac{c_t^a}{c_w^a} \tag{6.18}$$

where l_m is the membrane thickness.

Substituting (6.18) into (6.16), solving the resulting equation with the boundary condition $c^a|_{x=0} = c_h^a$ and substituting $x = l_b^a$ (Figure 6.5) into the solution we get

$$c_t^a = c_h^a \left(1 - \frac{j}{j_{\lim}^a}\right)\left(1 + \beta + n_d \frac{j}{j_w}\right)^{-1} \tag{6.19}$$

where c_h^a is methanol concentration in the anode channel,

$$\beta = \frac{D_m l_b^a}{D_b^a l_m} \tag{6.20}$$

$$j_w = F \frac{D_b^a c_w^a}{l_b} \tag{6.21}$$

and

$$j_{\lim}^a = 6F \frac{D_b^a c_h^a}{l_b^a} \tag{6.22}$$

is limiting current density: if $j = j_{\lim}^a$, methanol concentration in the ACL is zero.

With (6.19) the crossover flux (6.18) can be written as

$$N_{cross} = \frac{j_{\lim}^a}{6F} \left(\frac{\beta + n_d \dfrac{j}{j_w}}{1 + \beta + n_d \dfrac{j}{j_w}} \right) \left(1 - \frac{j}{j_{\lim}^a}\right) \tag{6.23}$$

The drag terms in (6.23) are small as compared to β. To show this we note that typically $l_m \simeq l_b^a$ and $D_m \simeq D_b^a$; this gives $\beta \simeq 1$. Further, with $D_b^a = 2 \times 10^{-5}\,\mathrm{cm^2 s^{-1}}$, $c_w^a \simeq 2 \times 10^{-2}\,\mathrm{mol/cm^3}$ and $l_b^a = 1 \times 10^{-2}\,\mathrm{cm}$ we get $j_w \simeq 5\,\mathrm{A/cm^2}$. Since $n_d \simeq 2$ and current density in DMFC usually does not exceed $0.2\,\mathrm{A/cm^2}$, we find that $n_d j / j_w \ll \beta$. Therefore, the terms with n_d

in (6.23) may be omitted. Physically it means that if current density in the cell is not large, the dominating mechanism of methanol transport through the membrane is diffusion.

Equation (6.23) then takes a form

$$N_{cross} = \frac{\beta_\star j^a_{\lim}}{6F}\left(1 - \frac{j}{j^a_{\lim}}\right) \tag{6.24}$$

where

$$\beta_\star = \frac{\beta}{1+\beta} \tag{6.25}$$

It is convenient to introduce equivalent crossover current density according to

$$j_{cross} \equiv 6FN_{cross} = \beta_\star j^a_{\lim}\left(1 - \frac{j}{j^a_{\lim}}\right) \tag{6.26}$$

Clearly, at $j = j^a_{\lim}$ crossover is zero, since there is no methanol in the ACL.

Equation (6.26) shows that j_{cross} linearly decreases with the current density from the maximum value $j_{cross}(0) = \beta_\star j^a_{\lim}$ at $j = 0$ to zero at $j = j^a_{\lim}$. This behavior correlates with many experimental data [3,27,28].

6.2.2.2 Oxygen

Following Ref. [29] we assume that all methanol permeated through the membrane is consumed on the cathode side in a direct catalytic burning with oxygen:

$$2CH_3OH + 3O_2 \rightarrow 2CO_2 + 4H_2O$$

The diffusion flux of oxygen in the cathode backing layer is thus a sum of fluxes required to consume proton current j and to burn methanol:

$$D^c_b \frac{\partial c^c}{\partial x} = \frac{j}{4F} + \frac{3}{2}N_{cross} \tag{6.27}$$

Substituting (6.18) into (6.27), solving the resulting equation with the boundary condition $c|_{x=x_2} = c^c_h$, substituting $x = x_1$ into the solution and taking into account that $x_2 - x_1 = l^c_b$ Figure 6.5), we get oxygen concentration in the CCL:

$$c^c_t = c^c_h\left(1 - \frac{j + j_{cross}}{j^c_{\lim}}\right) \tag{6.28}$$

where c_h^c is oxygen concentration in the cathode channel and

$$j_{\lim}^c = 4F \frac{D_b^c c_h^c}{l_b^c} \tag{6.29}$$

is the limiting current density due to imperfect oxygen transport through the backing layer: if $j + j_{cross} = j_{\lim}^c$, oxygen concentration in the catalyst layer is zero.

6.2.3 Polarization Curve

Relations (6.19) and (6.28) allow us to write down 1D polarization curve of DMFC. c_t appears in (6.14) and (6.15) through $\tilde{k}_t$ (6.11). Neglecting the term with n_d in (6.19) and using this equation in (6.14) and (6.15), we obtain polarization voltage of the anode side in the limiting cases of small and large current densities:

$$\frac{\eta^a}{b^a} = \ln\left(\frac{j}{j_\star^a}\right) - \ln \tilde{k}_h^a - \ln\left(1 - \frac{j}{j_{\lim}^a}\right) + \ln(1+\beta) \qquad j \ll j_\star^a \tag{6.30}$$

$$\frac{\eta^a}{b^a} = 2\ln\left(\frac{j}{j_\star^a}\right) - \ln \tilde{k}_h^a - \ln\left(1 - \frac{j}{j_{\lim}^a}\right) + \ln(1+\beta) \qquad j \gg j_\star^a \tag{6.31}$$

where

$$j_\star^a = \frac{2\sigma_t^a b^a}{l_t^a} \tag{6.32}$$

$$\tilde{k}_h^a = \frac{l_t^a i_\star^a}{j_\star^a}\left(\frac{c_h^a}{c_{ref}^a}\right) \tag{6.33}$$

Using (6.28) in (6.14) and (6.15), we obtain polarization voltage of the cathode side:

$$\frac{\eta^c}{b^c} = \ln\left(\frac{j}{j_\star^c}\right) - \ln \tilde{k}_h^c - \ln\left(1 - \frac{j + j_{cross}}{j_{\lim}^c}\right) \qquad j \ll j_\star^c \tag{6.34}$$

$$\frac{\eta^c}{b^c} = 2\ln\left(\frac{j}{j_\star^c}\right) - \ln \tilde{k}_h^c - \ln\left(1 - \frac{j + j_{cross}}{j_{\lim}^c}\right) \qquad j \gg j_\star^c \tag{6.35}$$

where

$$j_\star^c = \frac{2\sigma_t^c b^c}{l_t^c} \tag{6.36}$$

$$\tilde{k}_h^c = \frac{l_t^c i_\star^c}{j_\star^c}\left(\frac{c_h^c}{c_{ref}^c}\right) \tag{6.37}$$

and j_{cross} is given by (6.26).

Low- and high-current polarization voltages of the anode (6.30) and (6.31) differ only by a constant factor in the first term on the right side. The same is true for the cathode side (Eqs. (6.34) and (6.35)). It is convenient to combine each pair of these relations in one, approximately valid in the whole range of currents:

$$\frac{\eta^a}{b^a} = \phi\left(\frac{j}{j_\star^a}\right)\ln\left(\frac{j}{j_\star^a}\right) - \ln\tilde{k}_h^a - \ln\left(1 - \frac{j}{j_{\lim}^a}\right) + \ln\ (1+\beta) \tag{6.38}$$

$$\frac{\eta^c}{b^c} = \phi\left(\frac{j}{j_\star^c}\right)\ln\left(\frac{j}{j_\star^c}\right) - \ln\tilde{k}_h^c - \ln\left(1 - \frac{j + j_{cross}}{j_{\lim}^c}\right) \tag{6.39}$$

Here the matching function

$$\phi(\gamma) = 1 + \frac{\gamma}{1+\gamma} \tag{6.40}$$

varies from 1 to 2 as its argument changes from small to large values.

Equations (6.38) and (6.39) need some comments. The characteristic current densities at the anode (6.32) and at the cathode (6.36) do not coincide. For example, if $j_\star^c \gg j_\star^a$ we may have the situation when $j_\star^a \ll j \ll j_\star^c$. In other words, the cell may operate in a low-current regime on one side and in a high-current regime on the other.

The effect of crossover on η^a reduces to a constant shift by $\ln(1+\beta)$ (Eq. (6.38)). Physically, on the anode side crossover simply diminishes amount of methanol available for current production.

On the cathode side the effect of crossover is less trivial. Crossover current j_{cross} increases the effective current density $j + j_{cross}$ in the last term on the right side of Eq. (6.39). Therefore, due to j_{cross} the cathode side operates closer to the limiting current. If for example $j_{cross} \lesssim j_{\lim}^c$, already at small j the cathode side operates near the limiting current. Below we will consider several consequences of this effect.

Equating the argument of the last logarithm in (6.39) to zero and taking into account (6.26), after simple manipulations we get the limiting current density of the cathode side, which takes into account crossover: $j^c_{\lim\star} = j^c_{\lim} - \beta(j^a_{\lim} - j^c_{\lim})$. The overall limiting current density of the cell is thus

$$j_{\lim} = \min\{j^a_{\lim},\ j^c_{\lim} - \beta(j^a_{\lim} - j^c_{\lim})\} \tag{6.41}$$

Maximal $j_{\lim}$ is attained when $j^a_{\lim} = j^c_{\lim}$. Note that when $j^a_{\lim} = j^c_{\lim}$ the value of $j_{\lim}$ does not depend on β, i.e., crossover does not affect cell operation. Below we will see that this "cross-linked" regime can be organized under more general conditions, when local current density depends on a distance along the feed channel.

6.2.3.1 Cell Voltage

The cell voltage is

$$V_{cell} = V^T_{oc} - \eta^a - \eta^c - Rj \tag{6.42}$$

where the last term on the right side accumulates all ohmic losses (contact resistance, membrane resistance, etc.) Relation (6.42) with (6.38) and (6.39) is a 1D model of a DMFC.

This model can be used for analysis of cell polarization curves in cases when special precautions are made to provide uniform distribution of reactants over the cell surface. In practice DMFC most often operates in conditions when these non-uniformities cannot be neglected. Equations (6.38) and (6.39) provide a basis for construction of a more realistic quasi-2D model of DMFC.

6.3 1D+1D Model: The Nature of Mixed Potential

6.3.1 Basic Assumptions

Quasi-2D model is a generalization of 1D model of the previous section and thus it inherits all assumptions made to describe the transport across the cell and the kinetics of electrochemical reactions. The following additional assumptions are utilized to account for the transport of reactants in the feed channels:

1. In this section we assume plug flow conditions in the channels on both sides of the cell, i.e., flow velocity in the channel is constant and reactant concentration is uniform across the channel. The growth of flow velocity

in the anode channel due to gaseous bubbles will be taken into account in Section 6.4.

2. Pressure in the channels is constant.

Following the idea of a quasi-2D approach, the solutions to 1D problem (6.38) and (6.39) are now considered as *local*. These solutions contain current density j and methanol and oxygen concentration in the channel c_h^a, and c_h^c, respectively. In the previous section j, c_h^a and c_h^c were assumed to be independent of a distance along the channel z. Now we will consider them as functions of z; what we need are equations, which relate these values.

6.3.2 Continuity Equations in the Feed Channels

Since total molar concentration of the anode and the cathode flows is constant, it is convenient to work with the molar fractions ψ and ξ of methanol and oxygen, respectively. In terms of molar fractions, methanol and oxygen mass balance in the respective channel is

$$h^a v^a C^a \frac{\partial \psi}{\partial z} = -\frac{j}{6F} - N_{cross} \tag{6.43}$$

$$h^c v^c C^c \frac{\partial \xi}{\partial z} = -\frac{j}{4F} - \frac{3}{2} N_{cross} \tag{6.44}$$

where axis z is directed along the channels (co-flow conditions, Figure 6.3), h is channel height above the plane, v is flow velocity, C is total molar concentration of the mixture in the channel, $j(z)$ is local current density and N_{cross} is given by Eq. (6.24).

It is convenient to introduce dimensionless variables

$$\tilde{z} = \frac{z}{L} \qquad \tilde{\psi} = \frac{\psi}{\psi^0} \qquad \tilde{\xi} = \frac{\xi}{\xi^0} \qquad \tilde{j} = \frac{j}{j_{\lim}^{a0}} \tag{6.45}$$

where L is channel length,

$$j_{\lim}^{a0} = \frac{6FD_b^a C^a \psi^0}{l_b^a} \tag{6.46}$$

and the superscript "0" indicates the values at the channel inlet (at $\tilde{z} = 0$).

DMFC is usually run at constant oxygen and methanol stoichiometries λ^a and λ^c, respectively. By the definition

$$\lambda^a = \frac{6Fh^a v^a C^a \psi^0}{LJ} \qquad \lambda^c = \frac{4Fh^c v^c C^c \xi^0}{LJ}$$

where J is the mean current density in a cell.

With these relations (6.43) and (6.44) transform to

$$\lambda^a \tilde{J} \frac{\partial \tilde{\psi}}{\partial \tilde{z}} = -(\tilde{j} + \tilde{j}_{cross}) \tag{6.47}$$

$$\lambda^c \tilde{J} \frac{\partial \tilde{\xi}}{\partial \tilde{z}} = -(\tilde{j} + \tilde{j}_{cross}) \tag{6.48}$$

where

$$\tilde{j}_{cross} = \frac{6FN_{cross}}{j^{a0}_{\lim}} = \beta_{\star}\left(\tilde{\psi} - \tilde{j}\right) \tag{6.49}$$

is dimensionless equivalent crossover current density (*cf.* Eq. (6.26)).

With (6.49), Eqs. (6.47) and (6.48) are

$$\lambda^a \tilde{J} \frac{\partial \tilde{\psi}}{\partial \tilde{z}} = -\left(\tilde{j} + \beta_{\star}(\tilde{\psi} - \tilde{j})\right) \quad \tilde{\psi}(0) = 1 \tag{6.50}$$

$$\lambda^c \tilde{J} \frac{\partial \tilde{\xi}}{\partial \tilde{z}} = -\left(\tilde{j} + \beta_{\star}(\tilde{\psi} - \tilde{j})\right) \quad \tilde{\xi}(0) = 1 \tag{6.51}$$

The system (6.50) and (6.51) contains three unknowns: $\tilde{\psi}$, $\tilde{\xi}$ and $\tilde{j}$. The equation for $\tilde{j}$ follows from the equipotentiality of cell electrodes. Indeed, assuming that DMFC electrodes are ideally conductive (equipotential), the sum E of voltage losses on both sides of the cell must be constant along $\tilde{z}$:

$$E = \eta^a(\tilde{z}) + \eta^c(\tilde{z}) \tag{6.52}$$

Polarization voltages of the anode and the cathode sides are given by (6.38) and (6.39), respectively. In this section we will assume that both sides of the cell operate in the low-current regime, i.e. ϕ in (6.38) and (6.39) equals 1.

In dimensionless variables (6.45), Eqs. (6.38) and (6.39) then take a form

$$\frac{\eta^a}{b^a} = \ln\left(\frac{\tilde{j}}{q\tilde{\psi}}\right) - \ln\left(1 - \frac{\tilde{j}}{\tilde{\psi}}\right) + \ln(1 + \beta) \tag{6.53}$$

$$\frac{\eta^c}{b^c} = \ln\left(\frac{\tilde{j}}{\alpha q\tilde{\xi}}\right) - \ln\left(1 - \frac{\tilde{j} + \tilde{j}_{cross}}{\gamma\tilde{\xi}}\right) \tag{6.54}$$

where

$$q = \frac{l_t^a i_\star^a l_b^a}{6FD_b^a c_{ref}^a} \qquad \alpha = \frac{l_t^c i_\star^c c_{ref}^a}{l_t^a i_\star^a c_{ref}^c}\left(\frac{C^c \xi^0}{C^a \psi^0}\right) \qquad \gamma = \frac{2D_b^c l_b^a}{3D_b^a l_b^c}\left(\frac{C^c \xi^0}{C^a \psi^0}\right) \tag{6.55}$$

are parameters.

Introducing dimensionless potentials according to

$$\tilde{\eta} = \frac{\eta}{b^a} \qquad \tilde{E} = \frac{E}{b^a} \tag{6.56}$$

equation (6.52) takes a form $\tilde{\eta}^a + \tilde{\eta}^c = \tilde{E}$. With (6.53) and (6.54) we finally obtain

$$\begin{aligned} &\ln\left(\frac{\tilde{j}}{q\tilde{\psi}}\right) - \ln\left(1 - \frac{\tilde{j}}{\tilde{\psi}}\right) + \ln(1+\beta) \\ &+ p\left[\ln\left(\frac{\tilde{j}}{\alpha q \tilde{\xi}}\right) - \ln\left(1 - \frac{\tilde{j} + \tilde{j}_{cross}}{\gamma\tilde{\xi}}\right)\right] = \tilde{E} \end{aligned} \tag{6.57}$$

where

$$p = \frac{b^c}{b^a} \tag{6.58}$$

is the ratio of Tafel slopes. Note that to normalize potentials we used anodic Tafel slope b^a.

Equations (6.50), (6.51) and (6.57) is a 1D + 1D model of DMFC: the relation of local current density and voltage loss (6.57) couples two equations (6.50) and (6.51) for methanol and oxygen concentration in the respective channel.

Analysis of these equations for arbitrary set of parameters will be performed in Section 6.3.7. Here we will analyze the case of equal oxygen and methanol stoichiometries, when the system discussed has exact solution [30].

6.3.3 Solution for the Case of $\lambda^a = \lambda^c$

If $\lambda^a = \lambda^c$ solution to the system (6.50), (6.51) and (6.57) is

$$\tilde{\psi} = \tilde{\xi} \tag{6.59}$$

$$\tilde{j} = \kappa\tilde{\psi} \tag{6.60}$$

Indeed, under $\lambda^a = \lambda^c = \lambda$ Eqs. (6.50) and (6.51) coincide. The boundary conditions $\tilde{\psi}(0) = \tilde{\xi}(0) = 1$ also coincide; thus $\tilde{\psi} = \tilde{\xi}$. Substituting $\tilde{j} = \kappa\tilde{\psi}$ into (6.50) we get

$$\lambda \tilde{J} \frac{\partial \tilde{\psi}}{\partial \tilde{z}} = -(\kappa + \beta_\star(1-\kappa))\tilde{\psi} \tag{6.61}$$

Solving this equation we find

$$\tilde{\xi} = \tilde{\psi} = \exp\left(-\frac{\tilde{z}}{\mu}\right) \tag{6.62}$$

where μ is the characteristic length:

$$\mu = \frac{\tilde{J}\lambda}{\kappa + \beta_\star(1-\kappa)} \tag{6.63}$$

Using (6.62) in (6.60) we find

$$\tilde{j} = \kappa \exp\left(-\frac{\tilde{z}}{\mu}\right) \tag{6.64}$$

Parameter κ is hence local current density at the inlet: $\kappa = \tilde{j}(0)$. Substituting (6.62) and $\tilde{j} = \kappa\tilde{\psi}$ into (6.49) we get

$$\tilde{j}_{cross} = \beta_\star(1-\kappa)\exp\left(-\frac{\tilde{z}}{\mu}\right) \tag{6.65}$$

All variables thus decrease exponentially along $\tilde{z}$ with the same characteristic scale μ (6.63). Note that the exponential-like decay of local current density along the channel was obtained in 2D numerical calculations of Birgersson *et al.* [20,21].

Substituting $\tilde{j} = \kappa\tilde{\psi}$ into (6.53) and $\tilde{j} = \kappa\tilde{\xi} = \kappa\tilde{\psi}$ into (6.54) we find

$$\frac{\eta^a}{b^a} = \ln\left(\frac{\kappa}{q}\right) - \ln(1-\kappa) + \ln(1+\beta) \tag{6.66}$$

$$\frac{\eta^c}{b^c} = \ln\left(\frac{\kappa}{\alpha q}\right) - \ln\left(1 - \frac{\kappa + \beta_\star(1-\kappa)}{\gamma}\right) \tag{6.67}$$

We see that polarization voltages on both sides of the cell are constant along $\tilde{z}$.

Parameter κ is obtained from condition

$$\int_0^1 \tilde{j}\, d\tilde{z} = \tilde{J} \tag{6.68}$$

Using here (6.64) and calculating integral we find $\kappa\mu[1 - \exp(-1/\mu)] = \tilde{J}$. With (6.63), after simple manipulations we get equation for κ:

$$\frac{\kappa + \beta_\star(1-\kappa)}{\lambda \ln\left(1 - \dfrac{\kappa + \beta_\star(1-\kappa)}{\kappa\lambda}\right)} = -\tilde{J} \tag{6.69}$$

Equations (6.42), (6.66), (6.67) and (6.69) determine cell polarization curve. For given $\tilde{J}$, β and λ solution to (6.69) gives κ. Equations (6.66) and (6.67) then give η^a and η^c; cell voltage is calculated from (6.42).

Polarization curves for different β and λ are shown in Figure 6.6 (the other parameters are listed in Table 6.1). The increase in the rate of crossover β

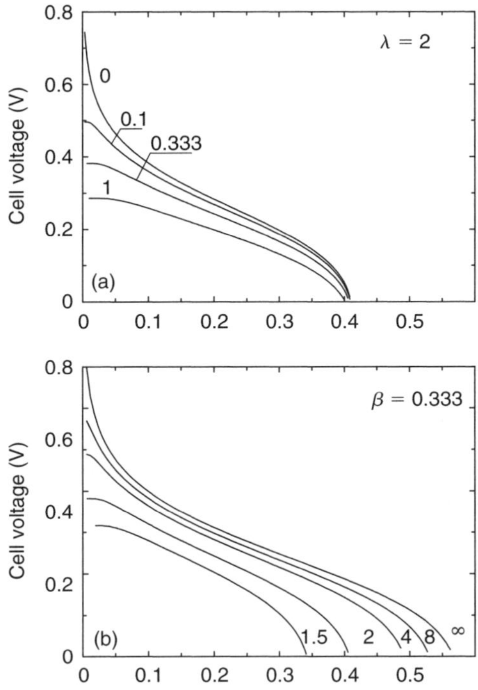

Figure 6.6 Voltage current curves of DMFC with equal stoichiometries of methanol and oxygen. (a) $\lambda = 2$; indicated are the values of crossover parameter β. (b) $\beta = 0.333$; indicated are the values of λ.

reduces cell open-circuit voltage V_{oc} (Figure 6.6(a)). Qualitatively, this is what we could expect.

Unexpectedly, however, V_{oc} depends on stoichiometry: with the growth of λ, V_{oc} increases (Figure 6.6(b)). At $\lambda = \infty$ cell open-circuit voltage reaches the value $V_{oc}^T - \tilde{E}_0(\lambda = \infty)$, where $\tilde{E}_0(\lambda = \infty) \lesssim 100\,\text{mV}$ (see below). Here E_0 is voltage loss at open circuit. The effect of V_{oc} lowering due to crossover (mixed potential) is well known in DMFC studies; the model enables to explain the origin of this effect.

6.3.4 Cell Depolarization at Zero Current: Mixed Potential

Analytical solutions of the previous section give exponentially decreasing with $\tilde{z}$, $\tilde{\psi}$, $\tilde{\xi}, \tilde{j}$, $\tilde{j}_{cross}$ and constant η^a and η^c. The function $\tilde{j}(\tilde{z})$ (6.64) is of the largest interest, since κ appears in polarization voltages (6.66) and (6.67).

Parameter κ is solution to Eq. (6.69); μ is given by (6.63). Behavior of κ and μ as $\tilde{J} \to 0$ determines cell voltage at zero current. To understand this behavior consider first the limiting cases.

Table 6.1 Parameters of a typical DMFC

	Anode side	Cathode side
Temperature (°C)	70	70
Pressure (atm)	2	2
D_b (cm^2/s)	2×10^{-5}	3×10^{-3}
$i_\star$ (A/cm^3)	10^{-2}	1
c_{ref} (mol/cm^3)	10^{-3}	7×10^{-5}
Oxygen molar fraction ξ^0	–	0.2
Methanol molar fraction ψ^0	0.018 (1M)	–
Tafel slope b (V)	0.05	0.05
Catalyst layer thickness l_t (cm)	10^{-3}	10^{-3}
Backing layer thickness l_b (cm)	2×10^{-2}	2×10^{-2}
Membrane thickness l_m (cm)	10^{-2}	
D_m (cm^2/s) for $\beta = 0.1, 0.333, 1$	$1 \times 10^{-6}, 3.33 \times 10^{-6}, 10^{-5}$	

In the case of zero crossover we put $\beta_\star = 0$ and (6.69) gives

$$\kappa = -\tilde{J}\lambda \ln\left(1 - \frac{1}{\lambda}\right) \tag{6.70}$$

The characteristic scale μ (6.63) then is

$$\mu = -\left[\ln\left(1 - \frac{1}{\lambda}\right)\right]^{-1} \tag{6.71}$$

We see that as $\tilde{J} \to$ zero, the "amplitude" κ of local current density (6.64) also tends to zero, whereas μ does not change. If, therefore, a cell without crossover is run at a constant $\lambda^a = \lambda^c = \lambda$, the $\tilde{z}$ shape of local current density is the same for all points of polarization curve. Variation of $\tilde{J}$ simply rescales the whole curve $\tilde{j}(\tilde{z})$.

Crossover, however, adds additional degree of freedom. If $\beta_\star \neq 0$, the characteristic scale of exponent μ varies with $\tilde{J}$. This can be shown explicitly for the case of large λ: the logarithm in (6.69) then can be expanded and we get $\kappa = \tilde{J}$. With this κ, from (6.63) we find

$$\mu = \frac{\tilde{J}\lambda}{\tilde{J} + \beta_\star(1 - \tilde{J})} \tag{6.72}$$

Now $\mu \to 0$ as $\tilde{J} \to 0$. Therefore, as mean current density tends to zero, both κ and μ in (6.64) tend to zero. Thus, at small $\tilde{J}$ local current $\tilde{j}$ decreases with $\tilde{z}$ faster than at large $\tilde{J}$.

Most interesting is the case of finite λ. Numerical solution to (6.69) for various λ is shown in Figure 6.7. As $\tilde{J} \to 0$ we again have $\mu \to 0$, whereas κ now tends to non-zero value $\kappa_0 \equiv \kappa|_{\tilde{J}=0}$ (Figure 6.7). The value of κ_0 is solution to (6.69) with $\tilde{J} = 0$. Clearly, as $\tilde{J} \to 0$ the expression under the logarithm in (6.69) should tend to zero. For κ_0 we thus have an equation

$$1 - \frac{\kappa_0 + \beta_\star(1 - \kappa_0)}{\kappa_0\lambda} = 0$$

The solution is

$$\kappa_0 = \frac{\beta_\star}{\lambda - 1 + \beta_\star} = \frac{\beta}{\lambda(1 + \beta) - 1} \tag{6.73}$$

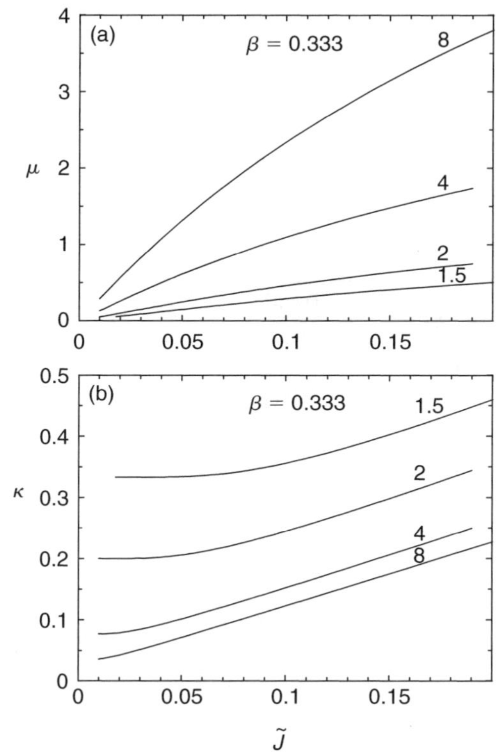

Figure 6.7 (a) Characteristic scale of the exponent μ and (b) amplitude factor κ in (6.64) as a function on mean current density $\tilde{J}$ for indicated values of λ. Crossover parameter $\beta = 0.333$.

With $\kappa = \kappa_0$ (6.63) gives

$$\mu = \frac{\lambda(\lambda(1+\beta)-1)\tilde{J}}{\beta(1+\lambda)} \tag{6.74}$$

and thus $\mu \to 0$ as $\tilde{J} \to 0$.

Therefore, as $\tilde{J} \to 0$ near the inlet forms a narrow "bridge" of local current with finite amplitude κ_0. The width of the bridge μ vanishes with $\tilde{J}$, whereas its amplitude κ_0 does not!

This is illustrated in Figure 6.8. In the case of zero crossover κ (6.70) decreases with $\tilde{J}$ and μ (6.71) remains constant (Figure 6.8(a)). In the logarithmic scale the curves in Figure 6.8(a) are parallel straight lines.

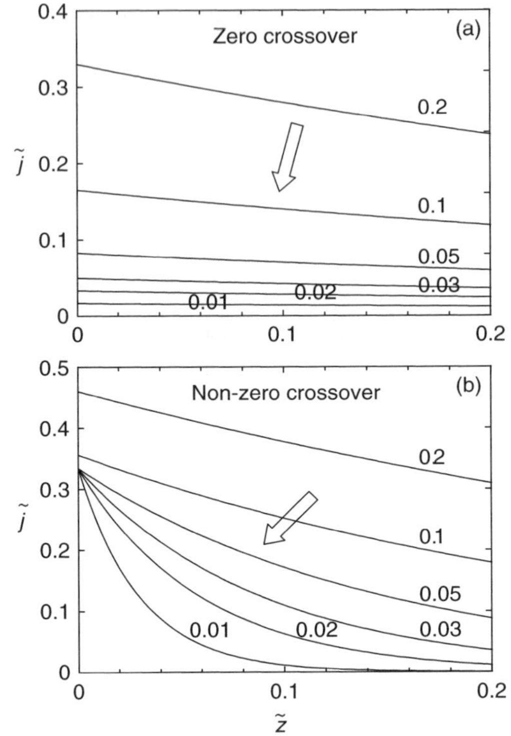

Figure 6.8 The shape of local current density along the channel for indicated values of mean current density $\tilde{J}$ in a cell. (a) The case of zero crossover ($\beta = 0$) and (b) non-zero crossover ($\beta = 0.333$). In both cases $\lambda = 1.5$. Arrows indicate evolution of the curve as $\tilde{J} \to 0$.

In the case of non-zero crossover, μ (6.74) tends to zero with $\tilde{J}$, while κ tends to a constant value κ_0 (6.73) (Figure 6.8(b)). Furthermore, *mean current density* in the bridge $\tilde{j}_{br}$ remains finite as $\tilde{J} \to 0$. Indeed,

$$\tilde{j}_{br} \simeq \frac{1}{\mu} \int_0^1 \kappa_0 \exp\left(-\frac{\tilde{z}}{\mu}\right) d\tilde{z} = \kappa_0 \left[1 - \exp\left(-\frac{1}{\mu}\right)\right]$$

Since $\mu \to 0$ as $\tilde{J} \to 0$, we have

$$\tilde{j}_{br} \to \kappa_0 = \frac{\beta_\star}{\lambda - 1 + \beta_\star} \tag{6.75}$$

Polarization voltages (6.66) and (6.67) depend on the local current density rather than on the total current in the system. Formation of the bridge

Table 6.2 Cell voltage drop (mV) at zero current for indicated values of β and λ

β	0.1	0.333	1
E_0 ($\lambda = 8$)	520	637	738
E_0 ($\lambda \to \infty$)	13	43	135

with finite current density thus induces finite η^a and η^c. Furthermore, since η^a and η^c are constant along z, the bridge is immediately "seen" everywhere along the cell surface.

Total voltage loss at open circuit is a sum $E_0 = \eta^a + \eta^c$, calculated with $\kappa = \kappa_0$. Using here (6.66) and (6.67) we get

$$\begin{aligned} E_0 &= b^a \left[\ln\left(\frac{\kappa_0}{q}\right) - \ln(1-\kappa_0) + \ln(1+\beta) \right] \\ &\quad + b^c \left[\ln\left(\frac{\kappa_0}{\alpha q}\right) - \ln\left(1 - \frac{\kappa_0 + \beta_\star(1-\kappa_0)}{\gamma}\right)\right] \end{aligned} \tag{6.76}$$

where κ_0 is given by (6.73).

With the parameters in Table 6.1 and $\lambda = 8$ we find the values of E_0, shown in the second row of Table 6.2. The bridge reduces DMFC open-circuit voltage by 500–700 mV, which agrees with the experiments [31, 32,27].

It should be emphasized that V_{oc} decreases as soon as the cell is connected to any load resistance. The bridge forms already at very small (strictly speaking, at infinitesimal) total current in the load. Care thus should be taken to prevent leakage current in a wet environment: this current allows bridge to form. If the cell is polarized, fuel and oxygen in a wet environment will be consumed even in the absence of a "real" load.

Equation (6.73) does not contain kinetic parameters of electrochemical reactions. This means that the bridge arises regardless of these kinetic details; the nature of the bridge is determined by electrostatic and transport phenomena. However, the value of voltage loss E_0 (6.76) due to the bridge includes parameters of reactions.

Relation (6.73) shows that $\kappa_0 \to 0$ as $\lambda \to \infty$. Therefore, the voltage loss E_0 can be reduced increasing λ at small currents (Figure 6.6(b)). Physically, the increase in λ increases the characteristic size of the bridge μ (6.74) (Figure 6.7(a)), "smearing" the bridge over larger surface.

6.3.5 The Limit of Large λ: 1D Cell

If $\lambda \gg 1 + \beta_\star(1/\kappa - 1)$, the logarithm in (6.69) can be expanded and we get $\kappa = \tilde{J}$. Substituting $\kappa = \tilde{J}$ into (6.66) and (6.67) and using the resulting expressions to calculate voltage loss $E = \eta^a + \eta^c$, we find

$$E = b^a \left[\ln\left(\frac{\tilde{J}}{q}\right) - \ln\left(1 - \tilde{J}\right) + \ln(1 + \beta) \right] + b^c \left[\ln\left(\frac{\tilde{J}}{\alpha q}\right) - \ln\left(1 - \frac{\tilde{J} + \beta_\star(1 - \tilde{J})}{\gamma}\right) \right] \quad (6.77)$$

The same relation can be obtained if we put in (6.53) and (6.54) $\tilde{\psi} = \tilde{\xi} = 1$ (no variation along the channel), $\tilde{j} = \tilde{J}$ (local current equals mean current) and sum up the resulting expressions. Equation (6.77) thus coincides with the voltage loss resulted from 1D model of DMFC. Clearly, in the case of large λ the variation of all parameters along the channel is negligible and we come to the relation for 1D cell.

In this limit the bridge does not form ($\kappa_0 = 0$) and E_0 is given by non-vanishing terms in (6.77) as $\tilde{J} \to 0$. However, one has to be careful when passing to the limit $\tilde{J} \to 0$ in (6.77). The formal limit cannot be calculated, since reaction activation terms in (6.77) originate from Tafel equation, which is not valid in the limit of zero $\tilde{J}$. At small currents Tafel equation should be replaced with the Bultler–Volmer formula, which gives zero activation voltage at zero current[2].

To pass to the limit $\tilde{J} \to 0$ in (6.77) we thus have to drop out the activation terms $\ln(\tilde{J}/q), \ln(\tilde{J}/(\alpha q))$ and put $\tilde{J} = 0$ in the other terms. This gives

$$E_0(\lambda \to \infty) = b^a \ln(1 + \beta) - b^c \ln\left(1 - \frac{\beta}{(1 + \beta)\gamma}\right) \quad (6.78)$$

With the parameters from Table 6.1 we get the values E_0 ($\lambda \to \infty$), shown in the third row of Table 6.2. For all β "1D values" E_0 ($\lambda \to \infty$) are significantly smaller, than E_0 ($\lambda = 8$). We see that in 1D approximation the effect of methanol crossover does not explain detected in experiments reduction of cell open-circuit voltage.

6.3.6 Cross-Linked Feeding

Consider cell operation at the oxygen-limiting current density. Formally, operation at the limiting current is equivalent to infinite voltage loss $\tilde{E}$. Thus, the

[2]Calculation of the limit $\tilde{J} \to 0$ in the previous section is justified, since $\kappa \simeq \kappa_0$ already at finite $\tilde{J} \simeq \tilde{J}_\star$. (Figure 6.8(b)). For all curves shown in Figure 6.8(b) Tafel equation at $\tilde{J} = \tilde{J}_\star$ is valid.

expression under the last logarithm on the left side of (6.57) should tend to zero and we get $\tilde{j} + \tilde{j}_{cross} = \gamma\tilde{\xi}$ or $\tilde{j} + \beta_\star(\tilde{\psi} - \tilde{j}) = \gamma\tilde{\xi}$. Using here $\tilde{j} = \kappa\tilde{\psi}$ and $\psi = \xi$ (6.60) we find

$$\kappa + \beta_\star(1 - \kappa) = \gamma$$

If $\gamma = 1$ we get $\kappa = 1$ and thus $\tilde{j}_{cross} = \beta_\star(1 - \kappa)\psi = 0$ *regardless of* $\beta_\star$. This case is of particular interest.

$\gamma = 1$ is equivalent to $j_{\lim}^{a0} = j_{\lim}^{c0}$ or $6FD_b^a c_M^0/l_b^a = 4FD_b^c c_{ox}^0/l_b^c$. In other words, at the inlet the flux of oxygen across the cathode backing layer equals the flux of methanol across the anode backing layer. Both fluxes are maximal, since they provide limiting current density. The concentrations of oxygen and methanol in the respective catalyst layer thus tend to zero.

The fact that $\tilde{j}_{cross}(\tilde{z}) = 0$ regardless of β means that under $\lambda^a = \lambda^c$ and $\gamma = 1$ the equality of oxygen and methanol through-plane fluxes at $\tilde{z} = 0$ provides their equality at any $\tilde{z}$. The condition

$$\lambda^a = \lambda^c \quad \text{and} \quad j_{\lim}^{a0} = j_{\lim}^{c0} \tag{6.79}$$

thus describes the regime of *cross-linked feeding*: at any $\tilde{z}$ available methanol and oxygen are consumed, so that regardless of β crossover is exactly zero and one does not need the membrane.

Equation (6.69) with $\kappa = 1$ gives

$$\tilde{J} = -\left[\lambda \ln\left(1 - \frac{1}{\lambda}\right)\right]^{-1} \tag{6.80}$$

Thus, in this regime the limiting current density in a cell is a function of λ only. Substituting this $\tilde{J}$ and $\kappa = 1$ into (6.63) we find

$$\mu = -\left[\ln\left(1 - \frac{1}{\lambda}\right)\right]^{-1} \tag{6.81}$$

This value coincides with those obtained for $\beta_\star = 0$ (*cf.* (6.71)). Physically, the regime of cross-linked feeding is equivalent to the case of ideal membrane with zero crossover. Using (6.81) in (6.62) and (6.64), we get

$$\tilde{\psi} = \tilde{\xi} = \tilde{j} = \left(1 - \frac{1}{\lambda}\right)^{\tilde{z}} \tag{6.82}$$

Thus, all variables in this regime are one–parameter functions of $\tilde{z}$.

6.3.7 The General Case of Arbitrary λ^a and λ^c

6.3.7.1 Equation for local current

Formation of the bridge is caused by methanol crossover. To understand the nature of the effect it is advisable to consider the case of large methanol stoichiometry. For further calculations it is convenient to convert (6.57) into differential equation for $\tilde{j}$.

Differentiating (6.57) with respect to $\tilde{z}$ and using (6.50) and (6.51) to exclude derivatives $\partial\tilde{\psi}/\partial\tilde{z}$ and $\partial\tilde{\xi}/\partial\tilde{z}$, after simple algebraic manipulations we come to

$$\frac{\partial\tilde{j}}{\partial\tilde{z}} = \frac{A_J}{B_J} \qquad \tilde{j}(0) = \kappa \tag{6.83}$$

where

$$\begin{aligned} A_J = \tilde{j}(\tilde{j} + \beta\tilde{\psi})\Big[\gamma(1+\beta)\lambda^c\tilde{\xi} + \Big(p\gamma(1+\beta)\lambda^a - \beta(1+p)\lambda^c\Big)\tilde{\psi} \\ + \Big((p\beta - 1)\lambda^c - p\gamma(1+\beta)\lambda^a\Big)\tilde{j}\Big] \end{aligned} \tag{6.84}$$

$$\begin{aligned} B_J = \lambda^a\lambda^c\tilde{J}(1+\beta)\Big[\Big(p\gamma(1+\beta)\tilde{\xi} + (1-p\beta)\tilde{\psi}\Big)\tilde{j} \\ + (1+p)\Big(\beta\tilde{\psi} - \gamma(1+\beta)\tilde{\xi}\Big)\tilde{\psi}\Big] \end{aligned} \tag{6.85}$$

Equation for κ is obtained if we substitute inlet values $\tilde{j} = \kappa$ and $\psi = \xi = 1$ into (6.57):

$$\begin{aligned} \ln\left(\frac{\kappa}{q}\right) - \ln(1-\kappa) + \ln(1+\beta) \\ + p\left[\ln\left(\frac{\kappa}{\alpha q}\right) - \ln\left(1 - \frac{\kappa + \beta_\star(1-\kappa)}{\gamma}\right)\right] = \tilde{E} \end{aligned} \tag{6.86}$$

Note that (6.83) does not contain α and q. These parameters appear in the solution only through the value of κ (6.86). In other words, α and q simply re-scale the curve $\tilde{j}(\tilde{z})$ as a whole, not affecting its shape.

In the general case of arbitrary λ^a and λ^c DMFC performance is governed by the system (6.50), (6.51) and (6.83). Parameters $\tilde{J}$ and $\tilde{E}$ determine the point on polarization curve. Mean current density in a cell is $\tilde{J}$; solution to (6.83) must, therefore, obey the relation

$$\int_0^{\tilde{z}_{\min}} \tilde{j}\, d\tilde{z} = \tilde{J} \tag{6.87}$$

where $\tilde{z}_{\min} = \min\{\tilde{z}_j, 1\}$ (see below). Iterations are required to determine $\tilde{E}$, which provides fulfillment of (6.87). Cell voltage may be calculated from (6.42) with $\tilde{j} = \tilde{J}$.

6.3.8 Numerical Solution

Numerical solution to the system (6.50), (6.51) and (6.83) is obtained by the standard Runge–Kutta method. Figure 6.9 shows total voltage loss $\tilde{E}$ as a function of $\tilde{J}$ for $\lambda^a = 4$, $\lambda^c = 2$ and the indicated values of parameter β. At small currents all curves form the plateau (Figure 6.9), which indicates formation of the bridge. The length of the plateau increases with β (Figure 6.9).

The shapes of local current density in the several points on polarization curve for $\beta = 1$ (Figure 6.9) are depicted in Figure 6.10. In contrast to the

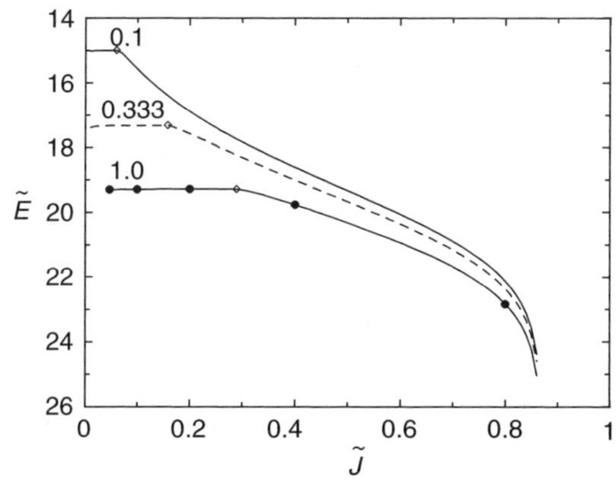

Figure 6.9 Total voltage loss $\tilde{E}$ versus mean current density $\tilde{J}$ for indicated values of crossover parameter β. Stoichiometries are $\lambda^a = 4$, $\lambda^c = 2$. The other parameters are $\alpha = 20.04$, $\gamma = 2.803$, $q = 1.727 \times 10^{-5}$. The profiles of local current density in the black points are shown in Figure 6.10. Diamonds: the points, where the bridge forms ($\tilde{z}_j = 1$, see the text).

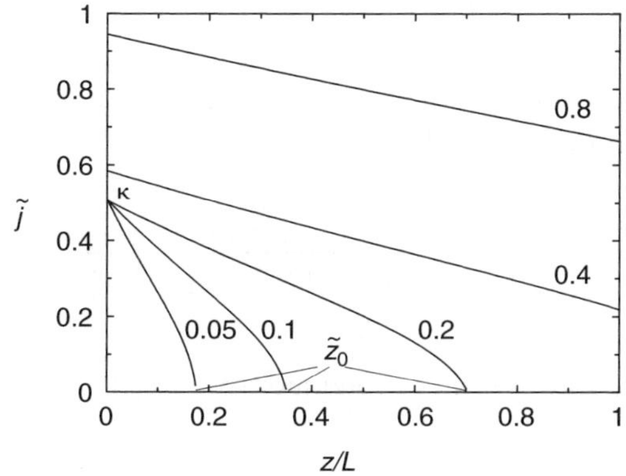

Figure 6.10 The profiles of local current density for indicated values of dimensionless mean current density in a cell and $\beta = 1$. The other parameters are the same as in Figure 6.9. The polarization curve for $\beta = 1$ is shown in Figure 6.9.

exponential shape of local current at $\lambda^a = \lambda^c$ (Figure 6.8(b)), in the case of $\lambda^a > \lambda^c$ the bridge has finite spread over the cell surface (Figure 6.10). At certain $\tilde{z}_j$ local current vanishes (Figure 6.10) and to the right of $\tilde{z}_j$ the solution does not exist.

As $\tilde{J}$ decreases, the point $\tilde{z}_j$ moves to zero, whereas κ remains constant (Figure 6.10). This means bridge "shrinking", quite analogous to that considered in the previous section.

To understand the effect of a localized current on cell performance in the next sections we derive "low–current" solutions to the system (6.50), (6.51) and (6.83) in the limiting cases of large and small methanol stoichiometry.

6.3.9 Large Methanol Stoichiometry, Small Current

6.3.9.1 The Shape of the Bridge

In the limit of small current we expand the right side of Eq. (6.83) over $\tilde{j}$ and retain only linear term. This yields

$$\frac{\partial \tilde{j}}{\partial \tilde{z}} = \frac{\beta\left[\lambda^c\gamma(1+\beta)\tilde{\xi} + \left(p\lambda^a\gamma(1+\beta) - \lambda^c\beta(1+p)\right)\tilde{\psi}\right]\tilde{j}}{\lambda^a\lambda^c\tilde{J}(1+\beta)(1+p)\left[\beta\tilde{\psi} - \gamma(1+\beta)\tilde{\xi}\right]} \tag{6.88}$$

Large methanol stoichiometry means that $\tilde{\psi} \simeq 1$. Putting in (6.88) $\tilde{\psi} = 1$ we find

$$\frac{\partial \tilde{j}}{\partial \tilde{z}} = \frac{\beta\left[\lambda^c\gamma(1+\beta)\tilde{\xi} + p\lambda^a\gamma(1+\beta) - \lambda^c\beta(1+p)\right]\tilde{j}}{\lambda^a\lambda^c\tilde{J}(1+\beta)(1+p)\left[\beta - \gamma(1+\beta)\tilde{\xi}\right]} \tag{6.89}$$

Since $\gamma \gtrsim 1, p \gtrsim 1, \tilde{\xi} \leqslant 1$ and $\lambda^a \gg \lambda^c$, we can omit terms with λ^c in the numerator of (6.89). This leads to

$$\lambda^c\tilde{J}\frac{\partial \tilde{j}}{\partial \tilde{z}} = \frac{\beta p\gamma\tilde{j}}{(1+p)\left[\beta - \gamma(1+\beta)\tilde{\xi}\right]} \tag{6.90}$$

Equation for $\tilde{\xi}$ in this limit is obtained from (6.51) if we put $\psi = 1$ there and neglect $\tilde{j}$ on the right side:

$$\lambda^c\tilde{J}\frac{\partial \tilde{\xi}}{\partial \tilde{z}} = -\beta_\star \qquad \tilde{\xi}(0) = 1 \tag{6.91}$$

This equation is valid provided that $\tilde{j} \ll \beta_\star$. Physically, $\tilde{j}$ should be much less than crossover current density; oxygen consumption then is determined mainly by crossover.

Solution to (6.91) gives linear decrease of oxygen fraction with the distance:

$$\tilde{\xi} = 1 - \frac{\tilde{z}}{\tilde{z}_{ox}} \tag{6.92}$$

Here

$$\tilde{z}_{ox} = \frac{\lambda^c \tilde{J}}{\beta_\star} \tag{6.93}$$

is a point where $\tilde{\xi}$ vanishes.

Substituting (6.92) into (6.90) and solving the resulting equation we find

$$\tilde{j} = \kappa \left(1 - \frac{\tilde{z}}{\tilde{z}_j} \right)^{\frac{p}{1+p}} \tag{6.94}$$

where

$$\tilde{z}_j = \lambda^c \tilde{J} \left(\frac{1}{\beta_\star} - \frac{1}{\gamma} \right) \tag{6.95}$$

Local current density (6.94) for several values of $\tilde{J}$ is shown in Figure 6.11 together with the numerical solutions of the full system of equations (6.50), (6.51) and (6.83). Again, as in Figure 6.10, current is localized in the domain $0 \leqslant \tilde{z} \leqslant \tilde{z}_j$ (Figure 6.11).

This localization explains the nature of the bridge. According to Eq.(6.49), crossover current $\tilde{j}_{cross}$ increases with the decrease in $\tilde{j}$. Suppose that we are moving along the cathode/anode channel from the inlet towards the outlet. Local current $\tilde{j}$ decreases with $\tilde{z}$ due to oxygen consumption. Methanol is in excess ($\tilde{\psi} = 1$); thus, according to (6.49) local crossover current increases with $\tilde{z}$. As we move along $\tilde{z}$ more oxygen is spent for burning of permeated methanol and thus less oxygen is left for current production. At $\tilde{z}_j$ (6.95) local current is zero, whereas $\tilde{j}_{cross}$ is maximal; starting from that point all available oxygen is consumed in the reaction with permeated methanol.

This picture takes place when $\lambda^a \gg \lambda^c$, when everywhere along the channel $\tilde{\psi} = 1$. In the case of $\lambda^a = \lambda^c$ local fraction of methanol $\tilde{\psi}$ decreases along $\tilde{z}$ and thus $\tilde{j}_{cross}$ increases with $\tilde{z}$ slower. Certain amount of oxygen is then available for current production everywhere; this leads to an exponential decay of local current along $\tilde{z}$.

According to (6.92), to the right of $\tilde{z}_{ox}$ there is no oxygen at all. Note that the length of the domain, where all oxygen is consumed in the reaction with permeated methanol $\Delta\tilde{z} = \tilde{z}_{ox} - \tilde{z}_j = \lambda^c \tilde{J}/\gamma$ increases with the mean current in the system $\tilde{J}$.

Current density at the inlet κ is obtained from condition $\int_0^{\tilde{z}_j} \tilde{j}\, d\tilde{z} = \tilde{J}$. Integrating (6.94) we find

$$\kappa = \frac{\gamma\beta_\star(1 + 2p)}{\lambda^c(\gamma - \beta_\star)(1 + p)} \tag{6.96}$$

Thus, κ does not depend on $\tilde{J}$, whereas $\tilde{z}_j$ (6.95) is proportional to $\tilde{J}$. Figure 6.11 confirms this result.

As before, mean current density in the bridge $\tilde{j}_{br} = \frac{1}{\tilde{z}_j}\int_0^{\tilde{z}_j} \tilde{j}\, d\tilde{z}$ remains constant as $\tilde{J} \to 0$. Using here (6.94) and calculating integral we find

$$\tilde{j}_{br} = \frac{(1 + p)\kappa}{1 + 2p} = \left[\lambda^c\left(\frac{1}{\beta_\star} - \frac{1}{\gamma}\right)\right]^{-1} \tag{6.97}$$

We see that $\tilde{j}_{br}$ does not depend on $\tilde{J}$.

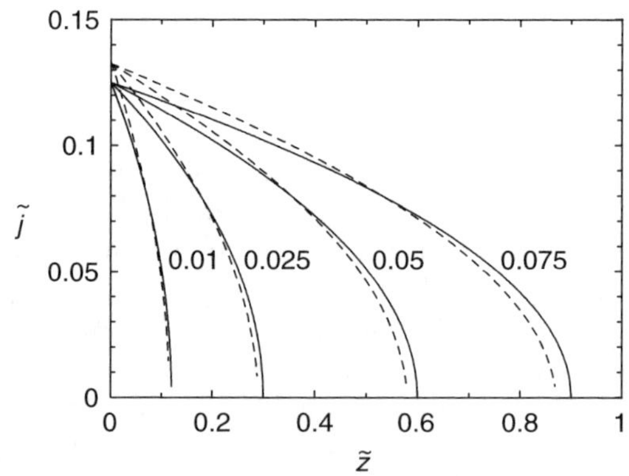

Figure 6.11 Evolution of local current density profile upon variation of the total current $\tilde{J}$ (indicated) in the cell. Solid lines: analytical solution (6.94) of reduced system and dashed lines: numerical solution of the full system of equations (6.50), (6.51) and (6.83). Parameters are $\beta = 1$, $\lambda^a = 100$, $\lambda^c = 8$, $p = 1$, $\gamma = 2$. The other parameters required for numerical solution are $\alpha = 20.04$, $q = 1.727 \times 10^{-5}$.

Parameter γ is proportional to oxygen concentration and to the mass transfer coefficient of oxygen through the cathode backing layer. Equations (6.95) and (6.97) show that when $\gamma = \beta_\star$ the bridge has zero thickness and infinite local current density. Therefore, at $\gamma < \beta_\star$ the cell does not work at all. Since $\beta_\star \leqslant 1$, we conclude that with $\gamma > 1$ the cell generates current at any rate of crossover.

What physically means small current? In this section all currents are normalized to the methanol-limiting current density at the inlet $j_{\text{lim}}^{a0} = 6FD_b^a C^a \psi^0 / l_b^a$. If methanol concentration is large enough, current in the cell is limited by the oxygen side. Small J then may be on the order of the oxygen-limiting current density in the system. The solution in Figure 6.11 then gives a picture of DMFC operation in a quite large range of operating current densities.

6.3.9.2 Plateau

Cell polarization curves for $\lambda^a \gg \lambda^c$ are qualitatively similar to the curves for $\lambda^a > \lambda^c$ shown in Figure 6.9. Suppose that we are moving along any of the curves in Figure 6.9 from large to small $\tilde{J}$. As soon as $\tilde{z}_j$ equals 1, mean current density in the current-carrying domain $[0, \tilde{z}_j]$ ceases to change (Eq. (6.97)). Since that moment this domain (bridge) supports constant mean current density.

Equation (6.95) allows us to calculate mean current density $\tilde{J}_\star$ when the bridge forms. Equating $\tilde{z}_j$ to 1 we get

$$\tilde{J}_\star = \left[\lambda^c \left(\frac{1}{\beta_\star} - \frac{1}{\gamma}\right)\right]^{-1} \tag{6.98}$$

As it should be, $\tilde{J}_\star$ coincides with the mean current density in the bridge (6.97).

For all $\tilde{J} \leqslant \tilde{J}_\star$ the value of κ does not vary with $\tilde{J}$ and thus the cell potential also does not change. $\tilde{J}_\star$ is hence the point where polarization curve enters the plateau (Figure 6.9).

As in the case of $\lambda^a = \lambda^c$, even at vanishingly small total current in the load the bridge supports finite current density (6.97) and thus it short-circuits DMFC electrodes. The respective loss in cell open-circuit voltage is given by (6.86) with κ (6.96). Indeed, since the sum of polarization voltages $\tilde{E}$ is constant along $\tilde{z}$, we may calculate $\tilde{E}$ at $\tilde{z} = 0$, where $\tilde{\psi} = \tilde{\xi} = 1$ and $\tilde{j} = \kappa$. Note that (6.86) and (6.96) give also the dependencies of $\tilde{E}$ on λ^c and β.

The shapes of the bridge for different λ^c are depicted in Figure 6.12. The increase in λ^c decreases κ and "smears out" the bridge over the larger domain. In view of (6.86) lower κ means higher cell open-circuit voltage.

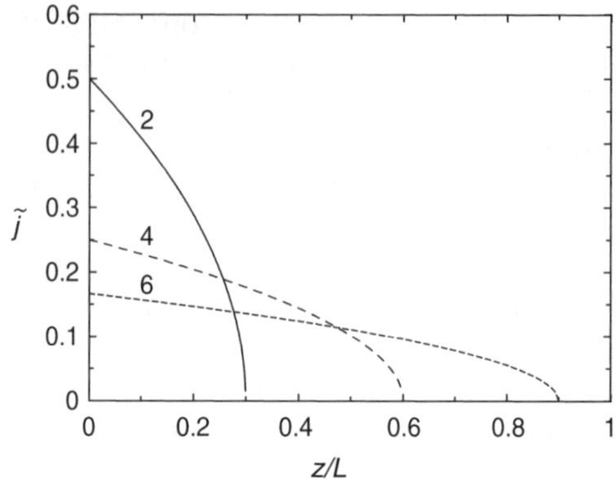

Figure 6.12 The shape of the bridge for indicated values of oxygen stoichiometry λ^c. $\beta = 1, \tilde{J} = 0.1, \lambda^a = 100$; parameters α, γ and q are the same as in Figure 6.9.

6.3.10 How to Determine $\beta_\star/\gamma$ Experimentally

The key parameters, which determine the regime of DMFC operation are crossover parameter $\beta_\star$ and transport parameter γ. The effect of the bridge provides a simple means for experimental determination of the ratio $\beta_\star/\gamma$.

Equation (6.98) can be transformed to

$$\frac{1}{\beta_\star} = \frac{1}{\gamma} + \frac{1}{\lambda^c \tilde{J}_\star} \tag{6.99}$$

This relation suggests the method for measurement of $\beta_\star/\gamma$. What we need is two polarization curves measured at the same $\lambda^c \ll \lambda^a$ but different inlet oxygen molar fractions ψ^0 and $\psi^0_k = k\psi^0$. From (6.99) we then get

$$\frac{1}{\beta_\star} = \frac{1}{\gamma} + \frac{1}{\lambda^c \tilde{J}_\star}$$

$$\frac{1}{\beta_\star} = \frac{1}{\kappa\gamma} + \frac{1}{\lambda^c \tilde{J}_{\star k}}$$

where $\tilde{J}_{\star k}$ is measured in experiment with $k\psi^0$. Here we have used relation $\gamma(k\psi_0) = k\gamma(\psi_0)$, which follows from the definition of γ (6.55).

Solving this system we find

$$\beta_\star = \lambda^c (k-1) \left(\frac{k}{\tilde{J}_{\star k}} - \frac{1}{\tilde{J}_\star} \right)^{-1} \tag{6.100}$$

$$\gamma = \frac{\lambda^c (k-1)}{k} \left(\frac{1}{\tilde{J}_{\star k}} - \frac{1}{\tilde{J}_\star} \right)^{-1} \tag{6.101}$$

These relations contain normalization factor j^{a0}_{lim}, which is usually not known. However, the ratio $\beta_\star/\gamma$ does not depend on j^{a0}_{lim}. From (6.100) and (6.101) we get

$$\frac{\beta_\star}{\gamma} = k \left(\frac{J_\star - J_{\star k}}{kJ_\star - J_{\star k}} \right) \tag{6.102}$$

where all current densities are in dimension form. The right side of (6.102) contains directly measurable values.

6.3.11 Large Oxygen Stoichiometry, Small Current

In that case in (6.88) we put $\tilde{\xi} = 1$ and neglect the term with λ^a in numerator. This yields

$$\lambda^a \tilde{J} \frac{\partial \tilde{j}}{\partial \tilde{z}} = \frac{\beta \left[\gamma(1+\beta) - \beta(1+p)\tilde{\psi} \right] \tilde{j}}{(1+\beta)(1+p) \left[\beta\tilde{\psi} - \gamma(1+\beta) \right]} \tag{6.103}$$

Equation for $\tilde{\psi}$ in this limit is obtained from (6.50) if we neglect $\tilde{j}$ on the right side of this equation:

$$\lambda^a \tilde{J} \frac{\partial \tilde{\psi}}{\partial \tilde{z}} = -\beta_\star \tilde{\psi} \qquad \tilde{\psi}(0) = 1 \tag{6.104}$$

Quite analogous to (6.91), this equation is valid provided that $\tilde{j} \ll \beta_\star$. Physically, small current means that methanol consumption is determined mainly by crossover.

Solution to (6.104) is

$$\tilde{\psi} = \exp\left(-\frac{\tilde{z}}{\mu_\psi} \right) \tag{6.105}$$

where the characteristic scale of the exponent is

$$\mu_\psi = \frac{\lambda^a \tilde{J}}{\beta_\star} \tag{6.106}$$

Substituting (6.105) into (6.103) and solving the resulting equation we find

$$\tilde{j} = \kappa \exp\left(-\frac{\tilde{z}}{\mu_j}\right) \tag{6.107}$$

where

$$\mu_j = \frac{\mu_\psi}{1+p} = \frac{\lambda^a \tilde{J}}{\beta_\star(1+p)} \tag{6.108}$$

and

$$\kappa = \beta_\star \left[\lambda^a(1+p)\left(1-\exp\left(-\frac{1}{\mu_j}\right)\right)\right]^{-1} \tag{6.109}$$

We see that local current decreases exponentially with the characteristic scale (6.108). Note that this scale is $(1 + p)$ times smaller than the scale of methanol (6.106).

As before, mean current density of the bridge does not depend on $\tilde{J}$:

$$\tilde{j}_{br} = \frac{1}{\mu_j}\int_0^1 \kappa \exp\left(-\frac{\tilde{z}}{\mu_j}\right) = \kappa\left(1-\exp\left(-\frac{1}{\mu_j}\right)\right) = \frac{\beta_\star}{\lambda^a\left(1+p\right)} \tag{6.110}$$

In the limit of $\tilde{J} \to 0$, κ (6.109) reduces to (6.110).

6.3.12 General Properties of the Bridge

Comparing (6.75), (6.97) and (6.110) we see that in all considered cases ($\lambda^a = \lambda^c$, $\lambda^a \gg \lambda^c$ and $\lambda^a \ll \lambda^c$) local current density in the bridge increases with $\beta_\star$ and decreases with the increase in stoichiometry of the "rate-determining" flow. This suggests that the bridge forms regardless of the relationship between λ^a and λ^c. Presumably, for $\lambda^c < \lambda^a$ the bridge has finite spread over the cell surface; for $\lambda^c \geqslant \lambda^a$ it has exponential shape along $\tilde{z}$.

6.3.13 Experimental Verification

Analysis above shows that the bridge is best seen when methanol stoichiometry largely exceeds oxygen stoichiometry. Cell polarization curves calculated for three values of λ^c under excess of methanol ($\lambda^c \ll \lambda^a$) are shown in Figure 6.13. As discussed above, the bridge manifests itself as a plateau at small current density. Furthermore, Eq. (6.98) shows that for various λ^c the

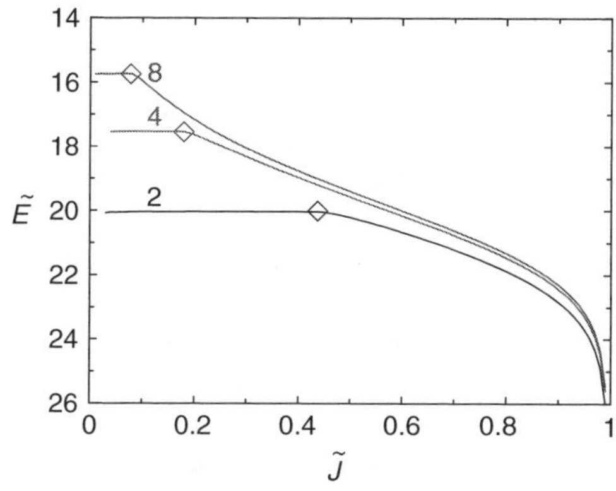

Figure 6.13 Model polarization curves of DMFC for indicated values of oxygen stoichiometry λ^c and $\lambda^a = 100$. Shown is the dimensionless total voltage loss $\tilde{E}$ as a function of mean current density in a cell $\tilde{J}$. Crossover parameter $\beta = 1$, the other parameters are the same as in Figure 6.9.

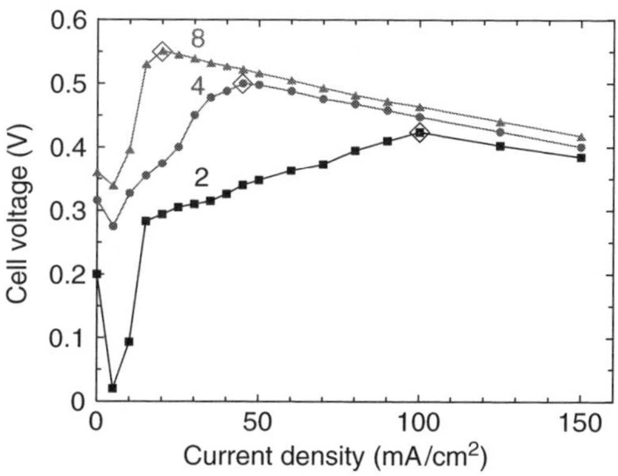

Figure 6.14 Experimental polarization curves of DMFC for indicated values of oxygen stoichiometry λ^c and $\lambda^a \gg \lambda^c$. Diamonds: the points, where the bridge forms (*cf.* Figure 6.13).

product $\tilde{J}_\star \lambda^c$ must remain constant. Numerical curves in Figure 6.13 confirm this result.

To verify these predictions the measurements of DMFC polarization curves in conditions $\lambda^c \ll \lambda^a, \tilde{J} \ll 1$ were performed [33]. For experimental details please see Ref. [33]. The results are shown in Figure 6.14. Comparing Figures 6.14 and 6.13 we see that the experimental polarization curves remarkably well reproduce two main features of the model curves: drastic change in the slope at a current density $J_\star$, where the bridge forms (large diamonds in Figure 6.14) and decrease of $J_\star$ with the increase in oxygen stoichiometry. Furthermore, Figure 6.14 indicates that the relation $J_\star \sim 1/\lambda^c$ (6.98) approximately holds: $J_\star$'s for $\lambda^c = 8, 4$ and 2 roughly are related as 1:2:4.

For $J \leqslant J_\star$ the model predicts constant cell voltage (Figure 6.13); the experimental curves, in contrast, exhibit decrease in cell voltage with the decrease in J to the left of $J_\star$ (Figure 6.14). This suggests that in the real cell local current density at the inlet κ *increases* as $J \to 0$. This may be due to the non-uniform flooding of the CCL by water produced in a reaction of methanol and oxygen. Since crossover is maximal where local current is minimal, the flooding of the cathode side is larger in the region of small local current. The effective current-generating domain may thus be narrower than that one determined by model equations.

If the bridge shrinks faster than prescribed by the model, local current density at the inlet must increase in order to provide required total current in a cell. The increase in κ inevitably leads to the increase in total voltage loss, or, equivalently, to the decrease in cell voltage.[★]

6.3.14 Oxygen and Methanol Utilization, Mean Crossover

Mass balance equations (6.47), (6.48) enable to formulate integral relations, which are valid regardless of the distribution of local current density in a cell. Equating the left sides of (6.47), (6.48) and integrating the result over $[0, \tilde{z}]$ we get

$$\lambda^a(\tilde{\psi} - 1) = \lambda^c(\tilde{\xi} - 1) \tag{6.111}$$

Introducing oxygen and methanol utilizations $u^c(\tilde{z}) = 1 - \tilde{\xi}(\tilde{z}), u^a(\tilde{z}) = 1 - \tilde{\psi}(\tilde{z})$ we immediately find that

$$\frac{u^a(\tilde{z})}{u^c(\tilde{z})} = \frac{\lambda^c}{\lambda^a} \tag{6.112}$$

Though u^a and u^c depend on $\tilde{z}$, their ratio does not.[3] In the case of $\lambda^c = \lambda^a$ oxygen and methanol utilizations are equal: $u^a = u^c$. Putting $\tilde{z} = 1$ in (6.112) we get the ratio of total methanol and oxygen utilizations in a cell: $u^a(1)/u^c(1) = \lambda^c/\lambda^a$.

Integrating (6.48) over $\tilde{z} \in [0,1]$ and taking into account that $\int_0^1 \tilde{j}\, d\tilde{z} = \tilde{J}$ after simple calculations we find

$$\frac{\tilde{J}_{cross}}{\tilde{J}} = \lambda^c u^c(1) - 1 = \lambda^a u^a(1) - 1 \tag{6.113}$$

where $\tilde{J}_{cross} = \int_0^1 \tilde{j}_{cross}\, d\tilde{z}$ and the last equality was obtained using (6.112).

[3] A more general form of Eq. (6.112), which takes into account gaseous bubbles in the anode channel is derived in Section 6.4.4.

★ The effect in fact is caused by formation of electrolytic domain; for the details please see the references quoted in the abstract (added upon proof reading).

The flux of methanol through the membrane is usually calculated measuring the amount of CO_2 in the cathodic exhaust [34,27]. This method, however, is not reliable since CO_2 permeates through the membrane from the anode to the cathode side [34]. Equation (6.113) provide a simple means for evaluation of the overall crossover current in DMFC: measuring oxygen and/or methanol concentration at the outlet one can calculate J_{cross} with (6.113).

When crossover is zero Eq. (6.113) gives

$$u^c(1) = \frac{1}{\lambda^c} \qquad u^a(1) = \frac{1}{\lambda^a} \tag{6.114}$$

These relations do not depend on the type of feed molecules and are thus valid also for hydrogen fuel cells.

6.4 DMFC with Gaseous Bubbles in the Anode Channel

Methanol oxidation gives one CO_2 molecule per each methanol. At typical working temperatures 70–90°C the amount of dissolved CO_2 is negligible and the molar flux of gaseous CO_2 in the anode channel may be comparable to the flux of liquid methanol there. In this section we develop a simple model of the flow with bubbles in DMFC anode channel [35]. The model is based on experimental findings of Yang et al. [36].

6.4.1 Flow Velocity

Figure 6.15 shows a picture of the flow in the anode channel at the mean current density 100 mA/cm^2 [36]. Close to the inlet the flow contains small bubbles, which do not dam the cross section of the channel (Figure 6.15). However, at the second and third turns of the meauder large bubbles form, which span the channel cross section. Due to the flux of gaseous CO_2 incoming into the channel, the length of large bubbles rapidly increases as they move towards the outlet (Figure 6.15).

This qualitative picture enables to formulate simple mass balance equation of the two-phase flow. Since amount of water largely exceeds amount of methanol, we may assume that the anode backing layer is always filled with the liquid. Gaseous methanol in the channel then does not contribute to current production, since gaseous transport to the catalyst sites is hindered. Moreover, the flux of gaseous CO_2 from the ACL to the channel even greater retards diffusion transport of gaseous methanol to the catalyst sites.

Large bubble blocks supply of liquid methanol to the adjacent porous volume. Local current density then partially consumes methanol from this volume. Estimate show that in typical conditions current does not consume

Figure 6.15 Anode flow field with bubbles (from Ref. [36]). Mean current density is $100\,\mathrm{mA/cm^2}$. The contrast is increased to represent the details.

all available in the backing layer methanol [35]. Next to the bubble arrives liquid plug, which partially restores local concentration of liquid methanol in the backing layer.

The concentration of methanol at any point of the channel provides the "boundary condition" for methanol transport across the cell. This concentration is a rapidly varying in time function, which jumps from methanol concentration in the liquid plug down to practically zero in the bubble. To derive the governing equation consider the following *gedankenexperiment*.

Suppose that we watch plugs and bubbles as they pass trough the channel cross section at z_0 (Figure 6.16). Each time when the plug crosses the plane z_0 we measure the respective time interval τ_p. Similarly, for every bubble crossing z_0 we measure time interval τ_{bb}. Let the total time of measurements is T.

Introducing mean over T lengths of plug $\overline{l}_p$ and bubble $\overline{l}_{bb}$ we can write

$$
\begin{aligned}
r \equiv \frac{\sum \tau_p}{T} &= \frac{\overline{l}_p}{\overline{l}_p + \overline{l}_{bb}} \\
\frac{\sum \tau_{bb}}{T} &= \frac{\overline{l}_{bb}}{\overline{l}_p + \overline{l}_{bb}} = 1 - r
\end{aligned}
\tag{6.115}
$$

where r and $1 - r$ are mean linear fractions of plugs and bubbles, respectively.

Now we can define time-average local molar concentration of the flow $\overline{C}$ as

$$
\overline{C} = rC_p + (1 - r)C_{bb} \tag{6.116}
$$

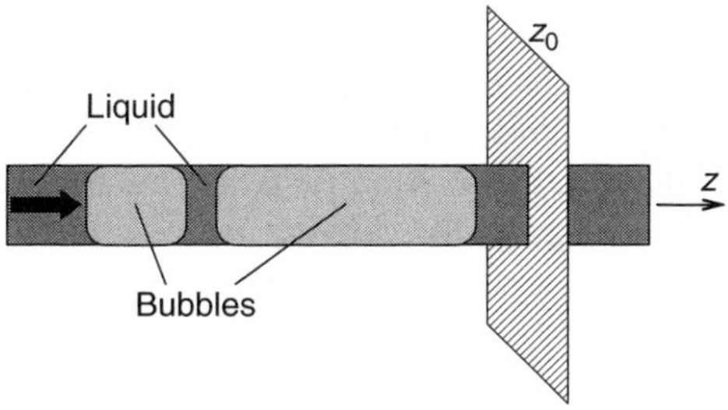

Figure 6.16 Sketch of the *gedankenexperiment* (see the text).

where C_p and C_{bb} are total molar concentrations of the plug and bubble, respectively. Since C_{bb} is roughly three orders of magnitude smaller than C_p, for any reasonable fraction of bubbles we have

$$\bar{C} \simeq rC_p \simeq rC^0 \tag{6.117}$$

The latter equality implies that methanol constitutes a small fraction in the mixture, so that methanol consumption does not change total molar concentration of the liquid.

Time-average mass flux of the flow must be continuous. Neglecting inertia of plugs we may assume that local velocity of plugs and bubbles is the same. For any z we then have $\nu^0 C^0 = \nu\bar{C} = \nu r C^0$, or

$$r\tilde{\nu} = 1 \tag{6.118}$$

Here

$$\tilde{\nu} = \frac{\nu}{\nu_0}$$

is dimensionless flow velocity. Qualitatively, due to the growth of bubbles parameter r decreases with z; the velocity of the flow thus must increase with the distance.

The "mean" bubble grows with the distance z due to the flux of CO_2 from the backing layer. For time average local concentration of carbon dioxide $\bar{c}_{CO_2}$ in the channel we may write

$$\frac{\partial(\nu\bar{c}_{CO_2})}{\partial z} = \frac{j}{6Fh} \tag{6.119}$$

Since all CO_2 is in the gas phase, it is reasonable to assume that the local fraction of bubbles is proportional to the local amount of CO_2 in the channel

$$1 - r = k\frac{\bar{c}_{CO_2}}{c^0} \tag{6.120}$$

Here c^0 is inlet methanol molar concentration, which is used here just as a convenient normalization factor. Simple Eq. (6.120) substitutes a detailed kinetics of bubble growth and coalescence in the channel.

Using Eq. (6.120) and introducing methanol stoichiometry $\lambda^a = 6Fhv^0c^0/(LJ)$, Eq. (6.119) can be transformed to

$$\lambda^a \tilde{J} \frac{\partial(\tilde{v}(1-r))}{\partial \tilde{z}} = k\tilde{j} \tag{6.121}$$

where

$$\tilde{j} = \frac{j}{j_{\lim}^{a0}} \qquad \tilde{J} = \frac{J}{j_{\lim}^{a0}} \qquad \tilde{z} = \frac{z}{L} \tag{6.122}$$

and $j_{\lim}^{a0}$ is given by (6.46).

In view of (6.118) $\partial(r\tilde{v})/\partial\tilde{z} = 0$ and Eq. (6.121) takes a form

$$\lambda^a \tilde{J} \frac{\partial \tilde{v}}{\partial \tilde{z}} = k\tilde{j} \qquad \tilde{v}(0) = 1 \tag{6.123}$$

Equation (6.123) shows that bubbles accelerate the flow at a rate proportional to local current density. Physically, j generates local flux of CO_2 into the channel. This flux increases local fraction of bubbles and thus flow velocity in the channel; the larger j the faster increases $\tilde{v}$ with $\tilde{z}$.

6.4.2 Mass Balance of Methanol and Oxygen

Time-average methanol concentration in the channel $\bar{c}$ obeys the equation

$$\frac{\partial(v\bar{c})}{\partial z} = -\frac{(j + j_{cross})}{6Fh} \tag{6.124}$$

where j_{cross} is given by (6.26). In dimensionless variables Eq. (6.124) takes a form

$$\lambda^a \tilde{J} \frac{\partial(\tilde{v}\tilde{c})}{\partial \tilde{z}} = -(\tilde{j} + \beta_\star(\tilde{c} - \tilde{j})) \qquad \tilde{c}(0) = 1 \tag{6.125}$$

where

$$\tilde{c} = \frac{\bar{c}}{c^0}$$

Equation (6.125) is the generalization of Eq. (6.50) to the case of variable flow velocity. Indeed, at $v = v^0$ we have $\tilde{v} = 1$. Furthermore, $\tilde{c} = \bar{c}/c^0 = \psi_p \bar{C}/(\psi^0 C^0) = \psi_p r C^0/(\psi^0 C^0) = \psi_p r/\psi^0 = \bar{\psi}/\psi^0 = \tilde{\bar{\psi}}_p$. Here ψ_p is methanol molar fraction in the liquid plug. Thus, in the absence of bubbles Eq. (6.125) reduces to (6.50).

Oxygen mass balance in the cathode channel is given by (6.51), where we should replace $\tilde{\psi}$ with $\tilde{c}$:

$$\lambda^c \tilde{J} \frac{\partial \tilde{\xi}}{\partial \tilde{z}} = -(\tilde{j} + \beta_\star(\tilde{c} - \tilde{j})) \qquad \tilde{\xi}(0) = 1 \tag{6.126}$$

As before, velocity of the flow in the cathode channel is assumed to be constant.

6.4.3 Potentials

The system of equations (6.123), (6.125) and (6.126) is completed by condition of equipotentiality of cell electrodes (6.57), where $\tilde{\psi}$ should be replaced with $\tilde{c}$:

$$\begin{aligned} &\ln\left(\frac{\tilde{j}}{q\tilde{c}}\right) - \ln\left(1 - \frac{\tilde{j}}{\tilde{c}}\right) + \ln(1+\beta) \\ &\quad + p\left[\ln\left(\frac{\tilde{j}}{\alpha q \tilde{\xi}}\right) - \ln\left(1 - \frac{\tilde{j} + \beta_\star(\tilde{c} - \tilde{j})}{\gamma \tilde{\xi}}\right)\right] = \tilde{E} \end{aligned} \tag{6.127}$$

All other parameters here are the same as in (6.57).

The system of equations (6.123), (6.125), (6.126) and (6.127) determines cell polarization curve. Solution of this system for certain $\tilde{E}$ and $\tilde{J}$ gives $\tilde{j}(\tilde{z})$. As before, iterations are required to determine $\tilde{E}$, which provides condition $\int_0^1 \tilde{j}\, d\tilde{z} = \tilde{J}$.

6.4.4 Conservation Laws

6.4.4.1 Limiting Flow Velocity in the Anode Channel

In the limit of negligible crossover ($\beta_\star \simeq 0$) Eqs. (6.123) and (6.125) determine the conservation law. Dividing (6.125) with $\beta_\star = 0$ by (6.123) we get

$$\frac{\partial(\tilde{v}\tilde{c})}{\partial \tilde{v}} = -\frac{1}{k} \tag{6.128}$$

Integrating this equation with the boundary condition $\tilde{c}|_{\tilde{v}=1} = 1$ we find

$$\tilde{v}\tilde{c} = 1 - \frac{\tilde{v} - 1}{k} \tag{6.129}$$

We see that for $\tilde{v} = 1 + k$ methanol concentration vanishes. Evidently, downstream from the point where $\tilde{c} = 0$ local current and CO_2 are not generated and flow velocity remains constant. In other words, $1 + k$ is the limiting velocity of the flow in the case of zero crossover. Clearly, crossover only reduces this value, since in the presence of crossover less methanol is available on the anode side to produce CO_2. We conclude that $1 + k$ is the upper limit of flow velocity in DMFC channel.

6.4.4.2 Bubbles and Feed utilization

Another integral of motion follows from Eqs. (6.125) and (6.126). Dividing (6.125) by (6.126) we get

$$\left(\frac{\lambda^a}{\lambda^c}\right)\frac{\partial(\tilde{v}\tilde{c})}{\partial\tilde{\xi}} = 1 \tag{6.130}$$

Taking into account that $\tilde{v}(1) = \tilde{c}(1) = \tilde{\xi}(1) = 1$ and integrating this equation from $\tilde{\xi} = 1$ to $\tilde{\xi}$ we find relation

$$\lambda^a(1 - \tilde{v}\tilde{c}) = \lambda^c(1 - \tilde{\xi}) \tag{6.131}$$

which holds for any $\tilde{z}$. This relation is the generalization of Eq. (6.111) to the case of anodic flow with variable velocity.

Introducing methanol and oxygen utilizations $u^a(\tilde{z}) = 1 - \tilde{c}(\tilde{z})$ and $u^c = 1 - \tilde{\xi}(\tilde{z})$, respectively, Eq. (6.131) transforms to

$$\lambda^c u^c = \lambda^a u^a - \lambda^c(\tilde{v} - 1)(1 - u^a) \tag{6.132}$$

The second term on the right side represents the effect of bubbles on utilizations. Since $\tilde{v} \geqslant 1$ and $u^a \leqslant 1$, the product $\lambda^c(\tilde{v} - 1)(1 - u^a)$ is positive. Comparing Eqs. (6.132) and (6.112) we see that bubbles always reduce oxygen utilization. The consequences of this effect will be discussed below. Note that (6.132) is valid for all $\tilde{z}$; the relation for total utilizations is obtained upon substitution $\tilde{z} = 1$ into (6.132).

6.4.5 Estimate of k

In view of (6.123) flow velocity monotonically increases with $\tilde{z}$; k thus determines maximal flow velocity at the outlet. To estimate a range of k variation the data [36] can be used. Measured outlet velocity in Ref. [36] is roughly 20 times larger, than the inlet velocity (at $J = 150\,\text{mA/cm}^2$ inlet and outlet values are 0.21 and 5 cm/s, respectively). With $k_{\max} = 30$ calculations give velocity at the outlet 10 times larger than at the inlet; we thus can take 30 as the characteristic maximal value of k corresponding to a flow with large fraction of bubbles.

k_{max} can also be estimated using another arguments. Physically, k depends on solubility of CO_2 in water, which is a strong function of temperature. To obtain k_{max} we assume zero solubility, i.e., all CO_2 produced in the reaction enters the channel in a gas phase. This assumption is fulfilled at typical for DMFC operating temperatures above 60°C, when solubility of CO_2 in water is low [37]. Molar concentration of pure CO_2 at atmospheric pressure is 3×10^{-5} mol/cm^3. With 1 M concentration of methanol at the inlet, Eq. (6.120) gives $1 - r \simeq 3 \times 10^{-2}\, k$. At $r = 0$ (negligible fraction of liquid plugs) we find $k \simeq 30$. This estimate coincides with that one performed above using the data on flow velocity in the anode channel. At room temperature solubility of CO_2 is high and the bubbles are not generated. This corresponds to $k = 0$; the range of k variation is thus $0 \leqslant k \lesssim 30$.

6.4.6 Numerical Polarization Curves

Equations (6.123) and (6.125)–(1.127) form a system of four equations for four unknowns: $\tilde{v}, \tilde{c}, \tilde{\xi}$ and $\tilde{j}$. Technically, to solve this system it is convenient to convert (6.127) into differential equation for $\tilde{j}$. Differentiating (6.127) with respect to $\tilde{z}$ and using (6.123), (6.125) and (6.126) to exclude derivatives $\partial\tilde{v}/\partial\tilde{z}$, $\partial\tilde{c}/\partial\tilde{z}$ and $\partial\tilde{\xi}/\partial\tilde{z}$, respectively, we get the desired equation (Appendix A, Eq. (6.230)).

Solution to the system (6.123), (6.125), (6.126) and (6.230) can be obtained by the standard Runge–Kutta method. Parameters for calculation (caption to Figure 6.17) correspond to typical DMFC operating conditions [38] (Table 6.1).

Figures 6.17 and 6.18 show the results for several k in the range $k = 0$ (no bubbles) to $k = 30$ (intensive bubbles formation). Figure 6.17 shows the profiles of methanol concentration, local current density and velocity down the channel. Figure 6.18 displays the respective polarization curves.

Under the conditions specified (caption to Figure 6.17) current in the cell is limited by the anode side. The maximal value of dimensionless limiting current density $\tilde{j}_{\lim}$ is thus 1 (i.e., in dimension form $j_{\lim} = j^{a0}_{\lim}$). Physically, $\tilde{j}_{\lim} = 1$ corresponds to uniform methanol concentration along the channel. This regime is realized if $\lambda^a \gg 1$ and $k = 0$ (no bubbles). Current in the cell is then limited only by the rate of methanol transport through the backing layer. Finite λ^a and/or formation of bubbles reduce this limit.

In the case of $k = 0$ (no bubbles) flow velocity is constant and the limiting current density is reduced only by finite λ^a. The reduction factor is $f_\lambda = -\lambda^a \ln(1 - (1/\lambda^a))$ [39]. For $\lambda = 4$ we get $f_\lambda \simeq 1.15$ and hence the limiting current density is $1/1.15 = 0.87$, in accordance with Figure 6.18(a).

Bubbles accelerate the flow and reduce local methanol concentration in the channel (Figure 6.17). Rapid decrease of $\tilde{c}$ with $\tilde{z}$ dramatically reduces limiting current: for $k = 30, \tilde{j}_{\lim}$ is 4 times smaller, than for $k = 0$

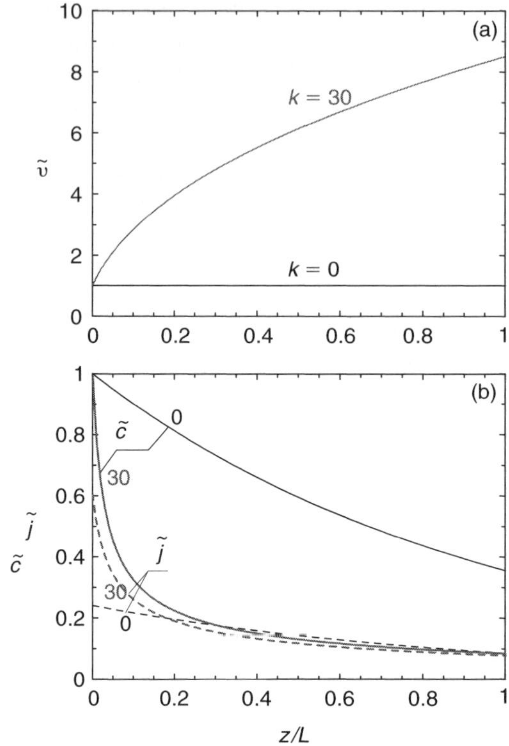

Figure 6.17 (a) Dimensionless flow velocity for $k = 0$ (no bubbles) and $k = 30$ (high rate of bubbles formation). (b) Local current density and mean methanol concentration down the channel for indicated values of k. The other parameters are $\alpha = 20.04$, $\gamma = 2.803$, $q = 1.727 \times 10^{-5}$. The respective polarization curves are shown in Figure 6.18.

(Figure 6.18(a)). Therefore, under specified conditions CO_2 bubbles dramatically lower DMFC performance.

The reduction of $\tilde{j}_{\lim}$ due to bubbles qualitatively is similar to that reduction due to finite stoichiometry of the flow. For example $\tilde{j}_{\lim} = 0.175$ (Figure 6.18) would exhibit the cell with pure liquid flow under $\lambda^a = 1.00337$ (the root of equation $-\lambda^a \ln(1 - 1/\lambda^a) = 1/0.175$). Thus, formation of bubbles may be thought of as a reduction of methanol stoichiometry. To partially compensate for the negative effect of bubbles larger methanol stoichiometries might be helpful. The results in Figure 6.18(b) confirm this conclusion.

Similar trend is detected in experiments [36]: the increase in inlet flow rate from 0.5 to 1 ml/min leads to better performance (Figure 13 in Ref. [36]). However, further increase in inlet flow rate up to 8 ml/min leads to performance degradation [36].

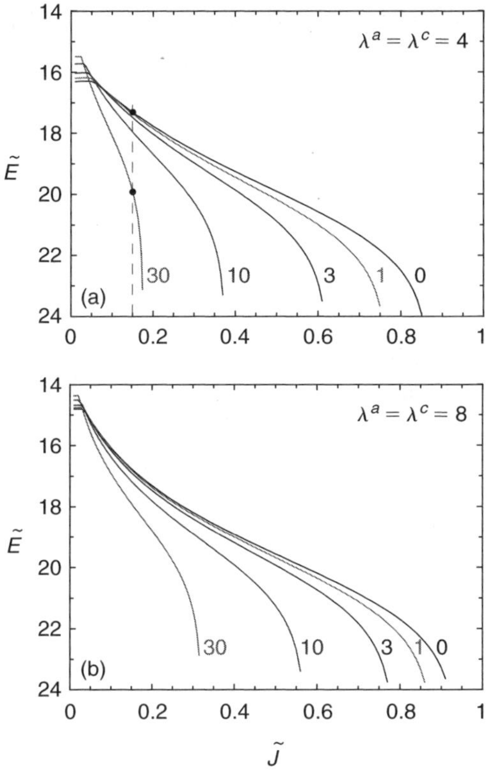

Figure 6.18 Total voltage loss $\tilde{E}$ versus mean current density $\tilde{J}$ (cell polarization curve) for (a) methanol and oxygen stoichiometries $\lambda^a = \lambda^c = 4$ and (b) $\lambda^a = \lambda^c = 8$. Indicated are the values of parameter k ($k = 0$: no bubbles, $k = 30$: intense bubbles formation). The other parameters are given in the caption to Figure 6.17. The profiles of parameters along the channel in the black points in (a) are shown in Figure 6.17.

The model above includes seven parameters. To understand the trends in a space of seven parameters it is advisable to construct approximate analytical solution to model equations. This can be done in the limit of small k, as described in the next section.

6.5 DMFC with Bubbles: Asymptotic Solutions

In the case of purely liquid flow we have $k = 0$ and thus we may put $\tilde{v} = 1$ in Eq. (6.125). The model of the previous section reduces then to the model of "bubbles-free" DMFC in Section 6.3. Under equal methanol and oxygen stoichiometries this latter model has exact solution. This makes it possible to develop a simple asymptotic model of a flow with bubbles, valid for small k [40]. The procedure is as following.

6.5.1 Basic Equations and Solution for Purely Liquid Flow

In the case of $\lambda^a = \lambda^c = \lambda$, Eqs. (6.123), (6.125) and (6.126), reduce to

$$\lambda J_0 \frac{\partial \tilde{v}}{\partial \tilde{z}} = k\tilde{j} \qquad \tilde{v}(0) = 1 \tag{6.133}$$

$$\lambda \tilde{J}_0 \frac{\partial (\tilde{v}\tilde{c})}{\partial \tilde{z}} = -(\tilde{j} + \beta_\star(\tilde{c} - \tilde{j})) \qquad \tilde{c}(0) = 1 \tag{6.134}$$

$$\lambda \tilde{J}_0 \frac{\partial \tilde{\xi}}{\partial \tilde{z}} = -(\tilde{j} + \beta_\star(\tilde{c} - \tilde{j})) \qquad \tilde{\xi}(0) = 1 \tag{6.135}$$

Equipotentiality of cell electrodes gives equation for $\tilde{j}$ (6.127), which completes the model.

With $k = 0$ (no bubbles in the flow) Eqs. (6.133)–(6.135) and (6.127) have analytical solution (6.62) and (6.64), where we should replace $\tilde{\psi}$ with $\tilde{c}$:

$$\tilde{v}_0 = 1 \tag{6.136}$$

$$\tilde{c}_0 = \tilde{\xi}_0 = \exp\left(-\frac{\tilde{z}}{\mu}\right) \tag{6.137}$$

$$\tilde{j}_0 = \kappa \exp\left(-\frac{\tilde{z}}{\mu}\right) \tag{6.138}$$

Here $\kappa \equiv \tilde{j}_0(0)$ is a solution to Eq. (6.69) with $\tilde{J} = \tilde{J}_0$:

$$\frac{\kappa + \beta_\star(1-\kappa)}{\lambda \ln\left(1 - \dfrac{\kappa + \beta_\star(1-\kappa)}{\kappa\lambda}\right)} = -\tilde{J}_0 \tag{6.139}$$

and the characteristic scale μ is given by (6.63) with $\tilde{J} = \tilde{J}_0$:

$$\mu = \frac{\tilde{J}_0 \lambda}{\kappa + \beta_\star(1-\kappa)} \tag{6.140}$$

In this section the subscript "0" denotes zero-order (bubbles-free) solution.

Substituting $\tilde{c}_0 = \tilde{\xi}_0$ and $\tilde{j}_0 = \kappa\tilde{c}_0 = \kappa\tilde{\xi}_0$ into (6.127) we come to:

$$\ln\left(\frac{\kappa}{q}\right) - \ln(1-\kappa) + \ln(1+\beta) + p\left[\ln\left(\frac{\kappa}{\alpha q}\right) - \ln\left(1 - \frac{\kappa + \beta_\star(1-\kappa)}{\gamma}\right)\right] = \tilde{E}_0 \tag{6.141}$$

where $\tilde{E}_0$ is the total voltage loss.

The second term on the left side of (6.141) shows that κ must be less than 1:

$$\kappa < 1 \tag{6.142}$$

Physically, local current density at the inlet cannot exceed dimensionless methanol-limiting current density there, which is 1.

The second term in the square brackets in (6.141) has sense if

$$\gamma > \gamma_\star \equiv \kappa + \beta_\star(1-\kappa) \tag{6.143}$$

In view of (6.139) κ is a function of $\tilde{J}_0$; (6.143) thus says that γ (which is proportional to oxygen concentration at the inlet) must be large enough to provide given $\tilde{J}_0$.

6.5.2 Equations for Corrections Due to Bubbles

To obtain solutions to (6.133)–(6.135) and (6.127) for small k we will exploit the following expansion:

$$\tilde{\nu} = 1 + k\tilde{\nu}_1 \tag{6.144}$$

$$\tilde{c} = \tilde{c}_0 + k\tilde{c}_1 \tag{6.145}$$

$$\tilde{\xi} = \tilde{\xi}_0 + k\tilde{\xi}_1 \tag{6.146}$$

$$\tilde{j} = \tilde{j}_0 + k\tilde{j}_1 \tag{6.147}$$

which tends to zero-order (bubbles-free) solution as $k \to 0$. Here $\tilde{\nu}_1, \tilde{c}_1, \tilde{\xi}_1$ and $\tilde{j}_1$ are first-order corrections due to bubbles.

Substituting (6.144) and (6.147) into (6.133) we get

$$\lambda\tilde{J}_0\left(k\frac{\partial\tilde{\nu}_1}{\partial\tilde{z}}\right) = k\tilde{j}_0 + k^2\tilde{j}_1 \tag{6.148}$$

Neglecting term with k^2 we find

$$\lambda \tilde{J}_0 \frac{\partial \tilde{\nu}_1}{\partial \tilde{z}} = \tilde{j}_0 \qquad \tilde{\nu}_1(0) = 0 \tag{6.149}$$

Using here (6.138) and solving we find

$$\tilde{\nu}_1 = \frac{\kappa\mu}{\lambda \tilde{J}_0}\left[1 - \exp\left(-\frac{\tilde{z}}{\mu}\right)\right] \tag{6.150}$$

Substituting (6.144)–(6.147) into (6.134) and (6.135), neglecting the terms with k^2 and equating the terms with k we find

$$\lambda \tilde{J}_0 \frac{\partial \tilde{c}_1}{\partial \tilde{z}} + \beta_\star \tilde{c}_1 = -(1-\beta_\star) j_1 + \kappa \tilde{c}_0 (1 - 2\tilde{c}_0) \qquad \tilde{c}_1(0) = 0 \tag{6.151}$$

$$\lambda \tilde{J}_0 \frac{\partial \tilde{\xi}_1}{\partial \tilde{z}} + \beta_\star \tilde{c}_1 = -(1-\beta_\star)\tilde{j}_1 \qquad \tilde{\xi}_1(0) = 0 \tag{6.152}$$

Here we used (6.149), (6.150) and relations $\partial \tilde{c}_0/\partial \tilde{z} = -\tilde{c}_0/\mu$, $\tilde{\nu}_1 = \kappa\mu(1-\tilde{c}_0)/(\lambda \tilde{J}_0)$, which follow from (6.137) and (6.150), respectively.

Equation for $\tilde{j}_1$ is obtained from (6.127). Without the loss of generality we will assume that the cell potential is fixed, that is on the right side of (6.127) $\tilde{E} = \tilde{E}_0$ does not depend on k (that is the reason why in this section $\tilde{E}$ is equipped with the subscript "0"). The bubbles then reduce mean current density (see below). Substituting (6.145)–(6.147) into (6.127), expanding the result over k and neglecting the terms with k^2 we get

$$\begin{aligned} &\ln\left(\frac{\kappa}{q}\right) - \ln(1-\kappa) + \ln(1+\beta) \\ &\quad + p\left[\ln\left(\frac{\kappa}{\alpha q}\right) - \ln\left(1 - \frac{\kappa + \beta_\star(1-\kappa)}{\gamma}\right)\right] + kf_1 = \tilde{E}_0 \end{aligned} \tag{6.153}$$

Here

$$f_1 = \frac{A_f \tilde{j}_1 + B_f \tilde{c}_1 + C_f \tilde{\xi}_1}{\tilde{c}_0 \kappa (1-\kappa)(\gamma - \gamma_\star)} \tag{6.154}$$

where

$$A_f = -\left[\gamma - \gamma_\star + p(1-\kappa)(\gamma - \beta_\star)\right] \tag{6.155}$$

$$B_f = \kappa\left[\gamma - \gamma_\star - p\beta_\star(1-\kappa)\right] \quad (6.156)$$

$$C_f = p\gamma(1-\kappa)\kappa \quad (6.157)$$

and we took into account that $\tilde{j}_0 = \kappa\tilde{c}_0$ and $\tilde{\xi}_0 = \tilde{c}_0$. Comparing (6.153) and (6.141) we see that zero-order terms cancel and (6.153) reduces to $f_1 = 0$, or

$$A_f\tilde{j}_1 + B_f\tilde{c}_1 + C_f\tilde{\xi}_1 = 0 \quad (6.158)$$

The system of equations (6.151), (6.152) and (6.158) determines the first-order corrections $\tilde{c}_1$, $\tilde{\xi}_1$ and $\tilde{j}_1$.

6.5.3 Solutions

Subtracting (6.152) from (6.151) we get equation for $y_1 = \tilde{c}_1 - \tilde{\xi}_1$:

$$\lambda\tilde{J}_0\frac{\partial y_1}{\partial\tilde{z}} = \kappa\tilde{c}_0(1-2\tilde{c}_0) \qquad y(0) = 0 \quad (6.159)$$

Using here $\tilde{c}_0$ (6.137) and integrating we find

$$\tilde{c}_1 = \tilde{\xi}_1 - \frac{\mu\kappa}{\lambda\tilde{J}_0}\left[\exp\left(-\frac{\tilde{z}}{\mu}\right) - \exp\left(-\frac{2\tilde{z}}{\mu}\right)\right] \quad (6.160)$$

Note that $\tilde{c}_1 - \tilde{\xi}_1$ does not depend on $\tilde{j}_1$: up to now we did not use Eq. (6.158) for $\tilde{j}_1$.

With (6.160) in hand we have to solve only one of the equations (6.151) and (6.152). Equation (6.158) allows us to express $\tilde{j}_1$ through $\tilde{c}_1$ and $\tilde{\xi}_1$:

$$\tilde{j}_1 = \frac{\left[\gamma - \gamma_\star - p\beta_\star(1-\kappa)\right]\kappa\tilde{c}_1 + p\gamma(1-\kappa)\kappa\tilde{\xi}_1}{\gamma - \gamma_\star + p(1-\kappa)(\gamma - \beta_\star)} \quad (6.161)$$

Substituting this $\tilde{j}_1$ into (6.152), using $\tilde{c}_1$ (6.160) and solving the resulting equation we find

$$\tilde{\xi}_1 = K_\xi\exp\left(-\frac{z}{\mu}\right)\left[\exp\left(-\frac{z}{\mu}\right) - \left(1 - \frac{z}{\mu}\right)\right] \quad (6.162)$$

where

$$K_\xi = \frac{\kappa(\gamma - \gamma_\star)\left[\gamma_\star + p\beta_\star(1-\kappa)\right]}{\gamma_\star^2\left[\gamma - \gamma_\star + p(1-\kappa)(\gamma - \beta_\star)\right]} \quad (6.163)$$

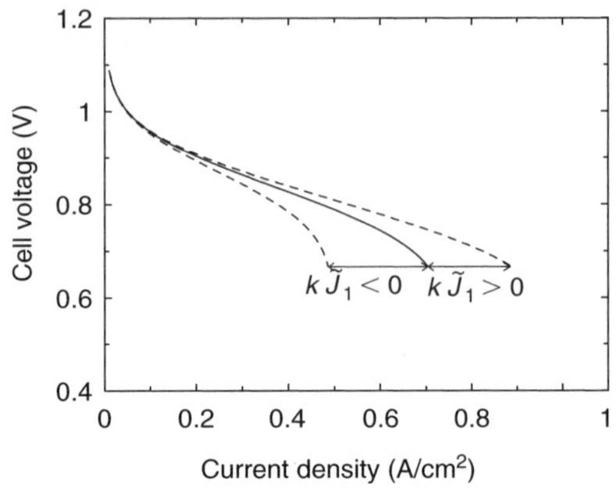

Figure 6.19 Sketch of the change of cell polarization curve due to bubbles. Depending upon the sign of $\tilde{J}_1$ bubbles shift mean current density to the right or to the left.

Expression in the square brackets in (6.162) is nonnegative. The sign of $\tilde{\xi}_1$, therefore, coincides with the sign of K_ξ; the latter is positive (see Appendix B). Physically, the bubbles always reduce time-average methanol concentration in the anode channel. More oxygen is then available on the cathode side, i.e., $\tilde{\xi}_1 > 0$.

Integrating (6.147) over $\tilde{z}$ we find

$$\tilde{J} = \tilde{J}_0 + k\tilde{J}_1 \tag{6.164}$$

where $\tilde{J}_{0,1} = \int_0^1 \tilde{j}_{0,1}\, d\tilde{z}$. Fixed E_0 means that the cell potential V_{cell} is fixed.[4] Formation of bubbles then shifts the point on the polarization curve, as illustrated in Figure 6.19. Positive $\tilde{J}_1$ shifts the curve to the right, whereas $\tilde{J}_1 < 0$ shifts it to the left (Figure 6.19).

Substitution of $\tilde{c}_1$ (6.160) and then $\tilde{\xi}_1$ (6.162) into (6.161) yields explicit dependence $\tilde{j}_1(\tilde{z})$. Integrating then $\tilde{j}_1$ over $\tilde{z}$ we find correction to mean current density due to bubbles

$$\tilde{J}_1 = \frac{1}{K_J}\left[A_J + B_J \exp\left(-\frac{1}{\mu}\right) + C_J \exp\left(-\frac{2}{\mu}\right)\right] \tag{6.165}$$

where

$$A_J = \lambda \tilde{J}_0 \gamma p \beta_\star \kappa^2 (1-\kappa) \tag{6.166}$$

$$B_J = 2\kappa^2 \gamma_\star \left[(\gamma - \gamma_\star)\left(\lambda \tilde{J}_0 - \gamma_\star - p\beta_\star(1-\kappa)\right) - \lambda \tilde{J}_0 p \beta_\star (1-\kappa)\right] \tag{6.167}$$

[4]To simplify the analysis in this section we neglect ohmic terms.

$$C_J = \lambda \tilde{J}_0 \kappa^2 \left[\gamma p \beta_\star (1-\kappa) - 2(\gamma - \gamma_\star)\left(\gamma_\star + p\beta_\star(1-\kappa)\right)\right] \quad (6.168)$$

$$K_J = 2\gamma_\star^3 \left[\gamma - \gamma_\star + p(1-\kappa)(\gamma - \beta_\star)\right] \quad (6.169)$$

μ and $\gamma_\star$ are given by (6.140) and (6.143), respectively, and κ is solution to Eq. (6.139). Note, that (6.165)–(6.169) do not contain α and q: as before, these parameters determine voltage loss $\tilde{E}_0$.

6.5.4 Profiles and Polarization Curves

Figure 6.20 shows analytical and numerical profiles of $\tilde{\nu}, \tilde{c}, \tilde{\xi}$ and $\tilde{j}$ along the channel. The numerical results were obtained by direct numerical integration of the basic system of equations (6.133)–(6.135) and (6.127). We see that for the set of parameters indicated in caption to Figure 6.20, the asymptotic formulas reproduce all shapes with quite reasonable accuracy even for $k = 2$.

The perturbation model was developed for small k; why then it works well for $k = 2$? Rigorous answer to this question requires accurate estimate of truncation terms. This analysis leads to very cumbersome relations and is not performed. Qualitatively, equations for the first-order corrections were obtained neglecting the terms proportional to k^2. Presumably, up to $k \simeq 1$ it is the inequality $k^2|\tilde{j}_1| \ll k\tilde{j}_0$ (or, equivalently, $k|\tilde{J}_1| \ll \tilde{J}_0$), which

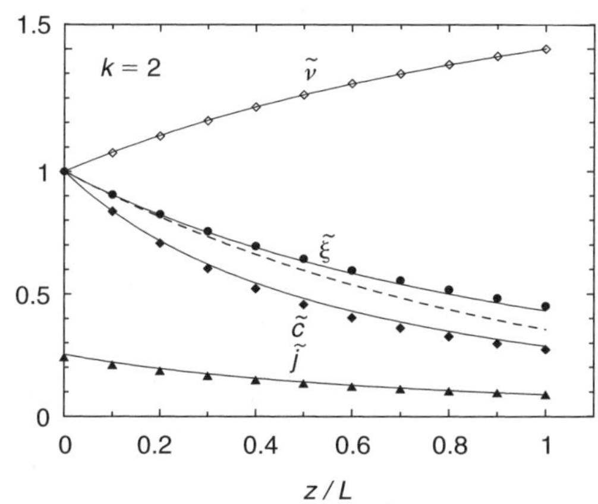

Figure 6.20 Profiles of velocity $\tilde{\nu}$ in the anode channel, time-average methanol concentration $\tilde{c}$, oxygen fraction in the cathode channel $\tilde{\xi}$ and local current density $\tilde{j}$. Points: asymptotic solutions and solid lines: numerical solutions of the basic system of equations (6.133)–(6.135) and (6.127). Dashed line: zero-order solution $\tilde{c}_0 = \tilde{\xi}_0 = \exp(-\tilde{z}/\mu)$. Mean current density $\tilde{J} = 0.15$. The other parameters are $k = 2$, $p = 1$, $\lambda = 4$, $\beta = 1$ ($\beta_\star = 0.5$), $\gamma = 2.805$. All variables are dimensionless.

determines validity of expansion (6.144)–(6.147). For the case shown in Figure 6.20 this inequality is fulfilled. In the following we will use $k = 1$ to emphasize the effects due to bubbles. Note that strictly speaking all the results below are valid for small k; at $k = 1$ they should be considered as qualitative.

Function $\tilde{J}_1(\tilde{J}_0)$ (6.165) is shown in Figure 6.21 for several values of parameters β and γ. Below we will see that the regime of cell operation is determined by γ: $\gamma > 1$ means that the current in a cell is limited by a lack of methanol, whereas at $\gamma < 1$ it is limited by a lack of oxygen.

Figure 6.21(a) shows that if oxygen is in large excess ($\gamma = 2.8$), the bubbles have negative effect on performance ($\tilde{J}_1$ is negative) almost in the whole range of currents. In the case of relatively low oxygen content and large crossover ($\gamma = 1.2, \beta = 5$) the bubbles improve cell performance ($\tilde{J}_1$ is positive) in a quite large range of currents; furthermore, the larger crossover the larger this improvement (Figure 6.21(b)). If crossover is small the bubbles diminish cell performance (the curve $\beta = 0.1$ in Figure 6.21(b)).

The situation, however, changes if the cell is run under the lack of oxygen ($\gamma = 0.9 < 1$). The bubbles dramatically improve cell performance under large and medium crossover (Figure 6.21(c), the curves for $\beta = 5$ and 1). When crossover is small the bubbles slightly decrease performance for almost all currents ($\beta = 0.1$, Figure 6.21(c)).

These results can be understood if we take into account that the bubbles always reduce time-average methanol concentration in the anode channel and hence diminish crossover. When oxygen is in large excess ($\gamma = 2.8$) reduction of crossover has minor effect on cell performance. However, under the shortage of oxygen ($\gamma = 1.2$ and $\gamma = 0.9$) lowering of crossover has strong impact on cell performance, since it saves oxygen for current generation. Figures 6.21(b) and (c) illustrate this effect: the larger β, the larger the gain in mean current due to bubbles formation. Figures 6.21(a) and (b) show that for $\gamma > 1$ there is optimal current density, when the gain in current is maximal.

To calculate cell polarization curve it is convenient to express all required values through κ. $\tilde{E}_0(\kappa)$ is given by (6.141), $\tilde{J}_0(\kappa)$ is given by Eq. (6.139), perturbed mean current density is $\tilde{J}(\kappa) = \tilde{J}_0(\kappa) + k\tilde{J}_1(\tilde{J}_0(\kappa))$, where $\tilde{J}_1$ is given by (6.165). Varying κ we thus can calculate the respective $\tilde{J}_0, \tilde{J}$ and $\tilde{E}_0$.

Figure 6.22 displays polarization curves $\tilde{E}_0(\tilde{J}_0)$ (purely liquid flow, dashed lines) and $\tilde{E}_0(\tilde{J}_0 + k\tilde{J}_1)$ (the flow with bubbles, solid lines) for $\gamma = 1.2, 0.9$ and the three values of crossover parameter β. Consider first the case of $\gamma > 1$ (Figure 6.22(a)). If crossover is small, the bubbles simply diminish limiting current density ($\beta = 0.1$, Figure 6.22(a)). However, for medium and large β the effect of bubbles is more complicated. Consider the curve for $\beta = 5$ (Figure 6.22(a)), where the features are best seen. In the range of small and medium currents the bubbles slightly improve cell performance due to the lowering of crossover. At higher currents the effect of

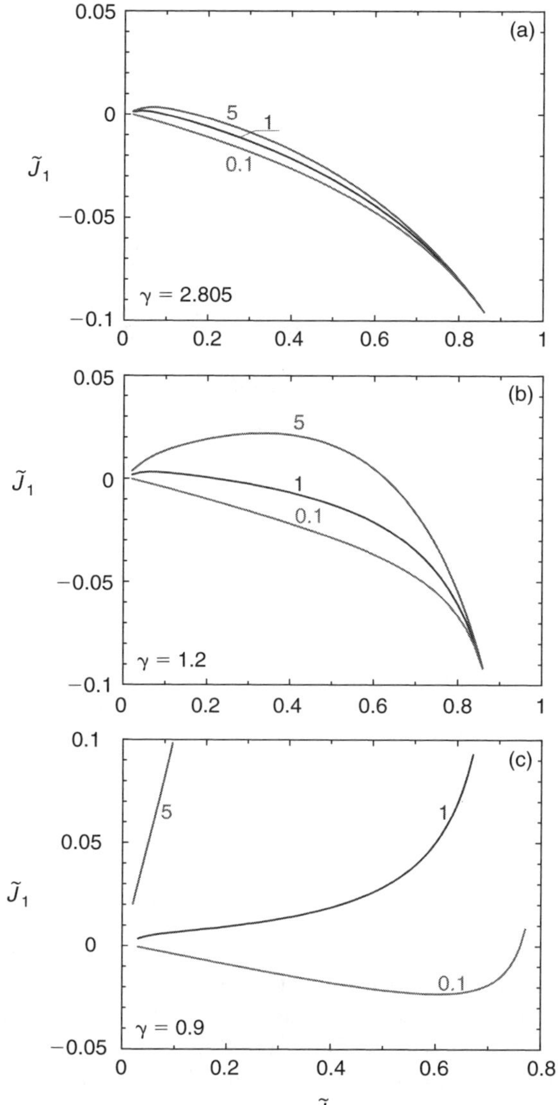

Figure 6.21 First-order correction $\tilde{J}_1$ to mean current density due to bubbles formation for (a) $\gamma = 2.805$, (b) $\gamma = 1.2$ and (c) $\gamma = 0.9$. Indicated are the values of β. The other parameters: $p = 1$, $\lambda = 4$.

bubbles is negative (Figure 6.22(a)). Thus, at $\gamma > 1$, close to the limiting current density the bubbles always reduce performance.

In contrast, when $\gamma < 1$ the bubbles increase the limiting current (Figure 6.22(b)). For $\beta = 1$ and 5 the bubbles improve cell performance in

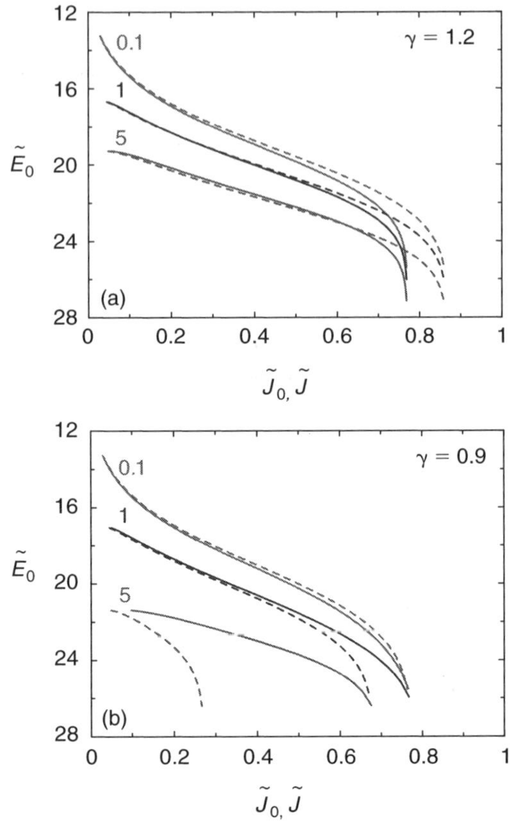

Figure 6.22 Total voltage loss (cell polarization curves) for (a) $\gamma = 1.2$ and (b) $\gamma = 0.9$. Indicated are the values of β. Dashed lines: purely liquid flow and solid lines: the flow with the bubbles. The other parameters are: $p = 1$, $\lambda = 4$, $\alpha = 20$, $q = 1.7e - 5$, $k = 1$.

the whole range of currents (Figure 6.22(b)). If crossover is small, the bubbles have negative effect on performance, excluding a small domain close to the limiting current. It is, therefore, advisable to consider the expression for the limiting current density, which is derived in the next Section.

6.5.5 Limiting Current Density and Optimal λ

6.5.5.1 The Two Regimes of Cell Operation

Formally, the limiting current density is attained when $\tilde{E}_0 \longrightarrow \infty$. Equation (6.141) shows that $\tilde{E}_0 \longrightarrow \infty$ if either $\kappa = 1$ or $\kappa + \beta_\star(1 - \kappa) = \gamma$. When $\kappa = 1$ the second term on the left side of (6.141) tends to infinity, which physically means that the current is limited by available amount of methanol (methanol-limiting regime). If $\kappa + \beta_\star(1 - \kappa) = \gamma$ the second term in the square brackets in (6.141) tends to infinity, i.e., current is limited by the lack of oxygen (oxygen-limiting regime).

In the case of oxygen limitation we have $\kappa + \beta_\star(1 - \kappa) = \gamma$. Since $\kappa \leqslant 1$ and $\beta_\star < 1$, it is easy to show that $\kappa + \beta_\star(1 - \kappa) \leqslant 1$. Therefore, if $\gamma > 1$ we have $\kappa + \beta_\star(1 - \kappa) < \gamma$ and oxygen side does not limit the current.

Thus, the nature of current limitation (the regime of cell operation) depends solely on the value of γ: when $\gamma > 1$ the cell works in the methanol-limiting regime, if $\gamma < 1$ it works in the oxygen-limiting regime. For $\gamma = 1$ we get $\kappa = 1$, i.e., methanol- and oxygen-limiting current densities coincide.

6.5.5.2 Methanol-Limiting Case ($\gamma > 1$)

Consider first the methanol-limiting case ($\gamma > 1$, $\kappa = 1$). Substituting $\kappa = 1$ into (6.139) we find

$$\tilde{J}_0^{\text{lim}} = -\left[\lambda \ln\left(1 - \frac{1}{\lambda}\right)\right]^{-1} \tag{6.170}$$

Using $\kappa = 1$ and $\tilde{J}_0 = \tilde{J}_0^{\text{lim}}$ (6.170) in Eq. (6.165) we find

$$\tilde{J}_1^{\text{lim}} = -\frac{(\lambda - 1)}{\lambda^2}\left(\lambda + \frac{1}{\ln\left(1 - \dfrac{1}{\lambda}\right)}\right) \tag{6.171}$$

Thus, methanol-limiting current density $\tilde{J}_0^{\text{lim}} + k\tilde{J}_1^{\text{lim}}$ does not depend on crossover. Physically, close to the limiting current concentration of methanol in the ACL tends to zero and crossover is negligible.

Function (6.171) is depicted in Figure 6.23(a). We see that the correction due to bubbles is always negative, i.e., bubbles reduce limiting current density; near $\lambda \simeq 1.856$ the reduction is maximal (Figure 6.23(a)). Note also that $\tilde{J}_1^{\text{lim}} \to 0$ as $\lambda \to 1$ or $\lambda \to \infty$ i.e., the limiting current tends to unperturbed value as $\lambda \to 1, \infty$. Physically, close to $\lambda = 1$ all methanol is consumed regardless of the presence of bubbles in the flow; in the opposite limit of $\lambda \to \infty$ the bubbles do not affect methanol concentration in the flow.

6.5.5.3 Oxygen-Limiting Case ($\gamma < 1$)

For the oxygen-limiting case we have $\kappa + \beta_\star(1 - \kappa) = \gamma$. Solving this for κ we find $\kappa = (\gamma - \beta_\star)/(1 - \beta_\star)$. This implies that $\beta_\star \leqslant \gamma$, otherwise κ is negative. Physically, $\beta_\star = \gamma$ is the limiting rate of crossover; if crossover exceeds this rate the cell does not generate current, since all available oxygen is consumed in the reaction with permeated methanol. Combining this

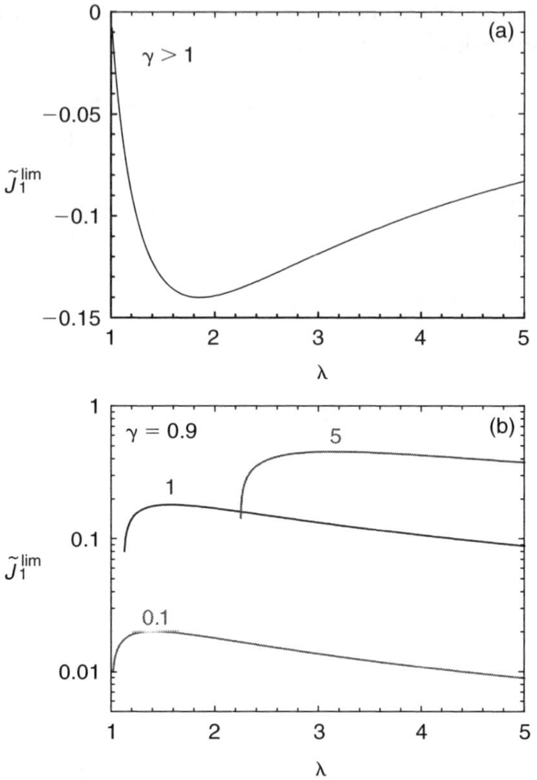

Figure 6.23 First-order correction to the limiting current density due to bubbles. (a) Methanol-limiting case ($\gamma > 1$) and (b) oxygen-limiting case ($\gamma < 1$), indicated are the values of β.

with $\gamma < 1$ we see that in the oxygen-limiting regime the following inequalities must be fulfilled:

$$\beta_\star \leq \gamma < 1 \tag{6.172}$$

Substituting $\kappa + \beta_\star(1 - \kappa) = \gamma$ and $\kappa = (\gamma - \beta_\star)/(1 - \beta_\star)$ into Eq. (6.139) we get

$$\tilde{J}_0^{\text{lim}} = -\frac{\gamma}{\lambda \ln\left(1 - \dfrac{\lambda_{\text{min}}}{\lambda}\right)} \tag{6.173}$$

where

$$\lambda_{\text{min}} = \frac{\gamma(1 - \beta_\star)}{(\gamma - \beta_\star)} \tag{6.174}$$

Equation (6.173) has sense if

$$\lambda > \lambda_{\min} \tag{6.175}$$

otherwise the argument of the logarithm is negative. We see that in the oxygen-limiting regime there is minimal λ, which provides generation of limiting current. At $\lambda \leqslant \lambda_{\min}$ much oxygen is consumed in the reaction with permeated methanol and the limiting current is not generated.

Substituting $\kappa + \beta_\star(1 - \kappa) = \gamma$, $\kappa = (\gamma - \beta_\star)/(1 - \beta_\star)$ and $\tilde{J}_0 = \tilde{J}_0^{\lim}$ (6.173) into (6.165) we find the correction to the limiting current density due to bubbles:

$$\tilde{J}_1^{\lim} = -\gamma\beta_\star \left[2\lambda^2 (\gamma - \beta_\star) \ln\left(1 - \frac{\lambda_{\min}}{\lambda}\right)\right]^{-1} \tag{6.176}$$

This function is depicted in Figure 6.23(b). Under lack of oxygen the effect of bubbles is always positive. Furthermore, Figure 6.23(b) shows that there exists optimal stoichiometry λ_1^{opt}, which provides maximal improvement of performance. λ_1^{opt} is a solution to equation $\partial\tilde{J}_1^{\lim}/\partial\lambda = 0$, which can be transformed to

$$\lambda_1^{opt} = \lambda_{\min} \left[1 - \frac{1}{2\ln\left(1 - \dfrac{\lambda_{\min}}{\lambda_1^{opt}}\right)} \right] \tag{6.177}$$

Note that λ_1^{opt} is a function of only one parameter $\lambda_{\min}$ (6.174), which is a combination of γ and $\beta_\star$. By direct substitution it is easy to verify that solution to Eq. (6.177) is $\lambda_1^{opt} = a\lambda_{\min}$, where a is solution to equation

$$a = 1 - \frac{1}{2\ln\left(1 - \dfrac{1}{a}\right)} \tag{6.178}$$

Calculation gives $a = 1.39795\ldots$ We, therefore, get a simple formula

$$\lambda_1^{opt} = \frac{a\gamma(1 - \beta_\star)}{(\gamma - \beta_\star)} \qquad \beta_\star < \gamma < 1 \tag{6.179}$$

From practical point of view a more important value is λ^{opt}, which provides maximal limiting current density of the cell in the oxygen-limiting

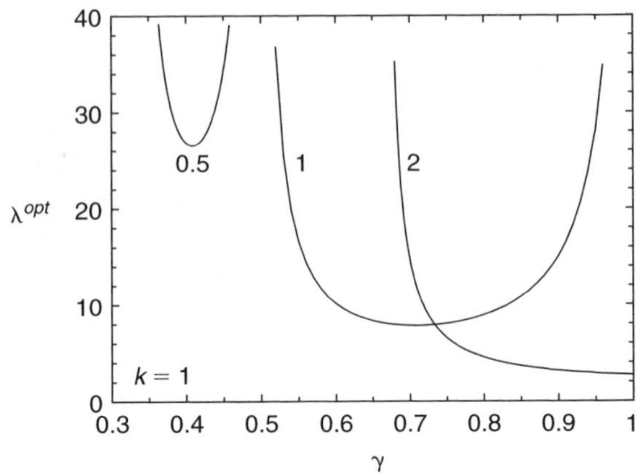

Figure 6.24 Optimal stoichiometry of the flows (solution to Eq. (6.180)), which provides the largest limiting current density in the oxygen-limiting regime. Indicated are the values of β. For all curves $k = 1$.

regime. λ^{opt} is a solution to equation $\partial(\tilde{J}_0^{\lim} + k\tilde{J}_1^{\lim})/\partial\lambda = 0$. With (6.173) and (6.176) this equation reduces to

$$\lambda^{opt} = -(\lambda^{opt} + \alpha)\left[1 - 2\left(1 - \frac{\lambda^{opt}}{\lambda_{\min}}\right)\ln\left(1 - \frac{\lambda_{\min}}{\lambda^{opt}}\right)\right] \tag{6.180}$$

where

$$\alpha = \frac{k\beta_\star}{\gamma - \beta_\star} \tag{6.181}$$

Solutions to Eq. (6.180) for several values of β are depicted in Figure 6.24. Note that Eq. (6.180) has solutions not for all parameters: if solution does not exist it means that λ^{opt} tends to infinity.

6.5.6 Remarks

The model shows that the central role in DMFC function plays parameter $\gamma = \dfrac{2D^c l^a}{3D^a l^c}\left(\dfrac{C^c \xi^0}{c^0}\right)$. The value $\gamma = 1$ separates the regimes of cell operation: at $\gamma > 1$ the cell current is limited by available amount of methanol, whereas at $\gamma < 1$ it is limited by the oxygen side. In the methanol-limiting case the bubbles dramatically decrease performance close to the limiting current density and have minor positive effect at small and medium currents. In the oxygen-limiting case the bubbles almost always improve performance.

γ is proportional to the product of inlet oxygen concentration by the mass transfer coefficient of oxygen through the cathode backing layer. Therefore, lack of oxygen or poor oxygen transport due to flooding may turn DMFC into oxygen-limiting regime. The bubbles in the anode channel then mitigate the negative effect of oxygen starvation. The effect of performance improvement due to bubbles may explain recent experimental result of Yang *et al.* [41], who reported that additional injection of CO_2 into the anode flow improves DMFC performance.

Physically, parameter k depends on temperature, type of backing layer, etc. For large k asymptotic solutions are not valid and direct numerical integration of (6.133)–(6.135) and (6.127) is necessary. Further efforts are needed to rationalize the dependence of k on the design parameters.

6.6 Kinetics of Methanol Oxidation and Cell Performance

One of the key problems on a way towards DMFC commercialization is sluggish kinetics of methanol oxidation [1]. In spite of 40 years history of the studies [42–44] reaction mechanism of methanol oxidation is still not fully understood.

In the previous sections kinetics of the electrochemical reactions on both sides of the cell were described by Tafel law. However, following [43] the mechanism of methanol oxidation includes relatively slow potential-independent step of methanol adsorption on the catalyst surface. This step in the chain of reaction events violates validity of Tafel law. In Refs. [44,15] performance of the ACL was studied numerically, taking into account non-Tafel effects. In this section we will study the effect of methanol adsorption on ACL performance analytically [45].

6.6.1 Model

6.6.1.1 Simplified Kinetics of Methanol Oxidation

To take into account adsorption step in the kinetics of methanol oxidation consider the following reaction scheme:

$$CH_3OH + \text{Site} \xrightarrow{k_a} CH_3OH_{ads} \tag{6.182}$$

$$CH_3OH_{ads} + H_2O \xrightarrow{k_\star \exp(\eta^a/b^a)} CO_2 + 6H^+ + 6e^- \tag{6.183}$$

Step (6.182) describes adsorption of methanol on the catalyst surface; step (6.183) represents the electrochemical conversion. Here k_a and $k_\star \exp(\eta^a/b^a)$

are the rate constants ($cm^3/mol/s$) of adsorption and conversion, respectively, $\eta^a = \varphi^a - \varphi_m$ is the local overpotential. As before, φ^a and φ_m are potentials of the carbon and electrolyte phase, respectively. Hereinafter the superscript "a" (anode side) will be omitted.

The fraction of occupied sites Θ on the catalyst surface is then subject to equation

$$A\frac{d\Theta}{dt} = k_a c_t A(1-\Theta) - k_\star \exp\left(\frac{\eta}{b}\right) c_w A\Theta \tag{6.184}$$

Here A is the molar concentration of active catalyst particles, c_t and c_w are the molar concentrations of methanol and water, respectively. Equation (6.184) implies that η is large enough, so that the electrochemical conversion is shifted towards oxidation and the backward reaction can be neglected. This condition is fulfilled for $\eta \gtrsim b$.

In the steady state $d\Theta/dt = 0$ and (6.184) yields

$$\Theta = \frac{k_a c_t}{k_a c_t + k_\star c_w \exp(\eta/b)} \tag{6.185}$$

The overall rate of electrochemical reaction Q (A/cm^3) is determined by the second term on the right side of (6.184). Using there (6.185) and multiplying the result by $6F$ we find

$$Q = \frac{6F k_a c_t k_\star c_w A \exp(\eta/b)}{k_a c_t + k_\star c_w \exp(\eta/b)} \tag{6.186}$$

Similar equation was derived by Meyers and Newman [12] from a much more detailed reaction scheme.

6.6.1.2 Basic Equations and the Conservation Law

As before, we will assume that methanol concentration does not vary significantly across the ACL. This condition is usually fulfilled due to a small thickness of the ACL and a high rate of methanol transport across the ACL by diffusion and electroosmosis.

Figure 6.25 illustrates typical profile of the overpotential $\eta(x)$ across the active layer. Let total proton current density generated in the ACL is j_1; voltage loss required to generate this current is $\eta(1) \equiv \eta_1$ (Figure 6.25). The subscripts "0" and "1" mark the values at the backing layer/ACL interface and at the membrane surface, respectively (Figure 6.25). Our goal is calculation of polarization curve $j_1(\eta_1)$. This function relates the boundary values in the problem of ACL performance, which is governed by two equations (*cf.* Section 6.2.1).

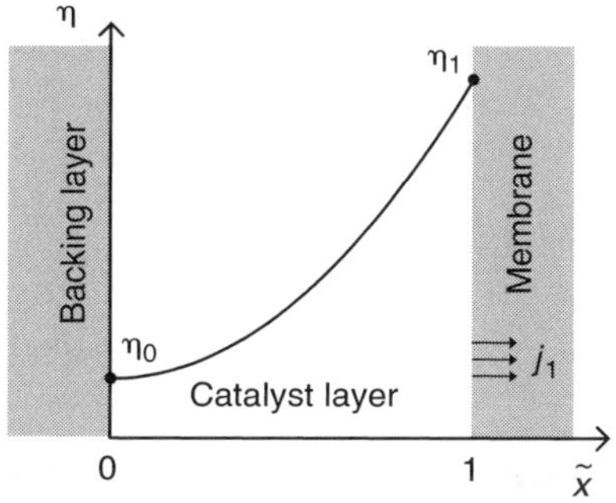

Figure 6.25 The system of coordinates and sketch of the membrane phase potential in the catalyst layer. The polarization curve relates the boundary values $j_1(\eta_1)$ in the problem of catalyst layer performance.

The first is equation of proton current conservation in the ACL

$$\frac{\partial j}{\partial x} = Q \tag{6.187}$$

which describes generation of protons at a rate $Q(x)$ (6.186). Here j is proton current density, x is the distance from the backing layer/ACL interface (Figure 6.25).

The second equation is Ohm's law

$$j = \sigma_t \frac{\partial \eta}{\partial x} \tag{6.188}$$

where σ_t is the conductivity of electrolyte phase. Equation (6.188) is valid provided that the variation of φ_c across the ACL is negligible and hence $\partial\varphi_m/\partial x \simeq -\partial\eta/\partial x$. This is justified since conductivity of carbon phase in the ACL is usually several orders of magnitude higher than σ_t.

Introducing dimensionless variables (6.6) and the characteristic current density $j_\star$ (6.7), Eqs. (6.186)–(6.188) take a form

$$\tilde{Q} = \frac{\exp(\tilde{\eta})}{\omega + p_t \exp(\tilde{\eta})} \tag{6.189}$$

$$\frac{\partial \tilde{j}}{\partial \tilde{x}} = \tilde{Q} \tag{6.190}$$

$$2\tilde{j} = \frac{\partial \tilde{\eta}}{\partial \tilde{x}} \tag{6.191}$$

The system is thus controlled by two dimensionless parameters:

$$\omega = \frac{j_\star}{6Fl_t k_\star c_w A} \tag{6.192}$$

is inverse characteristic current of electrochemical conversion and

$$p_t = \frac{j_\star}{6Fl_t k_a c_t A} \tag{6.193}$$

is inverse current of methanol adsorption. In the following we will also use parameter ψ, the ratio of conversion to adsorption currents

$$\psi \equiv \frac{p_t}{\omega} = \frac{k_\star c_w}{k_a c_t} \tag{6.194}$$

Either of the pairs ψ, p_t or ω, p_t can be used to characterize the system.

Multiplying together Eqs. (6.190) and (6.191) we get

$$2\tilde{j}\,\frac{\partial \tilde{j}}{\partial \tilde{x}} = \tilde{Q}\,\frac{\partial \tilde{\eta}}{\partial \tilde{x}}$$

or

$$\frac{\partial(\tilde{j}^2)}{\partial \tilde{x}} = \frac{1}{p_t}\frac{\partial H(\tilde{\eta})}{\partial \tilde{x}} \tag{6.195}$$

where

$$H(\tilde{\eta}) = \ln\left(\omega + p_t \exp(\tilde{\eta})\right) \tag{6.196}$$

Integrating Eq. (6.195) we find a conservation law [46]

$$H(\tilde{\eta}) - p_t\tilde{j}^2 = H(\tilde{\eta}_1) - p_t\tilde{j}_1^2 = H(\tilde{\eta}_0) \tag{6.197}$$

In the last equality we used $\tilde{j}(0) = 0$.

6.6.1.3 The General Form of Polarization Curve

Equation (6.197) enables to obtain polarization curve of the ACL $\tilde{\eta}_1(\tilde{j}_1)$. Taking into account (6.196), from (6.197) we get

$$\tilde{j}_1 = \sqrt{\frac{1}{p_t}\ln\left(\frac{1 + \psi\exp(\tilde{\eta}_1)}{1 + \psi\exp(\tilde{\eta}_0)}\right)} \tag{6.198}$$

Equation (6.198) contains $\tilde{\eta}_0$, electrolyte potential at $\tilde{x} = 0$. This parameter can be excluded using Eq. (6.191).

Taking the square of Eq. (6.191) and multiplying both sides by p_t we get

$$4p_t\tilde{j}^2 = p_t\left(\frac{\partial\tilde{\eta}}{\partial\tilde{x}}\right)^2 \tag{6.199}$$

From Eq. (6.197) we find

$$p_t\tilde{j}^2 = \ln\left(\frac{1+\psi\exp(\tilde{\eta})}{1+\psi\exp(\tilde{\eta}_0)}\right) \tag{6.200}$$

Using this in Eq. (6.199) we get equation for $\tilde{\eta}$:

$$p_t\left(\frac{\partial\tilde{\eta}}{\partial\tilde{x}}\right)^2 = 4\ln\left(\frac{1+\psi\exp(\tilde{\eta})}{1+\psi\exp(\tilde{\eta}_0)}\right) \tag{6.201}$$

Since $\partial\tilde{\eta}/\partial\tilde{x} > 0$ we may take a square root of this equation, which yields

$$\sqrt{p_t}\,\frac{\partial\tilde{\eta}}{\partial\tilde{x}} = 2\sqrt{\ln\left(\frac{1+\psi\exp(\tilde{\eta})}{1+\psi\exp(\tilde{\eta}_0)}\right)} \tag{6.202}$$

Separating variables and integrating we find

$$1-\tilde{x} = \left(\frac{\sqrt{p_t}}{2}\right)\int_{\tilde{\eta}}^{\tilde{\eta}_1}\left[\ln\left(\frac{1+\psi\exp(\tau)}{1+\psi\exp(\tilde{\eta}_0)}\right)\right]^{-1/2} d\tau \tag{6.203}$$

Putting here $\tilde{x} = 0$ we find relation of $\tilde{\eta}_0$ and $\tilde{\eta}_1$:

$$\left(\frac{\sqrt{p_t}}{2}\right)\int_{\tilde{\eta}_0}^{\tilde{\eta}_1}\left[\ln\left(\frac{1+\psi\exp(\tau)}{1+\psi\exp(\tilde{\eta}_0)}\right)\right]^{-1/2} d\tau = 1 \tag{6.204}$$

The latter equation allows us to exclude $\tilde{\eta}_0$ from (6.198), which then gives the desired relation $\tilde{j}_1(\tilde{\eta}_1)$.

6.6.2 Solutions

6.6.2.1 Small Variation of Overpotential Across the Active Layer

Let the variation of overpotential across the active layer is small: $\tilde{\delta} \equiv \tilde{\eta}_1 - \tilde{\eta}_0 \ll 1$. Conditions when this inequality is fulfilled will be derived below. In this case Eq. (6.204) can be simplified and solved for $\tilde{\eta}_1$.

Substituting $\tau = \tilde{\eta}_0 + \tilde{\delta}$ into (6.204), expanding logarithm over $\tilde{\delta}$ and performing integration over $0 \leqslant \tilde{\delta} \leqslant \tilde{\eta}_1 - \tilde{\eta}_0$ we get

$$\tilde{\eta}_1 = \tilde{\eta}_0 + \frac{\exp(\tilde{\eta}_0)}{\omega + p_t \exp(\tilde{\eta}_0)} \tag{6.205}$$

We see that

$$\tilde{\delta} = \frac{\exp(\tilde{\eta}_0)}{\omega + p_t \exp(\tilde{\eta}_0)} \tag{6.206}$$

For $\tilde{\eta}_0 \rightarrow \infty$ we find

$$\tilde{\delta}^{\lim} = \frac{1}{p_t} \tag{6.207}$$

Substituting (6.205) into (6.198) we obtain

$$\tilde{j}_1 = \sqrt{\frac{1}{p_t}\ln\left(\frac{\omega + p_t \exp\left(\tilde{\eta}_0 + \tilde{\delta}\right)}{\omega + p_t \exp(\tilde{\eta}_0)}\right)} \simeq \sqrt{\frac{1}{p_t}\ln\left(1 + p_t\tilde{\delta}^2\right)} \simeq \tilde{\delta} \tag{6.208}$$

where we have used $p_t\tilde{\delta}^2 \ll 1$ (see below). We, therefore, get

$$\tilde{j}_1 = \frac{\exp(\tilde{\eta}_0)}{\omega + p_t \exp(\tilde{\eta}_0)} \tag{6.209}$$

Since $\tilde{\eta}_1 - \tilde{\eta}_0 \ll 1$, we may write $\tilde{j}_1(\tilde{\eta}_1) \simeq \tilde{j}_1(\tilde{\eta}_0)$ and hence with the good accuracy (6.209) is the polarization curve of the ACL.

Equation (6.209) is trivial: when η is almost constant across the ACL, total proton current j_1 is simply a product of a constant rate of the reaction Q by the ACL thickness l_t. In dimensionless form this gives Eq. (6.209).

Solving (6.209) for $\tilde{\eta}_0$ we find

$$\tilde{\eta}_0 = \ln(\omega\tilde{j}_1) - \ln(1 - p_t\tilde{j}_1) \tag{6.210}$$

In terms of ψ Eq. (6.210) is

$$\tilde{\eta}_0 = \ln(p_t\tilde{j}_1) - \ln\psi - \ln(1 - p_t\tilde{j}_1) \tag{6.211}$$

The last logarithm in Eq. (6.210) shows that the limiting current density is

$$\tilde{j}_1^{\lim} = \frac{1}{p_t} \tag{6.212}$$

At $\tilde{\eta}_0 = 0$ Eq. (6.206) gives $\tilde{\delta} = 1/(\omega + p_t)$. Since $\tilde{\delta}(\tilde{\eta}_0)$ is monotonic, for arbitrary $\tilde{\eta}_0$ we have

$$\frac{1}{\omega + p_t} \leq \tilde{\delta} \leq \frac{1}{p_t} \qquad (6.213)$$

The inequality

$$\tilde{\delta} \ll 1$$

is thus guaranteed for any $\tilde{\eta}_0$ when

$$p_t \gg 1 \qquad (6.214)$$

With (6.213) we see that for $p_t \gg 1$ we have $p_t\tilde{\delta}^2 \leqslant 1/p_t \ll 1$, which justifies the expansion in terms of $p_t\tilde{\delta}^2$ in (6.208).

Polarization curves (6.211) for $p_t = 10$ are depicted in Figure 6.26 together with the exact numerical curves calculated from Eqs. (6.198) and (6.204) (see the next section). When $\psi \lesssim 10^{-2}$ the curves (6.211) coincide with the numerical ones and exhibit regular behavior: close to $\tilde{j}_1 = 0$ they have a Tafel-like shape $\tilde{\eta}_1 \sim \ln(\tilde{j}_1)$ and close to the limiting current they obey limiting law $\tilde{\eta}_1 \sim \ln(1 - p_t\tilde{j}_1)$.

Equation (6.211) shows that the variation of ψ shifts the polarization curve as a whole along the vertical axis. This is seen in Figure 6.26: the larger ψ the less the voltage loss. Physically, $k_\star$ is proportional to the exchange current density $i_\star$: $k_\star \sim i_\star$. Therefore, at a fixed p_t larger ψ corresponds to larger $i_\star$. The larger $i_\star$ the less overpotential is required for current production; this effect is accounted for in Eq. (6.211).

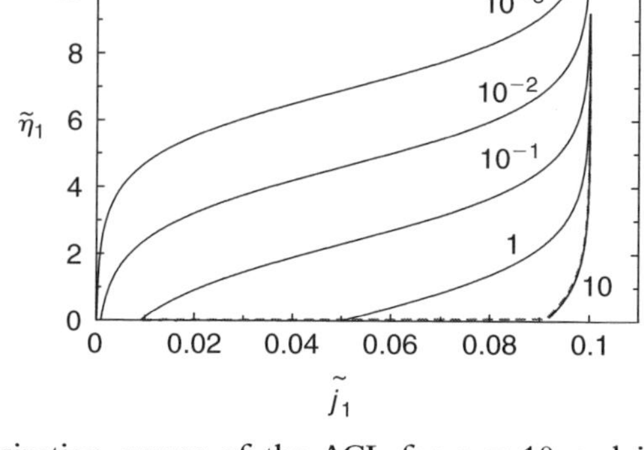

Figure 6.26 Polarization curves of the ACL for $p_t = 10$ and indicated values of $\psi = p_t/\omega$. Solid lines: Eq. (6.211) and dashed line: numerical solution to (6.204), (6.198). Above $\tilde{\eta}_1 \simeq 0.1$ the analytical and numerical curves are indistinguishable.

However, as ψ increases, in a region of small currents numerical polarization curves exhibit very rapid variation of $\tilde{j}_1$ with $\tilde{\eta}_1$ (dashed line in Figure 6.26), which is not described by Eq. (6.211). With the growth of ψ this region extends towards limiting current (Figure 6.26). When $\psi \gg 1$ the polarization curve "jumps" to a limiting current already at very small potentials $\tilde{\eta}_1$ (Figure 6.26). As ψ increases, the numerical curves in Figure 6.26 manifest a new mechanism of active layer operation.

Our simple model of methanol oxidation (Section 6.6.1) is valid when $\tilde{\eta} \gtrsim 1$. In the case of $p_t = 10$ shown in Figure 6.26 this new mechanism is seen at small $\tilde{\eta}_1$ and is thus beyond the scope of our model. However, at small p_t this mechanism appears at $\tilde{\eta}_1 \gtrsim 1$ and hence it can be analyzed.

6.6.2.2 The General Case: Active Layer of Variable Thickness

In the general case of arbitrary δ (arbitrary p_t) the procedure of calculation of polarization curve is as following. For given $\tilde{\eta}_1$ Eq. (6.204) has to be solved for $\tilde{\eta}_0$. Then the pair $(\tilde{\eta}_1, \tilde{\eta}_0)$ is substituted into (6.198) to calculate $\tilde{j}_1$.

The results of numerical calculations for $p_t = 0.1$ are shown in Figure 6.27. At very small $\psi = 10^{-3}$ we have qualitatively the same (as in the previous section) result: Tafel-like behavior $\tilde{\eta}_1 \sim \ln(\tilde{j}_1)$ at small currents and a limiting shape $\tilde{\eta}_1 \sim \ln(1 - p_t\tilde{j}_1)$ close to the limiting current $\tilde{j}_1^{\text{lim}} = 1/p_t$ (Figure 6.27).

However, as ψ increases, polarization curves at small currents become non-logarithmic. When $\psi = 10$ the whole curve changes its shape (Figure 6.27). The new mechanism provides much faster growth of current with the polarization voltage.

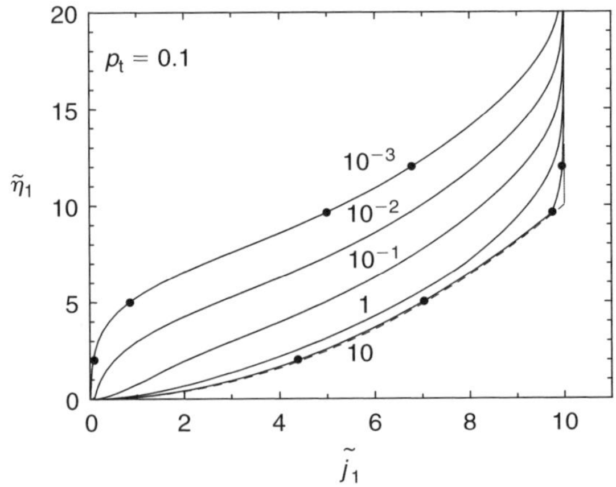

Figure 6.27 Polarization curves of the catalyst layer for $p_t = 0.1$ and indicated values of $\psi = p_t/\omega$. Dashed line: analytical solution for $\psi \gg 1$. Short dashed: limiting line $j_1^{\text{lim}} = 1/p_t$. Local overpotential and proton current density in the points indicated by filled circles are shown in Figures 6.28 and 6.29.

To understand this effect it is advisable to plot the shapes of local parameters $\tilde{\eta}(\tilde{x})$ and $\tilde{j}(\tilde{x})$ as we move along the polarization curve. Consider first these shapes in the several points on the "regular" curve for $\psi = 10^{-3}$ (filled circles in Figure 6.27). In the regular case local current is generated over the entire thickness of the catalyst layer (Figure 6.28(a)). At small currents local overpotential is almost constant along $\tilde{x}$; with the growth of current the curve $\tilde{\eta}(\tilde{x})$ shifts upward and the difference $\tilde{\eta}_1 - \tilde{\eta}_0$ increases (Figure 6.28(b)).

The shapes of $\tilde{\eta}(\tilde{x})$ and $\tilde{j}(\tilde{x})$ on the curve for $\psi = 10$ (filled circles in Figure 6.27) are quite different (Figure 6.29). As long as $\tilde{\eta}_1$ is below certain limiting value ($\tilde{\eta}_1 < \tilde{\eta}_1^\star \simeq 9.6$) the electrochemical reaction covers only part of the ACL. There is a point $\tilde{x}_0$, where local overpotential and current vanish ($\tilde{\eta}(\tilde{x}_0) = \tilde{j}(\tilde{x}_0) = 0$) (Figure 6.29). The growth of total current is provided by increase in the thickness of the current-generating domain (by the shift of $\tilde{x}_0$ to the left) at a cost of minimal growth of $\tilde{\eta}_1$. In other words, as $\tilde{j}_1$ increases the reaction occupies larger portion of ACL thickness, moving towards $\tilde{x} = 0$ (Figure 6.29). This is the regime with "variable thickness" of the ACL (VT regime).

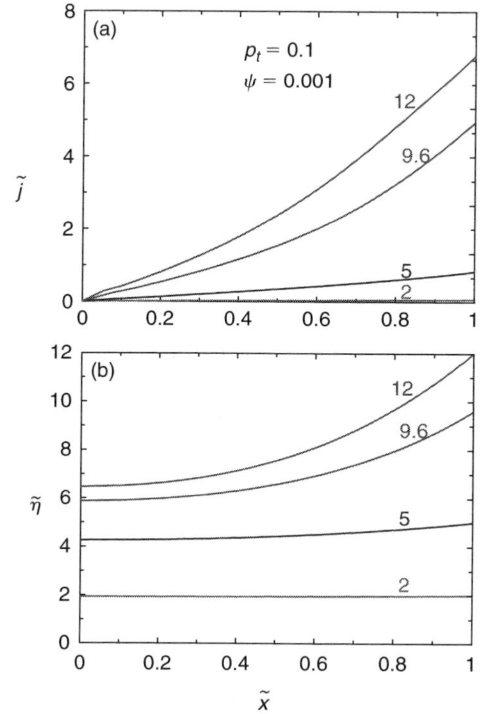

Figure 6.28 The distribution of (a) local current density and (b) local overpotential across the catalyst layer at different points on the polarization curve for $\psi = 0.001$ (Figure 6.27). Indicated are the values of $\tilde{\eta}_1$.

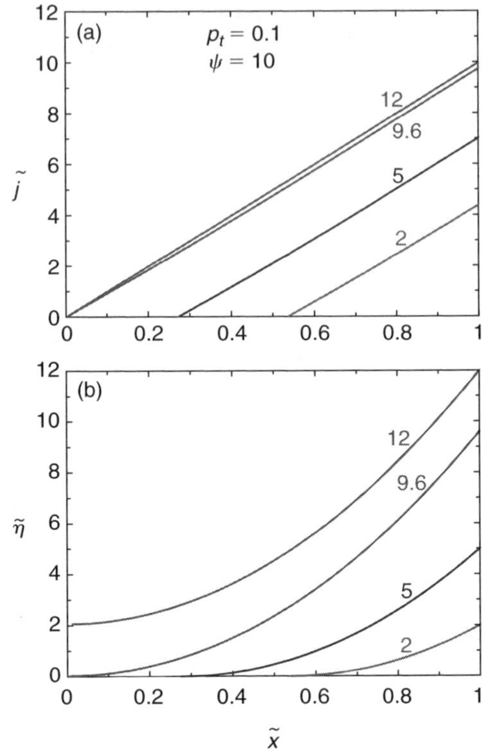

Figure 6.29 Same as in Figure 6.28 for the polarization curve with $\psi = 10$ (Figure 6.27).

When $\tilde{x}_0$ approaches zero, $\tilde{\eta}_1 = \tilde{\eta}_1^{\star}$ and the ACL enters the regular regime. The whole thickness of the ACL is now "covered" with the reaction (Figure 6.29). Further growth of total current is provided by the simultaneous increase in $\tilde{\eta}_0$ and $\tilde{\eta}_1$, which is more expensive in terms of voltage loss.

For the case of $\psi \gg 1$ the functions $\tilde{j}(\tilde{x})$ and $\tilde{\eta}(\tilde{x})$ can be obtained analytically. From the conservation law (6.197) we get relation of local $\tilde{j}$ and $\tilde{\eta}$:

$$\tilde{j} = \sqrt{\frac{1}{p_t} \ln\left(\frac{1 + \psi \exp(\tilde{\eta})}{1 + \psi \exp(\tilde{\eta}_0)}\right)} \tag{6.215}$$

Neglecting here both units under the logarithm sign and putting $\tilde{\eta}_0 = 0$ we find

$$\tilde{\eta} = p_t \tilde{j}^2 \tag{6.216}$$

or

$$\tilde{j} = \sqrt{\frac{\tilde{\eta}}{p_t}} \tag{6.217}$$

Using (6.217) in (6.191) and solving the resulting equation with the boundary condition $\tilde{\eta}(\tilde{x}_0) = 0$ we get

$$\tilde{\eta} = \frac{(\tilde{x} - \tilde{x}_0)^2}{p_t} \tag{6.218}$$

Equation (6.191) then gives

$$\tilde{j} = \frac{\tilde{x} - \tilde{x}_0}{p_t} \tag{6.219}$$

Local current density, therefore, increases with $\tilde{x}$ linearly and $\tilde{\eta}$ has parabolic shape along $\tilde{x}$. Numerical curves in Figure 6.29 confirm this result.

Substituting $\tilde{x} = 1$ into Eq. (6.219) we get the thickness $\tilde{\delta}_j = 1 - \tilde{x}_0$ of the current-generating domain as a function of total current:

$$\tilde{\delta}_j = p_t \tilde{j}_1 \tag{6.220}$$

Thus, $\tilde{\delta}_j$ increases with the current linearly.

To understand the character of the polarization curve $\tilde{\eta}_1(\tilde{j}_1)$ in the VT regime consider again the case of $\psi \gg 1$. From Eq. (6.216) we immediately obtain

$$\tilde{\eta}_1 = p_t \tilde{j}_1^2 \tag{6.221}$$

Therefore, instead of characteristic for regular regime logarithmic dependence (the first term on the right side of Eq. (6.210)), we have the parabolic law (6.221). This is seen in Figure 6.27: at $\tilde{\eta}_1 < \tilde{\eta}_1^\star \simeq 9.6$ the curve for $\psi = 10$ is very close to parabola.

Furthermore, for $\psi \gg 1$ the polarization curve does not depend on ψ. The curve for $\psi = 10$ is close to the limiting curve (dashed line in Figure 6.27). The latter curve is parabola (6.221) in the range $0 \leqslant \tilde{\eta}_1 < 1/p_t$; at $\tilde{\eta}_1 = 1/p_t$ it becomes a vertical line $\tilde{j}_1^{\lim} = 1/p_t$ (*cf.* Eq. (6.212)).

VT solutions arise when the exchange current density $i_\star \sim k_\star$ is large and/or the product $k_a c_t$ is small. The first term in denominator of Eq. (6.186) is then negligible and we get

$$Q \simeq 6Fk_a c_t A \tag{6.222}$$

which does not depend on overpotential.

Physically, Q (6.222) is the volumetric rate of methanol adsorption. VT-regime is thus the limit of fast electrochemical conversion: any methanol molecule which managed to sit onto the catalyst surface is immediately oxidized. In this regime methanol oxidation behaves like a "chemical" rather than electrochemical reaction. However, the reaction generates protons and it is activated close to the membrane, where the transport of protons is "cheaper".

It is advisable to write Eq. (6.220) in dimension form:

$$\delta_j = \frac{j_1}{6Fk_a c_t A} \tag{6.223}$$

where $\delta_j \equiv \tilde{\delta}_j l_t$. In the limit of Tafel kinetics, when the rate of methanol adsorption $6Fk_a c_t A$ is large, $\delta_j \to 0$ and VT regime disappears.

Using (6.222), Eq. (6.223) can be written as $j_1 \simeq \delta_j Q$. Physically, reaction rate Q in the current-generating domain is constant (6.222) and hence total proton current produced in this domain is simply a product of Q by domain thickness δ_j.

Voltage loss in the VT regime is smaller, than in the regular regime (Figures 6.26 and 6.27). The reason is that in the VT regime protons should be transported only through a current-generating domain, whereas in the regular regime reaction "covers" the whole thickness of the catalyst layer and expenses for proton transport are larger.

6.6.3 Optimal Catalyst Layer

Analysis above shows that optimal performance of the ACL is attained when

$$\tilde{j}_1^{\lim} \to \text{maximal} \tag{6.224}$$

$$\psi \to \text{maximal} \tag{6.225}$$

$$\tilde{\eta}_1 = p_t \tilde{j}_1^2 \to \text{minimal} \tag{6.226}$$

The first condition means that the reaction-limiting current density must be large. The second one warrants significantly large domain of currents where VT regime is realized and the third one requests the smallness of the polarization voltage in this regime.

Rewriting (6.224)–(6.226) in dimension form we get

$$j_1^{\lim} = 6Fl_t k_a c_t A \tag{6.227}$$

$$\psi = \frac{k_\star c_w}{k_a c_t} \tag{6.228}$$

$$\eta_1 = \frac{j_1^2}{12F\sigma_t k_a c_t A} \tag{6.229}$$

Consider the parameters which appear in (6.227)–(6.229). The following variation of three of them is unconditionally beneficial:

- Increase in the exchange current density $i_\star \sim k_\star$ increases ψ, thus extending the domain of currents where VT-regime exists.
- Increase in the thickness of the catalyst layer l_t increases $j_1^{\rm lim}$.
- Increase in proton conductivity of the catalyst layer σ_t lowers voltage loss η_1.

Parameters c_t and k_a appear in (6.227)–(6.229) only as a product $c_t k_a \equiv q_t$. The decrease in q_t gives larger ψ; at the same time smaller q_t decreases limiting current and increases voltage loss.[★]

The simplest way to organize VT regime in experiment (though at a cost of poor ACL performance) is to use small methanol concentration c_t. In the VT regime the slope of polarization curve near zero current density should be small (Figure 6.27). Furthermore, with the decrease in c_t the slope of the curve near $j_1 = 0$ should decrease.[5]

Control of c_t is particularly simple in experiments with the catalyst coated membranes, when methanol is supplied directly to the ACL. (Not through the backing layer as in the usual cell design. In the latter case c_t depends on local current density, which complicates the interpretation of results.)

If DMFC is designed to work at a constant current density, Eq. (6.223) gives optimal thickness of the catalyst layer. Together with the condition $\psi = k_\star c_w/(k_a c_t) \gtrsim 1$ Eq. (6.223) determines optimal composition and thickness of the ACL.

Non-Tafel kinetics limits the current in DMFC due to potential-independent step of methanol adsorption on the catalyst surface. However, the mechanism of variable thickness of the ACL provides a "cheap" route to the limiting current. Under high rate of electrochemical conversion the transport of protons becomes a "rate-determining step" and electrochemical reaction is activated close to the membrane, where protons are "cheaper". In this regime the thickness of current-generating domain is proportional to the total current.

[5] Note that similar effect causes formation of a bridge (Section 6.3).

[★] Interpretation of the effect of q_t variation on the ACL performance in Ref. [45] is erroneous.

6.7 List of Symbols

Symbol	Description
$\sim$	Marks dimensionless variables
$-$	Marks time-average values (Section 6.4)
A	Molar fraction of active catalyst particles in the anode catalyst layer (mol/cm^3)
A_a	Cell active area (cm^2)
b	Tafel slope (V)
C	Total molar concentration of the mixture in the channel (mol/cm^3)
c	Molar concentration (mol/cm^3)
c_{ref}	Reference molar concentration (mol/cm^3)
D_b	Diffusion coefficient of feed molecules in the backing layer (cm^2/s)
D_m	Diffusion coefficient of methanol in the membrane (cm^2/s)
E	Voltage loss, the sum of polarization voltages on the anode and the cathode sides (V)
F	Faraday constant (9.6495×10^4 Coulomb/mol)
$i_\star$	Exchange current density per unit volume (A/cm^3)
J	Mean current density in a cell (A/cm^2)
J_{cross}	Mean crossover current density in a cell (A/cm^2)
$J_\star$	Current density, at which the bridge forms (A/cm^2)
j	Local current density in a cell, local proton current density in the catalyst layer (A/cm^2) (Section 6.2.1)
j_0	Proton current density at the membrane/cathode catalyst layer interface (A/cm^2)
j_1	Proton current density at the anode catalyst layer/membrane interface (A/cm^2)
$j_\star$	Characteristic current density (A/cm^2)
$j_{\lim}$	Limiting current density (A/cm^2)
k	Parameter which determines the rate of bubbles formation (6.120)
k_t	Parameter (6.12)
k_h	Parameter (6.37)
h	Channel height (cm)
l_t	Thickness of the catalyst layer (cm)
l_b	Thickness of the backing layer (cm)
l_m	Thickness of the membrane (cm)
L	Channel length (cm)
N_{cross}	Molar flux of methanol through the membrane (crossover flux) ($mol/cm^2/s$)
n_d	Electroosmotic drag coefficient of water in the membrane
p	Ratio of Tafel slopes ($p = b^c/b^a$)
p_t	Inverse dimensionless current of methanol adsorption (6.193)
Q	Rate of electrochemical reaction (A/cm^3)
q	Dimensionless parameter

r	Linear fraction of liquid plugs in the anode channel
u	Feed molecules utilization
V_{cell}	Cell voltage (V)
V_{loss}	Total voltage loss, including ohmic losses (V)
V_{oc}^{T}	Thermodynamic open-circuit voltage (V)
V_{oc}	Cell open-circuit voltage (V)
v	Flow velocity in the channel (cm/s)
x	Coordinate across the cell (cm)
z	Coordinate along the channel (cm)
z_j	Coordinate, where local current vanishes (cm)
z_{ox}	Coordinate, where oxygen concentration vanishes (cm)

6.7.1 Superscripts

0	At the channel inlet (at $z = 0$)
a	Anode side
c	Cathode side

6.7.2 Subscripts

0	Cathode catalyst layer/membrane interface (Section 6.2.1); value at zero total current (Section 6.3); zero-order term (Section 6.5)
1	Membrane/anode catalyst layer interface; first-order correction (Section 6.5)
b	Backing layer
br	Bridge
lim	Limiting
M	Methanol
m	Membrane
ox	Oxygen
t	Catalyst layer
w	Water

6.7.3 Greek symbols

α	Dimensionless parameter (6.55)
β	Dimensionless parameter of crossover (6.20)
$\beta_\star$	$\beta/(1 + \beta)$
γ	Dimensionless parameter (6.55)
η	Polarization voltage (overpotential) (V)
κ	Dimensionless local current density at the oxygen channel inlet
λ	Stoichiometry of the feed flow
μ	Dimensionless characteristic scale of the exponent (6.63)

μ_ψ	Dimensionless characteristic scale of the exponent (6.106)
ξ	Oxygen molar fraction in the channel
σ_t	Proton conductivity of electrolyte phase in the catalyst layer (1/Ω/cm)
ϕ	Matching function (6.40)
$\varphi^{a,c}$	Local carbon phase potential in the catalyst layer (V)
φ_m	Local membrane phase potential in the catalyst layer (V)
ψ	Methanol molar fraction in the channel
ω	Inverse dimensionless current of methanol oxidation (6.192)

Appendix A: Equation for Local Current Density

Differentiating (6.127) with respect to $\tilde{z}$ and using (6.123), (6.125) and (6.126) to exclude derivatives $\partial\tilde{v}/\partial\tilde{z}$, $\partial\tilde{c}/\partial\tilde{z}$ and $\partial\tilde{\xi}/\partial\tilde{z}$ we get

$$\frac{\partial \tilde{j}}{\partial \tilde{z}} = -\frac{A}{B} \qquad \tilde{j}(0) = \kappa \tag{6.230}$$

where

$$\begin{aligned} A = {} & \left[k\lambda^c(1+\beta)(1-p\beta)\tilde{c} + \lambda^c(1-p\beta) + p\lambda^a\gamma(1+\beta)\tilde{\nu}\right]\tilde{j}^3 \\ & + \Big[k\lambda^c\beta(1+\beta)(1+p)\tilde{c}^2 + \Big(-k\lambda^c\gamma(1+\beta)^2\tilde{\xi} \\ & -p\lambda^a\gamma(1-\beta^2)\tilde{\nu} + \beta\lambda^c(2+p(1-\beta))\Big)\tilde{c} - \lambda^c\gamma(1+\beta)\tilde{\xi}\Big]\tilde{j}^2 \\ & + \left[(\beta^2\lambda^c(1+p) - p\lambda^a\gamma\beta(1+\beta)\tilde{\nu})\tilde{c}^2 - \lambda^c\gamma\beta(1+\beta)\tilde{\xi}\tilde{c}\right]\tilde{j} \end{aligned}$$

and

$$\begin{aligned} B = {} & \lambda^a\lambda^c J(1+\beta)\tilde{\nu}\Big[((1-p\beta)\tilde{c} + p\gamma(1+\beta)\tilde{\xi})\tilde{j} \\ & + \beta(1+p)\tilde{c}^2 - \gamma(1+p)(1+\beta)\tilde{\xi}\tilde{c}\Big] \end{aligned}$$

Equation for κ is (6.141). The standard algorithms of solution of nonlinear equations work better if we calculate the exponent of (6.141):

$$\frac{(1+\beta)\kappa}{q(1-\kappa)}\left(\frac{\kappa}{\alpha q}\right)^p\left(1-\frac{\kappa+\beta_\star(1-\kappa)}{\gamma}\right)^{-p} = \exp(\tilde{E})$$

and then transform it to

$$\frac{(1+\beta)\kappa}{\exp(\tilde{E})}\left(\frac{\kappa}{\alpha q}\right)^p = q(1-\kappa)\left(1-\frac{\kappa+\beta_\star(1-\kappa)}{\gamma}\right)^p$$

Appendix B

To show that K_ξ (6.163) is positive consider the right side of Eq. (6.163). In view of (6.143) the first factor in the numerator of (6.163) is positive ($\gamma - \gamma_\star > 0$). Since $\kappa < 1$, the factor in square brackets in the numerator is also positive. The sign of K_ξ is thus given by the sign of the expression in the square brackets in denominator. Equating it to zero we find: $\gamma(1 + p(1 - \kappa)) - \gamma_\star - p\beta_\star(1 - \kappa) = 0$; solving this for γ we get

$$\gamma = \frac{\gamma_\star + p\beta_\star(1-\kappa)}{1 + p(1-\kappa)} = \frac{\gamma_\star + a\beta_\star}{1+a} \tag{6.231}$$

where $a \equiv p(1 - \kappa) > 0$. We write $\gamma = \gamma_\star + \delta\gamma$; substituting this into (6.231) we find

$$\delta\gamma = \frac{a(\beta_\star - \gamma_\star)}{1+a} \tag{6.232}$$

Consider the sign of this expression. We have

$$\beta_\star - \gamma_\star = \beta_\star - \kappa - \beta_\star(1-\kappa) = -\kappa(1-\beta_\star)$$

Since $\beta_\star < 1$, we have $\delta\gamma < 0$. It means that Eq. (6.231) is satisfied when $\gamma < \gamma_\star$. Equation (6.143), however, prescribes that γ must exceed $\gamma_\star$. For $\gamma > \gamma_\star$ Eq. (6.231) is never satisfied, we have $K_\xi > 0$ and thus $\tilde{\xi}_1$ is also positive.

Acknowledgments

This work stemmed from numerical models of DMFC developed by the author in 1998–1999 together with Jiri Divisek and Alexei Kornyshev. Invaluable experimental data were provided by Katja Klafki and Klaus Wippermann. The author is grateful to all these people for their contribution and for many useful discussions.

REFERENCES

1. A. S. Aricò, S. Srinivasan, and V. Antonucci. DMFCs: from fundamental aspects to technology development. *Fuel Cells*, 1(2) (2001) 133–161.
2. R. Dillon, S. Srinivasan, A. S. Aricò, and V. Antonucci. International activities in DMFC R&D: status of technologies and potential applications. *J. Power Sources*, 127 (2004) 112–126.
3. X. Ren, P. Zelenay, S. Thomas, J. Davey, and S. Gottesfeld. Recent advances in direct methanol fuel cells at Los Alamos National Laboratory. *J. Power Sources*, 86 (2000) 111–116.
4. H. Dohle, H. Schmitz, T. Bewer, J. Mergel, and D. Stolten. Development of a compact 500 W class direct methanol fuel cell stack. *J. Power Sources*, 106 (2002) 313–322.
5. E. R. Choban, L. J. Markoski, A. Wieckowski, and P. J. A. Kenis. Microfluidic fuel cell based on laminar flow. *J. Power Sources*, 128 (2004) 54–60.
6. J. Larminie and A. Dicks. *Fuel Cell Systems Explained,* Chichester, Wiley, 2000.
7. K. Scott, P. Argyropoulos, and K. Sundmacher. A model for the liquid feed direct methanol fuel cell. *J. Electroanal. Chem.*, 477 (2000) 97–100.
8. S. F. Baxter, V. S. Battaglia, and R. E. White. Methanol fuel cell model: anode. *J. Electrochem. Soc.*, 146(2) (1999) 437–447.
9. H. Dohle, J. Divisek, and R. Jung. Process engineering of the direct methanol fuel cell. *J. Power Sources*, 86 (2000) 469–477.
10. K. Sundmacher, T. Schulz, S. Zhou, K. Scott, M. Ginkel, and E. D. Gilles. Dynamics of the direct methanol fuel cell (DMFC): experiments and model-based analysis. *Chem. Engineer. Sci.*, 56 (2001) 333–341.
11. J. Nordlund and G. Lindbergh. A model for the porous direct methanol fuel cell anode. *J. Electrochem. Soc.*, 149(9) (2002) A1107–A1113.
12. J. Meyers and J. Newman. Simulation of the direct methanol fuel cell. II. Modeling and data analysis of transport and kinetic phenomena. *J. Electrochem. Soc.*, 149(6) (2002) A718–A728.
13. J. Meyers and J. Newman. Simulation of the direct methanol fuel cell. III. Design and optimization. *J. Electrochem. Soc.*, 149(6) (2002) A729–A735.
14. G. Murgia, L. Pisani, A. K. Shukla, and K. Scott. A numerical model of a liquid-feed solid polymer electrolyte DMFC and its experimental validation. *J. Electrochem. Soc.*, 150 (2003) A1231–A1245.
15. J. Nordlund and G. Lindbergh. Temperature-dependent kinetics of the anode in the DMFC. *J. Electrochem. Soc.*, 151(9) (2004) A1357–A1362.
16. A. A. Kulikovsky, J. Divisek, and A. A. Kornyshev. Modeling the cathode compartment of polymer electrolyte fuel cells: dead and active reaction zones. *J. Electrochem. Soc.*, 146(11) (1999) 3981–3991.
17. A. A. Kulikovsky, J. Divisek, and A. A. Kornyshev. Two dimensional simulation of direct methanol fuel cell: a new (embedded) type of current collectors. *J. Electrochem. Soc.*, 147(3) (2000) 953–995.
18. J. Divisek, J. Fuhrmann, K. Gärtner, and R. Jung. Performance modeling of a direct methanol fuel cell. *J. Electrochem. Soc.*, 150 (2003) A811–A825.
19. Z. H. Wang and C. Y. Wang. Mathematical modeling of liquid-feed direct methanol fuel cells. *J. Electrochem. Soc.*, 150(4) (2003) A508–A519.
20. E. Birgersson, J. Nordlund, H. Ekström, M. Vynnycky, and G. Lindbergh. Reduced two-dimensional one-phase model for analysis of the anode of a DMFC. *J. Electrochem. Soc.*, 150 (2003) A1368–A1376.
21. E. Birgersson, J. Nordlund, M. Vynnycky, C. Picard, and G. Lindbergh. Reduced two-phase model for analysis of the anode of a DMFC. *J. Electrochem. Soc.*, 151 (2004) A2157–A2172.
22. A. A. Kulikovsky, T. Wüster, T. Egmen, and D. Stolten. Analytical and numerical analysis of PEM fuel cell performance curves. *J. Electrochem. Soc.*, 152(6) (2005) A1290–A1300.

23. A. A. Kulikovsky. Performance of catalyst layers of polymer electrolyte fuel cells: exact solutions. *Electrochem. Comm.*, 4(4) (2002) 318–23.
24. M. L. Perry, John Newman, and E. J. Cairns. Mass transport in gas-diffusion electrodes: a diagnostic tool for fuel-cell cathodes. *J. Electrochem. Soc.*, 145(1) (1998) 5–15.
25. M. Eikerling and A. A. Kornyshev. Modelling the performance of the cathode catalyst layer of polymer electrolyte fuel cells. *J. Electroanal. Chem.*, 453 (1998) 89–106.
26. X. Ren, T. Springer, T. Zavodzinski, and S. Gottesfeld. Methanol transport through the Nafion membranes. Electro-osmotic drag effects on potential step measurements. *J. Electrochem. Soc.*, 147(2) (2000) 466–474.
27. R. Jiang and D. Chu. Comparative studies of methanol crossover and cell performance for a DMFC. *J. Electrochem. Soc.*, 151 (2004) A69–A76.
28. V. Gogel, T. Frey, Z. Yongsheng, K. A. Friedrich, L. Jörissen, and J. Garche. Performance and methanol permeation of direct methanol fuel cells: dependence on operating conditions and on electrode structure. *J. Power Sources*, 127 (2004) 172–180.
29. W. Vielstich, V. A. Paganin, F. H. B. Lima, and E. A. Ticianelli. Non-electrochemical pathway of methanol oxidation at a platinum-catalyzed oxygen gas diffusion electrode. *J. Electrochem. Soc.*, 148 (2001) A502–A505.
30. A. A. Kulikovsky. On the nature of mixed potential in a DMFC. *J. Electrochem. Soc.*, 152(6) (2005) A1121–A1127.
31. M. K. Ravikumar and A. K. Shukla. Effect of methanol crossover in a liquid-feed polymer-electrolyte direct methanol fuel cell. *J. Electrochem. Soc.*, 143(8) (1996) 2601–2606.
32. Z. Qi and A. Kaufman. Open-circuit voltage and methanol crossover in DMFC. *J. Power Sources*, 110 (2002) 177–185.
33. A. A. Kulikovsky, K.Klafki, and K.Wippermann. Experimental verification of the effect of bridge formation in a direct methanol fuel cell. Electrochem. Comm., 7 (2005) 394–397.
34. H. Dohle, J. Divisek, H. F. Oetjen, C. Zingler, J. Mergel, and D. Stolten. Recent developments of the measurement of methanol permeation in a direct methanol fuel cell. *J. Power Sources*, 105 (2002) 274–282.
35. A. A. Kulikovsky. Model of the flow with bubbles in the anode channel and performance of a direct methanol fuel cell. *Electrochem. Comm.*, 7 (2005) 237–243.
36. H. Yang, T. S. Zhao, and Q. Ye. *In situ* visualization study of CO_2 gas bubble behavior in DMFC anode flow fields. *J. Power Sources*, 139 (2005) 79–90.
37. J. A. Dean. *Lange's Handbook of Chemistry*. McGraw-Hill, New York 1992.
38. A. A. Kulikovsky. 1D + 1D model of a DMFC: localised solutions and mixed potential. *Electrochem. Comm.*, 6 (2004) 1259–1265.
39. A. A. Kulikovsky. The effect of stoichiometric ratio λ on the performance of a polymer electrolyte fuel cell. *Electrochim. Acta*, 49(4) (2004) 617–625.
40. A. A. Kulikovsky. Bubbles in the anode channel and performance of a DMFC: asymptotic solutions. *Electrochim. Acta*, 2005 51 (2006) 2003–2011.
41. H. Yang, T.S. Zhao, and Q. Ye. Addition of non-reacting gases to the anode flow field of DMFCs leading to improved performance. *Electrochem. Comm.*, 6 (2004) 1098–1103.
42. V. S. Bagotzky and Yu. B. Vasilyev. Mechanizm of electrooxidation of methanol on the platinum electrodes. *Electrochim. Acta*, 12 (1967) 1323–1343.
43. H. A. Gasteiger, N. Markovic, Jr., P. N. Ross, and E. J. Cairns. Methanol electrooxidation on well-characterized Pt–Ru alloys. *J. Phys. Chem.*, 97 (1993) 12020–12029.
44. P. S. Kauranen, E. Skou, and J. Munk. Kinetics of methanol oxidation on carbon supported Pt and Pt + Ru catalyst. *J. Electroanal. Chem.*, 404 (1996) 1–13.
45. A. A. Kulikovsky. Active layer of variable thickness: the limiting regime of anode catalyst layer operation in a DMFC. *Electrochem. Comm.*, 7 (2005) 969–975.
46. A. A. Kulikovsky. Analytical model of the anode side of DMFC: effect of non-Tafel kinetics on cell performance. *Electrochem. Comm.*, 5 (2003) 530–538.

CHAPTER SEVEN

Methanol Reforming Processes

Arunabha Kundu*,[1], Yong Gun Shul* *and* Dong Hyun Kim**

Contents

Abstract

Hydrogen is seen by many as the energy carrier for the future, and development of the science and technology needed to produce, store and utilize hydrogen has emerged as an international research priority. Hydrogen has the highest energy density of any non-nuclear fuel and can be easily converted to electrical and thermal energy via highly efficient, non-polluting processes. Fossil fuels like natural gas and crude oil are obvious sources for the large quantities of hydrogen needed to initiate the transition to a "Hydrogen Economy."

Hydrogen can be produced either off-board or on-board way. Most of the studies for on-board hydrogen production for fuel cells are based on two types

* Department of Chemical Engineering, Yonsei University, 134 Shinchon Dong, Sudeamun-Ku, Seoul, South Korea 120-749, Republic of Korea

[1] Presently at Micro-fuel cell team, Electro-materials and Device Lab.(eMD Lab), Corporate R&D Center, Samsung Electro-Mechanics, Maetan3-Dong, Suwon, Korea

** Department of Chemical Engineering, Kyungpook National University, 1370 Sankyukdong, Daegu 702-701, Republic of Korea

Advances in Fuel Cells
0080453945

of compounds. One is oxygen containing compounds – methanol and ethanol, etc. The other option is hydrocarbons such as natural gas, gasoline and diesel fuel. Among them, methanol is the most attractive fuel for on-board hydrogen production because of its high H/C ratio, low reforming temperature and good miscibility with water. In addition, methanol is sulfur-free, eliminating the concerns with catalyst or electrode poisoning by sulfur. The absence of carbon–carbon (C—C) bonds in methanol comparatively reduces the risk of coking. The different reforming process of methanol (steam reforming of methanol (SRM), partial oxidation of methanol (POM) and oxidative methanol reforming (OMR)) has been discussed and compared.

Primarily owing to advances of the materials (e.g. catalysts) and reactors, the processing of hydrocarbons into H_2-rich gas has become fairly efficient. The chapter focuses on the different catalysts of recent literature and reactors by different workers for hydrogen production from methanol reforming process and operation of fuel cells from the production of hydrogen in this process. The commercial catalyst for SRM is $Cu/ZnO/Al_2O_3$. The other types of catalyst like ZrO_2 supported catalyst or addition of ZrO_2 and CeO_2 in $Cu/ZnO/Al_2O_3$, CuO/CeO_2 and Pd/ZnO have also been discussed.

The appropriate selection of reforming reactor is required for its suitability as on-board source of hydrogen in automobile industries as well as electronic equipment. The different approaches in this regard have been discussed. These are fixed bed reactor, monolithic reactor, wall-coated heat exchanger, micro-channel reactor and metallic foam reactor. Among them, micro-channel reactor seems to be most promising in portable electronics with respect to its fast response in transient behavior, higher heat (which will lead to better temperature control inside reactor) and mass transfer characteristics, less channeling of flow of reactant and less pressure drop. The further work on the stability of the coated catalyst with high catalytic performance is still required. The larger version of micro-channel reactor, i.e. wall-coated heat exchanger with channels can also be promising for automotive applications.

Studies on the integrated operation of methanol reformer with the fuel cell is very essential to apply it in on-board application. It is evident from the existing literature that the methanol reformer has a desirable fast response to achieve the reaction temperature and is highly suitable for use with fuel cells to power automobiles. However, more efforts are still needed on the cold start and transient characteristics of the reformer if it is to be used as an on-board vehicle fuel processor.

7.1 Introduction

7.1.1 Essentiality of Methanol Reforming

Currently there is major problem with the present transportation fuel. Stringent environmental regulations on the specifications of transportation fuels, general tendency of exploration of heavier crude oil throughout the globe and uncertainty over the future oil sources are forcing to look alternative sources of energy. The problems and its different approaches of solution for transportation fuels are given in Figure 7.1. The problems are covered in black color in the figure. Hydrogen-based transportation system is one of the best alternative.

Vehicles powered by proton-exchange membrane (PEM) fuel cells are approaching commercialization. Unlike internal combustion engines (ICEs), fuel cells are not limited by the thermal efficiency of the Carnot cycle. Hence, the fuel utilization is much more efficient than in conventional heat engines. Furthermore, fuel cells are silent during operation and have no moving parts, and the emissions of hazardous compounds to the atmosphere are low or even non-existing. The main feed for fuel cell is hydrogen. Compared with the gasoline and diesel fuel that presently dominate in the transportation

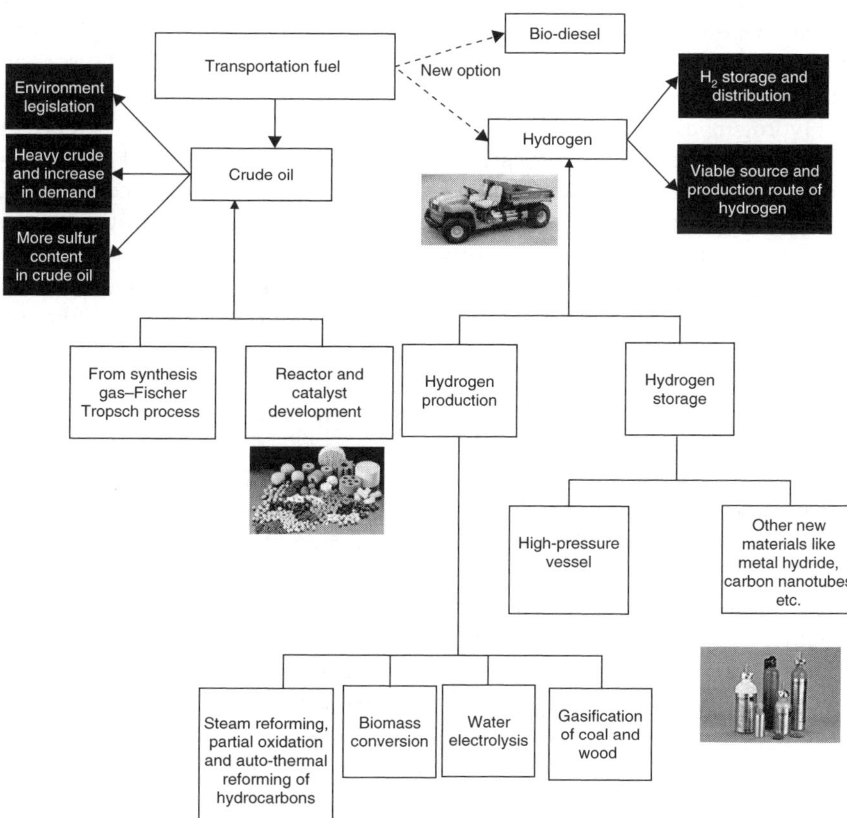

Figure 7.1 Problems in existing transportation fuels and the different approaches of its solution.

sector, hydrogen offers several significant potential advantages. First, hydrogen is in reality an "energy carrier" and not a primary fuel, and as such it can be produced from a wide range of potential feedstocks. Unlike gasoline and diesel, where there will be scarcity of oil especially light oils throughout the globe in a near future, hydrogen can be produced in many different ways and with resources that are abundant in nature. These resources include biomass, natural gas, landfill gas, oil and coal (Figure 7.1), as well as the wide range of additional resources that can be used to produce electricity. Second, hydrogen can be combusted cleanly with minimal air pollutant emissions and can be used in fuel cells to produce electrical energy with no emissions at all, except of water vapor. When used in fuel cell vehicles, hydrogen affords the opportunity of avoiding two key disadvantages of battery electric vehicles – short driving range and long refueling time – while providing many of the same private and public benefits. Hydrogen also offers opportunities for clean electricity production through "distributed generation" (DG). Under this concept, electrical power would be produced not through large power plants that typically transmit power relatively long distances, but rather through onsite generation that would produce power for local uses and it typically uses natural gas supplied by pipeline.

Hydrogen can be stored directly or produced on-board the vehicle by reforming methanol, or hydrocarbon fuels derived from crude oil (e.g., gasoline, diesel or middle distillates). Ogden *et al.* [1] compared the three leading options for fuel storage in on-board fuel cell vehicles in terms of the vehicle performance, fuel economy, weight and cost for various vehicle parameters, fuel storage choices and driving cycles. These are compressed hydrogen gas storage, on-board SRM and on-board partial oxidation of gasoline. The different possible fuel cell vehicle configurations have been shown in Figure 7.2 for these three processes. Vehicles with on-board SRM or partial oxidation of gasoline have about two-thirds the fuel economy of direct hydrogen vehicles. The efficiency is lower because of the conversion losses in the fuel processor, losses in making hydrogen from another fuel, reduced fuel cell performance on reformate, added weight of fuel processor components and effects of fuel processor response time. Ultimately they concluded that hydrogen is the preferred fuel compared to methanol.

Though hydrogen is the preferred fuel for fuel cell vehicles, for reasons of vehicle design, cost and efficiency, as well as potential energy supply and environmental benefits (e.g., the possibility for reduced total fuel cycle greenhouse gas emissions plus strictly zero tailpipe emissions), on-board hydrogen production is a promising alternative for commercialization of fuel cell at least in the early stages of commercialization till viable way of hydrogen production, transportation and storage is available. In other way, it can also work together with ICE. The possible combinations of on-board fuels and conversion technologies for personal transportation are summarized in Figure 7.3.

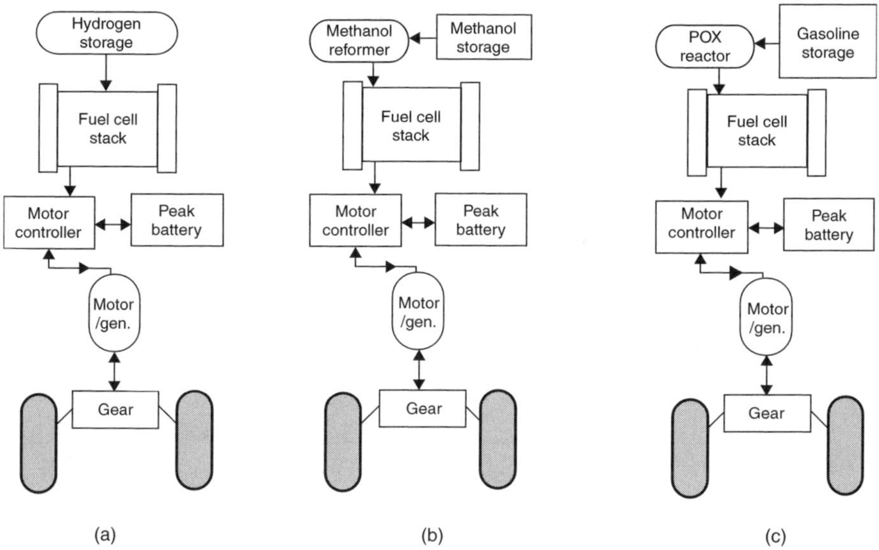

Figure 7.2 The three different configuration for fuel cell vehicle: (a) by off-board hydrogen production (b) and (c) on-board hydrogen production (adapted from Ogden *et al.* [1]).

Selection of the primary fuel for on-board hydrogen generation is a compromise that takes into account the energy value of the fuel, the temperature conditions of the reforming process, the composition of gases formed during reforming and cost. Table 7.1 gives the different aspects of reforming of different hydrocarbons. Among them, methanol reforming is the best in terms of the high energy density and comparatively low-temperature reaction. It should also be noted that volume % of carbon monoxide (CO) is very low (~1%) with the present reforming catalyst in contrast to the table due to the availability of most reliable technology and catalyst.

On the other hand, methanol, possibly in aqueous solution, appears to be leading as the fuel of choice for development of portable fuel cells. The conventional batteries lower power density and must be recharged periodically (Table 7.2), while fuel cell produces power continuously, as long as the fuel is supplied with high energy density. They can generate electrical energy of about 10 times that of a lithium-ion battery with the same cubic volume. Even at 5% system efficiency a fuel cell with a reformer would have higher energy density than a lithium polymer battery [3]. Though direct methanol fuel cell (DMFC) was one of the promising for use in laptop computers, digital cameras, Personal digital assistants and pocket television receivers, and also military purposes due to room temperature operation, it has some inherent

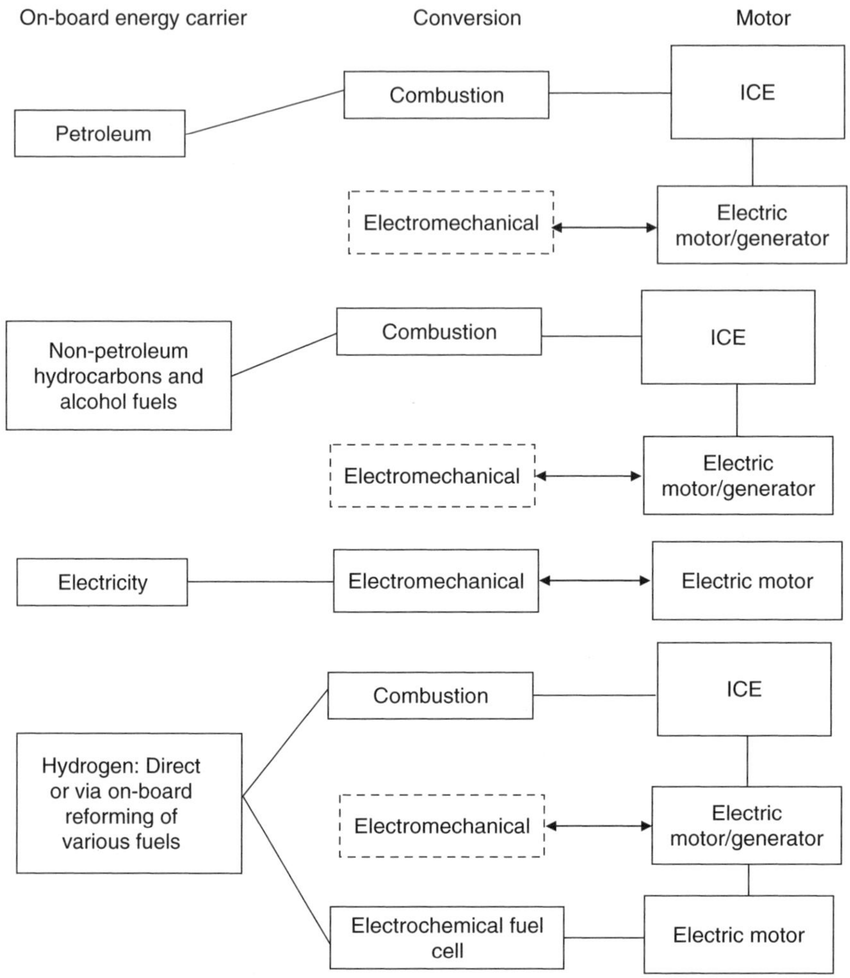

Figure 7.3 Possible combinations of on-board fuels and conversion technologies for personal transportation. (*Source*: The hydrogen economy-opportunities, costs, barriers and R&D needs (2004). The National Academic Press, Washington, DC.).

drawbacks. These are high cross over rate of methanol through the polymer electrolyte membrane and low reaction rate of methanol over an anode electrocatalyst [4]. Also DMFC gives less over-potential between cathode and anode due to higher oxidation potential of CH_3OH. Therefore, at 10 W and above, system manufacturing of DMFC is costly. In that case using on-board produced hydrogen from methanol is most promising to operate PEM fuel cell to use in these portables, but an integrated vaporizer and miniature methanol reformer with fuel cell is needed. The design of the fuel processor to reduce

Table 7.1 Different hydrocarbons used in reforming for production of hydrogen [2]

Raw material	Cost per 1 kg, dollar	Specific heat of combustion (kJ/g)	Reforming temperature (K)	Content in mixture (%)	
				H_2	CO
Methane	0.05–0.10	−50.1	1000	76.0	17.3
Propane	0.40–0.80	−46.5	700	74.0	7.2
Butane	–	−45.7	700	74.0	9.3
Octane	0.60–0.80	−44.7	>1000	70.8	20.4
Methanol	0.15–0.25	−19.7	533	74.4	4.3
Ethanol	0.20–0.30	−26.8	600	73.2	4.3
DME	–	−28.8	533	74.4	4.2

In terms of infrastructure facilities MeOH > EtOH > DME.

Table 7.2 Different battery technology

Technology	Energy density (kW_eh/l)	Energy density (kW_eh/kg)
Secondary cells		
Lead acid	0.07	0.035
Ni–Cd	0.170	0.055
Ni–metal hydride	0.250	0.070
Li–ion	0.350	0.120
Li–polymer	0.350	0.200
Fuel cell different fuel options		
Methanol	4.384	5.600
Butane	7.290	12.600
Iso-octane	8.680	12.340
Ethanol	5.900	7.500
Diesel	8.700	12.400
Hydrogen (gas, 24.8 MPa)	0.64	33.3
Hydrogen (Liquid, −253°C)	2.36	33.3

weight, volume and to respond quickly in a transient mode is important for the whole system.

7.1.2 Advantages of Methanol Reforming Compared to Reforming of Other Hydrocarbons

Methanol has a high H/C ratio (4:1). The absence of C—C bonds in methanol drastically reduces the risk for coke formation compared to other hydrocarbons like gasoline or diesel. It also makes the steam reforming process of methanol energetically favorable. Generally methanol reforming reaction occurs at comparatively lower temperature (200–300°C) as compared to the temperature (800–900°C) for gasoline reforming. Methanol steam reforming also requires relatively low water to carbon molar ratio. Moreover, as methanol can be produced from renewable sources (at present 90% production from natural gas), its reforming does not contribute to a net addition of CO_2 to the atmosphere. In addition, methanol is sulfur-free, eliminating the concerns with catalyst or electrode poisoning by sulfur, whereas logistic fuels such as jet and diesel fuels are readily available, but a compact and effective way to remove sulfur from these fuels is needed before entry to the hydrogen production unit. Methanol is also convenient for transportation and storage. This process produces a reformate with a high H_2 content and extremely low CO levels (it will not require water–gas-shift (WGS) reactor for lowering CO; preferential oxidation (PROX) reactor should be enough to get CO level below 10 ppm). CO is a fuel cell poison that deactivates the Pt-based catalyst at the anode already at concentrations exceeding a few parts per million. Methanol is the third most produced commodity chemical, after ethylene and ammonia, with a production capacity in excess of 25 million tons. If the entire 1995 methanol production capacity were dedicated to producing fuels for fuel cell cars, 7–8 % of total number cars worldwide can be fueled today [1]. The storage and refueling of methanol need little infrastructure change.

SRM has some drawbacks. Steam reforming has the reputation of being a kinetically slow reaction with residence times on the order of 1 s. Its large endothermic heat load exacerbated the consensus. However, there are some system-level advantages to steam reforming. First, because most PEM fuel cells run above atmospheric pressure, it is important that a fuel processor's primary conversion reaction run at a pressure higher than the fuel cell. Pressurizing liquid water and fuel for the steam reforming reaction is technologically straightforward and incurs a relatively small parasitic energy cost. Second, current fuel cell technologies do not consume all of the hydrogen produced. Thus unreacted hydrogen gas from the fuel cell could be combusted to provide a heat source for a steam reformer reactor. Third, the endothermic reaction makes the steam-reforming reactor a safer reactor.

7.1.3 Sources of Methanol

About 90% of methanol is produced from natural gas, although it would be possible to produce methanol via gasification of coal, heavy liquids, biomass or wastes.

MeOH is prepared from synthesis gas (which can be obtained from natural gas by partial oxidation (autothermal reforming), H_2O (steam reforming) and CO_2 (dry reforming)) containing H_2, CO, and CO_2 using MeOH synthesis catalyst ($Cu/ZnO/Al_2O_3$ or $Cu/ZnO/ZrO_2$ or sometimes with MnO_2 or MgO), and followed by distillation of MeOH. The process occurs at near 50 atm pressure in the temperature range of 550–650 K. This process corresponds to the following stoichiometry of the overall reaction:

$$CO + 2H_2 \rightleftarrows CH_3OH \qquad \Delta H^0 = -90.4\,\text{kJ/mol} \tag{7.1}$$

which is composed of two reactions

$$CO_2 + 3H_2 \rightleftarrows CH_3OH + H_2O \qquad \Delta H^0 = -49.4\,\text{kJ/mol} \tag{7.2}$$

and

$$CO + H_2O \rightleftarrows CO_2 + H_2 \qquad \Delta H^0 = -41\,\text{kJ/mol} \tag{7.3}$$

The process for the manufacture of methanol from coal comprises burning coal layers with oxygen-rich gases at 1300°C and 1–2 MPa in the presence of steam in a fixed-bed or fluidized-bed gasifier to obtain feed gas, dedusting, desulphurizing, dechlorinating, preparing synthesis gas with steam in the presence of a FeCrMg-based catalyst. Generally sulfur level is very high in the coal-derived synthesis gas. In that case copper-based catalyst should be poisoned as reported in Chinchen *et al.* [5].

7.1.4 Different Methanol Reforming Processes

Methanol is a fuel, which is readily available and can be catalytically converted into a H_2-rich gas at moderate temperature (200–350°C). The production of hydrogen from methanol is possible by three different ways. These are SRM, POM and OMR or autothermal methanol reforming. The comparisons of different reforming reactions with respect to process conditions and products produced are listed in Table 7.3. Hydrogen yields are significantly greater in SRM, rather than POM or autothermal reforming. OMR is advantageous with respect to no heat removal and supply, resulting in a simpler reactor geometry and faster responses. The main reason for the slow startup in SRM is poor heat conduction through the bed. But in OMR, controlling the heat generation by methanol oxidation is simple, and furthermore the bed is heated by the catalyst in the bed.

Table 7.3 Operating conditions and typical composition of produced gases in different reforming processes of methanol

	Steam reforming	Partial oxidation	Autothermal reaction/oxidative reforming
Water:methanol (molar ratio)	1.5–2	–	1.2–1.5
O_2/methanol	–	>0.5	0.1–0.25
Operating temperature	200–300°C	250–350°C	200–350°C
Operating pressure	3–5 bar	–	–
Catalysts used	Cu/ZnO/Al_2O_3, Cu/ZnO/Cr_2O_3, Pd/ZnO, CuO/CeO_2	Cu/ZnO/Al_2O_3, Cu/ZnO, Au/TiO_2, Cu/Zn/SiO_2, Cu/Cr	Cu/ZnO/Al_2O_3, Cu/ZrO_2/Al_2O_3, Cu/Cr_2O_3/Al_2O_3,
H_2 produced %	70–74	40–41	55–60
CO_2 produced %	22–25	20	22

N_2 produced %	–	21	21
CO produced %	<1	0.1–2	2
By-products	CO, coke, DME	CO, formaldehyde, methyl formate and DME	CO, formaldehyde, DME
Heat of reaction (ΔH^0), kJ/mol	+131	−154.8	−12
Reactions	$CH_3OH + H_2O \leftrightarrow CO_2 + 3H_2$ $CH_3OH \leftrightarrow CO + 2H_2$	$CH_3OH + \frac{1}{2}O_2 \leftrightarrow 2H_2 + CO_2$	$CH_3OH + \frac{1}{2}H_2O + \frac{1}{4}O_2 \leftrightarrow CO_2 + \frac{5}{2} H_2$
Remarks	1. High concentration of H_2 2. Comparatively less amount of CO 3. Most widely used	1. Less favorable H_2/CO_2 ratio 2. Large amount of CO	1. Net ideally heat required = 0 2. It is possible to get more compact design of the reactors

One drawback of OMR is formation of a hot-spot exceeding the Cu-based catalyst sintering temperature. This is solved in a new reactor design for OMR, not yet published, unfortunately.

7.1.4.1 Steam Reforming of Methanol

SRM refers to the chemical reaction between methanol and water vapor for the production of hydrogen gas. The key parameters of this reaction are given in Table 7.3. This process is typically carried out in the presence of metal oxide catalysts at temperatures ranging from 200°C to 300°C. The chemical reactions taking place during the reforming process are outlined as follows:

$$CH_3OH + H_2O \rightleftarrows CO_2 + 3H_2 \qquad \Delta H^0 = 49.4\ \text{kJ/mol} \tag{7.4}$$

$$CH_3OH \rightleftarrows CO + 2H_2 \qquad \Delta H^0 = 90.4\ \text{kJ/mol} \tag{7.5}$$

$$CO + H_2O \rightleftarrows CO_2 + H_2 \qquad \Delta H^0 = -41.4\ \text{kJ/mol} \tag{7.6}$$

Reaction (7.4) is the main reforming reaction which gives the stoichiometric conversion of methanol to hydrogen. It can be regarded as the overall effect of reactions of methanol decomposition (7.5) and the WGS reaction (7.6), the relative kinetics of which determine the amount of intermediate product CO formed in the process. In another group, it is claimed that CO is mainly formed from reverse water shift reaction and CO_2 is main product in the reaction [15, 54,56]. Proper control of methanol to water feed ratios and reformer operating temperature and pressure conditions can be used to minimize the amount of CO formed in the process, especially as higher CO formation lowers the hydrogen production rate and CO also acts as a poison for the fuel cell catalyst.

The thermodynamics of the reactions occurred in methanol steam reforming reaction states that both the reactions (eqs. (7.4) and (7.5)) are endothermic and their $\Delta H^0_{298\,\text{k}}$ values are 49.4 and 90.4 kJ/mol, respectively. With the temperature, the equilibrium hydrogen content decreases and the equilibrium CO content increases (Figure 7.4) [6]. The different equilibrium product compositions of the reformate gas ($P = 5$ Bar and $T = 200°C$) at different molar ratio of H_2O/MeOH have been given by Hohlein *et al.* [7] (Figure 7.5). Figures 7.4 and 7.5 are redrawn based on the computation by the method of constrained minimization of Gibbs free energy where the total pressure of the equilibrium is fixed at 2 atm. Water affects the selectivity toward hydrogen as a reactant in the methanol steam reforming, if it is not in excess of the stoichiometric ratio. Otherwise, the water effect is relatively small. At a definite temperature, if less water is present, then less fuel will be needed to feed into the combustion chambers and this can increase the efficiency of the system. Figure 7.5 also states that the equilibrium CO decreases with increasing steam to methanol ratios. Pan and Wang [8] have experimentally investigated the effect of H_2O/MeOH ratio on hydrogen selectivity, methanol conversion,

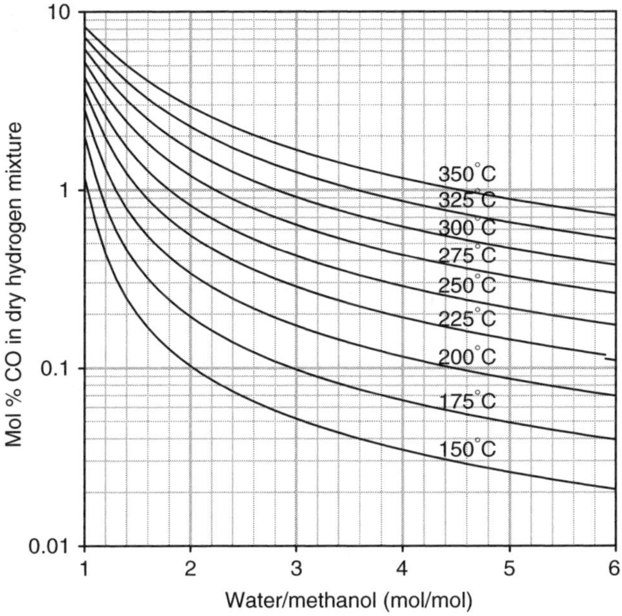

Figure 7.4 Theoretical CO concentration obtained by methanol steam reforming as a function of the molar water/methanol feed ratio and reaction temperature (total pressure 2 atm).

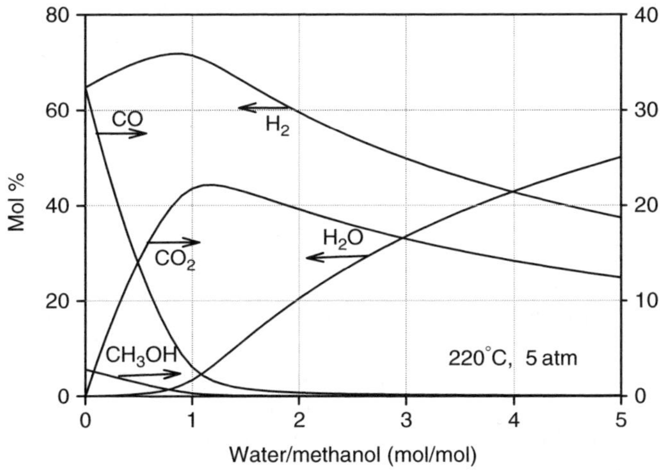

Figure 7.5 Equilibrium gas composition of reformate gas from methanol steam reforming.

concentration of CO and hydrogen in the plate-fin reformer. In their case, though CO composition lowers with the increase of H_2O/MeOH molar ratio, it also lowers the methanol conversion especially at high gas space velocities whereas the concentration of hydrogen almost remains constant (74.5%) irrespective of the H_2O/MeOH molar ratio (>1) and gas space velocity. Generally the H_2O/MeOH molar ratio of 1.5–2 is recommended for getting lower concentration of CO and higher conversion of methanol. The methanol reforming decreases with the increase of pressure, but remains above 98% even at pressures above 20 bar at H_2O/MeOH >1.5 [9]. Therefore within the range of operating pressure of methanol reforming, it has no effect on the performance.

Other reactions that can occur when catalysts other than those based on copper (such as noble metal catalyst) are used or when the steam to methanol ratio is too low, include methanation of CO and CO_2.

$$CO + 3H_2 \rightleftarrows CH_4 + H_2O \tag{7.7}$$

$$CO + 4H_2 \rightleftarrows CH_4 + 2H_2O \tag{7.8}$$

The above reactions are observed on catalyst containing iron e.g., $FeCr_2O_4$ and also on some Ni-containing catalysts [10]. The coke formation is also possible based on the following reactions. Though Amphlett *et al.* [11] have not observed the coke formation on commercial catalyst for SRM containing $Cu/ZnO/Al_2O_3$ or $Cu/ZnO/Cr_2O_3$, the recent investigations by Raimondi *et al.* [12] and Agarwal *et al.* [13] confirmed the possibility of coke formation by the presence of C—C bond and oxidized carbon species (as carbonate) in XPS study of deactivated catalyst surface. The coke formation may be also due to relatively low molar ratio of water: hydrocarbon (1.4), they used.

The coke formation may occur by these reactions:

$$CO + CO \rightleftarrows C + CO_2 \tag{7.9}$$

$$CO + H_2 \rightleftarrows C + H_2O \tag{7.10}$$

Small amount of production of dimethyl ether (DME) has also identified as a by-product of this reaction at high conversion [16]. DME formation is not directly related to methanol conversion. It is formed on the acid sites of the catalysts. Alumina containing catalysts usually have some acid sites on the surface. Hence DME formation depends on the acid sites generated by catalyst composition.

Table 7.4 shows the commercial status for the application of methanol steam reformer in automobile industries and portable power supply. There is a demonstration of fuel cell vehicle by different companies like Daimlerchrysler,

Table 7.4 Development/commercialization status of methanol steam reformers

Company/research laboratory	Characteristics
Haldor Topsoe From the cross reference of Ogden (2001), Report to the international energy agency	Small reformers for stand alone hydrogen production
Argonne National Laboratory From the cross reference of Ogden (2001), Report to the international energy agency	Methanol steam reformers for PEM fuel cells The speed range of 50–100 miles per hour and the range of 250–310 miles
Daimlerchrysler, Toyota, Daihatsu, Honda, Mazda, Mitsubishi, Volkswagen, Ford, GM/Opel and Nissan (US Fuel Cell Council report, 2005)	They have demonstrated fuel cell vehicles, which have developed based on on-board steam reformers for methanol
Northwest Power Systems (now called IdaTech) From the cross reference of Ogden (2001), Report to the international energy agency	They have developed a multi-fuel processor where pure hydrogen production via SRM, using a palladium membrane for the final purification step was used

(*Continued*)

Table 7.4 (*Continued*)

Company/research laboratory	Characteristics
Motorola (the report "Fuel Cells for Portable Power: Markets, Manufacture and Cost" for Breakthrough Technologies & US Fuel Cell Council Submitted January 13, 2003 by Darnell Group Inc. Corona, California, USA)	They have demonstrated an integrated vaporizer and miniature methanol steam reformer, and separately a miniature chemical heater – three of the key components in the reformed methanol to hydrogen fuel cell system. Eventually, Motorola plans to integrate all of these components, along with a fuel cell, into a single device producing 1 W or greater power
Battelle, Pacific Northwest Division [63]	A complete system incorporates two vaporizers/preheaters, a heat exchanger, a catalytic combustor and a catalytic methanol reformer in a volume $<0.25\,cm^3$ and weighing <1 g. These are used for microsensors and other micro-electro-mechanical systems MEMS to provide <1 W of power

Toyota, Daihatsu, Honda, Mazda, Mitsubishi, Volkswagen, Ford, GM/Opel and Nissan in addition to the application in portable electronics.

7.1.4.2 Partial Oxidation of Methanol

POM occurs in the presence of air (oxygen). The characteristics of this reaction are given in Table 7.3. Being exothermic ($\Delta H^0_{298} = -192.2\,\text{kJ/mol}$), the partial oxidation route is energetically favorable.

$$CH_3OH + \Delta O_2 \rightleftarrows 2H_2 + CO_2 \qquad \Delta H^0 = -192.2 \text{ kJ/mol} \quad (7.11)$$

The POM reaction to produce H_2 offers some advantages over the steam reforming with respect to the exothermicity of the reaction and the use of O_2 (air) instead of steam as oxidant, by which it decreases the total heat load. Another incentive comes from the simplicity of the POM reaction, which allows the development of compact hydrogen generators, particularly for car engine purposes. It also provides a fast response to transients. Its disadvantage lies on the fact that it produces less favorable H_2/CO_2 ratio compared to that of methanol steam reforming. As it is highly exothermic, temperature control can also be difficult inside the reactor.

The oxygen/methanol molar ratio should exceed 0.5 in order to get complete conversion; otherwise there is a chance of producing CO by methanol decomposition. Generally Cu/ZnO, Cu/ZnO/Al_2O_3 catalysts are used for this reaction. Pd/ZnO and Pd/ZrO_2 have also been used [17–19]. Au/TiO_2 as a catalyst for this reaction has also been reported to reduce the CO level in the resulting gas [20]. They have produced CO-free hydrogen with Au/TiO_2 catalyst at O_2/CH_3OH ratio of 0.3. Agrell *et al.* [14] have shown that the reaction pathway for POM over Cu/ZnO consists of consecutive combustion and steam reforming. This is followed by methanol decomposition, leading to CO formation.

7.1.4.3 Oxidative Methanol Reforming or Autothermal Methanol Reforming

OMR is a combination of partial oxidation and steam reforming (sometimes it is also called combined methanol reforming) at such a ratio that the overall reaction is thermally neutral or modestly exothermic, which means that the heat necessary to maintain SRM is supplied by the POM reaction. In reality, the OMR reaction cannot be completely heat balanced ($\Delta H^0 = 0$) because of heat loss. Therefore the OMR reaction is usually made slightly exothermic to account for some heat loss in the process. The reactions occurring in OMR are:

$$CH_3OH + \tfrac{1}{2} O_2 \rightleftarrows CO_2 + 2H_2 \qquad \Delta H^0 = -192.2 \text{ kJ/mol} \quad (7.11)$$

$$CH_3OH + H_2O \rightleftarrows CO_2 + 3H_2 \qquad \Delta H^0 = 49.4 \text{ kJ/mol} \quad (7.12)$$

In addition to these reforming reactions, the reverse reactions, the reverse shift reaction takes place simultaneously to some extent. Thus the product gas also contains a small amount of CO:

$$CO_2 + H_2 \rightleftarrows CO + H_2O \quad \Delta H^0 = 41\,\text{kJ/mol} \tag{7.13}$$

The OMR reaction can deliver relatively high hydrogen concentrations at moderate response rates. Hot spot also occurs in OMR because the exothermic and endothermic reactions are not balanced locally to be thermally neutral. The exothermic part proceeds at the inlet and the heat is carried by the reaction mixture is consumed subsequently by the endothermic steam reforming reaction. This has been one of the difficulties in OMR since the hot spot sinters the catalysts. Increasing the oxygen/steam ratio will increase the exothermic nature of the OMR reaction, while lowering the maximum theoretical hydrogen concentration. OMR is commonly operated with an excess of steam, 20–30%. Excess steam inhibits reverse WGS reaction since CO_2 is the primary product.

Early studies on OMR were reported by Huang and Wang [21] and Huang and Chren [22]. They studied the effect of O_2 addition to SRM reaction and the kinetics of OMR reaction over $Cu/ZnO/Al_2O_3$ catalysts. These authors proposed that the rate of the combined OMR process could be calculated by the sum of the rates of POM and SRM reactions. Recently Murcia-Mascaros *et al.* [23] reported the performance of $Cu/ZnO/Al_2O_3$ catalyst for OMR process prepared by first homogenous precipitation of chloride salts in the presence of urea to obtain hydrotalcite and oxychloride precursors followed by thermal decomposition in air. Lindstrom *et al.* [24] used Cu/Zn, Cu/Cr and Cu/Zr on Al_2O_3 for SRM reaction at near autothermal and under exothermic operating conditions separately in a monolithic reactor. For SRM reaction Zn-containing catalyst exhibited the highest conversion, whereas for the OMR, the chromium containing catalyst showed the highest conversion. Liu *et al.* [25] studied Pd/ZnO catalyst for OMR keeping in mind of the sintering effect of Cu-based catalyst at above 300°C. It was claimed that Pd/ZnO is highly active for OMR reaction to the same extent as commercial Cu–Zn catalyst and is also stable at higher Pd loading prepared by co-precipitation method.

7.2 Steam Reforming of Methanol

7.2.1 Catalyst Development

Methanol steam reforming takes place over Cu or Pd/Zn alloy catalysts at low temperatures, typically above 200°C. Low temperature reduces the equilibrium CO selectivity via the WGS reaction. Copper-based materials have been extensively studied under SRM reaction conditions. The commercial catalyst for SRM is $Cu/ZnO/Al_2O_3$. Copper is very susceptible to thermal sintering via a surface migration process at temperature above 330°C. Therefore, copper-based catalysts typically contain one or more oxides that enhance the stability mainly by physically separating the copper crystallites. Further synergetic effects have been observed, e.g. alloy formation, and morphology changes or the induction of strain in the crystallite lattice of copper metal in the case of Cu supported on ZnO. To obtain these promoting effects various preparation techniques have been tested. For copper supported on zinc oxide, mostly co-precipitation method is applied [26,27]. As Cu-based catalysts at the reduced state are phyrophoric in nature, there is a try to look alternative non-pyrophoric noble metal based like Pd, Pt and Rh catalyst supported on ZnO and CeO_2 for SRM [33]. The recently used different catalyst and its method of preparation, characteristics and operating conditions in SRM are given in Table 7.5.

In the 1980s, Takezawa *et al.* [41] and Takahashi *et al.* [42] reported on studies about Cu/SiO_2 catalysts. They investigated the effect of copper loading and calcinations temperature on the reaction and tried to understand the reaction pathway by comparison with a Pt/SiO_2 catalyst. It was suggested that, dependent on the catalyst material, the SRM proceeds via different intermediate reactions of formaldehyde, resulting in better SRM selectivities with the Cu/SiO_2 system. Cu/SiO_2 catalysts with higher copper loadings (10 wt%) or calcined at higher temperatures turned out to be highly active and selective due to the presence of CuO clusters and bulk CuO, which readily transformed into metallic copper during reaction [19]. Later they compared the SRM performance over copper and group VIII metals supported on SiO_2, but also on Pd with different supports (alumina, zirconia, ZnO, La_2O_3 and others). They assigned the high performance of copper-based and Pd/ZnO catalysts to these differences in the reactivity of formaldehyde and proposed that the modification of the Pd upon the formation of Pd/Zn alloys similarly influenced the structure of adsorbed HCOH as it was suggested for Cu/SiO_2. The relatively low performance of Cu/SiO_2 catalyst in the presence of Al_2O_3 may be due to the reaction with alumina in the catalyst to give acidic sites, leading to wax or carbon formation [44]. For all the other tested Pd, Ni and Pt catalysts (supported on various oxides) inferior hydrogen production rates and selectivities were observed. Pd/Zn alloy catalysts exhibit excellent thermal stability and are non-pyrophoric which are advantages for portable hydrogen production [63].

As sulfur is a common impurity in syngas, which is the primary raw material for the production of methanol, the studies on the effect of sulfur poisoning on the SRM catalyst is very important. Lindstorm and Pettersson [40] shows that the catalyst activity decreased almost 65% of its original activity by adding 2 ppm sulfur to the methanol. The selectivity of CO_2 also lowered considerably. They used $Cu/ZnO/Al_2O_3$ catalyst from Sud Chemie. The sintering effect also observed with this catalyst at above 340°C.

Table 7.5 Some recent studies on development of catalyst in the SRM

Composition of catalyst/ type of reactor used	Authors	Preparation method	Experimental set-up and operating conditions	Comments
$Cu/ZnO/Al_2O_3$ Fixed bed reactor	Kawamura *et al.* [4]	Prepared by co-precipitation	8 mm i.d., GHSV = 100,000 h, steam/methanol = 2:1, temperature = 200–300°C	The catalytic activity with the catalyst prepared at pH = 8.8 during precipitation shows superiority than commercial one (Sud Chemie, MDC-3). XRD analysis shows that the minimum crystallite sizes for CuO and malachite at pH = 8.8
$Cu/CeO_2/Al_2O_3$ Micro-channel reactor	Men *et al.* [28, 29]	Wet impregnation on γ-Al_2O_3-coated layer	steam/methanol = 1–1.1:1, temperature = 200–300°C, liquid flow rate = 1.5 g/h	Low Cu/Ce ratio of 0.1 and 0.2 shows higher activities with respect to methanol conversion. It may be due to the strong tendency of copper present in small amounts to associate with cerium oxide and the excess copper formed bulk and large copper oxide. DME has also been identified as by-product

$Cu/ZnO/Al_2O_3$ Micro-channel reactor	Yu *et al.* [30]	Wet impregnation on γ-Al_2O_3-coated layer	Reactor dimension: 40 mm × 40 mm × 8 mm Channel dimension: 1000 μm wide, 600 μm deep and 32 mm long, WHSV = 8.27–32.54 h, steam/methanol = 1.3:1, temperature = 220–300°C	14 wt% Cu and 6 wt% Zn shows highest activity which confirmed by TPR patterns and copper surface area. With the increase of thickness of the coating, CO increases.
$Cu/ZnO/Al_2O_3$ and $Cu/ZnO/Al_2O_3[ZrO_2]$ monlithic catalyst	Lindstrom and Pettersson [31]	Wet impregnation on γ-Al_2O_3-coated layer	Tube I.D. = 25 mm GHSV = 10,000 h temperature = 200–300°C, steam/methanol = 1.3:1	The 60 wt% Cu and 40 wt% Zn shows highest activity for SRM in the case of $Cu/ZnO/Al_2O_3$ where as 80 wt% Cu and 20 wt% Zn in the ZrO_2 doped $Cu/ZnO/Al_2O_3$ catalyst shows the highest activity. The ZrO_2 doped $Cu/ZnO/Al_2O_3$ catalyst were less active with respect to hydrogen yield, but they are more selective toward CO_2. Brunauner–Emmett–Teller (BET) surface area is also higher side with ZrO_2 doped $Cu/ZnO/Al_2O_3$ catalyst

(Continued)

Table 7.5 *(Continued)*

Composition of catalyst/ type of reactor used	Authors	Preparation method	Experimental set-up and operating conditions	Comments
$Cu/ZnO/Al_2O_3[ZrO_2]$ fixed bed reactor	Yong-Feng *et al.* [32]	Co-precipitation method at pH = 8–9	Temperature = 210–270°C, steam/ methanol = 1–1.3:1 WHSV = 3.56/h	The addition of ZrO_2 promoter in $Cu/ZnO/Al_2O_3$ system improves the overall catalytic activity. XRD and XPS confirms the improvement of the dispersion of CuO and Cu crystals and the protection from the formation of $CuAl_2O_4$ spinel-type compound by adding ZrO_2
Pd/ZnO and Pd/CeO_2 fixed bed reactor	Ranganathan *et al.* [33]	Thermal decomposition of zinc and cerium oxalate (prepared by the precipitation by the reaction between the corresponding nitrate and oxalic acid) followed by wet impregnation of $Pd(NO_3)_2$ on the support	5 mm i.d. U-tube quartz reactor. 30 mg of the catalyst, steam/ methanol = 1:1 temperature = 185–230°C	The formation of Pd–Zn alloys was observed when the reduction temperature was equal to or greater than 300°C The Pd/ZnO catalysts had lower SRM rates but were more selective to the production of CO_2 than the Pd/CeO_2 catalysts

Pd/ZnO Fixed bed reactor	Iwasa *et al.* [34]	Heat treatment of the physical mixture of Pd and Zn powders in a N_2 stream.	Temperature = 220°C GHSV = 7600/h	The Pd/Zn with a molar ratio of 1/8, the selectivity increases to 87.5%. Among the alloys, PdZn and $Pd_{3.9}Zn_{6.1}$ exhibit uniquely high performance for SRM
$Cu/SnO_2/SiO_2$	Lee *et al.* [35]	Co-precipitation method	Fixed bed reactor steam/methanol = 1.5:1 temperature = 250°C, contact time = 3.22 g-cat h/mol	The different copper precursors of $Cu(NO_3)_2$, $CuSO_4$ and $CuCl_2$ has been used for the preparation of catalyst. Though the highest activity is obtained with catalyst prepared from $Cu(NO_3)_2$. H_2 selectivity is also low for the catalysts prepared from $CuSO_4$ and $CuCl_2$. The increasing amount of acid on the surface of $Cu/SnO_2/SiO_2$ is the reason for low selectivity
CuO/ZrO_2	Purnama *et al.* [36]	Sol–gel nanocoating	10-mm i.d. fixed bed reactor	The activity of CuO/ZrO_2 catalyst was increased by temporary addition of oxygen to the feed. The catalytic activity is higher in CuO/ZrO_2 than the commercial one ($Cu/ZnO/Al_2O_3$). It increased long-term stability probably because the macroporous zirconia support is more effective than ZnO/Al_2O_3 in preventing copper sintering

(*Continued*)

Table 7.5 *(Continued)*

Composition of catalyst/type of reactor used	Authors	Preparation method	Experimental set-up and operating conditions	Comments
CuO/CeO_2 Fixed bed reactor	Lee and Kim [37]	Co-precipitation method	¼ inch s.s. tube and 50 cm height, temperature = 160–280°C, steam/methanol = 1–2	The catalyst showed a slow but steady deactivation during the course of reaction over 220 h
$Cu/CeO_2/ZrO_2$ Fixed bed reactor	Mastalir *et al.* [38]	Co-precipitation method	10-mm i.d. s.s tube, temperature = 230–300°C, catalyst weight = 0.1–0.77 g steam/methanol = 1, flow rate = 0.07 cm^3/min for time on stream measurement, flow rate = 0.00333–0.67 cm^3/min for kinetic experiments	Kinetic model suggested for the transformation involved the RWGS and methanol decomposition in addition to the SRM reaction. The initial deactivation was found and after 5 days of operation it showed constant activity
	Oguchi *et al.* [39]	Co-precipitation method	Catalyst weight = 0.1 g, flow rate of the mixture of water and methanol = 16.0, 4.5 and 6.5 ml/min	Both Cu_2O and Cu metal are active for SRM. The presence of Cu_2O seems to lead the durability of the catalysts

RWGS: reverse water–gas shift ; TPR: temperature-programmed reduction.

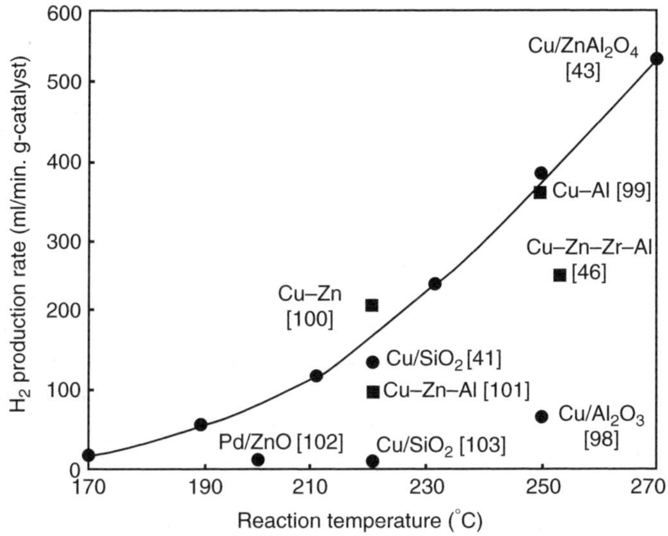

Figure 7.6 Comparisons of catalytic performances for SRM reactions: (●) Impregnation catalyst, (■) co-precipitation method [43].

Figure 7.6 shows the comparison of catalytic performance for SRM with two preparation methods (impregnation and co-precipitation methods) at different temperatures (adapted from [43]). It gives the relative idea of performance with different catalyst combination for SRM.

7.2.2 Effect of Different Promoters on the Catalytic Activity of Cu-based Catalyst in SRM

The Cu-based catalyst on Al_2O_3 base with Zn shows highest activity and high selectivity of CO_2 compared to that of Cr and Zr. But the addition of ZrO_2 on this $Cu/Zn/Al_2O_3$ further increases the activity and stability.

Lindström and Pettersson [45] compared copper-based catalysts (where nitrates of Cu, Zn, Cr and Zr were impregnated on alumina pellets, followed by drying and calcination) in the SRM reaction (molar ratio CH_3OH/H_2O of 1). For temperatures above 513 K the activity of Cu/ZrO_2 catalysts was comparable to copper-chromia and Cu/ZnO catalyst, while their selectivity toward CO_2 was higher. For the case of ternary mixture, Zr has more positive promotional effect on Cu than Zn at low temperatures with high and low Cr contents. Higher copper contents yield increased hydrogen concentrations in the product gas, except for Cu/Zn where the activity is virtually unchanged. They have given the explanation that zinc cannot influence the dispersion of copper in a positive way at high copper concentrations. Also high copper content can cause the formation of agglomerates, which leads to a decrease in the metal surface area and, hence, a lower catalyst activity. Chromium promotes catalyst activity, especially at high temperatures.

This group [31] also investigated the performance of copper-based monoliths (SRM with 30% excess of water) and observed a superior selectivity toward CO_2 for zirconia doped $Cu/Zn/Al_2O_3$ catalysts over the entire investigated temperature interval and for all copper loadings.

Concerning the stability of copper-based catalysts, Cu/ZrO_2 was compared with a commercial $Cu/ZnO/Al_2O_3$ catalyst under methanol steam reforming conditions for 250 h. For both systems, where the conversion is decreasing exponentially in an initial period, is followed by a linear behavior. However, after 150 h time on stream the conversion appears to be constant for the binary catalyst, while it continues to decrease for the ternary system [36].

Breen and Ross [46] performed a study to investigate the influence of the preparation method and the zirconia addition to copper-based catalysts on the SRM. Various types of catalysts were tested, i.e. binary materials with Cu supported on ZnO or ZrO_2, as well as ternary and quaternary systems with additional alumina (sequential and co-precipitation with the metal nitrate salts). Most of the catalysts were prepared by sequential precipitation, whereas the Cu–Zr material was prepared by co-precipitation. The activity of the Cu/ZrO_2 catalysts increased with increasing copper content under SRM conditions (H_2O/CH_3OH, 1.3:1). Moreover, no significant amounts of CO were detected for all catalysts up to temperatures of 553 K. Despite the low hydrogen production of the Cu–Zr systems, they showed higher turn over frequency than the Cu–Zn catalysts. The addition of zinc oxide to Cu–Zr materials, and to a lesser extent also the addition of alumina to Cu–Zn–Zr catalysts increased the copper dispersion, as well as the activity. The best performance was found for the Cu–Zn–Zr–Al catalysts. The further confirmation for the improvement of performance and stability has been provided by Yong-Feng *et al.* [32,47]. They found that the effect of the addition of Zr plays an important role in the rates of methanol conversion and H_2 production, with a respective increase of 15% and 10% where as the complete substitution of Al by Zr increases the amount of CO with the increase of methanol conversion and hydrogen production.

Cr_2O_3 is another candidate as catalytic promoter of Cu-based catalyst in SRM reaction. Ma *et al.* [49] observed little effect of Cr_2O_3 on the catalytic activity in Cu-based catalyst for SRM though they found a increase on total surface area and on copper surface area by the addition of chromium.

CeO_2 exhibits good ability in storing and releasing oxygen, so it is often used in redox catalysts to change the valence states of the active core ions. Yong-Feng *et al.* [47] investigated that CeO_2 did not have any improvement with respect to hydrogen production and methanol conversion though hydrogen selectivity and CO_2 selectivity increases and the complete substitution of Al_2O_3 by CeO_2 decreased the rates of methanol conversion and hydrogen production.

7.2.3 Reaction Mechanism and Kinetics of SRM Reactions

The different reaction pathways are given in Figure 7.7(a) for SRM [33]. The most acceptable pathway for steam reforming is through dehydrogenation to formaldehyde when using copper-based catalyst. Formaldehyde then hydrolyses to formic acid, which decomposes to carbon dioxide

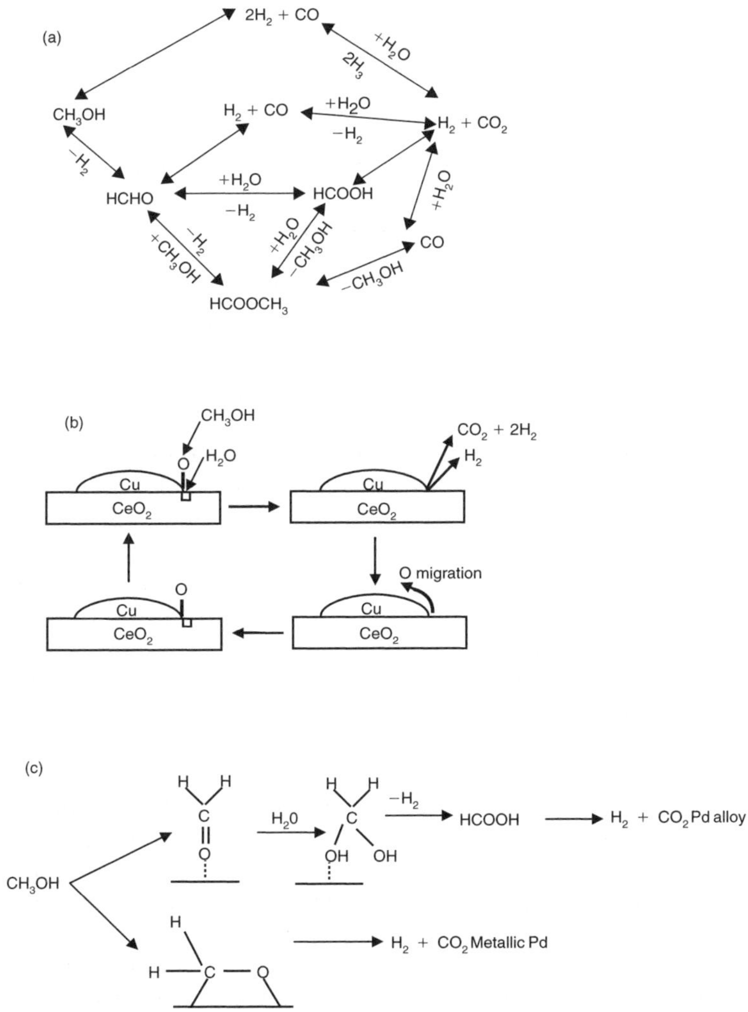

Figure 7.7 (a) The different reaction pathways for SRM [33]. (b) The reaction mechanism of SRM proposed in $Cu/CeO_2/Al_2O_3$ system by Men *et al.* [28,29]. (c) Reaction mechanism of SRM over Pd alloys and metallic Pd [113].

and hydrogen. The mechanism of SRM has been widely discussed in the literature [16, 48, 50–53]. The reaction is believed to proceed via four steps on Cu-based catalysts, as described below:

1. Dissociative adsorption of methanol to form surface methoxy groups

$$CH_3OH \longrightarrow CH_3O^a + H^a \tag{7.14}$$

2. Dehydrogenation of methoxy to formaldehyde or methylformate

$$CH_3O^a \longrightarrow H_2CO + H^a \tag{7.15}$$

$$CH_3O^a + H_2CO \longrightarrow CH_3OCHO + H^a \tag{7.16}$$

3. Hydrolysis of formaldehyde or methylformate

$$H_2CO + OH^a \longrightarrow HCOO^a + H_2 \tag{7.17}$$

$$CH_3OCHO + OH^a \longrightarrow HCOO^a + CH_3OH \tag{7.18}$$

4. Decomposition of the surface intermediates

$$HCOO^a \longrightarrow CO_2 + H^a \tag{7.19}$$

$$H_2CO \longrightarrow CO + H_2 \tag{7.20}$$

Where ^{a} denotes the adsorbed species. Detailed studies of the kinetics of the reactions show that methanol dehydrogenation controls the rate of steam reforming [16, 50]. It is more acceptable that CO is not a primary product and does not take part in the rate expression [15,54,56]. Since the reforming catalysts are also highly active in the WGS reaction, CO has been thought to be produced by the reverse WGS reaction (Eq. (7.21)). Breen *et al.* [54] have also recently reported that CO is not a primary but a secondary product of the steam reforming over $CuO/ZnO/ZrO_2/Al_2O_3$ catalyst by the reverse WGS reaction. But the nature of the active sites, namely whether Cu^0 or Cu^+ is more active for the SRM reaction and the role of the support, is still an open question.

With the high selectivity of Pd/ZnO for SRM, the decomposition of the surface formate through dehydrogenation to CO_2 and H_2 (Eq. (7.19)) should be dominant, and that CO is mainly formed via the decomposition of formaldehyde in the case of Pd/ZnO catalyst. The mode of further conversion of formaldehyde to either the surface formate or simple decomposition to form CO is probably related to its adsorption state on the catalyst surface. It is speculated that the surface monodentate H_2CO is stabilized on either Cu or the Pd–Zn alloy and undergoes further hydroxylation and dehydrogenation to form the surface formate, while the bidentate H_2CO tends to decompose directly to CO and H_2 (Figure 7.7(c) taken from [113]). However, under the oxyreforming condition, other reactions such as methanol decomposition (Eq.(7.5)) and reverse water gas shift (RWGS, Eq. (7.21)), which are very similar from a mechanistic point of view to the decomposition of surface formate through "dehydration" to CO and OH(a) (Eq. (7.22)), may also contribute to CO formation, depending on the surface characteristics and the composition of the catalysts. Liu *et al.* [55] confirmed this by the evidence of the significant increase in CO formation on their catalyst, HCP45, in which elemental Pd was detected by XRD

$$CO_2 + H_2 \longrightarrow CO + H_2O \tag{7.21}$$

$$HCOO^a \longrightarrow CO + OH^a \tag{7.22}$$

Recently Ranganathan *et al.* [33] have shown experimentally that the ZnO and CeO_2 supported Pd catalyst does not follow the reaction of CH_3OH decomposition followed by the WGS, and that the CO_2 formed during reactions involving HCHO and $HCOOCH_3$ was not due to the WGS reaction. It is also inactive in reverse WGS reaction.

Diverse power-law and Langmuir–Hinselwood rate expressions have been proposed in the literature to express the kinetics of the SRM. The proposed rate expressions are listed in Table 7.6 (partially taken from [56]). Every expression for the kinetics of SRM reaction in the existing literature shows that an increasing reforming rate with an increasing methanol partial pressure. Except for methanol, there is little agreement regarding the components affecting the reaction rate. Some rate expressions show that CO_2 inhibits the reforming [57]. Some rate expressions predict infinite reaction rate in the absence of hydrogen [50,58]. Lee *et al.* [56] carried out a kinetic study of methanol steam reforming over a commercial catalyst $Cu/ZnO/Al_2O_3$. They have found that the reaction rate increased with an increase in the methanol partial pressure but decreased with an increase in the hydrogen partial pressure indicating hydrogen inhibition. Partial pressure of carbon dioxide and water did not affect the reaction rate when it was in excess of methanol partial pressure. The one of other kinetic models is proposed by Peppley *et al.* [53], where reversible Langmuir–Hinselwood rate expressions for each of the overall reactions involved in SRM have been derived by a kinetic analysis of the elementary surface reactions occurring on the catalyst. The final form for methanol–steam reaction is in Table 7.6.

Men *et al.* [28,29] proposed the reaction mechanism for SRM with $Cu/CeO_2/Al_2O_3$ catalyst (Figure 7.7(b)). In the mechanism, they proposed the adsorption of methanol with metallic copper through the presence of surface oxygen created by CeO_2 which is strong reducing agent. Water is activated by the reduced ceria. The production of CO_2 consumes one surface oxygen and accordingly one oxygen is migrated to copper-ceria interface to regenerate the depleted copper surface oxygen.

Table 7.6 Rate expressions for SRM in the literature (partially taken from [56])

References	Rate expression	Catalyst	Activation energy (kJ/mol)
[108]	$-rM = \sqrt{(P_M P_W)}\ P_H^{-1.3}$	$Cu/ZnO/Cr_2O_3/Al_2O_3$	116.1
[109]	$-r_M = (kK_M P_M)/(1 + K_M P_M + K_W P_W)$	$Cu/ZnO/Al_2O_3$ (BASF)	102.6
[110]	$-r_M = (kK_M P_M - k' P_{CO} P_M^2)/(1 + KP_{CO})$	Cu/ZnO (Girdler G66B)	96
[16]	$-r_M = (kK_1 PM/\sqrt{K_3 P_H})/(1 + K_1 P_M/\sqrt{K_3 P_H} + 1/\sqrt{K_3 P_H})^2$	$Cu/ZnO/Al_2O_3$ (BASF S3-85)	110
[50]	$-r_M = kP_M^{0.26}\ P_W^{0.03}\ P_H^{-0.2}$	$Cu/ZnO/Al_2O_3$ (BASF S3-85)	105
[57]	Below 190°C	$Cu/MnO/Al_2O_3$	
	$-r_M = kP_M^{0.04}$		79.7
	$r_M = (k(P_M - P_C P_H^3/K_E P_W))/(1 + KP_M)$		77.3
	Above 200°C		
	$r_M = kP_M^{0.28} P_C^{-0.99-}$		77.7
	$r_M = (k(P_M P_W/P_H - P_C P_H^2/K_E))/(1 + KP_M)^4$		116.6

[51,53]	$-r_M = kK_1(P_M/\sqrt{P_H})(1-(P_H{}^3P_C/K_EP_MP_W))/DEN$	$Cu/ZnO/Al_2O_3$ (BASF K3-110)	102.8
[111]	$-r_M = kP_M{}^{0.4}(1-(P_CP_H{}^3/K_EP_MP_W)$	$Cu/ZnO/Al_2O_3$	83
[58]	$-r_M = kP_M{}^{0.63}P_W{}^{0.39}P_H{}^{-0.23}P_C{}^{-0.07}$	$Cu/ZnO/Al_2O_3$ (BASF K3-110)	74
[58]	$-r_M = kK_1(P_M/\sqrt{P_H})(1-(P_H{}^3P_C/K_EP_MP_W))/DEN$	$Cu/ZnO/Al_2O_3$ (BASF K3-110)	122.4
[112]	$-r_M = kP_M{}^{0.6}P_W{}^{0.4}$	$Cu/ZnO/Al_2O_3$ (Sud Chemie)	76
[56]	$-r_M = kK_1(P_M/\sqrt{P_H})/(1+K_1(P_M/\sqrt{P_H}))(1+\sqrt{K_2P_H})$	$Cu/ZnO/Al_2O_3$ (Synetix 33-5)	111
[37]	$-r_M = kK_1(P_M/\sqrt{P_H})/(1+K_1(P_M/\sqrt{P_H}))(1+\sqrt{K_2P_H})$	$CuO\text{-}CeO_2$	106
[73]	$r_M = kC_{MeOH}{}^{n1}C_{H20}{}^{n2}C_{H2}{}^{n3}$	$Cu/ZnO/Al_2O_3$ (G-66MR, Süd Chemie)	77
$n1 = 0.7, n2 = 0.1, n3 = -0.2$		Wall-coated catalyst	

C: carbon dioxide; H: hydrogen; M: methanol; W: water; $DEN = [1 + K_1(P_M/\sqrt{P_H}) + K_2P_C\sqrt{P_H} + K_3(P_W/\sqrt{P_H})](1+\sqrt{K_4P_H})$.

$K_1 = K_1'/\sqrt{K_2}$ and $k = k'C_{T,s1}C_{T,S2}$ where $C_{T,s1}$ and $C_{T,S2}$ are the total concentration of two kinds of active sites catalyst denoted as s1 and s2, respectively.

7.2.4 Reactor Development

7.2.4.1 Different Types of Reactors Used in This Process

Four different types of reactors were used for SRM. These are fixed bed reactor, monolithic reactor, wall-coated heat exchanger, micro-channel reactor and metallic foam reactor.

Fixed bed reactor Fixed bed reactors are mostly used in the methanol reforming reaction not only for production of hydrogen but also for the catalyst development for this process. The catalyst particle size is an important parameter for pressure drop inside the catalytic bed and intra-particle mass transfer resistance. In packed-bed reactors, the catalyst particle size cannot be too small to avoid excessive pressure drop. Due to intra-particle diffusional effect, however, smaller catalyst size is favorable for high apparent activity. A certain optimum catalyst size for reasonable pressure drop and high apparent activity may exist depending on the operating conditions and reactor capacity. Figure 7.8 plots the effectiveness factors of various catalyst sizes for a typical reforming mixture and reaction temperatures. The effectiveness factor was computed following the detailed procedure described in [56]. For a typical catalyst size of 1–4 mm, the figure shows that the intra-particle diffusional effect can be significant and the effectiveness factor can be considerably smaller than 1.0 in the usual reaction temperatures of 200–300°C.

De wild and Verhaak [59] studied the comparison of the performance with fixed bed reactor, catalyst-coated heat exchanger and coated aluminium foam bed structure for SRM. Though the temperature required to reach 95% of methanol conversion for fixed bed reactor is higher (280°C)

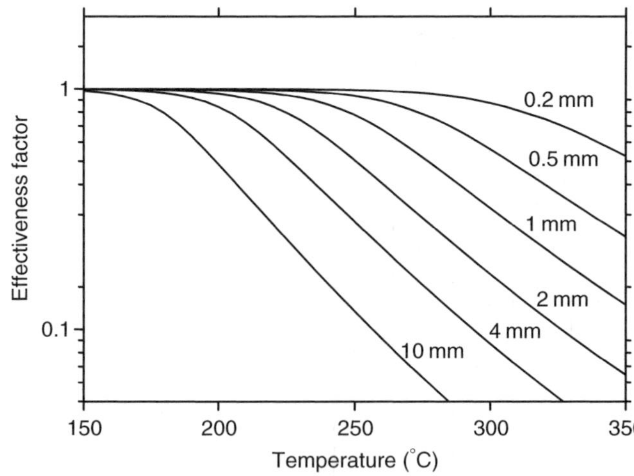

Figure 7.8 Effectiveness factor of catalyst in methanol steam reforming as a function of reaction temperature and catalyst diameter [56].

than foam type reactor (250°C) and heat exchanger (230°C), the deactivation rate in fixed bed reactor is almost 10 times less than other types of reactors. This may be due to the low amount of catalyst coated on heat exchanger (18 g catalyst) and foam type reactor (17 g catalyst) compared to that in fixed bed reactor (74 g catalyst).

Monolithic reactor Monoliths are uniform extruded structure composed of parallel flow through channels and are usually based upon ceramic materials or aluminium containing metals. The catalyst coated on the monolithic substrate for SRM is a good alternative especially in automotive environment due to its high tolerance of the mechanical stress caused by vibrations and constant alterations of the operating conditions, high activity of the catalyst and low-pressure drop compared to packed bed reactors.

One obvious disadvantage with monolithic catalysts is their low content of active material per unit reactor volume. Also for endothermic reactions, such as methanol steam reforming, the use of monoliths can be problematic as heat transfer is poor between channels due to the low heat conductivity of the ceramic material.

Lindstrom and Pettersson [31] studied monolithic reactors in the application of SRM. The monolithic reactors were dipped into γ-Al_2O_3 slurry to wash coat. The calcined γ-Al_2O_3 layer was impregnated by metal nitrate solution to get $Cu/ZnO/Al_2O_3$ and zirconium doped $Cu/ZnO/Al_2O_3$.

Lindstorm *et al.* [24] studied the steam reforming and combined methanol reforming for hydrogen generation over monolithic catalysts. Cu-based catalysts with three different promoters (Zn, Zr and Cr) were prepared on Al_2O_3-coated monolithic substrate in the same ratio. They obtained highest conversion with zinc containing catalyst for SRM whereas chromium containing catalyst shows better performance for OMR. The zirconium containing catalyst generated the lowest CO concentration. This may be due to different surface structure of the Zr-based catalyst in which the copper crystallites were found to be smaller than the zirconium crystallites (confirmed by XRD) and dispersed over the entire surface around single Zr particles compared to the zinc- and chromium-based catalyst. The temperature-programmed reduction studies also confirmed that the particle size is in the increasing order of $Cu/Cr < Cu/Zr < Cu/Zn$.

Wall-coated heat exchanger, micro-channel reactor and micro reactors Micro-channel reactors are (generally, but not exclusively) one type of micro reactors with three-dimensional structures, typically have dimensions of the channels in the sub-millimeter range (Figure 7.9), the effect being a reduction in diffusive transport limitations (a summary of advantages are given in Table 7.7). It therefore follows that the main feature of these reactors, in comparison to conventional chemical reactors, is the high surface area to volume ratio. Specific surface areas of micro-structures lie between

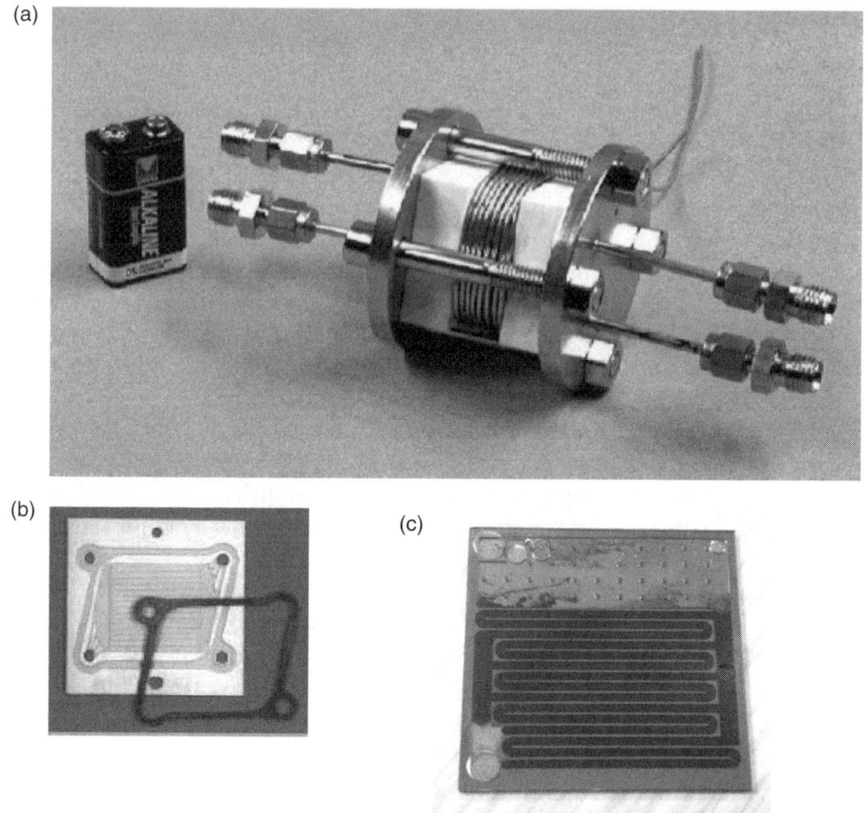

Figure 7.9 A multiplate-stacked reactor with an end plate: (a) gas phase microreactor with two steel end caps enclosing a stack of micro-structured steel platelets manufactured by wet chemical etching; for better insulation of this stack, two ceramic blocks are placed between the end caps and the stack [97]; (b) single plate with gasket of IMM [60]; (c) MEMS Based micro-reformer with micro-vaporizer [114].

10,000 and 50,000 m^2/m^3, while those of traditional reactors are generally about 100 m^2/m^3 and in rare cases reach 1000 m^2/m^3 [60]. High heat transfer coefficient is also obtained compared to traditional heat exchangers since the heat-transfer coefficient is inversely proportional to the channel diameter. This high heat-exchanging efficiency allows for fast heating and cooling in reaction mixtures and the elimination of hot spots or the accumulation of reaction heat within micro-structures so that undesirable side reactions and fragmentations are hindered. Therefore higher selectivity, yield and product quality is obtained by using these micro-channel reactors in many reactions irrespective of whether it is endothermic or exothermic reaction.

In addition to heat transport, mass transport is also considerably improved in micro-structured reactors. Mixing times in micromixers (down to several

Table 7.7 Advantages of micro-channel reactors

1. Enhanced heat transfer
2. Enhanced mass transfer
3. Economics to mass production
4. High surface area to volume ratio
5. Low-pressure drop through channels
6. Fast response in transient behavior
7. Easy and safe operation
8. Unit can be more compact with this type of reactor (reduction of 1/5 the size of the conventional unit is possible)

milliseconds) are generally smaller than in conventional systems and due to the small dimensions the diffusion times are very short, thus the influence of mass transport on the speed of a reaction can be considerably reduced.

A third aspect of micro-structured reactors is the hydrodynamic flow in the micro-channels, which is an important feature of microfluidics. The flows are mostly laminar, directed and highly symmetric. Very narrow residence times can be achieved through micro-channels which minimizes the back-mixing. For a given catalyst loading, the pressure drop in wall-coated micro-channels is substantially lower than that in packed-bed reactors; the losses in wall-coated micro-channels would be primarily skin-frictional losses, while in packed beds the form-frictional losses around the packed particulates would be dominant in addition to skin friction.

Process parameters such as pressure, temperature, residence time and flow rate are more easily controlled in reactions that take place in small volumes. The hazard potential of strongly exothermic or explosive reactions can also be drastically reduced. The possibility of reduction in dimensions with small volumes of reaction zone would allow application of high temperature or concentration with significant ease of process control and thermal management. It provides another advantage of meeting the demand of end-products by connecting more micro-channeled plates in parallel ("numbering-up") without revamping the unit a lot [61].

The micro-channels can be filled with catalyst powder forming micro-packed bed reactor or coated with the catalyst material forming wall-coated micro-channel reactor for the application of SRM. The walls of the micro-channels can be coated with catalytic material by a number of methods including sputtering, evaporation, slurry wash-coating, sol–gel coating or electroless plating. The different substrate materials (aluminium, stainless steel, silicon wafer) of the micro-channeled plate are used for this purpose. The most widely used substrate material is stainless steel for its high resistance of corrosion though it has less thermal conductivity (15/Wm/K) compared to copper (401/Wm/K) and aluminium (236/Wm/K). The silicon wafer has great potentiality to use as micro-channel reactor as methanol reformer for the application in electronic devices. To get the high resistivity of the materials for micro-channeled plate with high thermal conductivity, Pfeifer *et al.* [62] used aluminium alloys like $AlMg_3$. The micro-channels can be made by LIGA (lithography, electroplating and molding developed by IMM), wet and dry etching process and micro-machining. The different materials used for micro-channels, the selection criteria of the materials chosen and the details of the fabrication technique have been explained in the review of Holladay *et al.* [63].

The use of micro-packed bed reactor for the reforming of methanol to produce hydrogen has been focused a lot recently. Pattekar *et al.* [64] made use of a silicon chip for hydrogen production from methanol steam reforming. The chip-like reactors as a future part of a fuel processor/fuel cell system were regarded as an alternative to portable power supply such as batteries. The reactor housing was made of stainless steel, electrically heated and sealed with graphite. The micro-channel was a long serpentine of 1000 μm width and 230 μm depth fabricated by photolithography and KOH etching. Cu catalyst was sputtered to a thickness of 33 nm on to the top. Preliminary simulations revealed a non-uniform temperature distribution in the reactor housing, pointing to the importance of proper insulation especially in lower power systems. At a feed composition of 76 mol% methanol in steam, less than 7% conversion was achieved at 250°C. Selectivity to CO was higher than that to carbon dioxide and 7 mol% hydrogen was found in the product. Later Pattekar and Kothare [65] presented a silicon reactor fabricated by deep reactive ion etching (DRIE). Micro-channels of depth ranging from 200 to 400 μm and width of 1000 μm were fabricated in 1000-μm thick silicon substrates using photolithography followed by DRIE. A photoresist coating of up to 10 μm thickness (Shipley 1045, single/dual coat) was used as the etch mask for DRIE. The catalyst used was commercial one (Sud Chemie, $Cu/ZnO/Al_2O_3$ catalyst) and was loaded by passing the water-based suspension of catalyst particles (size range: 50–70 μm) through the micro-channel device. By this process, the particles got trapped in the micro-channels due to the presence of the filter near the outlet which allows only particles smaller than 20 μm to pass through. This resulted a micro-packed bed of catalyst particles. The temperature sensor is a platinum resistance-temperature device (RTD) with a linear temperature versus resistance characteristic. The heater is a platinum line meandering along the micro-channels (also on the backside of the substrate). They have obtained 88% methanol conversion at 1:1.5 molar ratio of methanol and water and the feed rate was 5 ml/h.

Table 7.8 Micro-packed bed reactor versus micro-channeled reactor

Micro-Packed bed catalytic Reactor	Micro-channeled catalyst-coated reactor
1. High-pressure drop, channeling of flow of reactants	1. Low-pressure drop, high heat and mass transfer rates
2. Comparatively high axial mixing, hot spot formation, low selectivity	2. Narrow RTD, better temperature distribution, high selectivity
3. Though high catalyst loading, very fast rate of catalyst losses	3. There is a tendency of losing catalytic activity for long-term operation, even in high catalyst loading.
4. Poor response to transient behavior	4. Better response to transient behavior

Most recently, Pattekar and Kothare [66] developed a radial flow micro-packed bed reactor which posseses less pressure drop compared to conventional one due to variable (increasing) flow cross section along the reaction path.

The catalyst-coated micro-channel reactor and plate type reactor is more attractive than micro-packed bed reactor due to low pressure drop, less channeling of gas and good response in transient behavior in catalyst-coated reactor (Table 7.8). The different workers were tried with the catalyst coating. Details are given in Table 7.9. The methanol steam reforming reactions using micro-channeled stainless steel plate has been carried out by Zapf *et al.* [68], Park *et al.* [69], and Men *et al.* [28,29]. The micro-channeled aluminium plate for the SRM has been carried out by Pfeifer *et al.* [75,76]. This reaction using wall coated-catalyst on the aluminum material has been carried out by De wild and Verhaak [59] and Fukuhara *et al.* [71]. To the best of our knowledge, Kwon *et al.* [77] has only performed the SRM reaction in micro-channel reactor made of silicon wafer with catalyst coating. The other user of catalyst-coated micro-channeled silicon wafer is Srinivas *et al.* [78] for the application of preferential CO oxidation. They coated the micro-channeled silicon wafer with alumina sol as undercoating followed by wash-coating of Pt/Al_2O_3 catalyst. They have also compared the catalytic performance using the catalyst coated on the micro-channels and in packed bed reactor. The larger version of micro-channel reactor, wall-coated heat exchanger has been used for integrated methanol steam reformer/PROX reactor for 20 KW_e Proton exchange membrane fuel cell (PEMFC) application [79].

The major concerned point for the catalyst-coated reactor is the low activity and durability of the catalyst, insufficient amount of catalyst loading in spite of the above advantages. The low amount of catalyst loading comes

Table 7.9 Summary of previous work of methanol steam reforming in plate-type reactor

Authors	Plate dimensions	Support and catalysts	Flow rate, water/methanol ratio and composition of products
1. Wild and Verhaak [59]	Heat exchanger	18 g wash-coated catalyst	3.5–14 lpm (1000–4000/h), 1.5–2 and 400–650 l/h H_2. Though the temp. can be controlled easily in the heat exchanger, the decrease in methanol conversion (catalyst deactivation) occurs after 150 h of about 15%
2. Haas-santo *et al.* [67]	Coating studies on stainless steel plate	No catalyst-only characterization with different types of alumina, silica and TiO_2 coating Two types of alumina sols prepared: (a) Al_2O_3 sol from ALSB in ethanol with HNO_3 as catalyst (b) Al_2O_3 sol from ALSB in ethanol and acetyl acetone as stabilizer	Al_2O_3 sol prepared from ALSB in ethanol with HNO_3 as catalyst produced cracked and non-adhesive coating. The reason behind the bad adhesion of these coatings is the high condensation rate in the sol leading to granular fine structure of the gel film. Spontaneous precipitation of the sol has to be avoided and gelling of the sol has to be enhanced in order to achieve a more polymer like structure film. The Al_2O_3 sol modified adding acetyl acetone gives more stable coating
3. Zapf *et al.* [68]	s.s. plates with three different micro-channels: 500 μm × 300 μm, 750 μm × 300 μm and 500 μm × 70 μm (width × depth)	γ alumina wash coats prepared by binder method (γ alumina, deionized water, PVA and acetic acid). Catalyst ($CuO/Cr_2O_3/Al_2O_3$) prepared by wet impregnation of $Cu(NO_3)_2$ and $Cr(NO_3)_2$ of 10 wt% aqueous solution on this calcined wash coats	Catalyst weight (Cu + Cr) = 1.5 mg, $H_2O/CH_3OH = 2$, flow rate = 1.5 g/h

4. Park *et al.* [69]	s.s. plate: micro-channels: 500 μm width, 200 μm in deep and 33 mm long	Alumina sol (15 wt% solid contents). Catalyst: $Cu/ZnO/Al_2O_3$, the catalyst coating was calcined at 350–400°C.	12 cm^3/h (feed flow rate), $H_2O/CH_3OH = 1.1$, reforming temperature = 260°C, composition of products: 73.4% H_2, 25% CO_2, 1.6% CO
5. Bravo *et al.* [70]	Quartz capillaries with inner diameters of 4.1 and 2 mm	Alumina sol prepared from boehmite powder with acid as catalyst. Catalyst: $Cu/ZnO/Al_2O_3$ catalyst was made powder in a mortar and pestle followed by ball milling (in the dispersion step). Slurry was made of catalyst powder and boehmite. Catalyst weight = 0.075 and 0.1 g and operating temperature = 230°C	0.2 ml/h (feed flow rate), $H_2O/CH_3OH = 1.1$
6. Fukuhara *et al.* [71]	Aluminium plate (thickness = 0.4 mm): 21 mm in diameter and 120 mm in length.	Catalyst was prepared by electroless method. The procedure involves the displacement of Al by Zn followed by the intermediate deposition of various metals (Fe, Ni, Co and Sn) followed by the Cu deposition by chemical reduction. Operating temperature = 200–350°C	Flow rate of methanol (0.0057 mol/min), $H_2O/CH_3OH = 1.1$, LHSV (based on the net volume of plated layer) = 10.1 h The activity of the catalyst decreases about 40% in 4800 min and restored its initial activity when the catalyst was oxidized by flowing oxygen. Before reaction: Electroless plating plate was dried in air for 12 h. After the plated catalyst was placed in the reactor, the catalyst was reduced in a hydrogen stream (100 ml/min) at 300°C for 1 h or oxidized in air stream (100 ml/min) at 300°C for 1 h

(*Continued*)

Table 7.9 *(Continued)*

Authors	Plate dimensions	Support and catalysts	Flow rate, water/methanol ratio and composition of products
7. Men *et al.* [28,29]	s.s. plate (10 micro-channels)	Alumina wash coating on micro channels prepared by binder method (exactly same as in [68]). The wash-coated plates were calcined at 600°C for 2 h. catalyst: $Cu/CeO_2/Al_2O_3$. By wet impregnation method, the catalyst was prepared. The impregnated wash coats were calcined for 2 h at 350°C	Catalyst weight = ~2 mg, H_2O/CH_3OH = ~1/1.1
8. Catillon *et al.* [72]	Copper metal foam	Alumina slurry was prepared by adding γ-Al_2O_3 into acidified 10 wt% α-Al_2O_3 in water wash-coated foam was dried at 100°C for 1 h and calcined at 300°C for 5 h. Catalyst: $Cu/ZnO/Al_2O_3$. The deposition of copper and zinc precursors on wash-coated foam was carried out by deposition–precipitation method. Operating temp. = 280°C; the catalyst was reduced without former calcinations in a 10% H_2 in Ar flow up to 400°C at 5°C/min	0.01–10 ml/min (feed rate), H_2O/CH_3OH = 1.5

9. Reuse *et al.* [73]	Aluminium containing stainless steel (FeCr alloy) For kinetics studies: 25 plates and each has 100 micro-channels each channel 40 mm long, 100 μm deep and 200 μm large. For coupled SRM/total oxidation: S-shaped channels: 17 rounded channels to give 34 straight channels with a total length of 30 mm long and each channel is 320 μm large and 100 μm deep	The commercial catalyst (Sud Chemie) micro-milled to a particle size in the nanometer range; 10 mg catalyst in each plate and a catalyst layer of 5 μm	Kinetics of SRM reaction was studied
10. Lim *et al.* [74]	s.s. plate and comprised of six plates with 40 channels each (45 mm length, 1 mm width and 0.5 mm depth)	Zirconia sol was prepared by adding HNO_3 (HNO_3:Zr = 1:2) to a Zirconium isopropoxide isopropanol complex. 1.33 g of zirconia powder was mixed with 0.07 g of zirconia sol to give ratio of 95:5 and then 10 ml isopropyl	0.01–0.1 ml/min (feed rate of liquid mixture), H_2O/CH_3OH = 1, 1.5, 73–75% H_2, 24–26% CO_2 and 0.0–1.2 % CO

(*Continued*)

Table 7.9 (*Continued*)

Authors	Plate dimensions	Support and catalysts	Flow rate, water/methanol ratio and composition of products
		alcohol was added to obtain viscosity. The catalyst slurry was obtained by mixing 0.2 g catalyst, 1 ml isopropyl alcohol and ball milling for 12 h. After coating with zirconia solution, the plate was dried at 70°C for 6 h. The catalyst slurry coated was dried at 70°C and calcined at 400°C for 2 h	
11. Pfeifer *et al.* [75,76]	Aluminium foils Micro-channels: 100 μm wide and 100 μm deep and 64 mm length and width of fin: 50 μm	Coating with the help of hydroxy ethyl cellulose. Pd/ZnO catalyst. ZnO nanoparticles (average particle diameter: 77 nm) was coated on channels and impregnated with palladium acetate in toluene after drying the coating (POST-IMP). In PRE-IMP method, ZnO nanoparticles were added in the heated palladium acetate solution after complete evaporation of the liquid and the dried powder after calcinations at 250°C was put into hydroxy ethyl cellulose solution after complete evaporation of toluene. Then the coating was applied on the micro-channels	0.011 ml/min (feed rate of liquid mixture), $H_2O/CH_3OH = 2$, LHSV = 6.1 h (based on total catalyst mass). The palladium content of the pre-impregnated samples increased the activity and stability upto 10 wt% Pd

12. Yu *et al.* [30]	Aluminium containing stainless steel Reactor: 40 mm × 40 mm × 8 mm and Channel: 1000 μm wide, 600 μm deep and 32 mm long	Wet impregnation on γ-Al_2O_3 coated layer. γ-Al_2O_3 layer was coated on the micro-channels by dipping in Al_2O_3 slurry. Catalyst: $Cu/ZnO/Al_2O_3$	WHSV = 8.27–32.54/h, steam/methanol = 1:1.3 temperature = 220–300°C
13. Kwon *et al.* [77]	Rectangular-shaped micro-channel on (110) silicon wafer by anisotropic wet etching with KOH solution micro-channels were 600 μm in width and 240 μm in depth	Undercoating with alumina sol prepared by Yoldas method followed by drying at room temperature for a day and then calcinations at 500°C. Two types of catalyst slurry were prepared with solvents of deionized (DI) water and alumina sol by using a ball-milling method. The alumina-coated channels were filled with either type of slurry, then dried for a day and calcined at 300°C	1 cm^3/h methanol solution, steam/methanol = 1:1 temperature = 280°C

PVA: polyvinyl alcohol.

from the adjustment of the coating with more porosity and less binding to the adhered surface and the affect of channel's dimension (which affects flow behavior).

Recently another type of reactor called metallic foam reactor has been used for SRM reaction [72]. Metallic foams have some advantages over ceramic monoliths such as higher thermal conductivity and mechanical shock resistance. However the coating of catalysts onto metallic substrate is not as easy as for ceramics, since the smooth metallic surface leads to a weak adhesion of the catalyst.

7.2.4.2 Different Coating Processes for Wall-Coated Heat Exchanger and Micro-Channel Reactor and Its Comparison in Terms of Stability, Performance and Durability

There are mainly three types of coating procedure of catalyst on micro-channels or walls. (1) One method involves the preparation of catalyst by conventional one (or use of commercial catalyst), then making very fine powder by milling followed by making catalyst slurry with different types of binders (Alumina/Zirconia sol, hydroxyethyl cellulose, hydroxyl propyl cellulose, etc.). The prepared slurry was coated on the undercoated channels/walls by dipping or spray method. The undercoating is required to improve the adhesion between the substrate and the catalyst layer. The undercoating can be made with wash-coating of prepared Al_2O_3 or ZrO_2 sol (by conventional way or adding acetyl acetone as stabilizer), anodic oxidation (if the substrate is aluminium), making alumina binder with Polyvinyl alcohol (PVA), electrophoretic deposition and chemical vapor deposition (CVD). The characteristics of different coating procedures have been listed in Table 7.10. Among them sol–gel coating has been used widely. (2) In another method of coating of catalyst, one comparatively thick layer of support material (like Al_2O_3 or ZnO or ZrO_2) was coated on the substrate by the above mentioned method followed by wet impregnation of the metal solution in nitrate form or other form. (3) Another method is made by elctroless plating [71]. In this method, a copper-based catalyst was prepared on an aluminium plate by the electroless plating, which consisted of displacement plating of zinc, an intermediate plating of metal (iron, nickel, cobalt or tin) and chemical plating of copper. There is also a potentiality of coating of Pd-based catalyst based on this method.

Recently Kwon *et al.* [77] compared the performance of micro-channel reactor based on silicon substrate coated with commercial catalyst, Cu/ZnO/Al_2O_3 (Sud Chemie) by making two types of catalyst slurry after undercoating the micro-channels with alumina sol. Two types of catalyst slurry were prepared with solvents of deionized (DI) water and alumina sol by using a ball-milling method.

They had shown that DI water-based catalyst coated by the fill-and-dry coating method on the micro-channels exhibited superior catalytic activity than alumina sol-based catalyst due to less surface area of the catalyst exposed

Table 7.10 Different coating procedure for undercoating the substrate to improve the adhesion between substrate and catalyst layer (partially taken from http://www.fzk.de/imvt/englisch/beschichtung.htm)

Name of supported coating	Characterization	Reactions carried out
(a) Anodic oxidation of aluminium	Surface enlargement upto 2400 m^2/m^2 (BET surface area/geometrical surface area) with 20-m layer thickness	
(b) Sol–Gel coating (HNO_3 as catalyst) with excess water	Surface enlargement upto 400 m^2/m^2 with 3 μm layer thickness	Lim *et al.* [74] used plate-type reactor for SRM using commercial catalyst. ZrO_2 sol was used
(c) Sol–Gel coating (acetyl acetone as stabilizing agent in the molar ratio of 1:1 and 1:2 molar ratio of metal:water)	Surface enlargement upto 150 m^2/m^2 with 1 μm layer thickness [67]	
(d) Alumina binder (mixture of alumina powder, PVA and acetic acid)	Surface enlargement upto 75 m^2/m^2 [68]	$CuO/Cr_2O_3/Al_2O_3$ catalyst coated in micro-channeled s.s. plate for methanol steam reforming
(e) Sol–Gel coating (acetyl acetone as stabilizing agent in the molar ratio of 1:1 and acetic acid (1:4 molar ratio)	Surface enlargement is very less and used for membrane purposes	
(f) Washcoating with nanoparticles and hexa propyl cellulose	Surface enlargement upto 250 with 20 μm layer thickness [75,76]	Pd/ZnO catalyst for methanol steam reforming
(g) Electrophoretic deposition of nano-particles	Surface enlargement up to 100- with 3-μm layer thickness	
(h) CVD deposition of Al_2O_3	Surface enlargement up to 100- with 10-μm layer thickness	

to reactants is available. It is to be noted that they have not compared the post-reaction stability of the coated layer. Therefore it is still an open question which method should be applied to coat the catalyst with stability and durability in addition to high performance in the micro-channel reactor.

Brandner *et al.* [80] studied the effect of the temperature shock on the performance and stability of the catalyst coating in micro-channel reactor. Though they used the oxidation of CO to CO_2 as the case studies on a platinum catalyst, the study is very important on the future direction of coated catalyst on micro-channel reactors. Yu *et al.* [30] performed the stability test of catalyst coating with 40 μm layer with their prepared catalyst Cu/ZnO/Al_2O_3 (14/6 wt% ratio of Cu/Zn) by operating continuously for 100 h. Though there was initial deactivation of the catalyst, they obtained constant and stable conversion of methanol after 20 h of operation.

7.2.4.3 Numerical Modeling of the Reactors

Numerous modeling and simulation for fixed bed catalytic reactors can be seen in the literature. As methanol reforming occurs in gas phase, the modeling approach makes it easy in the case of fixed bed catalytic reactor. Rabou [81] has developed a semi-empirical numerical model for the kinetics of the methanol reforming process. The modeling for hydrodynamics and mass and heat transfer characteristics in these reactors has been studied extensively. Therefore the brief discussion of numerical modeling of micro-channel reactor with special emphasizes to flow distribution and heat transfer characteristics has been presented.

An understanding of the fluid distribution in the parallel process structures required for chemical production in micro-channel systems is critical for industrial scale-up. It is therefore of interest to develop a tool to estimate quickly the quality of the distribution and the maximum velocity difference attained in the reactor. Commenge *et al.* [82] has analyzed the influence of the geometrical dimensions of the reactor micro-structure on the velocity distribution between channels. The study compared two approaches: an approximate pressure drop model and more detailed finite-volume calculations. The results obtained by the approximate model, based on pressure drop calculations through a resistive network of ducts, had shown to be in good agreement with those obtained with the finite-volume method. The approximate model provided rapid calculation of the fluid distribution and offers a qualitative understanding of the influence of geometric parameters and characteristic dimensions on the quality of fluid distribution among the channels of a micro-structured plate.

The fluid flow phenomena in micro-channels have been studied with two types of micro-channels-with smooth micro-channels and micro-channels with wall roughness. List of different experimental investigations are reported in Ref. [104]. They had attempted to reveal the actual reasons of disparity between the theoretical predictions and data of measurements obtained for single-phase flow in micro-channels. For this they had considered the influence of different factors (roughness, energy dissipation, etc.) on flow characteristics. Recently Pfeifer *et al.* [83] validated the simulation results of the flow distribution in micro-channel reactors by CFD (using FLUENT) by experimental results carried out by thermal anemometry.

In the fixed bed reaction system, heat transfer in the catalyst bed takes place by convection and this transfer becomes the rate limiting process. In plate type reaction system, heat exchange occurs mainly by conductive mode. The heat transfer characteristics of water flowing in micro-channels has been studied by Tiselj *et al.* [105]. They have used triangular silicon micro-channels with hydraulic diameter of 160 μm. It was mainly focused on the effect of axial heat flux on heat transfer in micro-channel for low Reynolds numbers. Hetsroni *et al.* [106] presented data of heat transfer in circular, triangular, rectangular and trapezoidal micro-channels with hydraulic diameters ranging from 60 to 2000 μm. The effects of geometry, axial heat flux due to thermal conduction through the working fluid and channel walls, as well as the energy dissipation were reported. Fukuhara and Igarashi [107] studied theoretically the heat transfer and reaction characteristics in a plate-type reactor consisting of one endothermic and one exothermic channel stacked in two tiers or several endothermic and exothermic channels stacked in a multi-tier configuration and compared the performance and heat transfer characteristics with fixed bed reactor. In their simulation, though the conversion in endothermic channel is little less with wall-type reactor compared to that in fixed bed reactor due to a difference in the total surface area of loaded catalyst, the temperature fluctuation between inlet and outlet of the channel is very minimum (~5°C), whereas the fixed bed reactor showed the temperature difference of about 30–40°C between inlet and outlet for both endothermic and exothermic reactions at the same length of the reactor.

7.3 Fuel Cell Operation with Methanol Reforming Process

The majority of the world's leading automotive manufacturers now have PEM fuel cell vehicle research and development (R&D) programs underway and a number of companies have publicly stated their aim of launching pre-production fuel cell vehicles and some companies are in the way of commercialization. However, the choice of fuel for the fuel cell stacks continues to be a key issue regarding the commercialization of fuel cell vehicles. As it is already stated that methanol reforming is one of the best method to be chosen for on-board hydrogen supplying in fuel cell. The key issues of integrated operation of fuel cell with methanol reformer are the selection of catalysts and reactors for methanol reforming since temperature cycling and mechanical vibration are inevitable to supply on-board hydrogen for fuel cells in automotive application. Reduction of CO is essential before feeding to the fuel cell. Methanol processor must convert all the input methanol to hydrogen. Otherwise the unreacted methanol contacts the anode catalyst and will react to form CO which deactivates the electrode. With methanol steam reformer, there is possibility of omitting the humidifier section in the anode side of the fuel cell (which helps to get more compactness of the system) if we operate the methanol steam reformer with excess water/methanol ratio. Also the design of fuel processor will depend on the operating characteristics (the fuel, startup/shutoff frequency and power demand) of the fuel cell and on the ability of the fuel cell stack to tolerate reaction products such as CO.

Recent years have seen the rapid development of methanol reformers in particular applications involving cars and portable applications. Figure 7.10 shows the different parts of fuel processor for the applications of car and portable electronics.

Numerous papers have been published by many researchers investigating the effect of cold start and the dynamic characteristics of methanol reformers. Hohlein *et al.* [7] manufactured a methanol steam reformer system operated at 240, 260 and 280°C, respectively, and noted the intimate relationship between the heating mode and the concentration of the emitted CO. They also found that the characteristics of the starting process were less than expectation. Using their newly constructed methanol reformer, Emonts *et al.* [84] discovered that under low load, the conversion rate of methanol was significantly improved when the operating temperature was increased from 260 to 280°C, with 100% conversion efficiency of methanol. Further increase in temperature, however, resulted in less evident improvement. They further revealed that the higher the temperature, the higher was the concentration of the emitted CO, the level of which was reduced as the load was increased. Wiese *et al.* [85] investigated the quasi-static and dynamic behavior as well as the start up condition of a methanol reformer and noted that reaction first occurred after 10 min from start up. As the reaction rate of the fuel supply system was found to influence the dynamic response of the reformer significantly, they recommended that an injection method be adopted to enhance performance. Han *et al.* [86] built a 3 kW methanol reformer, which provided a thermal efficiency beyond 89% and attained maximum hydrogen production after 15 min from cold start up. Nagano *et al.* [87] investigated catalytic methanol steam reforming by simulation and experiment. They found that there was a trade off relationship between methanol conversion and CO concentration. The trade off relationship was improved by using an internal corrugated metal heater and an external catalytic combustion heater to enhance the heat transfer. Optimal reaction parameters were also obtained. With a $Cu/ZnO/Al_2O_3$ catalyst, Choi *et al.* [88] studied the hydrogen emission characteristics with and without adding water to a methanol reformer. When no water was added, 100% reaction efficiency of the methanol could be achieved, while the presence of water resulted in a high concentration of

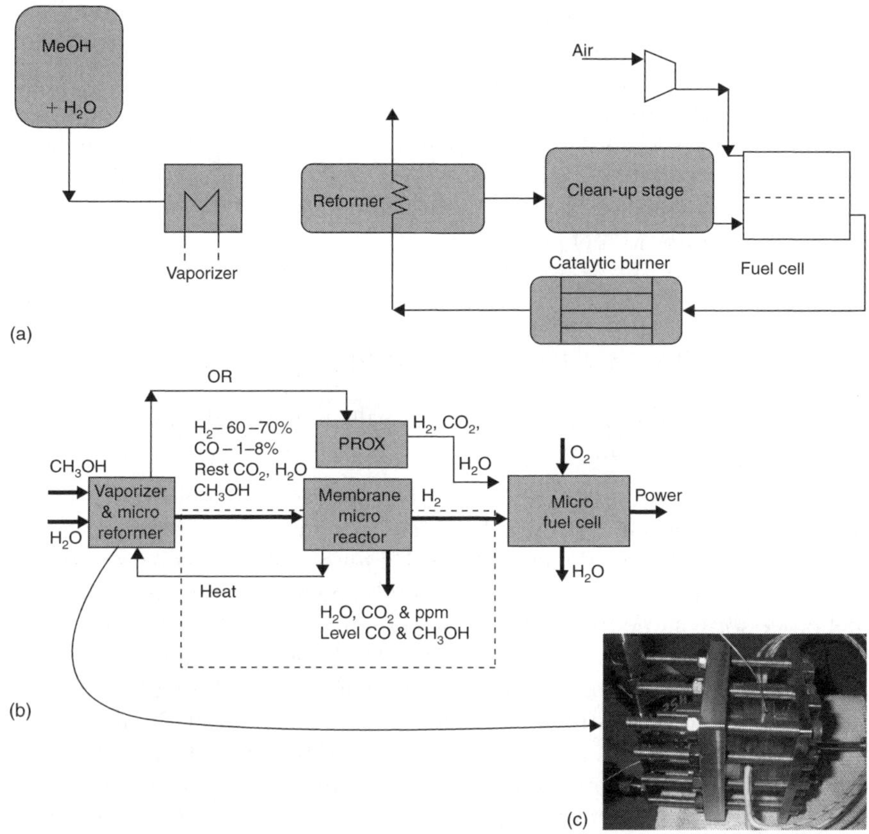

Figure 7.10 Integrated fuel processor with PEMFC: (a) for automobile purpose, (b) for portable's application and (c) image of reactor and vaporizer with SRM reaction for portable application [115].

hydrogen and a low level of CO. Holladay *et al.* [89] fabricated a micro scale methanol reformer and discovered that the temperature of 360°C resulted in a conversion efficiency of 99% using the steam reforming method for a water/methanol ratio of 1.8. Holladay *et al.* [90,91] have manufactured 23 mW fuel cell with phosphoric acid doped polybenzimidazole (PBI) membrane with integrated fuel processor at the operating temp of 50–200°C. The volume of fuel processor is less than 0.3 cm^3 and weight of less than 1 g. The fuel processor consists of the reactor for SRM followed by methanation reactor to reduce CO level in the resulting gas. Lindstrom and Pettersson [92] developed a methanol reformer fitted in cars for Volvo. The catalyst in the reformer under a combined partial oxidation and steam reforming method commenced operation after 4–6 min. Loffler *et al.* [93] included a pre-reformer in their reformer system, which promoted achieving the activation temperature of the catalyst through fuel oxygenation and the main reformer attained the reaction condition by the steam reforming method. This system resulted in a high thermal efficiency and a high purity of the hydrogen produced. Shul *et al.* [94] studied the integrated fuel processor with PEMFC. They successfully operated the fuel cell with the fuel processing with $Cu/ZnO/Al_2O_3$ as methanol steam reforming catalyst and Cu/CeO_2 catalyst as PROX catalyst. The fuel cell used consists of Pt/Co/Cr as cathode catalyst and Pt/Ru alloy as anode catalyst. Lattner and Harold [95] developed 50 KW using three options in fuel processor:

1. SRM, with heat supplied by catalytic combustion, followed by CO removal in a PROX reactor.
2. Adiabatic autothermal reforming of methanol, followed by CO removal in a PROX reactor.
3. Adiabatic autothermal reforming in a palladium membrane reactor with countercurrent steam sweep.

The SRM and autothermal reforming fuel processors coupled with a PEM fuel cell achieve about 50% overall efficiency (LHV basis), with roughly equal fuel processor volumes of about 29 and 22 l for 50 kW net power generation, respectively, whereas the autothermal membrane reactor is a reduction of the fuel processor volume to 13 l, at the expense of a more complex steam system and a small reduction in overall efficiency. In all cases, $Cu/ZnO/Al_2O_3$ catalyst had been used for reforming purpose. Pan *et al.* [96] operated PEMFC with acid doped PBI as membrane at higher temperature (125 and 200°C) by the direct resulting gas from methanol reformer (operating temperature:185–260°C) by eliminating CO reducing process (PROX or other reactors). The PEMFC performance at 125°C is significantly decreased at the CO concentration of 0.5–10% in the reformate gas whereas at 200°C, there is no effect on the performance of PEMFC with the CO concentration of 3% in the reformate gas. Dudfield *et al.* [79] performed integrated methanol steam reformer/PROX reactor for 20 KW_e PEM fuel cell applications. The volume of the fuel processor is 1.85 l having 2.46 g weight with 8.5 g of catalyst loading. This study is important in the view of automotive application that they used compact reactor design by using plate type of reactor with wash-coated catalyst.

7.4 Concluding Remarks

Fuel cell can be operated efficiently by on-board hydrogen production. Methanol seems to be one of the most suitable candidates for on-board hydrogen production. The application of methanol reforming process for fuel cell in automobile purpose and electronic application has been mainly focused. The different catalysts and reactors for methanol reforming have been discussed. Among the catalysts, CuO/CeO_2, Pd/ZnO and CuO/ZrO_2 catalysts are used apart from the commercial catalyst $Cu/ZnO/Al_2O_3$. The reaction mechanism and kinetics of the reaction has been explained. The improved performance of ZrO_2 doped $Cu/ZnO/Al_2O_3$ catalyst has also been reported. The reaction rate of steam reforming is fairly fast and the developed catalyst is in a pretty mature stage after a long research and development.

Among the reactors, micro-channel and wall-coated reactors seems to be promising both for automobile and electronic purposes due to its inherent advantages with respect to hydrodynamics and flow configuration and fast response in transient behavior. The high surface to volume ratio, efficient heat and mass transfer characteristics, vastly improved fluid mixing, etc, in this reactor allow precision control of reaction with improved conversions, selectivities and yields of desired products. The literature studies on PEMFC operation with fuel processor which includes methanol reformer followed by purification stage have also been reported. The lack on studies of long-term stability and thermal cycling testing is there on both the fuel processor and the fuel cell. It is expected further improvement in size and performance of a steam reformer reactor to become attractive for use in a fuel processor.

ACKNOWLEDGMENT

One of the authors (A. Kundu) wishes to thank Dr. P. K. Mukhopadhayay (Former Chairman, Indian Oil Corporation Ltd., India) who motivated him toward this research.

REFERENCES

1. J. M. Ogden, M. M. Steinbugler, and T. G. Kreutz, *J. Power Sources*, 79 (1999) 143.
2. A. V. Makunin, M. S. Granovskii, E. B. Ivanov, V. M. Fomin, *Chem. Petrol. Eng.*, 39 (11–12) (2003) 704.
3. J. Hu, Y. Wang, D. VanderWiel, C. Chin, D. Palo, R. Rozmiarek, R. Dagle, J. Cao, J. Holladay, and E. Baker, *Chem. Eng. J.*, 93 (2003) 55.
4. Y. Kawamura, K. Yamamoto, N. Ogura, T. Katsumata, and A. Igarashi, *J. Power Sources*, 150 (2005) 20.
5. G. C. Chinchen, P. J. Denny, J. R. Jennings, M. S. Spencer, and K. C. Waugh, *Appl. Catal.*, 36 (1988) 1.
6. A. T. Larson, US 2425625 (1947).
7. B. Hohlein, M. Bee, J. B. Hansen, P. Brockerhoff, G. Colsman, B. Emonts, R. Menzer, and E. Riedel, *J. Power Sources*, 61 (1996) 143.
8. L. Pan, and S. Wang, *Int. J. Hydrogen Energy*, 30 (2005) 973.
9. J. B. Hansen, W. Vielstich, H. A. Gasteiger, and A. Lamm, in Handbook of Fuel Cells – Fundamentals, Technology and Applications, (Eds. W. Vielstich, A. Lamm and H. Gasteiger), John Wiley & Sons, Wiley-VCH Verlag GmbH & Co. KGaA, Weinheim 3 (2003) 141.

10. B. A. Peppley, J. C. Amphlett, R. F. Mann, W. Vielstich, H. A. Gasteiger, and A. Lamm, in Handbook of Fuel Cells – Fundamentals, Technology and Applications (Eds. W. Vielstich, A. Lamm and H. Gasteiger), John Wiley & Sons, Wiley-VCH Verlag GmbH & Co. KGaA, Weinheim 3 (2003) 131.
11. J. C. Amphlett, M. J. Evans, R. A. Jones, R. F. Mann, and R. D. Weir, *Can. J. Chem. Eng.*, 59 (1981) 720.
12. F. Raimondi, K. Geissler, J. Wambach, and A. Wokaun, *Appl. Surf. Sci.*, 189 (2002) 59.
13. V. Agarwal, S. Patel, and K. K. Pant, *Appl. Catal. A: Gen.*, 279 (2005) 155.
14. J. Agrell, G. Germani, S. G. Jaras, and M. Boutonnet, *Appl. Catal. A*, 242 (2003) 233.
15. J. Agrell, H. Birgersson, and M. Boutonnet, *J. Power Resources*, 106 (2002) 249.
16. C. J. Jiang, D. L. Trimm, M. S. Wainwright, and N. W. Cant, *Appl. Catal. A: Gen.*, (1993) 145.
17. M. L. Cubeiro, and J. L. G. Fierro, *J. Catal.*, 179 (1998a) 150.
18. M. L. Cubeiro, and J. L. G. Fierro, *Appl. Catal.*, 168 (1998b) 307.
19. N. Takezawa, and N. Iwasa, *Catal. Today*, 36 (1997) 45.
20. F.-W. Chang , H.-Y. Yu, L. S. Roselin, and H.-C. Yang, *Appl. Catal. A: Gen.*, 290 (2005) 138.
21. T. J. Huang, S. W. Wang, *Appl. Catal.*, 24 (1983) 287.
22. T. J. Huang, and S. L. Chren, *Appl. Catal.*, 40 (1988) 43.
23. S. Murcia-Mascaros, R. M. Naavarro, L. Gomez-Sainero, U. Costantino, M. Nocchetti, and J. L. G. Fierro, *J. Catal.*, 198 (2001) 338.
24. B. Lindström, J. Agrell, and L. J. Pettersson, *Chem. Eng. J.*, 93 (2003) 91.
25. S. Liu, K. Takahashi, and M. Ayabe, *Catal. Today*, 87 (2003) 247.
26. K. Klier, V. Chatikavanij, R. G. Herman, and G. W. Simmons, *J. Catal.*, 74 (1982) 343.
27. M. M. Günter, T. Ressler, R. E. Jentoft, and B. Bems, *J. Catal.*, 203 (2001) 133.
28. Y. Men, H. Gnaser, R. Zapf, V. Hessel, and C. Zeigler, *Catalysis Comm.*, 5 (2004a) 671.
29. Y. Men, H. Gnaser, R. Zapf, V. Hessel, C. Ziegler, and G. Kolb, *Appl. Catal. A: Gen.*, 277 (2004b) 83.
30. X. Yu, S.-T. Tu, Z. Wang, and Y. Qi, *J. Power Sources*, 150 (2005) 57.
31. B. Lindström, L. J. Pettersson, *J. Power Sources*, 106 (2002) 264.
32. L. Yong-Feng, D. Xinfa, and L. Weiming, *Int. J. Hydrogen Energy*, 29 (2004) 1617.
33. E. S. Ranganathan, S. K. Bej, and L. T. Thompson, *Appl. Catal. A: Gen.*, 289 (2005) 153.
34. N. Iwasa, T. Mayanagi, S. Masuda, and N. Takezawa, *React. Kinet. Catal. Lett.*, 69 (2) (2000) 355.
35. E. K. Lee, H.-S. Kim, K.-D. Jung, O.-S. Joo, and Y.-G. Shul, *React. Kinet. Catal. Lett.*, 81(1) (2004) 177.
36. H. Purnama, F. Girgsdies, T. Ressler, J. H. Schattka, R. A. Caaruso, R. Schomacker, and R. Schlogl, *Catal. Lett.*, 94 (1–2) (2004) 61.
37. J. K. Lee, and D.-H. Kim, *Korean Chem. Eng. Res.*, 42 (4) (2004) 387.
38. A. Mastalir, B. Frank, A. Szizybalski, H. Soerijanto, A. Deshpande, M. Niederberger, R. Schomacker, R. Schlogl, and T. Ressler, *J. Catal.*, 230 (2005) 464.
39. H. Oguchi, T. Nishiguchi, T. Matsumoto, H. Kanai, K. Utani, Y. Matsumura, and S. Imamura, *Appl. Catal. A: Gen.*, 281 (2005) 69.
40. B. Lindström, and L. J. Pettersson, *Catal. Today*, 74 (2001) 27.
41. N. Takezawa, H. Kobayashi, A. Hirose, M. Shimokawabe, and K. Takahashi, *Appl. Catal.*, 4 (1982) 127.
42. K. Takahashi, H. Kobayashi, and N. Takezawa, *Chem. Lett.*, 6 (1985) 759.
43. T. Takeguchi, Y. kani, M. Inoue, and K. Eguchi, *Catal. Lett.*, 83 (2002) 49.
44. M. V. Twigg, and M. S. Spencer, *Topics Catal.*, 22 (2003) 191.
45. B. Lindström, and L. J. Pettersson, *Int. J. Hydrogen Energy*, 26 (2001) 923.
46. J. P. Breen, and J. R. H. Ross, *Catal. Today*, 51 (1999) 521.
47. L. Yong-Feng, D. Xinfa, and L. Weiming, *J. Nat. Gas Chem.*, 12 (2003) 71.

48. S. P. Tonner, D. L. Trimm, M. S. Wainwright, and N. W. Cant, *Ind. Eng. Chem. Res. Dev.*, 23 (1984) 3.
49. L. Ma, D. L. Trimm, and M. S. Wainwright, *Topics Catal.*, 8 (1999) 271.
50. C. J. Jiang, D. L. Trimm, M. S.Wainwright, and N. W. Cant, *Appl. Catal.*, 93 (1993) 245.
51. B. A. Peppley, Ph.D. Thesis, Royal Military College of Canada, Ontario, Canada (1997).
52. B. A. Peppley, J. C. Amphlett, L. M. Kearns, and R. F. Mann, *Appl. Catal.*, 179 (1999) 21.
53. B. A. Peppley, J. C. Amphlett, L. M. Kearns, and R. F. Mann, *Appl. Catal,*. 179 (1999) 31.
54. J. P. Breen, F. C. Meunier, and J. R. H. Ross, *Chem. Commun.* (1999) 2247.
55. S. Liu, K. Takahashi, K. Uematsu, and M. Ayabe, *Appl. Catal. A: Gen.*, 283 (2005) 125.
56. J. K. Lee, J. B. Ko, and D. H. Kim, *Appl. Catal. A: Gen.*, 278 (2004) 25.
57. R. O. Idem, and N. N. Bakhshi, *Chem. Eng. Sci.*, 51 (1996) 3697.
58. S. R. Samms, and R. F. Savinell, *J. Power Sources*, 112 (2002) 13.
59. P. J. de Wild, and M. J. F. M. Verhaak, *Catal. Today*, 60 (2000) 3.
60. K. Jahnisch, V. Hessel, H. Lowe, and M. Baerns, *Angew. Chem. Int. Ed.*, 43 (2004) 406.
61. V. Hessel, S. Hardt, H. Lowe, A. Muller, and G. Kolb, *Chemical Micro Process Engineering,* Wiley-VCH Verlag GmbH & Co. KgaA (2005).
62. P. Pfeifer, K. Schubert, M. A. Liauw, and G. Emig, *TransIChemE* 81 Part A (2003) 711.
63. J. D. Holladay, Y. Wang, and E. Jones, *Chem. Rev.*, 104 (2004) 4767.
64. A. V. Pattekar, M. V. Kothare, S. V. Karnik, M. K. Hatalis, in *Proceedings of the International Conference on Microreaction Technology (IMRET 5)*, Strasbourg, France, May 27–30 (2001).
65. A. V. Pattekar, and M. V. Kothare, *J. Microelectromec. Sys.*, 13(1) (2004) 7.
66. A.V. Pattekar, and M. V. Kothare, *J. Power Sources*, 147 (2005) 116.
67. K. Haas-Santo, M. Fichtner, and K. Schubert, *Appl. Catal. A: Gen.*, 220 (2001) 79.
68. R. Zapf, C. Becker-Willinger, K. Berresheim, H. Bolz, H. Gnaser, V. Hessel, G. Kolb, P. Lob, A.-K. Pannwitt, and A. Ziogas, *TransIChemE Part* A 81 (2003) 721.
69. G.-G. Park, D. J. Seo, S.-H. Park, Y.-G. Yoon, C.-S. Kim, and W.-L. Yoon, *Chem. Eng. J.*, 101 (2004) 87.
70. J. Bravo, A. Karim, T. Conant, G. P. Lopez, and A. Datye, *Chem. Eng. J.*, 101 (2004) 113.
71. C. Fukuhara, H. Ohkura, Y. kamata, Y. Murakami, and A. Igarashi, *Appl. Catal. A: Gen.*, 273 (2004) 125.
72. S. Catillon, C. Louis, and R. Rouget, *Topics Catal.*, 30/31 (2004), 463.
73. P. Reuse, A. Renken, K. Haas-santo, O. Gorke, and K. Schubert, *Chem. Eng. J.*, 101 (2004) 133.
74. M. S. Lim, M. R. Kim, J. Noh, and S. I. Woo, *J. Power Sources*, 140 (2005) 66.
75. P. Pfeifer, K. Schubert, M. A. Liauw, and G. Emig, *Appl. Catal. A.*, 270 (2004) 165.
76. P. Pfeifer, K. Schubert, M. A. Liauw, and G. Emig, *Appl. Catal. A.*, 286 (2005) 175.
77. O. J. Kwon, S.-M. Hwang, J.-G. Ahn, and J. J. Kim, *J. Power Sources*, 156 (2005) 253.
78. S. Srinivas, A. Dhingra, H. Im, and E. Gulari, *Appl. Catal. A: Gen.*, 274 (2004) 285.
79. C. D. Dudfield, R. Chen, and P. L. Adcock, *Int. J. Hydrogen Energy*, 26 (2001) 763.
80. J. J. Brandner, G. Emig, M.A. Liaw, and K. Schubert, *Chem. Eng. J.*, 101 (2004) 217.
81. L. P. L. M. Rabou, *Int. J. Hydrogen Energy*, 20(10) (1995) 845.
82. J. M. Commenge, L. Falk, J. P. Corriou, and M. Matlosz, *AIChE J.*, 48, (2002) 345.
83. P. Pfeifer, A. Wenka, K. Schubert, M. A. Liauw, and G. Emig, *AIChE J.*, 50 (2004) 418.
84. B. Emonts, J. B. Hansen, S. L. Jorgensen, B. Hohlein, and R. Peters, *J. Power Sources*, 71 (1998) 288.
85. W. Wiese, B. Emonts, and R. Peters, *J. Power Sources*, 84 (1999) 187.
86. J. S. Han, I. S. Kim, and K. S. Choi, *J. Power Sources*, 86 (2000) 223.
87. S. Nagano, H. Miyagawa, O. Azegami, and K. Ohsawa, *Energy Convers. Manage.*, 42 (2001) 1817.
88. Y. Choi, and H. G. Stenger, *Appl. Catal. B: Environ.*, 38 (2002) 259.
89. J. D. Holladay, E. O. Jones, M. Phelps, and J. Hu, *J. Power Sources*, 108 (2002) 21.

90. J. D. Holladay, J. S. Wainright, E. O. Jones, and S. R. Gano, *J. Power Sources*, 130 (2004) 111.
91. J. D. Holladay, E. O. Jones, R. A. Dagle, G. G. Xia, C. Cao, and Y. Wang, *J. Power Sources*, 131 (2004) 69.
92. B. L. Lindstrom, and J. Pettersson, *J. Power Sources*, 118 (2003) 71.
93. D. G. Loffler, K. Taylor, and D. Mason, *J. Power Sources*, 117 (2003) 84.
94. Y.-G. Shul, J.-M. Park, H.-J. Kim, H. Han, H. Kim, D.H. Kim, and S.-E. Yoo, *Third International Conference on Fuel Cell Science, Engineering and Technology*, Ypsilanti, MI, May 23–25 (2005).
95. J. R. Lattner, and M. P. Harold, *Appl. Catal. B: Environ.*, 56 (2005) 149.
96. C. Pan, R. He, Q. Li , J. O. Jensen, N. J. Bjerrum, H. A. Hjulmand, and A. B. Jensen, *J. Power Sources*, 145 (2005) 392.
97. V. Hessel, H. Löwe, and T. Stange, *Lab. Chip*, 2 (2002) 14N.
98. T. Inui, M. Suehiro, and Y. Takegami, *J. Jpn. Petrol. Inst.*, 25 (1982) 63.
99. M. Matsukata, S. Uemiya, and E. Kikuchi, *Chem. Lett.*, (1988) 761.
100. G. Chen, S. Fujita, S. Matsumoto, and N. Takezawa, *J. Mol. Catal. A*, 124 (1997) 123.
101. K. Miyao, H. Onodera, and N. Takezawa, *React. Kinet. Catal. Lett.*, 53 (1994) 379.
102. N. Iwasa, S. Kudo, H. Takahashi, S. Matsuda, and N. Takezawa, *Catal. Lett.*, 19 (1993) 211.
103. H. Kobayashi, N. Takezawa, and C. Minochi, *J. Catal.*, 69 (1981) 487.
104. G. Hetsroni , A. Mosyak, E. Pogrebnyak, and L. P. Yarin, *Int. J. Heat Mass Transfer*, 48 (2005) 1982.
105. I. Tiselj , G. Hetsroni, B. Mavko, A. Mosyak, E. Pogrebnyak, and Z. Segal, *Int. J. Heat Mass Transfer*, 47 (2004) 2551.
106. G. Hetsroni, A. Mosyak, E. Pogrebnyak, and L. P. Yarin, *Int. J. Heat Mass Transfer* (2005) (in press).
107. C. Fukuhara, and A. Igarashi, *Chem. Eng. Sci.*, 60 (2005) 6824.
108. J. Barton, and V. Pour, *Coll. Czech. Chem. Commun.*, 45 (1980) 3402.
109. E. Santacesaria, and S. Carra, *Appl. Catal.*, 5 (1983) 345.
110. J. C. Amphlett, M. J. Evans, R. F. Mann, and R. D. Weir, *Can. J. Chem. Eng.*, 63 (1985) 605.
111. K. Geissler, E. Newson, F. Vogel, T.-B. Truong, P. Hottinger, and A. Wokaum, *Phys. Chem. Chem. Phys.*, 3 (2001) 289.
112. H. Purnama, T. Ressler, R. E. Jentoft, H. Soerijanto, R. Schogl, and R. Schomacker, *Appl. Catal. A*, 259 (2003) 83.
113. N. Iwasa, and N. Takezawa, *Topics Catal.*, 22 (2003) 215.
114. A. Kundu, J. H. Jang, H. R. Lee, S.-H. Kim, J. H. Gil, C. R. Jung, and Y. S. Oh, *J. Power Sources*, 162 (2006) 572.
115. A. Kundu, J. M. Park, J. E. Ahn, S.S Park, Y. G. Shul, and H. S. Han, *Fuel* (in press).

INDEX

Instructions for Authors of Articles for *Advances in Fuel Cells*

ELSEVIER

1. General Statement

Advances in Fuel Cells intends to fill the information gap between regularly scheduled journals and university level textbooks by providing in-depth review articles over a broader scope than in journals or texts. Therefore, articles should be comprehensive, authoritative, critical, and readable reviews of recent research and development in the field of fuel cells. Preference will be given to creative reviews on timely, substantial topics and those likely to promote additional fuel cell research and development. The author's own work in the field should not be singled out for special emphasis, and all contributors to the subject are to be treated on equal footing when selecting material for in-depth discussion. The presentation should not be elementary; it should be addressed to readers who are competently trained in fundamentals but do not have extensive knowledge of the specialized topic. All articles must be comprehensive in the treatment of their subject matter, as defined in an introductory paragraph.

2. Preliminary Approval of Articles

Articles for *Advances in Fuel Cells* are prepared in one of two ways: (a) the Editors solicit suggestions from the Editorial Advisory Board and extend invitations to authors; (b) potential authors contact the Editors. In case (a), the author is assured that the manuscript, if prepared in accord with the guidelines, will be accepted for publication subject to changes that result from the peer-review process. In case (b), authors are requested to send the Editors (1) a two- or three-page detailed topical outline; (2) an estimate of the number of references; (3) an estimate of the number of doubles-paced typewritten pages; (4) a tentative submittal date. This material will be examined by the Editors and expert reviewers. A decision is usually reached in 2–3 weeks, and the author is notified in writing. If the proposal is approved, the author is expected to prepare the manuscript in accord with the proposal and these guidelines. *All manuscripts are peer reviewed, and final acceptance is contingent upon response by the author to the reviewers' comments.*

3. Guidelines for Manuscript Preparation

3.1 Format of Manuscript

All contributions should be typed, double spaced, with wide margins and on one side of the page only. They must be written in English. The length of an article is relatively flexible, but it must be adequate for comprehensive coverage of the subject:

(1) The *title* should be concise and accurately reflect the content.
(2) Full names of authors should be given. Indicate, with an asterisk, the author to whom all correspondence should be addressed.
(3) *Main divisions and subdivisions* must use the following designations: 1, 1.1, 1.2, 1.3, 2, 2.1, 2.2, 2.2.1, etc.
(4) The *Introduction*, usually one or two manuscript pages, should state precisely the purpose and goals of the review and its scope and limitations, as well as the years of literature covered. Historical material and

references to earlier pertinent reviews, as well as a limited amount of material of personal or anecdotal nature, may be included.

(5) Material in the *review proper* should be logically arranged and presented in a clear and concise style.

(6) *Acknowledgments* should be placed in a separate paragraph at the end of the review.

3.2 Tables

Tables should be kept to a minimum; no data should be recorded in both graphical and tabular form. All tables should be provided with headings and should be intelligible without reference to the text.

3.3 Nomenclature and Units

Each article should be consistent within itself as to abbreviations, symbols and units. Authors should use SI units wherever possible and when these are not used should provide a conversion factor to SI units. Axes for graphs and headings for tables should be given in quantity calculus form, e.g. times as t/s, potential as E/mV, current density as $i/\mathrm{A/cm^2}$. Any electrochemical abbreviations should be written in lower case without stops, e.g., emf, ac.

3.4 References

References should be indicated in the text in square brackets and listed at the end of the paper as follows:

1. R. A. Vargas, A. Garciá, and M. A. Vargas, *Electrochim. Acta*, 43 (1988) 1271.
2. J. Newman, *Electrochemical Systems*, 2nd edn, Prentice-Hall, Englewood Cliffs, NJ, 1991.
3. A. R. Hillman, in *Electrochemical Science and Technology of Polymers*, R. G. Linford (Ed.), Vol. 1, Elsevier, Amsterdam, 1987, Chapter 5.
4. B. Miller, *Proceedings of the 6th Australian Electrochemical Conference, Geelong, Vic.*, 19–24 February, 1984; *J. Electroanal. Chem.*, 168 (1984) 91.

Abbreviations of journal titles should follow those in World List of Scientific Periodicals (Fourth Edition).

4. Manuscript Submittal

All invited and unsolicited manuscripts as well as book chapter review proposals should be sent to:

Prof. Tim S. Zhao by email (metzhao@ust.hk)